Biological Magnetic Resonance
Volume 12

NMR of Paramagnetic Molecules

A Continuation Order Plan is available for this series. A continuation order will bring delivery of each new volume immediately upon publication. Volumes are billed only upon actual shipment. For further information please contact the publisher.

Biological Magnetic Resonance
Volume 12

NMR of Paramagnetic Molecules

Edited by

Lawrence J. Berliner

Ohio State University
Columbus, Ohio

and

Jacques Reuben

Hercules Incorporated
Research Center
Wilmington, Delaware

PLENUM PRESS • NEW YORK AND LONDON

The Library of Congress has cataloged the first volume of this series as follows:

Library of Congress Cataloging in Publication Data

Main entry under title:

Biological magnetic resonance:

Includes bibliographies and indexes.
1. Magnetic resonance. 2. Biology—Technique. I. Berliner, Lawrence, J. II. Reuben, Jacques.
QH324.9.M28B56   574.19′285   78-16035
                                                                                       AACR1

ISBN (Vol. 12) 0-306-44387-2

© 1993 Plenum Press, New York
A Division of Plenum Publishing Corporation
233 Spring Street, New York, N.Y. 10013

Printed in the United States of America

Contributors

S. Alam • Department of Chemistry, Washington State University, Pullman, Washington 99164-4630

Lucia Banci • Department of Chemistry, University of Florence, 50121 Florence, Italy

Robert G. Bryant • Department of Biophysics and Department of Chemistry, University of Rochester, Rochester, New York 14642. *Present address*: Department of Chemistry, University of Virginia, Charlottesville, Virginia 22901

Stefano Ciurli • Institute of Agricultural Chemistry, University of Bologna, 40127 Bologna, Italy

I. Constantinidis • Department of Radiology, Emory University School of Medicine, Atlanta, Georgia 30322

Jeffrey S. de Ropp • Nuclear Magnetic Resonance Facility, University of California, Davis, California 95616

J. E. Erman • Department of Chemistry, Northern Illinois University, De Kalb, Illinois 60115

Gerd N. La Mar • Department of Chemistry, University of California, Davis, California 95616

Cathy Coolbaugh Lester • Department of Biophysics and Department of Chemistry, University of Rochester, Rochester, New York 14642. *Present address*: Baker Laboratory of Chemistry, Cornell University, Ithaca, New York 14853

Jean-Marc Lhoste • Institut Curie, Section de Biologie, Centre Universitaire, 91405 Orsay Cedex, Paris, France

Claudio Luchinat • Institute of Agricultural Chemistry, University of Bologna, 40127 Bologna, Italy

Joël Mispelter • Institut Curie, Section de Biologie, Centre Universitaire, 91405 Orsay Cedex, Paris, France

S. J. Moench • Department of Chemistry, University of Denver, Denver, Colorado 80208

Michel Momenteau • Institut Curie, Section de Biologie, Centre Universitaire, 91405 Orsay Cedex, Paris, France

D. J. Russell • Bristol Meyers Squibb, P. O. Box 191, New Brunswick, New Jersey 08903-0191

J. D. Satterlee • Department of Chemistry, Washington State University, Pullman, Washington 99164-4630

Ursula Simonis • Department of Chemistry and Biochemistry, San Francisco State University, San Francisco, California 94132

F. Ann Walker • Department of Chemistry, University of Arizona, Tucson, Arizona 85721

Q. Yi • Department of Chemistry, Washington State University, Pullman, Washington 99164-4630

Preface

This volume is the first of a two-part series devoted to the magnetic resonance of paramagnetic molecules. The subject has not been treated specifically since the publication of a book by La Mar, Horrocks, and Holm (Academic Press, 1973). It is therefore very fitting that La Mar and de Ropp have contributed the opening chapter, which describes in general the NMR methodology as applied to paramagnetic proteins. Relaxation theory is treated in detail by Lucia Banci, while Lester and Bryant address the theory and applications specific to water relaxation in paramagnetic systems. Due to its relevance to contrast agents in NMR imaging (MRI), the importance of the latter subject has grown substantially.

In dealing with more biochemical systems, Walker and Simonis discuss model heme and heme–protein systems, while Satterlee addresses peroxidases. Mispelter, Momenteau, and Lhoste have contributed an insightful chapter on the application of heteronuclear NMR to model and biological systems. The unusual NMR aspects in polymetallic systems are illuminated by Luchinat and Ciurli in the final chapter.

We are proud that this volume contains contributions from researchers at the forefront of their fields. As always, we invite your comments, criticisms, and suggestions for future volumes.

Lawrence J. Berliner
Jacques Reuben

Contents

Chapter 2

Nuclear Relaxation in Paramagnetic Metalloproteins

Lucia Banci

Chapter 3

Paramagnetic Relaxation of Water Protons: Effects of Nonbonded Interactions, Electron Spin Relaxation, and Rotational Immobilization

Cathy Coolbaugh Lester and Robert G. Bryant

Chapter 4

Proton NMR Spectroscopy of Model Hemes

F. Ann Walker and Ursula Simonis

Chapter 5

Proton NMR Studies of Selected Paramagnetic Heme Proteins

*J. D. Satterlee, S. Alam, Q. Yi, J. E. Erman, I. Constantinidis,
D. J. Russell, and S. J. Moench*

Chapter 6

Heteronuclear Magnetic Resonance: Applications to Biological and Related Paramagnetic Molecules

Joël Mispelter, Michel Momenteau, and Jean-Marc Lhoste

Chapter 7

NMR of Polymetallic Systems in Proteins

Claudio Luchinat and Stefano Ciurli

NMR Methodology for Paramagnetic Proteins

Gerd N. La Mar and Jeffrey S. de Ropp

1. INTRODUCTION

1.1. Background

While it has long been recognized that both the hyperfine shifts and paramagnetic relaxation can provide unique and valuable information on the electronic, magnetic, and molecular structural properties of the active site of paramagnetic metalloproteins, access to this information has been thwarted by the absence of a systematic and reliable strategy for assigning the resonances to specific nuclei in the molecule. Thus the initial excitement about and interest in NMR spectroscopy of paramagnetic proteins in the late 60s to middle 70s faded somewhat on the full realization of the apparent barriers to unambiguous signal assignment. The early strategies for assigning hyperfine-shifted resonances in a paramagnetic protein relied on a number of methods, of which most had only a very limited applicability. On the one hand was the least definitive method, namely, the comparison of the shifts of model compounds with those of intact proteins; the selection of appropriate models often relied heavily on intuition (La Mar, 1979). While immensely useful as a starting point in the absence of other data, this method is unreliable except as a guide for the design of more definitive experiments. On the other extreme was the very direct approach represented by the specific isotope labeling of individual functional groups of native prosthetic groups. This, however, involves laborious synthetic methodology and is restricted to the class of b-type hemoproteins where

Gerd N. La Mar ● Department of Chemistry, University of California, Davis, California 95616. **Jeffrey S. de Ropp** ● Nuclear Magnetic Resonance Facility, University of California, Davis, California 95616.

Biological Magnetic Resonance, Volume 12: NMR of Paramagnetic Molecules, edited by Lawrence J. Berliner and Jacques Reuben. Plenum Press, New York, 1993.

the reversibly removable heme could be replaced by one with only ^2H for ^1H or ^{13}C for ^{12}C substitution at restricted positions (Mayer et al., 1974; La Mar et al., 1978; La Mar et al., 1980a, b, 1981a, b, 1985, 1988; Satterlee et al., 1983; de Ropp et al., 1984; Sankar et al., 1987). Both of the above strategies, moreover, are restricted by their nature to identifying signals solely from residues directly coordinated to the paramagnetic metal ion.

The third early strategy, and the one with the apparent widest applicability, was the analysis of paramagnetically induced electron–nuclear dipolar relaxation (Swift, 1973; Bertini and Luchinat, 1986), for which it was assumed the induced relaxivity was related to the inverse sixth power of the distance from the metal ion. This strategy, together with judicious use of model compounds, provided most of the assignments for nonheme paramagnetic proteins to the middle 1980s. The analysis of paramagnetically induced dipolar relaxation, moreover, not only appeared to provide quantitative determination of the relative distance of nonequivalent nuclei from the paramagnetic center, but was the only strategy for assigning the signals from protons on residues not coordinated to the metal. In fact, the prospect of using the paramagnetically induced dipolar relaxation was viewed as an advantageous trade-off for the loss of the conventional assignment strategy of diamagnetic systems. The advent of modern 1D and 2D NMR methodology in the late 70s and early 80s has provided an effective and reliable methodology for the assignment of resonances and the determination of the structure of diamagnetic biopolymers (Wüthrich, 1986; Robertson and Markley, 1990). The observation of the necessary spectral parameters in a diamagnetic system, however, depends on narrow lines (long T_2) and long spin–lattice relaxation times (T_1) dominated by diamagnetic dipole–dipole relaxation. Since many resonances in paramagnetic proteins are characterized by broad lines and very short T_1's dominated by paramagnetic influences, it was initially thought that neither multipulse 1D methods such as the nuclear Overhauser effect (NOE) nor any of the 2D methods would yield useful information on a paramagnetic system.

The recent renaissance of interest in, and productive results for, NMR in paramagnetic systems is based largely on significant improvements in the methodology for the assignment and determination of structure. The first report of an NOE for a hyperfine-shifted proton in a paramagnetic molecule was in 1971 (Redfield and Gupta, 1971) and was not followed by the next reports for several years (Keller and Wüthrich, 1978; Trewhella et al., 1979). Since that time, however, the 1D NOE has been systematically introduced for the whole range of hemoproteins in most accessible oxidation and spin states (Ramaprasad et al., 1984; Unger et al., 1985b; McLachlan et al., 1988; Yu et al., 1986; Lecomte and La Mar, 1986, 1989; Thanabal et al., 1987a, b, 1988a, b; Thanabal and La Mar, 1989; Satterlee and Erman, 1991; Dugad et al., 1990c). More recently, similar studies have been extended to other metalloprotein systems such as iron–sulfur cluster proteins (Dugad et al., 1990c; Banci et al., 1990a; Bertini et al., 1990, 1991a; Cheng et al., 1990; Cowan and Sola, 1990), superoxide dismutases (Banci et al., 1990a, b), nonheme iron proteins (Maroney et al., 1986; Scarrow et al., 1990) and Co(II)-substituted Zn proteins (Banci et al., 1992b). The first reported 2D NMR experiments involving hyperfine-shifted and relaxed protons demonstrated correlations between diamagnetic and paramagnetic cytochrome c derivatives, as facilitated by electron exchange (Santos et al., 1984; Boyd et al., 1984; Turner, 1985). In fact, several of the early 2D NMR studies focused on the transfer of assignments from a well-characterized diamagnetic to an isostructural paramagnetic system via 2D exchange spectroscopy (EXSY) in the belief that the complete assignments in the paramagnetic derivative itself were not directly obtainable by 2D NMR. The first successful 2D dipolar correlation (NOESY) (Yu *et al.*, 1986) and scalar or spin correlation (COSY) (Peters *et al.*, 1985; Jenkins and Lauffer, 1988a, b) experiments for resonances solely within paramagnetic systems were reported for inorganic complexes in the middle 80s and have been quickly followed by a flurry of extensions to the full range of paramagnetic systems that yield reasonably well-resolved ^{1}H NMR spectra. What is not known yet is the scope of and limits to the practical application of modern multipulse and multidimensional NMR experiments to paramagnetic systems. However, while it is now obvious that 2D NMR methodology has considerably greater utility in paramagnetic systems than originally thought, it is also becoming clear that interpretation of paramagnetically induced relaxivity in terms of distance is not as simple or unambiguous as earlier assumed.

1.2. Scope of This Review

In this article we assess the current understanding of experimental approaches to obtaining assignments and determining solution structures of the active sites of paramagnetic proteins. While many review articles and chapters (see, for example, Satterlee, 1986) and two monographs (La Mar *et al.*, 1973; Bertini and Luchinat, 1986) have appeared over the years dealing with the theoretical and interpretive bases of hyperfine shifts and paramagnetically induced relaxation in paramagnetic metalloproteins, only one has emphasized experimental approaches (Keller and Wüthrich, 1981). We will restrict ourselves to the areas where most of the advances have occurred in the past decade: 1D and 2D dipolar correlation and 2D scalar correlation. Also considered briefly are 1D and 2D exchange spectroscopy as they relate to assignments. The early interpretations of relaxation rates in paramagnetic systems solely in terms of the inverse sixth power of the distance from the metal made the implicit assumption that the protons of interest did not exhibit NOEs (Swift, 1973; Bertini and Luchinat, 1986). With the recent realization that not only do paramagnetic molecules exhibit NOEs, but large proteins also exhibit NOEs close to those of diamagnetic systems (Thanabal and La Mar, 1989; Dugad *et al.*, 1990c), the simplistic interpretations of paramagnetic relaxivity is invalidated (Granot, 1982). We will consider paramagnetic relaxation in some detail both because it can serve as a valuable structural probe and because appropriately measured relaxation rates are crucial to the interpretation of NOE data.

This article is not intended to be a review of the literature in this field. Much of that will appear in other chapters in this volume dealing with either the theoretical details of the paramagnetic shift and relaxation or the interpretive bases of these properties in terms of the structure and function of particular classes of molecules. Instead, we will restrict ourselves to reports that serve to define the current understanding of the scope and limitations of NMR methodology as applied to very strongly relaxed and shifted resonances. The 1D experiments are now mature, and the exposition of the factors contributing to the effectiveness of the experiments cites results only inasmuch as they directly address methodology. For the 2D experiments, clearly, the methodology is incompletely explored and developed, and our coverage will be more of a status report that may rapidly become dated. More extensive citations to 2D results will be made, using as a basis published results available as of the final draft of this chapter, April 1, 1992.

The principle of the NOE is considered in sufficient detail both to understand the limitations on interpreting steady-state NOEs and to appropriately measure and interpret paramagnetic relaxivity in terms of distance and electronic relaxation time. Essentially all of the NOE principles covered here are available in a more detailed manner in a recent monograph (Neuhaus and Williamson, 1989), where even some aspects of the influence of paramagnetism are anticipated. For the 2D experiments, the necessary equations are drawn from available monographs and are considered only to ascertain the optimal conditions for cross-peak detection in the presence of strong paramagnetic relaxation (Noggle and Schirmer, 1971; Neuhaus and Williamson, 1989). Moreover, we will usually restrict our discussion to ^{1}H NMR spectroscopy, which has provided most of the information on the active sites of paramagnetic metalloproteins. Extensions to heteronuclei are straightforward; references to heteronuclei are made where appropriate.

Most of our experience in this laboratory with the NOE and 2D experiments has been with hemoproteins and (to a lesser degree) iron–sulfur proteins (the structures of the active sites are illustrated in Fig. 1). The reasons for these choices, aside from the fact that these represent two classes of diverse and important metalloproteins, are twofold. The hemoproteins have played a particularly important role in extending the limits of the 1D and 2D NMR methodology because the assignments needed to test the efficacy of methods had previously been provided by extensive isotope labeling studies (Mayer et al., 1973; La Mar et al., 1978, 1980a, b, 1981a, b, 1985, 1988; Satterlee et al., 1983; de Ropp et al., 1984). The iron–sulfur proteins exhibit hyperfine shifts primarily for the coordinated Cys protons for which spatial and scalar connectivities can be anticipated to some degree (Phillips and Poe, 1973; Markley et al., 1986). Moreover,

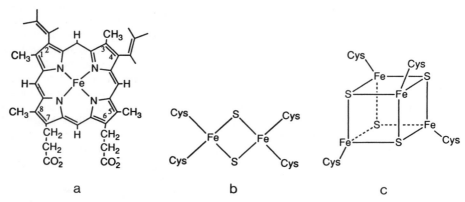

Figure 1. The molecular structure of the active sites of (a) the iron porphyrin in b-type hemoproteins, (b) the two-iron cluster of plant-type ferredoxins, and (c) the four-iron cluster of a bacterial-type ferredoxin, Fd, or high-potential iron protein, HiPiP.

both classes of proteins exhibit multiple paramagnetic states with variable relaxivity and come in a large range of molecular sizes (La Mar, 1979; Satterlee, 1986; Howard and Rees, 1991). We will consider primarily two properties of a protein to describe the efficacy of the various 1D and 2D experiments: the intrinsic relaxivity of a metal ion, which determines the magnitudes of paramagnetic relaxation, and the size of the protein, which determines the linewidths and ^1H–^1H cross relaxation.

1.3.   General Considerations

The paramagnetic relaxation rate ρ for a proton experiencing negligible contact shift for a magnetically isotropic metal ion is related to the spin state S of the metal, the proton–metal distance R_M, and the electron spin relaxation time T_{1e}, as (Swift, 1973; Bertini and Luchinat, 1986):

$$\rho(\text{para}) = \tfrac{2}{15} g^2 \beta^2 \gamma^2 S(S+1)(R_M)^{-6} f(T_{1e}) \tag{1}$$

where g is the Lande g-factor for the metal ion, β is the Bohr magneton, and γ is the nuclear magnetogyric ratio; for very short $T_{1e} < \omega_e^{-1}$, $f(T_{1e}) = 10T_{1e}$. For a given electronic structure, $S(S+1)f(T_{1e})$ is fixed and dictates that $\rho(\text{para}) \propto R^{-6}$. The measurement of nuclear $\rho(\text{para})$ for known R_M is the only method for determining T_{1e} in the absence of solution ESR spectra. The influences of paramagnetism (hyperfine shift, relaxation) can be viewed as advantageous or disadvantageous, inasmuch as they provide unique information but in themselves can seriously interfere with the ability to assign the influenced proton signals. However, neither the advantages nor disadvantages of paramagnetism are uniform throughout a paramagnetic system, but depend on the region of the protein relative to the central metal ion. Based on relaxation via Eq. (1), we define four ^1H zones of a system, as shown in Fig. 2. Zone I is closest to the metal and contains protons with paramagnetic influence such that $T_1 < 2$ ms; Zone II contains protons such that $2\text{ ms} \leq T_1 < 20\text{ ms}$; Zone III has $20\text{ ms} \leq T_1 < 200\text{ ms}$; and Zone IV has $T_1 \geq 200$ ms. In practice, the relaxation properties are not quantitatively described by these zones because of contributions to relaxation from delocalized spin density or anisotropy of the paramagnetic moment (see Section 2.6). The paramagnetic contribution to T_2 [linewidth $= \Delta = (\pi T_2)^{-1}$] is generally larger than to T_1; moreover, diamagnetic T_2's are much shorter than T_1's in biopolymers, so that the correlation between linewidth and T_1 in the various zones depends on the size of the protein. Clearly Zone IV is minimally perturbed as compared to an isostructural diamagnetic system and does not require a serious consideration of the influence of paramagnetism; hence, it is of little interest here. Zone I is by far the most difficult to address with respect to both detection and assignment. Most proteins investigated by ^1H NMR have few protons in Zone I. If a protein has a significant number of protons in Zone I, it is unlikely to be a suitable candidate for profitable NMR investigation of its active site. Zone III exhibits only moderate paramagnetic influence and has shown great potential for being successfully addressed by 2D methods. It is Zone II protons that pose the most interesting challenges to locate, assign and interpret.

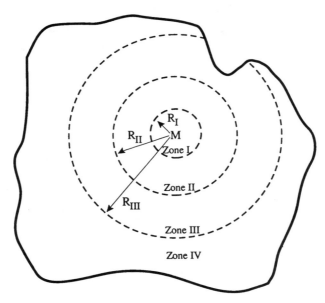

Figure 2. Partitioning of a given protein into zones, as defined by distance from the iron ion resulting in decade changes in the paramagnetic contribution to the spin–lattice relaxation rate $\rho(para)$. The outermost, Zone IV has a negligible contribution from paramagnetism with $\rho(para)$ $< 5\,\text{s}^{-1}$ or $T_1(\text{para}) > 200$ ms. The T_1 ranges for the other zones are I, <2 ms; II, $2 \leq T_1 < 20$ ms; III, $20 \leq T_1 < 200$ ms. These zones provide a useful way for generalizing the influence of paramagnetism on various experimental methodology for a variety of proteins. The sizes of the zones and the range of linewidths associated with them depend on the protein and are given in Table 1 for two spin states for two reference proteins, low-spin and high-spin ferric forms of 17-kDa myoglobin, Mb, and 42-kDa horseradish peroxidase, HRP.

Table 1 shows the actual sizes of the zones, the linewidth ranges associated with the zones, and the number of protons in each zone for two important reference proteins for this chapter, 17-kDa myoglobin (Mb) and 42-kDa horseradish peroxidase (HRP). For each protein, both the low-spin ferric ($S = \frac{1}{2}$) derivatives (metMbCN, HRP-CN) and the high-spin ferric ($S = \frac{5}{2}$) derivatives (metMbH$_2$O, resting-state HRP) are included; the intrinsic paramagnetic relaxivity of the $S = \frac{5}{2}$ state is greater than for the $S = \frac{1}{2}$ state by a factor ~ 10. Among a series of proteins with isoelectronic metal ions, $S(S + 1)f(T_{1e})$ is essentially invariant (i.e., heme proteins, iron–sulfur cluster proteins); hence, the radii of zones I–III remain fixed and the number of protons in Zone IV simply increases for larger proteins. For these systems the problem of identifying the protons in Zones I–III increases as the molecule gets bigger because Zone-IV protons can interfere with the detection of the protons of interest. For a given protein, the number of protons in Zones I, II, and III increases as $S(S + 1)f(T_{1e})$ increases because the radii of the zones become larger (more "problems"). The ^{1}H NMR spectra of 17-KDa myoglobin in the low-spin ferric metMbCN (Emerson and La Mar, 1990a)

TABLE 1
Size, Linewidth Ranges, and Proton Content of the Relaxation Zones of Various Proteins

Protein	Redox/spin state	Zone[a]	Outer radius R_j (Å)[b]	Linewidth range, Hz[c]	Number of protons in zone[d]
Myoglobin (Mb)	Fe(III), $S = \frac{1}{2}$	I	3.3	>350	2
(155 residues,		II	4.9	$350 > \Delta > 90$	8
17 kDa,		III	7.0	$90 > \Delta > 15$	~50
950 protons)		IV	>7.0	<15	9×10^2
	Fe(III), $S = \frac{5}{2}$	I	4.7	>350	9
		II	7.0	$350 > \Delta > 90$	52
		III	10	$90 > \Delta > 15$	~10^2
		IV	>10	<15	~8×10^2
HRP	Fe(III), $S = \frac{1}{2}$	I	3.3	>500	2
(308 residues,		II	4.9	$500 > \Delta > 120$	7
18% carbohydrate,		III	7.0	$120 > \Delta > 30$	~50
42 kDa		IV	>7.0	<30	3×10^3
~3.0×10^3 protons)	Fe(III), $S = \frac{5}{2}$	I	4.7	>10^3	9
		II	7.0	$10^3 > \Delta > 150$	~50
		III	10	$150 > \Delta > 30$	10^2
		IV	>10	<30	3×10^3

[a] Zones are defined by T_1 values: I (2 ms < T_1), II (2 < T_1 < 20 ms), III (20 < T_1 < 200 ms), IV (200 ms < T_1).
[b] Illustrated in Fig. 1.
[c] Based on experimental estimates; contains both paramagnetic and diamagnetic contributions.
[d] Estimated on the basis of crystal structures of MbCO (Kuriyan et al., 1986), and CcP (Finzel et al., 1984).

and high-spin ferric metMbH$_2$O (La Mar et al., 1980a) are illustrated in Figs. 3 and 4.

1.4. Experimental Considerations

Three properties of a paramagnetic system place more stringent demands on an NMR spectrometer and on the ability to extract the relevant NMR parameters of interest than in an analogous diamagnetic system: much faster nuclear relaxation rates (both T_1 and T_2) by factors of 10^2 to 10^4, a much larger range of nuclear relaxation rates in a given system, and a considerably expanded range of chemical shifts. Each of these properties can cause experimental problems for a particular spectrometer or demand care in setting up any experiment on even the most suitable spectrometer (Martin et al., 1980; Freeman, 1988). ^{1}H NMR spectra illustrating the wide range of shifts, linewidths, and T_1's are illustrated throughout this chapter as needed to describe the results for 1D and 2D experiments. The ^{1}H spectrum of low-spin ferric metMbCN collected under a variety of experimental conditions is illustrated in Fig. 3; T_1's range from 3 ms to 200 ms and linewidths from 400 Hz to 20 Hz.

1.4.1. Data Processing

The information on broad peaks in a free induction decay (FID) lies in the very beginning, so that apodization with an exponential suitable for resolution near the

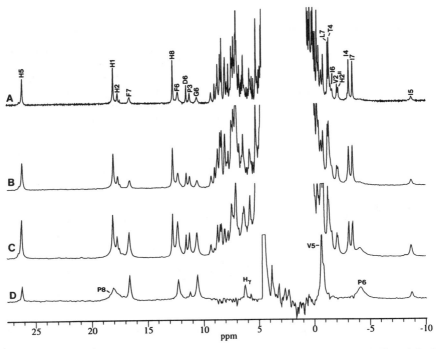

Figure 3. 500-MHz ^{1}H NMR spectra at 35 °C of sperm whale low-spin, cyanide-ligated ferric myoglobin, metMbCN, in ^{2}H$_2$O, pH = 8.6. (A) Repetition rate of 2 s^{-1}, exponential apodization with linebroadening of 3 Hz. (B) Same as (A), except linebroadening of 25 Hz. (C) Same as (B), except repetition rate of 20 s^{-1} and 10 times more scans. The tenfold enhancement in repetition rate allows 10 times as many scans to be collected in the same total time, 80 scans in trace (C) compared to eight scans in traces (A) and (B). (D) Same as (C) except that a SUPER-WEFT pulse sequence with a relaxation dalay of 30 ms was utilized. Repetition rate of 20 s^{-1} and exponential apodization with linewidth of 25 Hz was employed as in trace (C). Other parameters in traces (A) to (D) were kept constant. The labeled assignments are as follows: $H5$, $H1$, $H8$ = heme 5-, 1-, 8-CH$_3$. $H2$ = heme 2-H$_\alpha$. $H2''$ = heme 2-H$_{\beta\text{trans}}$. H$_\gamma$ = heme γ-meso H. The resolved amino acid signals are, for His F8: P8 = ring C$_4$H, P6 = ring C$_2$H; P3 = C$_\beta$H; for Phe CD1: F7 = C$_\zeta$H, F6 = C$_\varepsilon$Hs; for His E7: D6 = C$_\delta$H; for His FG3: G6 = C$_\delta$H; for Val E11: V2 = C$_\alpha$H, V5 = C$_{\gamma 2}$H$_3$; for Leu G5: L7 = C$_{\delta 2}$H$_3$; for Thr E10: T4 = C$_\gamma$H$_3$, and for Ile FG5: I4 = C$_\gamma$H$_3$, I5 = C$_\gamma$H′, I6 = C$_\gamma$H, I7 = C$_\delta$H$_3$. Note the increased signal-to-noise ratio of broad peaks in the progression of traces A–C, and the fact that there is suppression of the diamagnetic envelope at the faster repetition rate in trace (C). The SUPER-WEFT data in trace (D) provides a dramatic suppression of slowly relaxing peaks, allowing the ready detection of P8 and P6 as well as revealing H$_\gamma$ within the diamagnetic envelope. Assignments are from Emerson and La Mar (1990a) and Yu *et al.* (1990).

diamagnetic region (i.e., 3 Hz) leads to marginal sensitivity for the broad peaks for metMbCN (Fig. 3A). More severe apodization (i.e., 25 Hz) greatly improves the detectability of the broad peaks, but some resolution for the more weakly shifted upfield peaks is lost, as shown in Fig. 3B. The point to be made is that, in the processing of

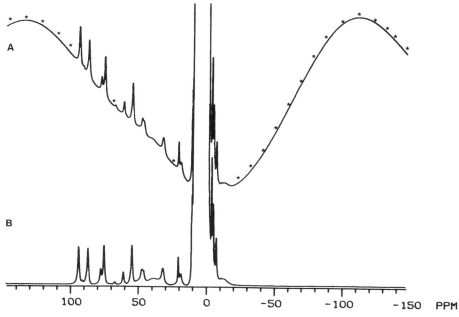

Figure 4. (A) 360-MHz ^{1}H NMR trace of 2-mM high-spin ferric 17-kDa metMbH$_2$O in ^{2}H$_2$O, pH = 6.2 at 25 °C, collected on a Nicolet NT-360 MHz spectrometer operating in the quadrature mode. The spectra consists of 5×10^3 scans collected at a rate of 20 s^{-1}, using 2048 points over a ±54-kHz bandwidth; the exponential apodization introduced 10-Hz line broadening. The low-field lines exhibit linewidths of ~300 Hz and T_1's of 2–10 ms. (B) The spectrum in (A) subjected to a fifth-order polynomial baseline-straightening routine. The points used to define the baseline in (A) are marked by asterisks.

FIDs for paramagnetic systems, a single window function will never optimize all the signals, and in general, several window functions must be used to examine the range of broadened signals. However, simple exponential multiplication is generally all that is necessary for a 1D trace; more complex window functions are needed for 2D experiments and will be considered in Section 4.

1.4.2. Bandwidth

The larger spread of shifts demands that a large spectral region be investigated. This requires a faster analog-to-digital converter (ADC). For 500-ppm peak spread, this requires a 250-KHz ADC on a 500-MHz instrument, but only a 150-KHz ADC on a 300-MHz instrument. Hence the maximum sampling rate of the ADC limits the spectral width that can be interrogated on any given instrument. The larger spectral width can require a larger number of data points, placing demands on the computer. However, if the paramagnetically relaxed signals are of primary interest, a threefold increase in bandwidth relative to a diamagnetic system (as for metMbCN in Fig. 3) does not necessarily demand more points, since the broader lines are adequately digitized with fewer points; often the number of points need not exceed that normally used in a diamagnetic system, and it can often be less.

The 90° pulse length t_{90} also limits the ability to detect peaks over a large spectral width. The effective excitation window, in Hz, is $\sim(4t_{90})^{-1}$. For a 10-μs t_{90}, this limits the effective spectral width to ± 25 KHz (or ± 50 ppm at 500 MHz) on either side of the carrier frequency (assuming quadrature detection). Effective excitation of a 400-ppm spectral width demands an $\sim$4-μs 90° pulse for a 300-MHz instrument. If there are very intense solvent lines, it may be necessary to keep the carrier in the diamagnetic region to avoid interfering artifacts, and this practice could increase pulse power demands since large hyperfine shifts are usually to low field. The effective spectral width for 1D experiments can be expanded to that desired simply by reducing the flip angle, i.e., a 10-μs 90° pulse effective over 25 KHz is effective over 50 KHz as a 5-μs 45° pulse (however, all signal intensities per pulse are reduced by a factor of $\sin 45° = 0.71$).

A third problem associated with both broad peaks and large spectral width is the severe baseline distortions which arise, in large part, due to defective points at the beginning of the FID. This problem also plagues diamagnetic systems, but is strongly exacerbated in paramagnetic systems because the defective points may represent a significant portion of the time domain for the broad signals. An example is given in the trace for high-spin ferric myoglobin, metMbH$_2$O, as shown in Fig. 4A. Several approaches to predicting the initial points can produce more stable baselines in diamagnetic systems (Tang and Norres, 1988; Tiredi and Martin, 1989; Marion and Bax, 1989; Gesmar and Led, 1988; Otting *et al.*, 1986), but these are much less effective in paramagnetic systems. Most spectrometer software packages produce baseline straightening routines based on fitting the distorted baseline to a third- to fifth-order polynomial. However, great care must be taken in selecting baseline reference points that are at least three linewidths from any resonance; otherwise, the routine will diminish intensities of broad lines in comparison to narrow lines. The influence of applying the baseline straightening for metMbH$_2$O is illustrated in Fig. 4B. The baseline points utilized in the routine are marked by asterisks in Fig. 4A.

1.4.3. Dynamic Range

The limits on the detectability of the weakest signal in the presence of a very strong signal are determined by the number of bits of the ADC. A 12-bit ADC would detect a signal in the FID up to 2.4×10^{-4} as intense as the strongest signal. Thus the detectability of a peak can be compromised as the intensity of the diamagnetic envelope gets very large for a protein with many protons in Zone IV, even in ^{2}H$_2$O solution. Modern spectrometers rely on 16-bit ADCs. While this provides a dynamic range that is unlikely to lose signals for even the largest paramagnetic protein studied to date, the dynamic range must not only allow detection of the signal itself but also detect the change in the intensity of a signal in a difference trace such as an NOE (see below). In the next section we consider ways to improve the dynamic range in a spectrum.

1.4.4. Repetition Rates

In contrast to diamagnetic systems, where T_1's have a relatively narrow range, the dynamic range for a paramagnetic system can be effectively controlled by the repetition rate selected for a particular experiment. Emphasizing information from relaxed and shifted lines dictates that the pulse recycle time can be decreased to a value comparable to that of the paramagnetically relaxed lines of interest. Thus, while a cycle time of 1 s may be necessary for a diamagnetic protein of 17 KDa, emphasis on lines for protons in Zone II in a paramagnetic derivative allows a recycle time of 10–100 ms without affecting these peak intensities. The more slowly relaxing protons (particularly those in Zone IV), on the other hand, are effectively steady-state saturated and exhibit strongly attenuated intensities in a resulting trace. The ^{1}H NMR trace of metMbCN collected at 20 s^{-1} is illustrated in Fig. 3C. The more effectively relaxed the lines, the shorter the optimal pulse recycle time. Since each pulse cycle takes less time, many more scans can be collected per unit time, significantly improving the signal–noise ratio for the broad resonances. An even stronger emphasis on the broad, quickly relaxing signals at the expense of the narrow, slowly relaxing diamagnetic envelope is given by the SUPERWEFT pulse sequence (Inubushi and Becker, 1983), where rapid pulsing conditions are combined with an inversion recovery sequence $(180°\text{-}\tau\text{-}90°)_t$, where t is selected to avoid saturation of the hyperfine shifted peaks of interest, and within each cycle a time τ is waited after the 180° pulse to null the slower-relaxing residual diamagnetic envelope while allowing the strongly relaxed signals to recover completely. An example is illustrated in Fig. 3D which not only strongly emphasizes the resolved broad lines but allows the detection of rapidly relaxing lines within the diamagnetic envelope (Emerson and La Mar, 1990a).

2. PARAMAGNETIC RELAXATION

2.1. Utility of Data

In general, the total spin–lattice and spin–spin relaxation rates are given by the sum of diamagnetic and paramagnetic contributions:

$$\rho(\text{total}) = \rho(T) = \rho(\text{dia}) + \rho(\text{para}) \tag{2a}$$

$$\pi\Delta(\text{total}) = \pi\Delta(T) = \pi\Delta(\text{dia}) + \pi\Delta(\text{para}) \tag{2b}$$

First and foremost, a measured $\rho_i(\text{para})$ together with a known R_{M_i} are required to characterize the electronic structure of the metal site in terms of T_{1e}, as obtained by Eq. (1). These determinations, however important, are of no concern to us in this article. In conjunction with the 1D and 2D NOE experiments of interest in Sections 3 and 4, the total relaxation rate (both paramagnetic and diamagnetic contributions) is crucial to both the setup of the experiment and the extraction of distance information.

The paramagnetically influenced nuclear spin–lattice, $\rho(\text{para})$, and spin–spin, $\pi\Delta(\text{para})$, relaxation rates enter into assignment and structural determination strategy based on their purported dependence on the inverse sixth power of the distance R_{M_i} of the nucleus i from the metal center M, as reflected in Eq. (1). Hence, for two nonequivalent nuclei i, j, one or the other of the following relations is assumed valid:

$$\frac{\rho_i(\text{para})}{\rho_j(\text{para})} = \frac{(R_{M_i})^{-6}}{(R_{M_j})^{-6}} \tag{3a}$$

$$\frac{\Delta_i(\text{para})}{\Delta_j(\text{para})} = \frac{(R_{M_i})^{-6}}{(R_{M_j})^{-6}} \tag{3b}$$

If one nucleus is identified and its $\rho_i(\text{para})$ (or Δ_i) measured and R_{M_i} known, $\rho_j(\text{para})$ [or $\Delta_j(\text{para})$] leads to R_{M_j}. This type of analysis relies on identifying $\rho(\text{para})$ or $\Delta(\text{para})$, which in general is not the only, and sometimes not even the dominant, contribution to a measured ρ or Δ. Moreover, the particular contribution(s) to the measured relaxation rate depend(s) on how the experiment is performed, and crucially on the time domain over which the data are analyzed.

We will consider the nature of the experiments in order to lay the groundwork for the expected experimental and interpretative problems (Granot, 1982) in correctly measuring $\rho(\text{total})$ and $\rho(\text{para})$ that have, in general, been ignored in the literature. In order to illustrate the basis of these potential pitfalls for measuring $\rho(T)$ and $\rho(\text{para})$, we will explore in a cursory fashion the experiments and the equations on whose basis T_1 determinations are made.

2.2. Relaxation Equations

We first consider an isolated single spin i whose recovery is given by the rate equation

$$\frac{dI}{dt} = -\rho_i(T)[I - I_0] \tag{4}$$

where I is the z-magnetization of spin i perturbed from its equilibrium value I_0 and ρ_i is the total or *intrinsic* relaxation rate. A plot of $\log(I_t - I_0)$, where I_t is the perturbed I after a time t, versus t yields a straight line with slope $\rho_i(T)$. In general, however, and particularly for protons in macromolecules, protons are not isolated, but interact strongly. Hence we extend the system to a pair of protons i, j with z-magnetizations I, J that interact but are isolated from other spins (two-spin approximation). For simplicity, we will assume equal equilibrium intensities, i.e., $I_0 = J_0$. Under these circumstances, the recovery of magnetizations I, J for the two protons is given by the coupled equations (Noggle and Schirmer, 1971; Neuhaus and Williamson, 1989)

$$\frac{dI}{dt} = -\rho_i(T)[I - I_0] - \sigma_{ij}[J - J_0] \tag{5a}$$

$$\frac{dJ}{dt} = -\rho_j(T)[J - J_0] - \sigma_{ij}[I - I_0] \tag{5b}$$

where $\rho_i(T)$, $\rho_j(T)$ are defined by Eq. (2a). We assume that the only diamagnetic relaxation contributions are from ^{1}H–^{1}H dipole–dipole interactions, a reasonable assumption for macromolecules, yielding the well-known equations

$$\rho_i(\text{dia}) = \frac{K^2\tau_r}{10r_{ij}^6}\left[\frac{6}{1+4\omega^2\tau_r^2} + \frac{3}{1+\omega^2\tau_r^2} + 1\right] \tag{6}$$

$$\sigma_{ij} = \frac{K^2\tau_r}{10r_{ij}^6}\left[\frac{6}{1+4\omega^2\tau_r^2} - 1\right] \tag{7}$$

with $K = \gamma^2\hbar$, ω the Larmor frequency, r_{ij} the interproton distance, and τ_r the reorientation time of the r_{ij} vector (assumed to be the molecular tumbling time), σ_{ij} is called the cross-relaxation rate. Note that while $\rho_i(\text{dia})$ is always positive, σ is positive only for $\tau < 1.12\omega^{-1}$ (small molecules); for $\tau > 1.12\omega^{-1}$ (large molecules), σ is negative. The correlation time, in the absence of internal mobility, can be estimated from the Stokes–Einstein equation

$$\tau_r = \frac{4\pi a^3\zeta}{3kT} \tag{8}$$

where ζ is the viscosity, a is the molecular radius of the protein, and k is the Boltzmann constant. The values of cross relaxation expected for a variety of common functional groups, as identified by characteristic r_{ij}, in proteins of variable size are listed in Table 2; σ_{ij} was estimated using Eqs. (7) and (8).

Inspection of Eq. (5a) reveals that the simple recovery of magnetization that directly yields $\rho_i(T)$ does not occur for the I magnetization *unless the second term in Eq. (5a) is zero*. In the general case, if σ is not zero, recovery of magnetization for either spin

TABLE 2
Predicted Cross-Relaxation Rate σ in s^{-1} for Various Proton Pairs in Proteins as a Function of τ_r[a]

Proton pair	r_{ij}, Å	3 ns 6[b] 4Fe-Fd,[c] HiPiP	5 ns 14 2Fe-Fd, cyto	8 ns 16 Mb	21 ns 42 HRP, LiP	48 ns 78 LPO	75 ns 155 MPO
Geminal	1.77	−5	−10	−14	−40	−67	−135
Vicinal *cis*	2.2	−1.3	−2.7	−3.8	−11	−18	−36
Vinyl *cis*	2.5	−0.65	−1.3	−1.8	−5.2	−8.2	−18
Vicinal *trans*	3.0	−0.21	−0.42	−0.58	−1.7	−2.8	−57
	3.5	−0.08	−0.17	−0.24	−0.68	−1.1	−2.3
	4.0	−0.04	−0.08	−0.10	−0.30	−0.50	−1.0
	5.0	−0.01	−0.02	−0.03	−0.08	−0.13	−0.27

[a] Cross-relaxation rates estimated *via* Eqs. (7), (8) in text.
[b] Molecular size in kDa.
[c] Representative proteins: Fe, ferredoxin; cyto, cytochrome c, b_5; HiPiP, high-potential iron protein; Mb, myoglobin; HRP, horseradish peroxidase; LiP, lignin peroxidase; LPO, lactoperoxidase; MPO, myeloperoxidase.

upon a perturbation of either or both spins is a complicated process, which is not single-exponential and for which an attempt to fit the recovery to a single exponential will (except under very limiting circumstance) not yield meaningful results (Noggle and Schirmer, 1971; Granot, 1982; Neuhaus and Williamson, 1989). Even the experimental approach to determination of these recovery rates for I, J now has several possibilities. The correct experiment must be done and interpreted critically in order to extract the appropriate parameters relevant to analysis of paramagnetic relaxation via R^{-6} or interpretation of a steady state NOE in terms of interproton distance.

2.3. The T_1 Experiment

For a coupled two-spin system, there are two direct experimental approaches to determining the recovery of magnetization of one of the spins (Noggle and Schirmer, 1971; Neuhaus and Williamson, 1989). We will consider here a qualitative picture of the observed recovery rates for those alternate approaches; both approaches are generally necessary if differential paramagnetic relaxation and NOE data are to be analyzed. A selective T_1 determination involves perturbing only the magnetization of the spin of i, I, and following its recovery; such a recovery rate is defined as $(T_{1i}^{-1})^{\text{sel}}$. The selective perturbation of I can involve either a soft $180°$ pulse on I ($I \rightarrow -I_0$) or saturation of I by the decoupler ($I \rightarrow 0$). Since the perturbation is selective, J is unperturbed from J_0 at $t = 0$ ($J_0 - J_0 = 0$), so the second term in Eq. (5a) vanishes initially. For the selective $180°$ pulse, therefore, we have an equation valid near $t = 0$:

$$\frac{dI}{dt}\bigg|_{t=0}^{\text{sel}} = -\rho_i(T)[-2I_0] \tag{9}$$

for which the initial recovery of I yields $\rho(T)$:

$$(T_{1i}^{-1})^{\text{sel}} = \rho_i(T) \tag{10}$$

Because $\sigma \neq 0$, J will receive magnetization from I shortly after $t = 0$, with the result that J and J_0 will differ, so that the second term in Eq. (5a) contributes to the recovery of I. Hence the recovery of I is no longer a single exponential but depends on ρ_i, ρ_j as well as σ_{ij}. When the T_1 experiment is carried out with a nonselective $180°$ pulse (as most commonly reported for paramagnetic systems), we have $I \rightarrow -I_0$ and $J \rightarrow -J_0$ at $t = 0$, and since $I_0 = J_0$, Eq. (5a) at $t = 0$ reduces to:

$$\frac{dI}{dt}\bigg|_{t=0}^{\text{nonsel}} = -\rho_i(T)[-2I_0] - \sigma[-2I_0] = -[\rho_i(T) + \sigma_{ij}][-2I_0] \tag{11}$$

Thus the initial recovery for a nonselective T_1 determination is a single exponential with a recovery rate

$$(T_{1i}^{-1})^{\text{nonsel}} = \rho_i(T) + \sigma_{ij} \tag{12}$$

Comparison of Eqs. (10) and (12) shows (since σ is negative for a macromolecule) that the selective T_1 will always be shorter than the nonselective T_1. For the purpose of

NOE interpretation, $\rho_i(T)$ is needed, and it is clear that the selective T_1 experiment must be performed. This will be discussed again in Section 3. To extract ρ(para) via Eq. (2a), ρ(dia) must be known. It is frequently not known, and, except in very small molecules, not negligible compared to ρ(para), as is often assumed. For small model compounds, σ, ρ(dia) $\ll \rho$(para), so the selective and nonselective T_1 experiments yield the same rate, namely ρ(para). The problem of ρ(dia) is serious only for macromolecules and becomes intractable for very large molecules because it *is not defined* for individual spins (spin diffusion) (Hull and Sykes, 1975; Kalk and Berendsen, 1976; Duben and Hutton, 1990).

For intermediate-size macromolecules, however, Eqs. (6) and (7) indicate that ρ_i(dia) $\rightarrow -\sigma_{ij}$ (the basis for the limiting steady state NOE $= \sigma_{ij}/\rho_i = -1$ in macromolecules). Hence, we can insert ρ_i(dia) $= -\sigma_{ij}$ and Eq. (2a) into Eq. (12) to obtain (Granot, 1982)

$$(T_{1i}^{-1})^{\text{nonsel}} = -\sigma_{ij} + \rho_i(\text{para}) + \sigma_{ij} = \rho_i(\text{para}) \tag{13}$$

Therefore the *nonselective* T_1 for spin i directly provides an estimate of ρ_i(para) for subsequent analysis in terms of Eq. (1) or (3a). We say *estimate* rather than *determine* the rates because a much more detailed analysis is required to assess the validity of the assumption made in the above qualitative analysis, i.e., how long is the single exponential range defined by $(dI/dt)|_{t=0}$ in Eqs. (10) and (12) valid?

2.4. Analysis of T_1 Data

A critical analysis of the problems to be encountered in the determination of ρ_i(para) in macromolecules has been reported by Granot (1982) and largely ignored in the subsequent literature. Granot demonstrated that a coupled spin system in a macromolecule experiencing paramagnetic relaxation generally exhibits nonexponential recovery for all z-magnetization. He showed that the description of the recovery of magnetization in terms of a single exponential is highly restricted and depends on the ratio of the cross-relaxation rate σ_{ij} to the difference $\Delta\rho$ in relaxation rates for the two coupled spins, in terms of the parameter,

$$\alpha = 2|\sigma_{ij}/[\rho_i(\text{para}) - \rho_j(\text{para})]| \tag{14}$$

We recommend turning to this elegant paper for details. However, the conclusions are that a T_1 is defined for a proton in a macromolecule ($\omega^2\tau_r^2 > 1$) only under the following stringent conditions:

Case i. Strong Cross Relaxation ($\alpha \gg 1$).   Only the nonselective experiment yields a single exponential, with

$$(T_{1i}^{-1})^{\text{nonsel}} = (T_{1j}^{-1})^{\text{nonsel}} = \tfrac{1}{2}[\rho_i(\text{para}) + \rho_j(\text{para})] \tag{15}$$

Both protons contribute equally to either relaxation rate so that *individual distances to the metal* R_{M_i}, R_{M_j} *cannot be determined.* Moreover, the intrinsic relaxation rates needed to interpret 1D NOEs are not defined.

Case ii. Weak Cross Relaxation ($a \ll 1$). Single exponential recovery is predicted for both experiments with $(T_{1i}^{-1})^{\text{sel}}$ and $(T_{1i}^{-1})^{\text{nonsel}}$ given by Eq. (10) and Eq. (13), respectively. However, since in practice $|\sigma| \ll |\Delta\rho|$ often means $|\sigma| \ll \rho_i(\text{para})$, $\rho_j(\text{para})$, the two experiments frequently yield essentially the same slopes $\sim\rho_i(\text{para})$ for proton pairs.

Case iii. Intermediate Cross Relaxation ($a \sim 1$). Most real cases do not fall into (i) or (ii) above, but in the broad range between the two. The first result of importance is that under no circumstances do Eqs. (5a, b) predict a single exponential behavior over even one lifetime of the exponential for either the selective or nonselective T_1 experiment. *Reliable estimates of meaningful relaxation times can be obtained only from the initial slope of a semilogarithmic plot of magnetization versus time, for which the measurement time t must be restricted to a very small fraction of the determined T_1* (Granot, 1982). (The single exponential is maintained over a longer time, however, for the selective experiment.) The results from the two experiments are the same as given in Eqs. (10) and (13). It is important to note that the experimental points in a T_1 determination must be closely spaced near $t = 0$, and the individual points must have extremely good signal-to-noise ratios in order to accurately represent the initial slope because the magnetization will change by only a small fraction.

Perhaps the most emphatic point to be made on the basis of the above discussion is that determining a nonselective or selective T_1 value by simply fitting experimental points over an arbitrary time range to a single exponential, as has frequently been reported, cannot be justified in any paramagnetic macromolecule; such T_1's are likely to be meaningless. The initial slope method must always be used and the recovery profile inspected for change in slope at the shortest times relevant to the system studied. The larger the protein and the closer a neighboring proton, the more crucial it is to use the initial slope method. Several reports where T_1 recovery data plots have been presented show clearly that the plots are anything but single exponential (Sletten *et al.*, 1983; Thanabal *et al.*, 1987b). Plots of $\ln(I_0 - I_\tau)$ versus τ for a nonselective inversion–recovery experiment for several hyperfine-shifted resonances in ferricytochrome c are illustrated in Fig. 5. Clearly the lines are not straight over a significant time, and the degree of curvature varies appreciably (Sletten *et al.*, 1983), being largest for a methylene proton (line e). For the methyl peak (line b), the comparison of the nonselective (unlabeled) and selective plots (marked SL) shows that the selective T_1 is shorter, as expected, and that the single exponential response is maintained over a longer time than for the nonselective experiment, as predicted (Granot, 1982). T_1 plots for low-spin HRP-CN (larger size and thus larger σ than ferricytochrome c) show even greater nonlinearity and confirm that there is a large difference between $(T_1^{-1})^{\text{nonsel}}$ and $(T_1^{-1})^{\text{sel}}$ when σ is large (Sette and La Mar, unpublished observations).

For a given proton in a given protein, increasing the paramagnetic relaxivity will "quench" the cross-relaxation effects that produce the curvature in the T_1 plots. Thus addition of increasing amounts of exogenous Gd^{+3} increases the $\rho_i(\text{para})$ of the methyl peak a of ferricytochrome c (Fig. 6) with the slope for the nonselective experiment becoming steeper and less curved as $[Gd^{+3}]$ increases (Sletten *et al.*, 1983). Thus T_1 plots will be more linear for a given proton in a molecule of a given size as the paramagnetic relaxivity increases (see Case ii above). Conversely, for a given functional group

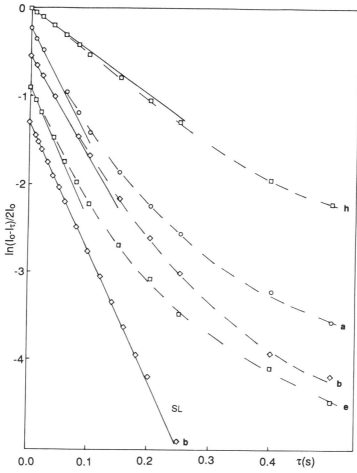

Figure 5. Plots of $\ln(I_0 - I_\tau)/2I_0$ versus τ for selected hyperfine-shifted resonances of ferricyto-chrome c in ^2H_2O. The peak labeling corresponds to that in (C) of Fig. 6. The solid lines reflect the initial linear portions for each peak: dashed curves indicate connectivity and have no theoretical significance. The curve labeled SL was obtained by a selective inversion–recovery experiment [5 ms 90° pulse, carrier at 33 ppm]; the other curves represent nonselective inversion–recovery data. All curves are plotted to the same scale but are displaced vertically for clarity (Sletten *et al.*, 1983).

with fixed paramagnetic relaxivity, the nonlinearity in the T_1 determination for even short t increases dramatically as the protein size σ increases. An indication that caution must be exercised in measuring T_1 is reflected in the degree that the apparent $(T_{1i}^{-1})^{sel}$ and $(T_{1i}^{-1})^{nonsel}$ differ, or by the magnitude and number of observed NOEs when spin i is saturated. The presence of NOEs and the difficulties in measuring ρ are intimately related and will be considered again in Section 3.2.

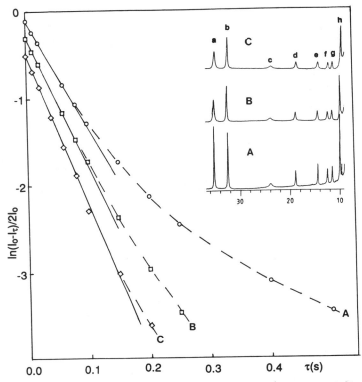

Figure 6. Effect of Gd^{+3} on the low-field portion of the 360-MHz ^{1}H NMR of ferricytochrome c in ^2H_2O, and on the initial slope of $\ln(I_0 - I_\tau)/2I_0$ for a nonselective T_1 determination of peak a. The curves A, B, C correspond to 0, 2.0, and 4.0 equivalents of Gd^{+3} per protein, with traces labeled A, B, C in the upper right-hand corner. The straight lines indicate the initial slope; dashed lines indicate data-point trends and have no theoretical significance. All curves are plotted to the same scale but are displaced vertically for clarity. Adapted from Sletten *et al.* (1983).

2.5. Relevance to Metalloproteins

There are therefore two limitations to the use of T_1's; the T_1 must be determined correctly, and the correctly determined T_1 may or may not allow the desired interpretation. The three cases above can be described as "easy" experiment and straightforward interpretation (Case ii), "easy" experiment and no interpretability (Case i), and "difficult" experiment and limited interpretability (Case iii). The problem that may be anticipated in extracting T_1 from an inversion–recovery experiment and interpreting it (Granot, 1982) can be summarized in a plot of $|\Delta\rho|$ versus $|\sigma|$, which marks the boundaries between the above cases i, ii, and iii [defined by $\alpha = 10$ or 0.1 in Eq. (14)], as illustrated in Fig. 7. Included in Fig. 7 are points that correspond to two types of functional groups, a methylene group ($r_{ij} = 1.77$ Å) and a vinyl cis or vicinal aromatic proton pair ($r_{ij} = 2.5$ Å) in a variety of proteins of variable size (τ_r) and paramagnetic

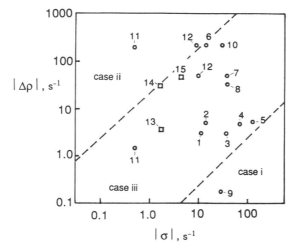

Figure 7. Plot of $\Delta\rho = |\rho_i - \rho_j|$ versus $|\sigma_{ij}|$ for the geminal methylene proton pair with $r_{ij} = 1.77$ Å ($\bigcirc$) and aromatic vicinal or vinyl *cis* proton pairs with $r_{ij} = 2.5$ Å ($\square$) for a variety of metalloproteins. The ρ_i, ρ_j were taken from the literature and σ_{ij} estimated *via* Eqs. (7) and (8). The borders between the three cases defined in the text are shown; the dashed lines correspond to $\alpha = 10$ and 0.10 [Eq. (14)]. The data points correspond to the following: 1, H93 $C_\beta H_2$ of metMbCN (Emerson and La Mar, 1990a); 2, H102 $C_\beta H_2$ of ferricytochrome b_{562} (Wu *et al.*, 1991); 3, H170 $C_\beta H_2$ of HRP-CN (de Ropp *et al.*, 1991b); 4, 8-CH_2 of lactoperoxidase-CN (Thanabal and La Mar, 1989); 5, 8-CH_2 of myeloperoxidase-CN (Dugad *et al.*, 1990c); 6, propionate $C_\alpha H_2$ of metMbH$_2$O (Unger *et al.*, 1985b); 7, propionate α-CH_2 of resting-state horseradish peroxidase (Thanabal *et al.*, 1988b); 8, propionate α-CH_2 of horseradish peroxidase compound I (Thanabal *et al.*, 1988b); 9, Tyr β-CH_2 of uteroferrin (Scarrow *et al.*, 1990); 10, His β-CH_2 of Cu$_2$Co$_2$-superoxide dismutase (Banci *et al.*, 1989); 11, two Cys $C_\beta H_2$ of oxidized HiPiP (Bertini *et al.*, 1991a); 12, two Cys $C_\beta H_2$ of 2-Fe reduced ferredoxin (Dugad *et al.*, 1990b); 13, vinyl H_α–$H_{\beta c}$ in metMbCN (Emerson and La Mar, 1990a); 14, Phe 43 *o*-H–*m*-H in metMbCN (Emerson and La Mar, 1990a); 15, vinyl H_α–$H_{\beta c}$ in horseradish peroxidase–cyanide (de Ropp *et al.*, 1991b).

relaxivity. It is clear that even strong paramagnetic relaxation leads to problems in extracting T_1 if the protein is large enough (i.e., high-spin ferric HRP or lactoperoxidase, LPO). More importantly, Fig. 7 shows that methylene groups are clearly within the intermediate (Case iii) to strong (Case i) cross-relaxation limits for many of the low-spin ferric hemoproteins. A generalization may be drawn from Fig. 7 that methylene protons T_1 must be extracted from the initial slope for any protein and are unlikely to provide structural information from their differential paramagnetic relaxivity. Another corollary is that large NOEs and simplistic analysis or interpretation of measured T_1 in terms of distance to the metal is mutually exclusive.

2.6. Interpretation of Differential Relaxivity

Having dealt with the scope and limitations of determining a T_1 which can be (at least semiquantitatively) related to ρ_i(para), we consider here briefly the limitations of the simplistic interpretation in terms of $(R_{M_i})^{-6}$. First, significant relaxation of a proton signal in a protein can be observed and measured only if the signal is either contact or dipolar shifted outside the diamagnetic envelope. In fact, practical ^{1}H studies are largely restricted to paramagnetic proteins where the relaxed resonances are also hyperfine shifted (Jesson, 1973; Bertini and Luchinat, 1986). Unfortunately, the very interactions that allow the resolution of the peaks and the quantification of ρ(para) can also seriously interfere with extracting quantitative information on T_{1e} or $\,^{-}(R_{M_i})^{-6}$. The details of relaxivity are treated elsewhere in this volume. However, we need to consider the form of the expression to understand the limitations of the interpretation. For ρ_i(para) we have (Swift, 1973; Bertini and Luchinat, 1986):

$$\rho_i(\text{para}) = B[(R_{M_i})^{-6} + \rho_L^2 R_{L_i}^{-6}]f(\tau_e) + C(A/\hbar)^2 f'(\tau_e) \qquad (16)$$

where $B = \frac{2}{15} g^2 \beta^2 \gamma^2 S(S + 1)$, $C = \frac{2}{3} S(S + 1)$, A is the Fermi contact coupling constant, and $f(\tau_e)$, $f'(\tau_e)$ are functions of the electron spin correlation times whose form need not be specified for the analysis of differential dipolar relaxation via Eq. (3a). The contribution to the first term in Eq. (16) results from electron spin–nuclear dipole relaxation. The second term in Eq. (16) arises from scalar relaxation, makes a negligible contribution to ρ_i(para), and can be neglected. It is seen, however, that the dipolar (first) term, in general, does not depend only on R_{M_i}. If the proton of interest is on a ligand covalently attached to the metal and experiences a contact shift, then there is nearby significant delocalization spin density ρ_L on atom L of the ligand, which is a distance R_{L_i} from the nucleus i of interest (Waysbort and Navon, 1975; Koenig, 1982; Unger et al., 1985a). This is shown schematically in Fig. 8A. While this spin density is small, $\rho_L \ll 1$, the distance of this delocalized spin density center L to proton i, R_{L_i}, is likely very small, i.e., $R_{L_i} \ll R_{M_i}$, and hence can be extremely effective in relaxing the nucleus. From this it follows that only where ρ_L is sufficiently small that $\rho_L^2(R_{L_i})^{-6} \ll (R_{M_i})^{-6}$ for all nuclei to be compared, does Eq. (3a) follow, because of the cancellation of common constants and $f(\tau_e)$ in Eq. (16). Unfortunately, if large contact shifts are observed, $\rho_L^2(R_{L_i})^{-6}$ may not be insignificant compared to $(R_{M_i})^{-6}$, and relative relaxivity can give a poor quantitative indication of relative distance to the metal (Unger et al., 1985a). Therefore, distance to the metal determined for a proton with a substantial contact shift appears shorter than it really is.

A case in point is the low-spin ferric complexes of hemin (Fig. 1A), for which vinyls are variably substituted. The various heme methyls (each 6.1 Å from the central iron), reflect ρ(para) varying over a factor of 4 (Unger et al., 1985a). However, the variation in ρ(para) could be quantitatively correlated with the square of the contact shift ($\propto \rho_L^2$) on the adjacent aromatic carbon, as shown in Fig. 8B. The magnitude of the ρ_L needed to cause the excess relaxation, moreover, was shown to be consistent with the spin density calculated from the observed contact shift (Unger et al., 1985a). This may be an extreme case because of the short T_{1e} and very large delocalized spin density, but it illustrates the potential for very large errors in distance. Had it not been known that the methyls all have $R_{Fe} \sim 6.1$ Å, the differences in ρ(para) of a factor 4 could have led to the erroneous conclusion that R_{Fe} differs by $\sqrt[6]{4}$, or 25%. The relaxation properties of the coordinated Cys in iron–sulfur cluster proteins also appear to be particularly strongly influenced by delocalized spin density (Busse et al., 1991).

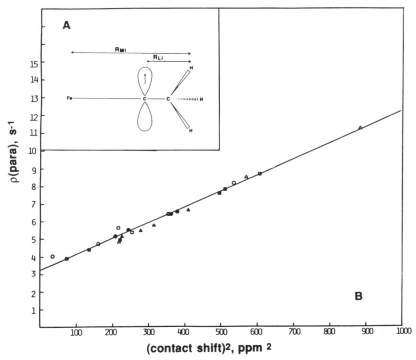

Figure 8. (A) Relaxation of a methyl group by unpaired spin density on the aromatic carbon to which it is attached and the paramagnetic metal center. The geometry of the interaction is characterized by distances R_{M_i}, the metal-center-to-proton distance, and R_{L_i}, the delocalized spin-density-to-proton distance. (B) Plot of observed $\rho(para)$ versus contact shift (delocalized spin density) squared for a variety of low-spin *bis*-cyano complexes of the ferric porphyrin shown in Fig. 1a, except with variable substituents at the 2,4 positions; 2,4-vinyl (●), 2,4-H (▲), 2,4-Br (■), 2,4-acetyl (○), 2-H, 4-acetyl (□), 2-acetyl, 4-H (△). Adapted from Unger *et al.* (1985a).

Well-resolved lines that are considerably relaxed also occur for residues near, but not coordinated to, the metal ion in cases when the metal magnetic moment is strongly anisotropic; low-spin Fe(III) and high-spin Co(II) appear to be the most prominent examples (Jesson, 1973; Bertini and Luchinat, 1986). For such nuclei, only the first term in Eq. (16) is operative, and $\rho_L = 0$, so, at first glance, relative relaxivity should be given by Eq. (3a). However, the magnetic anisotropy makes the term in B in Eq. (16) not a scalar, but a tensor ($g_{\parallel} \neq g_{\perp}$ for an axial system), so that its magnitude depends on the position of a proton in the magnetic axes of the system (Sternlicht, 1965; Vasavada and Nageswara Roa, 1989). At the same distance from the metal, a proton is more efficiently relaxed in the direction of the maximum g value than in the direction of minimum g value. The exact ratio depends on the particular system, but the difference in relaxivity due to anisotropy for the same R can vary considerably. Thus, in a low-spin iron (III) system, where $g_{\parallel} > g_{\perp}$, noncoordinated axial protons appear closer to the iron and equatorial protons appear when one uses the average magnetic moment, further from the iron than they really are. These errors may, or may not, be crucial in an assignment or structural determination. The bottom line is that considerable caution should be exercised and potential uncertainties assessed about any distances determined from paramagnetic relaxivity.

The influences of paramagnetism on linewidth are similar to ρ, with (Swift, 1973; Bertini and Luchinat, 1986):

$$T_2^{-1} = \pi\Delta_i(\text{para}) = \tfrac{1}{2}B[(R_{M_i})^{-6} + \rho_k^2(R_{k_i})^{-6}]f^*(\tau_e) + \tfrac{1}{2}C(A/\hbar)^2 f^\dagger(\tau_e)$$
$$+ ES^2(S+1)^2(R_{M_i})^{-6}B_0^2 T^{-2}\tau_r \qquad (17)$$

where $E = (4\gamma^2 g^4 \beta^4/45k^2)$ and $f^*(\tau_e), f^\dagger(\tau_e)$ are functions of the electron correlation times whose form is unimportant in analysis of differential relaxation. The important point is that not only can delocalized spin density contribute to the first term, as for $\rho(\text{para})$, but the scalar relaxation term involving C is not necessarily negligible. Hence linewidth analysis in terms of $(R_{M_i})^{-6}$ is not recommended in any protein. It is noted that Eq. (17) has an additional term, called the Curie spin term (Gueron, 1975; Vega and Fiat, 1976), which results from modulation not of the instantaneous S by electron spin relaxation but of the average S, $\langle S_z \rangle$, by the molecular reorientation τ_r. Since $\langle S_z \rangle \propto B_0/T$, the relaxation depends on B_0^2/T^2 and is readily shown to be present if the linewidth increases dramatically with field strength B_0. This effect is important with metal ions with short T_{1e} and large $S(S+1)$, such as high-spin Fe(II), Fe(III), and Co(II), and can be the dominant mechanism for large proteins (long τ_r) at high field strength ($\propto B_0^2$) or low temperatures ($\propto T^{-2}$). If it is assumed that the first two terms in Eq. (17) are independent of B_0 (this may or may not be justified), a plot of Δ versus B_0^2 for different protons in a given molecule should yield a series of straight lines whose relative slopes $\propto (R_{M_i})^{-6}$. Studies on superoxide dismutase using Curie spin relaxation have been used to glean assignments (Banci et al., 1987). It is likely that the utility of analyzing the Curie spin relaxation in any complicated real system is limited by the ability to make meaningful linewidth estimates for a series of broad and overlapping resonances.

3. 1D CORRELATION

3.1. Spin Decoupling

The double resonance experiment (pulse sequence in Fig. 9D) has seen very limited use for paramagnetic proteins, since the linewidths generally obscure any multiplet structure; it is included only for completeness. Prominent exceptions are the small, low-spin ferric hemoproteins with many protons close to the Zone III–IV boundary, where the components of a vinyl group were routinely identified in a difference trace upon irradiating the H_α signal (Keller and Wüthrich, 1981). A single case for identifying hyperfine-shifted amino acid signals has been reported (Lecomte et al., 1989). In contrast, ^{13}C labeling of hemoproteins in all oxidation and spin states has led to ^{1}H and ^{13}C spectra with either resolved multiplet structure or residual broadening that allowed

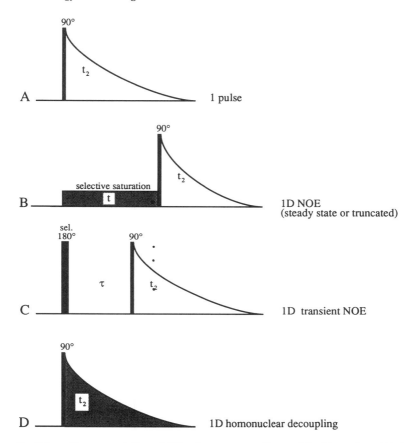

Figure 9. Schematic representation of 1D pulse sequences discussed in this section. (A) 1-pulse sequence. (B) Pulse sequence for steady-state or truncated NOE; in the latter case the saturation time t is varied to obtain the build-up curve of the NOE. (C) Transient NOE pulse sequence with delay τ after the selective pulse. (D) Homonuclear decoupling pulse sequence, with frequency-selective decoupling applied during the acquisition time t_2.

unambiguous assignment by spin decoupling (Sankar *et al.*, 1987). It is unlikely, however, that the 1D experiment can compete with the homonuclear ^{1}H 2D correlation experiments (see Section 4.2).

3.2. Nuclear Overhauser Effect

In treating the various aspects of the NOE, we rely on the same set of equations used to analyze relaxation rates, namely Eqs. (5a, b), making the same assumption that we have a pair of interactive nuclei i, j that are isolated from other nonequivalent protons. For the ρ(dia), we assume exclusive dipole–dipole relaxation such that ρ_i(dia) and σ_i are again given by Eqs. (6) and (7). We will also assume that the τ_r (which determines σ_i) is given by Eq. (8), i.e., there is no internal mobility. Considerations of internal mobility are largely the same as in a diamagnetic system (Noggle and Schirmer, 1971; Neuhaus and Williamson, 1989), and have been treated in some detail for a paramagnetic system (Duben and Hutton, 1990). An NOE for spin i is observed any time J is perturbed from J_0, provided $\sigma_{ij} \neq 0$. This perturbation of J, as in the case of a T_1 determination, can be effected in several ways, either saturation [$J \to 0$ in Eq. (5a)], or selectively inverting J ($J \to -J_0$). The general NOE for spin i upon saturating j is defined as:

$$\eta_i\{j\}(t) = \frac{I(t) - I_0}{J_0} \tag{18}$$

where I_0, J_0 are the equilibria magnetizations of i, j, and $I(t)$ is the i spin magnetization some time t after perturbing spin j. The experiment is always carried out in the difference mode, where one spectrum is collected with spin j saturated (the decoupler on-resonance for spin j), which will be subtracted from another spectrum for which the decoupler is placed in some portion of the spectral window so as not to influence any spin in the system. The difference trace will show changes in the intensity of spin i due to perturbing j. Pulse sequences appropriate for the various forms of the experiment are given in Fig. 9.

3.2.1. Steady-State NOE

Upon saturating spin j ($J = 0$) continuously (pulse sequence in Fig. 9B) until the NOE becomes time-independent [$(dI/dt) = 0$], Eqs. (5a) and (18) rearrange to yield

$$\eta_i\{j\} = \sigma_{ij}/\rho_i(T) \tag{19a}$$

Conversely, saturating i, setting $(dJ/dt) = 0$ in Eq. (5b) leads to

$$\eta_j\{i\} = \sigma_{ij}/\rho_j(T) \tag{19b}$$

Note that $\eta_i\{j\} \neq \eta_j\{i\}$ unless $\rho_i(T) = \rho_j(T)$. If the system is diamagnetic, Eqs. (6) and (7) for $\rho_i(\text{dia})$ and σ_{ij} yield the curve a in Fig. 10; a positive $\eta = 0.5$ for fast motion ($\omega^2\tau_r^2 \ll 1$) and $\eta = -1.0$ for slow motion ($\omega^2\tau_r^2 \gg 1$), with a crossover from positive to negative NOE at $\omega\tau_r = 1.12$ for an isolated system. In this isolated two-spin system, η is independent of interproton distance since σ_{ij} and $\rho_i(\text{dia})$ depend identically on r_{ij}^{-6}. The region $\omega^2\tau_r^2 \gg 1$ where $\eta \to -1$ is particularly problematic since under these circumstances (large τ_r or large ω) cross relaxation is so strong that T_1 for an individual spin is not defined and NOEs are essentially randomized over the whole molecule, which agrees with the definition of spin diffusion (Hull and Sykes, 1975; Kalk and Berendsen, 1976).

If $\rho_i(T)$ and $\rho_j(T)$ have both diamagnetic and paramagnetic contributions, i.e., Eq. (2a), we can rearrange the expression for the total relaxation rate for i as (Dugad et al., 1990a)

$$\rho_i(T) = \rho_i(\text{dia}) + \rho_i(\text{para}) = \rho_i(\text{dia})[1 + Q_i] \tag{20}$$

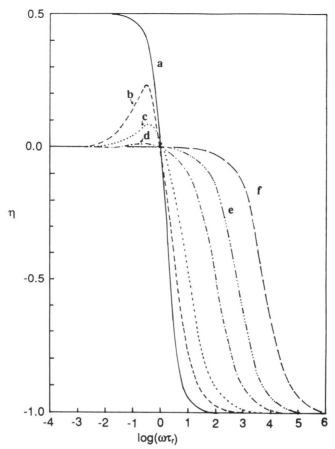

Figure 10. Plot of the steady-state NOE as a function of $\omega\tau_r$ and $\rho(para)$ for a pair of isolated spins. Curves a, b, c, d represent a proton pair separated by 1.77 Å experiencing $\rho(para) = 0, 2, 10$, and 100 s^{-1}, respectively; curves a, e, f represent a proton pair separated by 3.5 Å experiencing $\rho(para) = 0, 10$, and 100 s^{-1}, respectively (curve a is independent of r_{ij}).

where we define the parameter

$$Q_i = \rho_i(\text{para})/\rho_i(\text{dia}) \tag{21}$$

The NOE can then be rewritten:

$$\eta_i\{j\} = \frac{\sigma_{ij}}{\rho_i(\text{dia})}\left[\frac{1}{1 + Q_i}\right] = \frac{\eta_i\{j\}(\text{dia})}{1 + Q_i} \tag{22}$$

where $\eta_i\{j\}(\text{dia})$ is the NOE that would have been observed for the spin i in an isostructural diamagnetic system. Since for any $\rho_i(\text{para}) \neq 0$, $Q_i > 0$, the steady-state NOE for a pair of protons in a paramagnetic system will be smaller than in an isostructural diamagnetic system, with the degree of reduction determined by Q_i.

Q_i, however, is not a constant even for a given protein. Thus $\rho(\text{dia}) \propto \tau_r$ via Eq. (6), and for any given system can be modulated by changes in solvent viscosity [Eq. (8)]. $\rho(\text{para})$, however, is largely independent of molecular motion for metal ions which yield well-resolved ^{1}H NMR spectra (short T_{1e}), but does depend on R_{M_i} (Dugad et al., 1990a). Therefore Q_i will decrease with increased τ_r, ω, or R_{M_i} in any system (Duben and Hutton, 1990). Figure 10 shows the influence of $\rho(\text{para}) = 2 \text{ s}^{-1}$ (Zone IV), 10 s^{-1} (Zone III) and 100 s^{-1} (Zone II) on the steady-state NOE for a pair of isolated protons at separations of 1.77 Å and the influence of $\rho(\text{para}) = 10 \text{ s}^{-1}$, 100 s^{-1} for $r_{ij} = 3.5$ Å. The relaxivity values of 10 s^{-1} and 100 s^{-1} correspond to typical values in low-spin and high-spin ferric hemoproteins, respectively. Paramagnetic contributions have the effect of most strongly quenching the NOE for small molecules at low ω, with the effect strongly decreasing for large τ_r (large molecules or high viscosity) (Duben and Hutton, 1990). For any paramagnetic system, the reduction of the steady-state NOE can be minimized by sufficiently immobilizing the molecule (Yu et al., 1986; Dugad et al., 1990a). Note that for fixed $r_{ij} = 1.77$ Å (line a, b, c, d) or $r_{ij} = 3.5$ Å (line a, e, f) increased $\rho(\text{para})$ moves the asymptotic approach to $\eta = -1$ to larger τ_r (in the absence of paramagnetic relaxation (line a), η is independent of r_{ij}). Moreover, the approach is at larger τ_r for larger r_{ij}. This influence can be interpreted as a suppression of spin diffusion (Duben and Hutton, 1990). In other words, an intermethylene steady-state NOE at log $\omega\tau = 2$ is primary and (at least semiqualitatively) interpretable in a paramagnetic protein in Zone I and portions of Zone II, while in such a diamagnetic system, saturating one proton would lead to NOEs for all protons (spin diffusion). However, the degree of suppression of spin diffusion will be maximal in Zone I and sharply decrease as the protons are further removed from the metal.

Another advantage of steady-state NOE determination in a paramagnetic molecule is that, even if spin diffusion is not completely quenched, the increased $\rho_i(T)$ means that a steady state is reached for much shorter periods t of saturation, which makes less likely the appearance of secondary NOEs. Such primary NOEs have been observed in paramagnetic proteins to 155 kDa at 360 MHz (Dugad et al., 1990c). The interpretation of NOEs is still plagued by factors such as multiple spin effects and internal motions, and any quantitative interpretation of NOEs based on the two-spin approach is at risk if saturation of peak j gives NOEs to other protons beside i. In such a multispin system great care must be exercised in determining any interproton distance. Scenarios that can cause serious problems have been considered in detail elsewhere (Noggle and Schirmer, 1971; Neuhaus and Williamson, 1989).

Since $\rho(\text{para})$ is largely independent of τ_r for metallochromophores such as hemes whose electronic structure (and hence electronic T_{1e}) is largely independent of the size of the protein (Dugad et al., 1990a), it is predicted from Eqs. (21) and (22) that steady-state NOEs for a given functional group (fixed r_{ij}) in a paramagnetic system increase with molecular size. Such effects have been observed in a series of low-spin ferric hemo-proteins ranging in size from 16 to 160 kDa (Ramaprasad et al., 1984; Emerson et al., 1988; Thanabal et al., 1987a, b; Thanabal and La Mar, 1989; Dugad et al., 1990c). For small molecules, the paramagnetic suppression of NOEs is most effective (see Fig. 10 at negative log $\omega\tau_r$); moreover, many such molecules also have small negative NOEs due to being close to the region where NOEs change sign (Barbush and Dixon, 1985).

However, placement of these molecules into viscous solvents routinely leads to significantly increased NOEs to allow complete assignment for protons in Zone III such as low-spin ferric model hemes (Yu *et al.*, 1986; Chatfield *et al.*, 1988; Licoccia *et al.*, 1989). Estimation of ρ(para) is sufficient to determine the viscosity needed to yield detectable NOEs. Obviously the method will work better for compounds with small rather than large ρ(para). The addition of viscosity-increasing agents (ethylene glycol, glycerol, DMSO, glucose) to water may make NOEs more practical for small proteins with particularly effective paramagnetic relaxivities. The effect of increasing NOE with viscosity has been demonstrated for two small proteins (Dugad *et al.*, 1990a; Bertini *et al.*, 1990).

NOEs in themselves do not give distances. Although it is frequently claimed that paramagnetic relaxation introduces a r_{ij}^{-6} dependence to the steady-state NOE, this is true only if the ρ(para) is comparable for all protons (i.e., as induced by dissolved oxygen for small molecules). In a protein where ρ(para) depends on R_M^{-6}, the NOE itself still does not yield an estimate of σ; this requires measuring $\rho(T)$. It is crucial that the $\rho(T)$ be used, and, as shown in Section 2, *this requires the selective T_1 measurement* (Noggle and Schirmer, 1971; Granot, 1982; Neuhaus and Williamson, 1989). Use of a nonselective T_1 can introduce large errors in the σ calculated via Eq. (19), and hence large errors in r_{ij}. Only in a few cases in the literature have selective T_1's been determined to interpret η (Lecomte and La Mar, 1986; Lecomte *et al.*, 1989; Unger *et al.*, 1985b; Yamamoto *et al.*, 1989). The importance of using T_1^{sel} rather than T_1^{nonsel} can be appreciated by recasting the steady-state NOE using Eqs. (10), (12), i.e.,

$$\eta_i\{j\} = \frac{\rho_i^{nonsel} - \rho_i^{sel}}{\rho_i^{sel}} = \frac{\rho_i(T) + \sigma_{ij} - \rho_i(T)}{\rho_i(T)} = \frac{\sigma_{ij}}{\rho_i(T)} \qquad (23)$$

If the NOEs are significant, $\rho^{sel} \neq \rho^{nonsel}$, and ρ^{sel} must be measured. The problems encountered in determining r_{ij} are minimized by large ρ(para) and small σ_{ij}. Thus ρ^{nonsel} values of $\sim$200 and 500 s^{-1} (Zone II) for the heme methylene protons of high-spin ferric metMbH$_2$O are essentially the same as ρ^{sel}, since the expected difference, $\sigma_{ij} \sim -14$ s^{-1}, is within experimental error of each T_1 determination (Unger *et al.*, 1985b). However, for the same relaxivity, a methylene group in 42-kDa high-spin ferric heme peroxidase should show a difference between *correctly* determined ρ^{sel} and ρ^{nonsel} of $\sim$40 s^{-1}, which is not negligible for the slower relaxing peak. For a methylene group in Zone III, ρ^{nonsel} is likely a poor estimate for a protein as small as 16 kDa (Mb), and likely a disaster for a larger peroxidase. For a large enough protein, i.e., 150 kDa, even protons in Zone II could have significant differences ($\sim$150 s^{-1}) between ρ^{sel} and ρ^{nonsel}. Unfortunately, it is precisely these cases where ρ^{sel} and ρ^{nonsel} differ most that the ability to measure either is most problematical (see Sections 2.4 and 2.5).

3.2.2. Truncated Driven NOE (TOE)

The Ideal TOE. A more effective method for obtaining distance directly via the NOE, and at the same time minimizing secondary NOEs or spin diffusion, is to follow the time course of the buildup of the $\eta_i\{j\}(t)$ after a variable irradiation time of j; the pulse sequence for this experiment is given in Fig. 9B. Assuming instantaneous saturation of j at $t = 0$ (see below), setting $J = 0$ in Eq. (5a) and integrating, one obtains the *ideal* TOE equation (Noggle and Schirmer, 1971; Neuhaus and Williamson, 1989)

$$\eta_i\{j\}(t) = \frac{\sigma_{ij}}{\rho_i}[1 - \exp(-\rho_i t)] \qquad (24)$$

a similar equation holds for $\eta_j\{i\}(t)$ by interchanging i and j. In this equation t is the time that peak j is continuously saturated. The predicted ideal TOE profiles for both protons of a methylene group ($r_{ij} = 1.77$ Å) in Mb ($\sigma_{ij} = -14 \text{ s}^{-1}$) in Zone II ($\rho_i = 200 \text{ s}^{-1}$, $\rho_j = 500 \text{ s}^{-1}$), are illustrated in Fig. 11.

For very short irradiation times ($t \ll \rho_i^{-1}$), Eq. (24) simplifies to:

$$\eta_i\{j\}(t)|_{t=0} = \sigma_{ij}t \qquad (25)$$

The initial slope in the ideal TOE experiment directly yields σ_{ij} and, moreover, *is not influenced by paramagnetism*. By following the time profile of $\eta(t)$ as predicted by

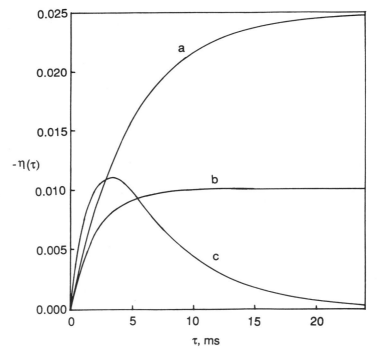

Figure 11. Plot of the steady-state NOEs (lines a, b) and transient NOE (line c) for two protons separated by 1.77 Å ($\sigma_{ij} = -14 \text{ s}^{-1}$) with $\rho_i(T) = 200$, $\rho_j(T) = 500 \text{ s}^{-1}$, which represents a heme propionate α-methylene group in Zone II of high-spin ferric metMbH$_2$O; $\eta_i\{j\}$, $\eta_j\{i\}$ correspond to curves a, b, respectively.

Eq. (24), a two-parameter fit would yield both σ_{ij} and ρ_i. The asymptotic TOE for long t, is, of course, the steady-state NOE [Eq. (19)]. If σ_{ij} can be obtained from the initial slope, $\rho_i(T)$ can be obtained from fitting the early portion of the curvature without entering the long t region where secondary NOEs interfere. The catch to this seemingly effective experiment for a paramagnetic system is twofold. First, the limit $t \ll \rho_i^{-1}$ is much shorter than for a diamagnetic system [by the ratio Q_i in Eq. (21), which can be very large], and there is a question whether the necessarily small NOEs at short t can be quantitated. The second problem, however, is much more severe. Equation (24) *assumes that spin j is saturated completely and instantly*, and this is not experimentally achievable (Lecomte *et al.*, 1991). While this problem also plagues diamagnetic systems, it is much more serious in paramagnetic systems.

The Real TOE. It has been shown in detail that complete saturation of a strongly relaxed signal is sometimes experimentally impossible due to power limitations, and frequently impractical in that induced artifacts obscure the desired NOEs. Using ^1H resonances in Zone II (i.e., a peak with $\Delta \sim 300$ Hz and $\rho \sim 200$ s^{-1}, a heme methyl in high-spin ferric Mb), Fig. 12A gives the degree of the methyl peak saturation as a function of the on-resonance decoupler power, 1000, 300, and 70 Hz, which produces 99, 95, and 50% steady-state saturation (Lecomte *et al.*, 1991). To saturate completely ($\sim$99%), oscillations in J are set up which will modulate $\eta_i\{j\}$. The highest power is not necessarily available in any given instrument; moreover, use of such power yields unacceptable artifacts (see below). The use of lower (reasonable) power to saturate j

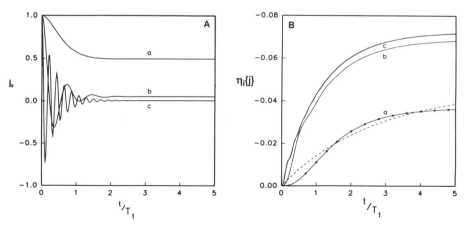

Figure 12. (A) Time dependence of the z magnetization of the irradiated spin j during an NOE experiment. Spin j is the 8-CH$_3$ line of sperm whale *met*-aquo myoglobin. The T_1 and T_2 values are 10 and 3 ms, respectively. The irradiation powers are 70 Hz (a), 300 Hz (b), 1000 Hz (c). (B) Time dependence of the NOE to spin i as spin j is irradiated. Spin i corresponds to one of the 6-α propionate protons of sperm whale *met*-aquo myoglobin and is characterized by $T_1 = 5.2$ ms and $\sigma_{ij} = -14$ Hz ($r_{ij} = 2.9$ Å). Power levels $\gamma B_2/2\pi$ are equal to 70 Hz (a), 300 Hz (b), 1000 Hz (c). The dotted line represents the time dependence of the NOE of the observed spin i for the ideal case of instantaneous saturation of spin j [Eq. (24)]. The dashed line represents the best fit of the simulated data of line a to Eq. (24) (Lecomte *et al.*, 1991).

leads to a delay in reaching equilibrium saturation, as well as $J \neq 0$ at any t, negating the use of the ideal Eqs. (24) or (25). Instead, a much more complicated analysis is required, as has been presented by Wagner and Wüthrich (1979). Simulations as to what the profile of $\eta_i\{j\}$ would be for saturation of a comparably relaxed proton ($\rho = \sim200\ s^{-1}$) of a methylene group ($\sigma = -14\ s^{-1}$) in the same protein under the power levels that yield 99%, 95%, and 50% saturation of j according to the time profiles in Fig. 12A are shown in Fig. 12B; also included is the "ideal response" according to Eq. (24). The 1000-Hz saturation power yields a response curve that closely resembles the ideal curve except for the oscillations (Torrey, 1949); these oscillations particularly interfere with quantitating the initial slope. For the more practical lower irradiation powers, the response curves for η_i show a sigmoidal behavior (curve a) due to incomplete saturation at short t. The "worst" portion of each curve is at the initial slope, thereby invalidating Eq. (25) for directly obtaining σ. The dotted line in Fig. 12B represents a fit to Eq. (24) using as "experimental" the simulation for a decoupler power of 70 Hz to achieve 50% saturation. The fit to Eq. (24) yields σ and ρ values that are off by a factor of 2 from the true values (Lecomte et al., 1991). These problems will be greatly diminished as ρ(para) is reduced. Thus, while a TOE may be effective for protons in Zone III, it is unlikely to be useful for protons in Zone II, although such studies have been reported.

3.2.3. Off-Resonance Effects

Common to both the steady-state and the truncated NOEs in paramagnetic systems is the need to saturate resonances with short T_1, T_2. For the TOE, extensive saturation is essential; for the steady-state NOE, the degree of saturation is not crucial since the η_i can be shown to be linear in the degree of saturation of peak j (Noggle and Schirmer, 1971; Neuhaus and Williamson, 1989). The need to use large decoupler power to completely saturate a peak is countered by the need to minimize off-resonance saturation of other peaks in the spectrum. While NOEs for protons in Zone II of a small to intermediate (16–42 kDa) protein are small (0.5–10%), complete saturation of any one peak can lead to off-resonance saturation of greater than 10% for all resonances within 10,000 Hz! The effect of decoupler power on off-resonance saturation of a peak 300 Hz wide with $\rho \sim 200\ s^{-1}$ as a function of the offset of the decoupler, in Hz, is shown in Fig. 13 (Lecomte et al., 1991); the three power levels used are the same as above, namely, those leading to 99, 95, and 50% saturation on resonance. The profiles show that for 99% on-resonance saturation, off-resonance peaks are saturated to 50% up to 2000 Hz away (6.7 ppm at 300 MHz), and to 10% up to 10 kHz (or 33 ppm) away. Such off-resonance effects exceed, and hence strongly obscure, any NOEs. For 95% on-resonance saturation, off-resonance saturation to 10% is predicted to 3000 Hz (10 ppm at 300 MHz), which is also unacceptable. The validity of these simulations is confirmed for the case of 50% on-resonance saturation, for which the experimental points for the described line fall precisely on the predicted profile, as shown from line c in Fig. 13.

3.2.4. Differentiating NOEs from Off-Resonance Effects

Since the degree of off-resonance saturation depends only on the amount of offset, and not on its direction, placement of the decoupler frequency in the off-resonance

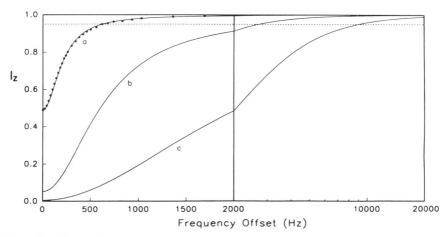

Figure 13. Plot of I_z versus RF offset for a line characterized by $T_1 = 4.7$ ms and $T_2 = 1.1$ ms. The line is irradiated with constant RF power $\gamma B_2/2\pi$ equal to 70 Hz (a), 300 Hz (b), 1000 Hz (c) during a time sufficient to reach steady state (≈ 20 ms). The data points represent experimental results on sperm whale *met*-aquo myoglobin 8-CH$_3$ with an actual decoupler power of 70 Hz. The dashed horizontal line represents the 5% saturation level (Lecomte *et al.*, 1991).

reference trace symmetrical to the peak which is a candidate for a NOE will lead to cancellation of off-resonance effects (provided the decoupler is very stable) and reveal only a true NOE. Should there be several candidates for NOEs, several different reference traces must be collected. A second procedure relies on the fact that a steady-state NOE is independent of the degree of saturation. The off-resonance effect, on the other hand, depends much more strongly on the degree of saturation, as shown clearly in Fig. 13 (Lecomte *et al.*, 1991). Thus the dependence of intensity in the difference trace on the degree of on-resonance saturation will identify NOEs; whether the NOE can be quantitated depends on the separation of the two peaks. These differentiation procedures are tedious, but necessary. Note that the failure to correct for off-resonance effects will indicate protons are closer than in reality for true NOEs and suggest the presence of NOEs where there are none.

A third time-consuming but effective strategy, particularly when symmetrical decoupler placement is not practical or when there is a very crowded spectral region, is to step the decoupler through a resonance and monitor the intensity in the difference trace for all peaks. A case in point is a portion of the high-spin ferric Mb trace for a series of resonances with $\Delta \sim 200$–300 Hz and T_1's 2–5 ms, as shown in Fig. 14 (Lecomte *et al.*, 1991). Saturating peak g gives a response to peaks d and f, and we wish to identify which of these is an NOE. By stepping the decoupler through resonance g at constant power, the intensity profile for peak d is that for off-resonance saturation, while that for f clearly maximizes when the decoupler is on g. An alternative is to saturate at constant decoupler power each of the peaks in a crowded region and for each case monitor the intensity of all nearby peaks in the difference trace. An example for another high-spin ferric Mb (Pande *et al.*, 1986) is shown in Fig. 15 for the intensity of peak d

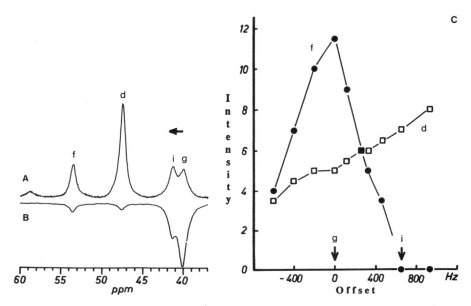

Figure 14. (A) Portion of the 360-MHz ^{1}H NMR spectrum of sperm whale myoglobin in ^{2}H$_2$O, pH = 6.2, at 25 °C. (B) Steady-state NOE difference spectrum resulting from 50% saturation of peak g. (C) The intensity of lines d and f in the difference spectra is plotted versus constant-power decoupler offset in Hz. The vertical scale is in arbitrary intensity units (Lecomte *et al.*, 1991).

(Fig. 15B) and peak h (Fig. 15C) upon saturating each of the other peaks (Lecomte *et al.*, 1991). The only NOE among these sets of peaks is between peaks f and h.

3.2.5. Transient NOEs

The form of this experiment eliminates saturation of any resonances and hence obviates many of the problems which characterize steady-state and truncated NOEs (Noggle and Schirmer, 1971; Neuhaus and Williamson, 1989). Instead, spin j is selectively inverted by a selective 180° pulse ($J = -J_0$ at $t = 0$) and the intensity of I monitored as a function of the time τ after the 180° pulse (the pulse sequence is illustrated in Fig. 9C), i.e.,

$$\eta_i\{j\}(\tau) = \eta_j\{i\}(\tau) = \frac{\sigma}{D}\{\exp[-(\bar{\rho} - D)\tau] - \exp[-(\bar{\rho} + D)\tau]\} \tag{26}$$

where

$$\bar{\rho} = \tfrac{1}{2}(\rho_i(T) + \rho_j(T)) \tag{27}$$

and

$$D = \sqrt{\Delta\rho^2/4 + \sigma^2} \tag{28}$$

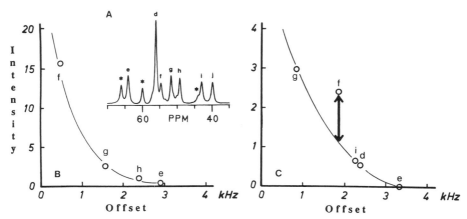

Figure 15. (A) Portion of the reference spectrum of *Aplysia limacina* metmyoglobin in ^2H_2O, pH 6.2, 25 °C. (B) The intensity of peak *d* in the steady-state NOE difference spectra obtained upon irradiation of transitions *f*, *g*, *h*, and *e* is plotted versus decoupler offset. The line is an approximate fit to the expected off-resonance saturation effect. (C) The intensity of peak *h* in the difference spectra obtained upon irradiation of transitions *g*, *f*, *i*, *d*, and *e* is plotted versus decoupler offset. The line is an approximate fit to the expected off-resonance saturation effect. The intensity change can be accounted for by off-resonance effects in all cases except peak *f*. The NOE contribution is marked by the double-headed arrow. The vertical scale is in arbitrary intensity units (Lecomte *et al.*, 1991).

with $\Delta\rho = |\rho_i(T) - \rho_j(T)|$. For $\rho_i(T) = \rho_j(T) = \rho$, Eq. (26) reduces to:

$$\eta_i\{j\}(\tau) = \eta_j\{i\}(\tau) = \exp[-(\rho - |\sigma|)\tau] - \exp[-(\rho + |\sigma|)\tau]$$
$$= \{1 - \exp(-2|\sigma|\tau)\}\{\exp[-(\rho - |\sigma|)\tau]\} \qquad (29)$$

The latter form of Eq. (29) represents an exponential buildup at a rate dependent on 2σ multiplied by an exponential decay rate dependent on $\rho - \sigma$. Note that, in this expression, τ refers to the time interval after the selective 180° pulse. Two facts are noteworthy about the transient NOE in comparison to TOE: the NOEs are identical for the two protons even if $\rho_i \neq \rho_j$, and the NOE builds up and decays, so the time interval must be selected carefully in order to maximize the sensitivity if only a single τ value is to be detected. The value of $\tau(\tau_{max})$ that gives the maximum transient NOE is

$$\tau_{max} = \frac{1}{2D} \ln[(\bar{\rho} - D)/(\bar{\rho} + D)] \qquad (30)$$

The comparison of the truncated and transient NOE responses for pairs of differentially relaxed geminal protons in Zones II ($\rho_i = 500$ s^{-1}) and III ($\rho_j = 200$ s^{-1}) of a 16-kDa protein ($\sigma_{ij} = -14$ s^{-1}) are compared in Fig. 11. For comparable relaxation rates, the maximum transient NOE is less than the steady-state NOE, the degree of reduction increasing with ρ. If $\Delta\rho$ is sufficiently large (i.e., $\rho_i > \rho_j$) the transient NOE can be larger than the steady-state NOE for *i*, but not for *j* (Fig. 11).

The initial slope of a transient NOE,

$$\frac{dI}{dt}\bigg|_{t=0} = 2\sigma t \tag{31}$$

is twice that of a TOE. This advantage, however, is generally lost for a paramagnetic protein because complete inversion of a signal is not attainable. This is due to the fact that the 180° pulse length needed to ensure reasonable selectivity is comparable (and often longer) than the ρ^{-1} for the peak to be measured. Limited selectivity of the inversion can introduce problems similar to but much less severe than those due to off-resonance saturation. Generally a balance between selectivity and inversion must be reached to optimize sensitivity (Paci et al., 1990). Because of the limitation on inversion and the weaker transient NOE response, the 1D transient NOE has limited applicability to paramagnetic molecules; a single case of its use has been reported and found less than satisfactory (Paci et al., 1990; Bertini et al., 1991b). The true potential of the transient NOE experiment is realized in its 2D formulation, the NOESY experiment, which will be considered in Section 4.

3.3. Magnetization Transfer Via Exchange

This phenomenon is related to the 1D NOE experiment both conceptually and experimentally (Sandström, 1982). It deals with the transfer of magnetization between two spins not by dipolar interaction but because the two alternate environments of a spin interconvert. Thus the identity of a given nucleus in the alternate chemical environments can be related, which leads not only to assignments from one form transferred to another but to a characterization of the rate process for the transfer. The method has played a key role in the early assignments of paramagnetic ferricytochrome by saturation transfer to diamagnetic ferrocytochrome, as modulated by electron self-exchange (Boyd et al., 1984; Santos et al., 1984; Turner, 1985; Keller and Wüthrich, 1981; Senn and Wüthrich, 1985). For a spin in two alternate environments i, j, with interconversion rates k_i, k_j,

$$i \underset{k_j}{\overset{k_i}{\rightleftharpoons}} j \tag{32}$$

The relevant equations for the magnetization vectors I, J in analogy to Eqs. (5a, b) are (Sandström, 1982):

$$\frac{dI}{dt} = -\rho_i(T)[I - I_0] - k_i I + k_j J \tag{33}$$

$$\frac{dJ}{dt} = -\rho_j(T)[J - J_0] - k_j J + k_i I \tag{34}$$

Saturating spin j ($J = 0$) for a time t (according to the pulse sequence in Fig. 9B), yields an analog of the truncated NOE, i.e.:

$$I(t) = I_0 \left[C \exp[-(\rho_i + k_i)t] + \frac{\rho_i}{\rho_i + k_i} \right] \tag{35}$$

where C depends on how fast spin j is saturated. For long saturation times ($t \to \infty$), the steady-state equation results, which defines a saturation factor F_i for spin i,

$$F_i = \frac{I_\infty}{I_0} = \frac{\rho_i}{\rho_i + k_i} \tag{36}$$

For $k_i \ll \rho_i$, $F_i = 1$ and no effect on I is observed; for $k_i \gg \rho_i$ ($F_i = 0$), the signal for i disappears. Since ρ is larger in a paramagnetic than an isostructural diamagnetic system, this experiment will require faster exchange rates to give the desired $0 < F_i < 1$. Moreover, since this experiment demands that the ^{1}H resonances of the alternate chemical environments be resolved, the chemical shift difference for the two environments, Δv_{ij}, must be larger than for a diamagnetic system in order to meet the resolution condition $2\pi\Delta v > k$. The large hyperfine shifts generally provide the necessary condition. A "transient" magnetization experiment can be performed where j is inverted at $t = 0$ ($J \to -J_0$) by a selective $180°$ pulse, leading to a time-dependent response in $I(t)$. This experiment is impractical for paramagnetic systems for the same reason as for the transient NOE; however, it forms the basis of the analogous 2D experiment, EXSY (Macura and Ernst, 1980; Boyd et al., 1984), which will be discussed briefly in Section 4.4.

4. 2D CORRELATION

4.1. General Considerations

It is only recently that 2D NMR has been successfully applied to paramagnetic molecules. While it is clear that much information can be obtained from these experiments, neither the scope nor the limitations has been fully explored. Thus the information available now should be considered a status report rather than a definitive appraisal. The potential problems are manifold, including the intrinsic efficacy of the experiment, the limitation of the spectrometer, the method of data processing, and the way that a specific experiment is programmed on a given instrument. As in the case of the 1D experiments, all responses in a paramagnetic system are weaker than in a diamagnetic system and occur on a much shorter time scale, placing greater demands on the fidelity of the spectrometer. We will not consider any theoretical aspects of the 2D experiments, drawing the relevant equations, when needed, either from the literature (Bax, 1982; Ernst et al., 1987; Neuhaus and Williamson, 1989; Martin and Zektzer, 1988) or by analogy with the 1D experiment.

Moreover, we will consider only a small fraction of the various 2D experiments described in the literature, namely those which are considered the workhorses for diamagnetic systems and those which have been found to be effective for paramagnetic systems. From this experience some generalizations are made as to which classes of 2D experiments are likely to be productive for paramagnetic systems. The considerations that dictate the productive experimental approach and processing strategy for paramagnetic systems are not all unique to paramagnetic systems. Indeed, similar considerations apply to 2D NMR involving quadrupolar nuclei (Domaille, 1984; Venable *et al.*, 1984; Gunther *et al.*, 1986; Chandrakumar and Ramamoorthy, 1992) and ^{1}H 2D studies *in vivo* (Delikatney *et al.*, 1991) in that both exhibit broad lines (short T_2). The conditions for a paramagnetic protein are somewhat more complicated than these because not only is the range of linewidth very large, but both T_1 and T_2 can be very short.

4.1.1. Pulse Sequences

Figure 16 shows pulse sequences of interest for homonuclear and heteronuclear 2D NMR (Bax, 1982; Ernst *et al.*, 1987). For our purposes, we group them into the following categories:

Class i. Pulse sequences where responses (scalar coherence for the experiments listed) develop during the acquisition time t_2, i.e., after the final or read pulse (Fig. 16A, B, I). For these pulse sequences choice of the acquisition times in the two dimensions are crucial for detecting responses.

Class ii. Pulse sequences where responses build (and decay) during the mixing time τ_m or other delay, prior to the read pulse and the t_2 acquisition time (Fig. 16C, D, E). In these experiments, choice of acquisition times is important but secondary to the selection of τ_m.

Class iii. Pulse sequences that include refocussing delays τ_D based on the magnitude of J, which generally stems from either proton homonuclear couplings $^3J_{HH}$ or heteronuclear couplings $^1J_{XH}$. Here the magnitude of τ_D relative to T_2 is of crucial concern (Fig. 16F, G, H, I). (Pulse sequences of class iii can include pulse sequences of class i.)

For all three classes, paramagnetic relaxation can strongly attenuate responses generated in each type of pulse sequence, and failure to anticipate strong relaxation in setting the parameters discussed above can obviate production of useful results. Specifically, for class i pulse sequences, t_1 and t_2 must be set based on the paramagnetically influenced T_2's. By choice of apodization, t_2 and t_1 can be reduced in postacquisition processing to account for shorter T_2's (see below). For class ii pulse sequences, useful results will be obtained only if τ_m is shortened appropriately for the paramagnetically influenced T_1 for NOESY (Section 4.3) or T_{1p} ($\sim T_2$) for TOCSY/HOHAHA (Section 4.2.4) experiments. In the case of class iii pulse sequences, useful results can be obtained only if paramagnetic relaxation does not destroy the magnetization during the required τ_D delays; T_2 cannot be short compared to τ_D. Hence, for each class of experiments, strong relaxation exerts profound effects on choice of experimental parameters and indeed on whether the experiment will succeed at all.

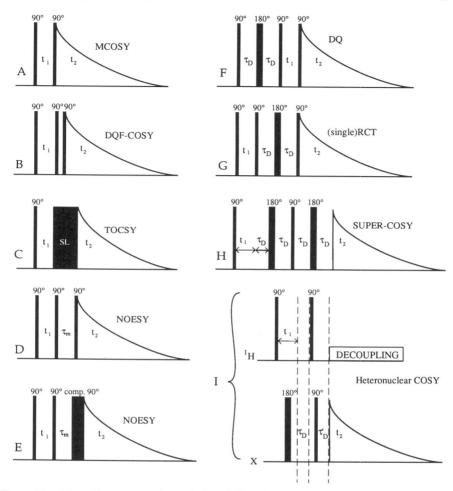

Figure 16. Schematic representations of selected 2D pulse sequences. In general, τ_D represents delays, which are functions of spin coupling J, τ_m is the mixing time for NOESY, t_1 the evolution time, t_2 the acquisition time, and SL the spin-lock time for TOCSY. It is important to note that for paramagnetic (unlike diamagnetic) systems, τ_D and SL may be large compared to T_2; and t_1 total, SL, τ_D, and τ_m may each be of magnitude similar to that of t_2. (A) Magnitude COSY (MCOSY). (B) Double quantum filtered COSY (DQF-COSY). (C) TOCSY. (D) NOESY. (E) NOESY with composite 90° pulse. (F) Double quantum homonuclear correlation DQ. (G) Relayed coherence transfer COSY (RCT), illustrated for a single relay step. (H) SUPER-COSY. (I) X-H heteronuclear COSY.

4.1.2. Pulse Power and Effective Bandwidth

In 1D pulse sequences, a 90° pulse is beneficial but often not essential. Such is not the case in many 2D experiments, whose success critically depends on uniform 90° pulses for all peaks in a spectrum (Bodenhausen *et al.*, 1984; Ernst *et al.*, 1987). A square pulse actually produces only a narrow frequency range of uniform excitation. For the study of diamagnetic proteins, the rolloff of pulse power at large resonance offsets is rarely a concern. This is not so for paramagnetic systems, where one of the most serious experimental constraints is lack of effective pulse power for peaks of large hyperfine shift. Lack of a uniform 90° pulse can give rise to various artifacts, such as mixing of scalar and dipolar correlations in a single experiment (Bodenhausen *et al.*, 1984) or simple reductions in intensity of diagonal and cross peaks with increasing resonance offset, resulting in missing correlations at large offsets from the carrier.

The reduction in pulse power at large carrier offsets is further compounded by the substitution of composite pulses (Bodenhausen *et al.*, 1984; Bax, 1985; Freeman, 1987) for simple 90° (or 180°) pulses in many common 2D pulse sequences. Composite pulses are popular and effective for studying diamagnetic systems, since they enhance B_1 homogeneity over a small bandwidth, correct for small errors in pulse phase, and in some cases help in water suppression (Bax, 1985; Freeman, 1987). However, the benefits gained over small bandwidths are in paramagnetic systems more than offset by the much narrower range of excitation, since a composite pulse is even less effective than a single pulse for a hyperfine shifted peak. It is useful to examine pulse sequences on any spectrometer and, if necessary, edit out composite pulses for studies on paramagnetic systems. If pulse power is still not sufficient for the bandwidths under consideration, moving the carrier should be considered. Most biomolecular 2D ^1H NMR spectra are acquired with the carrier on the water solvent, thereby placing at the same frequency two potential sources of artifacts: the residual solvent signal and any transmitter artifacts. However, if necessary, the transmitter frequency can be moved elsewhere, while irradiation of the solvent is still carried out on a second r.f. channel. Many paramagnetic systems have predominantly downfield shifts, and the carrier could be placed downfield to insure a more uniform excitation of the peaks of interest.

4.1.3. Acquisition Times and Digitization

The basic considerations for digitization and acquisition times for paramagnetic relative to diamagnetic systems are similar in 1D and 2D NMR, with requirements now for appropriate digitization in two dimensions. The sweep width needed to encompass all peaks determines the rate of digitization of the FID (assuming simultaneous acquisition on both channels of the ADC), whose inverse defines the dwell times DW_1, DW_2. The product of the dwell time and the number of points N_p in the t_2 dimension or the number of blocks N_b in the t_1 dimension sets the total acquisition time in each dimension.

$$\text{Acq}(t_2) = (DW_2)(N_p) \tag{37}$$

$$\text{Acq}(t_1) = (DW_1)(N_b) \tag{38}$$

The acquisition time determines the spectral resolution and is determined by N_p and N_b. Although $5T_2$ is defined as full decay of the FID, in practice the FID is monitored through $\text{Acq}(t_2) \sim 3T_2$. The same arguments apply in t_1, where N_b is now set by the number of t_1 blocks collected and is usually a minimum value sufficient to detect the responses of interest yet avoid prolonging the 2D data collection by unnecessary collection of additional N_b (see below). We find $\text{Acq}(t_2)$ as short as $\sim 2T_2$ to be sufficient. The total pulse-sequence recycle time T_R is given by the sum of $\mathrm{Acq}(t_2)$, the predelay, the pulses, and the interpulse delays.

In paramagnetic systems the much shorter T_1's and T_2's allow much shorter acquisition times than in diamagnetic cases, so that in practice (de Ropp and La Mar, 1991; Keating et al., 1991; Busse et al., 1991) the number of points and blocks are very similar to those in a diamagnetic system, i.e., 256 blocks × 1024 points. This, combined with larger sweep widths for paramagnetic systems, leads to (often much) lower spectral digital resolution than in diamagnetic cases; the larger linewidths, however, readily allow this. Zero-filling after apodization to increase digitization can be applied as needed. In t_1 the emphasis on keeping minimal N_b means that zero-filling is routinely applied to obtain adequate digital resolution, often more than once. Peaks not producing correlations (such as non–spin coupled resonances in COSY spectra) should be folded-in both in f_2 and in f_1 to increase digital resolution as well as decrease N_b for any needed $\mathrm{Acq}(t_1)$. Digital resolution of 10–30 Hz/point for Zone III and 30–100 Hz/point for Zone II appear reasonable.

4.1.4. Pulse Sequence Repetition Rates and Total Time for Data Collection

The total time TT for a 2D NMR experiment can be expressed as:

$$TT = (N_b)(N_s)(T_R) \tag{39}$$

where N_b and T_R are defined above and N_s is the number of scans per block. Under nonsaturating conditions, the signal-to-noise for a cross peak correlates with $\sqrt{N_s}$. The desired cross peaks for all 2D experiments are generally much weaker than in a diamagnetic system due to the enhanced T_1 relaxation and broader lines. The improvement in sensitivity with N_s alone, however, is very costly (and likely prohibitive) in TT. The very nature of the paramagnetically influenced protons, however, allows much of this "lost" sensitivity to be recovered by repeating the pulse sequence much more rapidly than in diamagnetic systems. This permits a much shorter T_R in a paramagnetic system and consequently a much larger value of N_s without a change in TT than for an isostructural diamagnetic system. For 2D homonuclear correlation in diamagnetic systems T_R is set ~1.5 T_1, with T_R typically 1–2 s. The much shorter T_1's in paramagnetic systems therefore indicate values of T_R of 3–30 ms for Zone II and 30–300 ms for Zone III. Because of the presence of a range of T_1's, no one data set is close to optimal for more than half of one of the zones, and several different T_R with compensating N_s to keep TT comparable may be necessary. In practice, T_R as short as 30 ms have been reported (see, for example, Keating et al., 1991; Busse et al., 1991). The improved sensitivity with much larger N_s due to much shorter T_R, however, does not come without a price. The enhanced repetition rates can lead to increased artifacts such as antidiagonals and t_1 streaks (see below) for intense narrow resonances. These artifacts are largely limited to the diamagnetic region and generally do not interfere with analysis of hyperfine-shifted cross peaks; some artifacts in COSY spectra can also be dealt with by symmetrization (see below).

An additional factor that needs to be addressed in an experimental setup is use of phase-sensitive or magnitude mode for 2D data collection. To achieve phase-sensitive data in f_1, twice as many t_1 blocks (real and imaginary) need to be accumulated as for magnitude mode data (States *et al.*, 1982; Marion and Wüthrich, 1983) to achieve equal t_1 acquisition time and spectral resolution. For some 2D experiments on paramagnetic systems, a given TT is more effectively used by doubling N_s and collecting half as many t_1 blocks than by collecting data in phase-sensitive mode with N_s halved and N_b doubled. Magnitude-mode homonuclear 2D experiments produce phase-twist lineshapes with broad tailing bases (Bax, 1982) and have been largely abandoned for diamagnetic studies due to the resultant poor resolution. In paramagnetic systems with high chemical shift dispersion, the drawbacks in lower resolution may well be outweighed by the twofold gain in N_s per given TT, at least in COSY studies (Section 4.2).

4.1.5. Signal Responses and Data Processing

One can visualize 2D experiments as producing two types of responses: in-phase and antiphase (Ernst *et al.*, 1987). In the former case, the magnetization produced by the 2D pulse train is at a maximum after the readout pulse and decays during the acquisition period. In the latter case, the resultant magnetization is divided into antiphase components, and the net response develops only after the readout pulse. The shape of the Fourier-transformed signals that result from an antiphase COSY cross peak as a function of linewidth is shown in Fig. 17. Class i experiments produce initial antiphase magnetization and class ii experiments yield in-phase magnetization. Class i and ii experiments also differ in the optimal apodization utilized, since for Class i a pseudoecho type function such as Gaussian or unshifted sine-bell/sine-bell-squared window is optimal, while for Class ii, strongly phase-shifted sine-bell/sine-bell-squared or exponential functions are optimal. Several of the most useful window functions are illustrated in Fig. 18. In either case the optimal apodization matches the natural time dependence of the response (Bax, 1982; Ernst *et al.*, 1987). The need for optimal sensitivity is balanced by the frequent need in 2D spectra for resolution enhancement to reduce

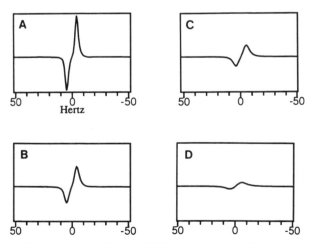

Figure 17.   Effect of increasing linewidth on COSY cross-peak intensity (for fixed J). (A) $J \gg \Delta$. (B) $J > \Delta$. (C) $J \sim \Delta$. (D) $J < \Delta$. Note strong antiphase cancellation of cross-peak intensity as Δ increases.

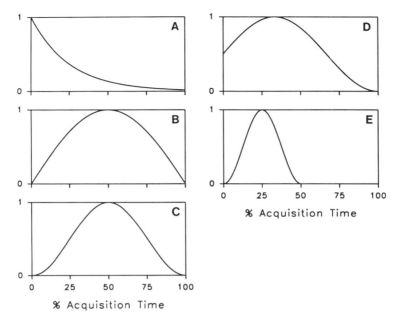

Figure 18. Selected window functions commonly utilized for apodization of paramagnetic biomolecules. (A) Exponential multiplication. (B) $0°$-shifted sine–bell. (C) $0°$-shifted sine–bell-squared. (D) $45°$-shifted sine–bell-squared. (E) $0°$-shifted sine–bell-squared applied over 50% of the acquisition time.

tailing of peaks; hence for class ii experiments functions such as $30–60°$ shifted sine-bell squared (Fig. 18D) have worked well since the initial part of the function reduces the broad base of peaks. For magnitude-mode COSY data, the combination of a phase-twist lineshape followed by a magnitude-mode calculation produces a broad base to peaks with long tails, which is best suppressed by application of $0°$ shifted sine-bell or sine-bell-squared window function (Bax, 1982).

   In addition to the control of the phase shift, the sine-bell and sine-bell-squared functions in most software packages offer easy control of the period of the function, i.e., control over how many points of the collected data the function is applied to before decaying to zero. For class i experiments, COSY in particular, the optimal choice of t_2 and t_1 is a function of the T_2 of the various peaks of interest (see Section 4.2). For a wide range of T_2's, one can collect numerous 2D data sets or more simply adjust the t_1, t_2 acquisition times artificially in data processing by varying the period of the apodization function, as shown in Fig. 18E. Figures 19B, C compare the effect of apodizing a magnitude COSY spectrum with $0°$-shifted sine-squared-bell over 256×256 (16 ms × 16 ms) and 128×128 (8 ms × 8 ms) points; notice that the shorter-period function significantly improves the detectability of the cross peaks for broad resonances, with the price of loss of resolution near the diagonal as the apparent T_2 is reduced; such a tradeoff characterizes all applications of paramagnetic systems with a wide range of T_1's and T_2's.

Section 1.4.2 discussed the problem of spectral baseline roll in 1D spectra of strongly paramagnetic systems because of distortions of the early points of the FID or large first-order phase corrections. An analogous situation exists in both dimensions of a 2D experiment (Marion and Bax, 1989). In the t_2 dimension, care should be taken to adjust the timing of the first sampled data point to produce the minimum first-order phase correction for the first block of data. To reduce initial distortions of the FID one can employ a slower sampling rate in t_2; i.e., a smaller sweep width to fold-in resonances of large hyperfine shift. Baseline distortions in f_1 are generally less troublesome than in f_2 if the delay between the pulses nesting the first t_1 delay (t_1 time of the first block) is kept short to minimize baseline roll. Failure to do so produces an analogous situation to distortion of the initial points of a 1D FID with resultant baseline curvature (Marion and Bax, 1989). Mathematical correction of the intensity of the first t_1 point is also routinely applied to correct the initial distortion of the response in t_1 (Otting et al., 1986). All modern NMR software packages provide routines that allow the user to choose baseline points to fit to a function, typically a third to fifth-order polynomial that corrects to a flat baseline. Baseline straightening in both f_2 and f_1 of 2D spectra can be employed. Generally, baseline straightening is done uniformly on all f_2, f_1 slices using points selected in the first transformed t_1 block (for f_2) or on summed slices (for f_1). The same precautions apply as for 1D spectra: care must be exercised in choice of baseline points in systems with broad lines.

Instrumental instabilities such as imperfections in pulse lengths and phase setting often produce artifacts known as t_1 streaks or ridges in 2D spectra (Mehlkopf et al., 1984). For the often weak correlations for paramagnetically relaxed lines, such artifacts can be especially troublesome. Rapid pulse repetition rates used in paramagnetic systems accentuate these ridges and can also produce antidiagonals in homonuclear 2D spectra (Turner and Patt, 1989; Derome and Williamson, 1990; Turner, 1992). To deal with these artifacts in magnitude 2D data, symmetrization can be employed (Baumann et al., 1981; Ernst et al., 1987; Wider et al., 1984), with the caution that false cross peaks can be produced or real ones annihilated; but false cross peaks in practice can be identified by studies done at different temperatures due to the strong and predictable temperature dependence of hyperfine shifted peaks (Section 4.2.5). Figure 19 shows magnitude COSY spectra of metMbCN without (Fig. 19A) and with (Fig. 19B) symmetrization. Symmetrization is effective in absolute mode spectra in rejecting artifacts from t_1 ridges and antidiagonals that do not obey the expected symmetry about the diagonal expected of homonuclear correlation cross peaks. Symmetrization is not recommended for phase-sensitive data and should always be used with caution. Inspection of symmetrized and nonsymmetrized data along with the aforementioned variable temperature studies aid in identifying real correlations.

4.2. Scalar Correlation

In this section we consider the experiments done to map spin–spin coupling networks, the major assignment tool of functional groups (side chains) in biopolymers. We consider primarily 3-bond ^1H–^1H data, for which J coupling is small, <15 Hz, and therefore not resolved for ^1H signals in Zones I–III of metalloproteins. One-bond heteronuclear couplings, on the other hand, i.e., ^1H–^{13}C, ^1H–^{15}N, are much larger. For two nuclei i, j coupled via $J_{ij} \neq 0$, the cross peak intensity (coherence) develops during

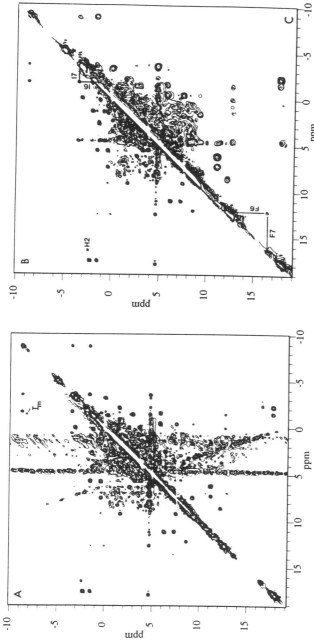

Figure 19. Magnitude COSY spectra of sperm whale metMbCN at 500 MHz, 35 °C, pH = 8.6 in ^2H_2O. All three spectra are from the same raw data but are processed differently. (A) Data processed without symmetrization, utilizing apodization by 0°-shifted sine–bell-squared over 16-ms acquisition time in both t_1 and t_2. (B) and (C) split diagonal representation of symmetrized COSY spectra with (B) apodized with 0°-shifted sine–bell-squared over 16-ms acquisition time in each dimension and (C) apodized with 0°-shifted sine–bell-squared over 8 ms in each dimension. Note that symmetrization removes the weak cross peak I_m (arising from a minor conformation Ile residue) present in A but not in B; however, symmetrization also eliminates troublesome t_1 ridges and artifacts. Comparing traces B and C, resolution is lost near the diagonal in C (16–17 cross peak is apparent in B but not C) and the intensity of cross peaks from narrow lines is diminished or lost (note absence of cross peak H2 in C). However, detectability of broad lines increases, most strikingly for the F6–F7 cross peak (Phe 43 p-H:m-H) only detected in C. Data was collected with a repetition rate of 3 s^{-1}, utilizing acquisition times of 64 ms (t_2) and 16 ms (t_1) with 1024 t_2 and 256 t_1 points, zero-filled to 1024 × 1024 points. Total data collection time was 2.5 h in a 10-mm tube. Full discussion of assignments is given in Emerson and La Mar (1990a) and Yu *et al.* (1990).

the acquisition time t in a sinusoidal fashion due to the coupling and at the same time decays exponentially as dictated by the linewidth πT_2^{-1}, according to Bax (1982):

$$a_{ij} = \sin(\pi J_{ij} t) \exp(-t/T_2) \tag{40}$$

with $T_2^{-1} = \frac{1}{2}(T_{2i}^{-1} + T_{2j}^{-1})$. The full development of coherence ($\pi J t = \pi/2$) requires $t = 1/2J$, for $^1\text{H}-^1\text{H}$ coupling of 7 (vicinal) and ~ 14 Hz (geminal); this requires 70 and 35 ms, respectively. The larger heteronuclear coupling ~ 125 Hz ($^{13}\text{C}-\text{H}$) and ~ 50 Hz ($^{15}\text{N}-\text{H}$) require only 4 and 10 ms, respectively.

4.2.1. Homonuclear ^{1}H Correlation or COSY

The relevant pulse sequences are shown in Fig. 16 (Aue *et al.*, 1976). The key points are that coherence develops completely during the detection period (Eq. 40), and the cross peak response is antiphase, as shown in Fig. 17. For protons in Zones I–III, the condition $T_2^{-1} \gg J$ applies, so coherence does not develop significantly, and has a pseudoecho form, as shown in Fig. 20. The COSY response in a paramagnetic system is going to be weaker than in an isostructural diamagnetic system due both to short-circuiting the full buildup of coherence (sin $\pi J t$) by T_2 relaxation (e^{-t/T_2}), and by substantial diminution of the detection of the weak coherence due to partial cancellation of the broad components of the antiphase cross peak response (Fig. 17) (Oschkinat and Freeman, 1984). However, there will always be some coherence buildup, and the antiphase cancellation will never be complete; the response, however, can be too weak to detect practically. To detect COSY cross peaks for paramagnetically relaxed ^{1}H signals simply requires that more care be exercised in searching for this weak response. Thus the emphasis in data collection is on sensitivity, i.e., maximum N_s (scans per block) per unit time, fastest repetition rates, and minimal acquisition time.

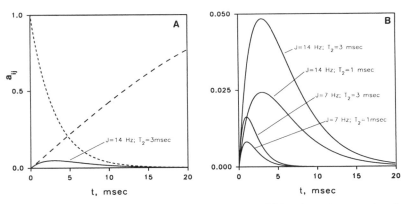

Figure 20. Plots of COSY cross-peak intensity a_{ij} versus acquisition time t_1 or t_2 as given by Eq. (40). (A) Plots of $\sin(\pi J t)$ (dashed line), e^{-t/T_2} (dotted line), and a_{ij} (solid line) for $J = 14$ Hz and $T_2 = 3$ ms. (B) Plots of a_{ij} for the specified combinations of J and T_2. The vertical axis is expanded $20\times$ in B relative to that in A. In B, the maximum intensity for each plot is at the point $t = T_2$. As T_2 decreases (larger Δ), the maximum in the curve shifts to shorter t; also, shorter T_2 and/or smaller J results in diminished cross-peak intensity.

The form of the coherence for the relevant case $T_2^{-1} \gg J$ leads to a maximum coherence for $t = T_2$ (Fig. 20), and this is optimally detected for $t_{aq} = 2T_2$ for both dimensions using a pseudoecho filter that profiles the coherence, i.e., a $0°$-shifted sine-bell or sine-bell-squared filter (as shown in Fig. 18B, C). The dilemma for a paramagnetic system is that T_2's can range over well more than an order of magnitude for protons of interest, demanding a large range of t_{aq} for optimally detecting the cross peaks for each pair of protons. In fact, however, a single $t_{aq} = 2T_2$ for the narrowest pair of the relaxed lines is sufficient and determines the recycle time (by its T_1) and N_s by the TT available. Since the coherence develops in the detection period, *the data collected for the longest $2T_2$ contains the coherence for all pairs of peaks with broader lines* (shorter T_2), but the optimal detection of each coherence will demand different processing of the data; i.e., the number of t_1 blocks and t_2 points of the total collected must be selected that correspond to a $t = 2T_2$ for the faster-relaxing proton of a pair and the pseudoecho filter applied to this fraction of the originally collected data (as in Fig. 18E). As in the case of 1D spectra, the highly differential relaxation rates in a paramagnetic system do not allow a single set of processing conditions that are optimal, but require a range of conditions suitable to limited subsets of the differentially relaxed lines. The apodization over a fraction of the collected data improves cross peak intensity for broad peaks, but the concomitant line broadening for all resonances leads to loss of resolution for the narrower peaks. This is illustrated in the MCOSY map of metMbCN for which many of the upfield cross peaks readily detected when $t_{aq} = 16$ ms for t_1, t_2 (Fig. 19B) are severely broadened when the larger linewidth (100 Hz, 80 Hz) Phe CD1 p–H:m–H (F6:F7) cross peak is detected by using only data points over $t_{aq} = 18$ ms (Fig. 19C).

The relative intensities of cross peaks in a system should not depend on the particular way the COSY experiment is executed, be it phase sensitive, magnitude, or DQF-COSY. However, since the phase-sensitive experiments (States *et al.*, 1982; Marion and Wüthrich, 1983) demand twice the t_1 blocks to provide the same inherent signal-to-noise ratio, as compared to MCOSY, MCOSY is clearly more sensitive. The advantage of the phase-sensitive COSY variants, namely, better resolution, is relatively unimportant for paramagnetically relaxed lines since they are usually also hyperfine shifted and hence resolution is less of a problem. The ease of setup of the MCOSY, together with its lack of required phasing, will likely make this experiment the simplest workhorse for detecting scalar connectivity for hyperfine-shifted, relaxed resonances. A comparison of the different variants of COSY has been carried out for metMbCN, for which MCOSY and phase-sensitive COSY gave essentially identical cross peak responses (Yu *et al.*, 1990). The DQF-COSY spectrum, conversely, exhibits weaker cross peaks than MCOSY selectively for severely broadened resonances, as illustrated in Fig. 21. The cross peaks for Ile 99(FG5), which involve a line 70 Hz broad (I5:I6, I5:I7) are selectively decreased in Fig. 21A relative to Fig. 21B, and the cross peak for the Phe 43(CD1) 100 Hz broad m–H to p–H (F6:F7) is missing in the DQF-COSY (Fig. 21C) relative to that in the MCOSY map (Fig. 21D). While the cross peaks that show selectively reduced intensity in the DQF-COSY map involve resonances with some of the largest offsets, varying the offset by moving the carrier did not influence the DQF-COSY peak intensities (not shown). The effect is not understood, but the DQF appears to discriminate against broad lines.

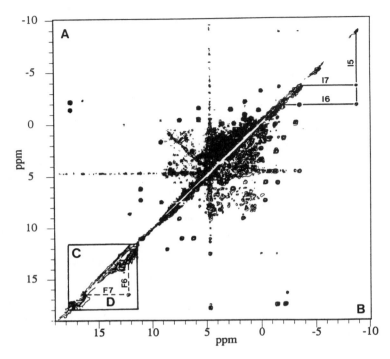

Figure 21. Split-diagonal comparison of DQF-COSY and MCOSY spectra of sperm whale metMbCN at 500 MHz, 35 °C, pH = 8.6, collected and processed under identical conditions except as noted below. (*A*) DQF-COSY processed with 0°-shifted sine–bell-squared apodization over 16 ms in both t_1 and t_2. (*B*) MCOSY spectrum with same apodization as trace *A*. (*C*) Insert of DQF-COSY spectrum processed with 0°-shifted sine–bell-squared apodization over 8 ms in both t_1 and t_2. (*D*) Insert of MCOSY spectrum with same apodization as (*C*). Note that while the DQF-COSY spectrum has slightly better resolution near the diagonal, the MCOSY spectrum provides much stronger cross peaks among signals of large Δ. This is particularly apparent for the Ile 99(FG5) spin system marked in trace *B* (I5 to I6, I7 cross peaks) compared to trace *A* and by the detection of the Phe 43(CD1) cross peak F7–F6 in trace *D* but not in trace *C*. Both data sets were collected identically with $t_1 = 16$ ms, $t_2 = 64$ ms, 1024 t_2 points, 256 t_1 points, sweepwidth 16 KHz, pulse repetition rate of 3 s^{-1}, and 96 scans per t_1 block in a 10-mm tube. Both data sets were processed identically with the apodization given above, zero-filled to 1024 × 1024 points, and displayed with a magnitude calculation in f_1. The only variance between the two data sets was that the MCOSY dataset was collected in half the total time of the DQF-COSY data set, which required twice as many t_1 blocks for hypercomplex data acquisition (de Ropp and La Mar, unpublished results).

4.2.2.   Homonuclear COSY-Type Experiments with Delays

The problem of antiphase cross peak cancellation has been addressed for small J in diamagnetic systems. The introduction of appropriate delays in the pulse sequence (of the order $1/J$), as shown in Fig. 16, allows the formation of in-phase cross peaks;

two such variants are SUPERCOSY (Kumar et al., 1984) and ISECR-COSY (Telluri and Scheraga, 1990). Both of these experiments have been shown to result in superior cross peak sensitivity in the case of the low-spin ferric hemin, particularly for the narrowest lines involved in very small spin couplings, i.e., protons in Zone IV (Bertini et al., 1991c). The delays needed for larger J values, however, are much larger than T_2 for lines in Zones III and II, and hence the T_2 relaxation within the necessary delays will likely annihilate the cross peaks.

Another early workhorse experiment for diamagnetic proteins is the relay coherence transfer or RCT-COSY (Eich et al., 1982; Bax and Drobny, 1985) which establishes sequential connectivity; the pulse sequence is shown in Fig. 16. The delays needed to implement this method, however, are again of the order of $(2J)^{-1}$ to $(4J)^{-1}$, which are much longer than T_2 for any proton but those in the outer ranges of Zone III and in the "diamagnetic" Zone IV. Other experiments in the category are the homonuclear multiquantum experiments (Bax et al., 1981; Rance et al., 1985), such as the double-quantum-pulse sequence shown in Fig. 16. Hence these 2D experiments with delays are unlikely to have much significant utility for the signal in Zones I and II or most of Zone III; cursory investigation of these methods appears to confirm this very limited utility (Satterlee et al., 1991; Yu and La Mar, unpublished observations).

4.2.3. Heteronuclear COSY

Most of the considerations addressed for homonuclear COSY are similarly relevant in heteronuclear COSY (Maudsley and Ernst, 1977; Bodenhausen and Freeman, 1977; Maudsley et al., 1977). However, since the one-bond ^1H-^{13}C and ^1H-^{15}N couplings are much larger than typical ^1H-^1H couplings, a given system for which $T_2^{-1} > J\,(^1H$-$^1H)$ can have $T_2^{-1} < J\,(^1H$-$^{13}C)$; hence, coherence is more strongly developed and antiphase cancellation is less of a problem for a given zone. This advantage is diminished somewhat by the need to introduce the delay of $(2J)^{-1}$ after the final pulse to yield decoupled spectra for the observed nucleus (see Fig. 16). Nevertheless, several successful 2D studies have been reported for the natural abundance ^{13}C signals of the heme methyls of low-spin ferric hemoproteins (Yamamoto, 1987; Yamamoto et al., 1988). An example for sperm whale metMbCN is illustrated in Fig. 22. The ability to observe the cross peaks at natural abundance for ^{13}C reflects the great promise of this technique if ^{13}C (or ^{15}N) labeling can be introduced. The sensitivity should also improve significantly for ^1H detection, since proton-detected versions of heteronuclear single-bond correlations utilizing either single or multiple quantum coherence (HMQC) are more sensitive than X-nucleus detection sequences (Bax and Marion, 1988; Bax et al., 1990). HMQC has recently been applied to low-spin heme with weakly relaxed Zone III signals and also to a small (82 amino acids) heme protein for the detection of ^{13}C-^1H single-bond correlations (Timkovich, 1991). However, for paramagnetically influenced resonances of a two-iron ferredoxin, no single-bond ^{15}N-^1H correlations were detected (Oh and Markley, 1990a, b). This could be due to the larger linewidths in this system or the smaller $^1J_{NH}$ (than $^1J_{CH}$).

Detection of long-range correlations involving protons and heteronuclei, such as $^2J_{CH}$ couplings by the HMBC sequence (Bax and Summers, 1986) will be difficult in strongly relaxed systems. The refocusing interpulse delays used in such pulse sequences

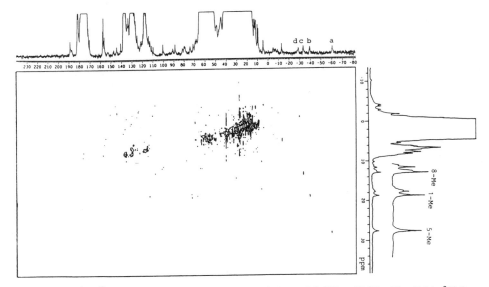

Figure 22. ^{1}H–^{13}C heteronuclear COSY of sperm whale metMbCN at 22 °C, pH = 7.5 in ^{2}H$_2$O. Reference one-dimensional ^{1}H and ^{13}C spectra are displayed along the f_1 and f_2 axes, respectively. From the known ^{1}H assignments, the ^{13}C assignments of signals *a, b, c* were made. The data was collected using carbon detection at 67.8 MHz with 2048 data points in t_2 over 30 KHz and 64 t_1 points with f_1 = 14 KHz. 1024 scans per t_1 block were utilized with a repetition rate of *circa* 1 s^{-1}. The data was processed with exponential apodization in both dimensions and zero-filled to 256 × 2048 and results are shown in absolute-value mode (Yamamoto, 1987).

are of the order of $1/2J$, which for $^2J_{CH}$ is on the order of 35 ms. The rapid spin–spin relaxation of the systems under consideration here will likely annihilate coherence formed during an interpulse delay which is long compared to T_2 and make application to paramagnetic biomolecules difficult. This has been confirmed in a study using HMBC, which detected $^2J_{CH}$ correlations for weakly relaxed resonances in low-spin heme but not for Zone III resonances in cytochrome c-551 (Timkovich, 1991). The same study did find weak $^2J_{CH}$ correlations in the heme protein by use of the heteronuclear multiple-quantum relay (HMQR) experiment (Lerner and Bax, 1986), which does not employ a delay based on $^2J_{CH}$ and hence may be the method of choice for such studies in paramagnetic proteins. Another method for detecting spin connectivities across portions of molecules without $^3J_{HH}$ (such as peptide bonds or pyrrole rings of heme groups) in diamagnetic systems is heteronuclear–heteronuclear correlation in molecules with isotopic enrichment. Such experiments include ^{13}C–^{15}N (Niemczura *et al.*, 1989) and ^{13}C–^{13}C scalar correlations, and their success in paramagnetic systems (if any) will be based on the fact that $^1J_{CC}$ is much larger than typical $^3J_{HH}$ and $^1J_{CN}$ is somewhat larger. No experimental data on these methods from paramagnetic systems is available to date. Overall, applications of heteronuclear scalar correlation to paramagnetic biomolecules to date have been few, but the potential benefits of large couplings and use of isotopic enrichment should ensure future endeavors in this area.

4.2.4. TOCSY/HOHAHA

The detection of scalar connectivities in the rotating frame has several advantages over COSY variants. The relevant pulse sequence, with isotropic mixing time τ_m as defined by a spin-lock field, is given in Fig. 16 (Braunschweiler and Ernst, 1983; Bax and Davis, 1985b). The spin-lock field is conveniently provided by an MLEV-17 or DIPSI pulse train. The cross peaks have purely in-phase absorptive character, leading to better resolution and obviating the antiphase cancellation of coherence present in COSY. Moreover, the same experiment carried out at different mixing times gives both primary and remote connectivities. This is the current workhorse experiment for defining total scalar connectivities in diamagnetic systems. The development of coherence takes the form of Eq. (40) but is more complicated and not subject to a single analytic form (Cavanagh *et al.*, 1990). What is crucial for the experiment in comparison to the COSY variants is that the coherence develops solely during the directly unobservable mixing time τ_m. This requires care in the selection of the value of τ_m in order to provide information on rapidly relaxed protons in the final data set, as well as consideration of the relaxation properties in subsequent data processing. Particularly for paramagnetic systems for which resolution of hyperfine-shifted peaks is unlikely to be a problem, the window function should maximally retain the data at short t for sensitivity, for which exponential or strongly phase-shifted sine-bell or sine-bell-squared functions are appropriate. Identical processing of COSY and TOCSY data does not lead to a meaningful comparison of the relative effectiveness of the two experiments.

Since the coherence builds up solely during τ_m, and during this time the relevant in-plane magnetization decays by $e^{-t/T_{1\rho}}$ ($T_{1\rho} \sim T_2$), the selection of τ_m must consider the T_2's for a particular pair of protons in order that information on that set can exist in the final data set. Therefore τ_m must be shorter to effectively detect primary cross peaks in a paramagnetic system than in an isostructural diamagnetic system, and the broader the lines (shorter T_2), the shorter τ_m. A reasonable test as to whether a cross peak can be expected for a given resonance in a TOCSY map at a given τ_m is to determine if the particular peak of interest is still present in the first slice of the TOCSY experiment. A series of such first slices can be quickly collected and processed. A set of these first slices as a function of τ_m is shown in Fig. 23 for low-spin metMbCN. Note that several peaks (Phe 43 p-H, Ile 99 C_γH) are essentially absent for $\tau_m > 20$ ms. The upfield portion of the 25-ms TOCSY and MCOSY spectra of metMbCN are compared in Fig. 24. Peaks lost in TOCSY with $\tau_m = 25$ ms relative to MCOSY include those involving the broadest upfield resonance, the Ile 99 $C_{\gamma 1}$H coupling to the geminal partner $C_{\gamma 2}$H. Also missing in this map, but not shown, is the cross peak for Phe 43 p-H:m-H, since both peaks have widths $\sim$100 Hz. However, both sets of peaks become detectable with $\tau_m = 10$ ms (Qin and La Mar, unpublished results). On the other hand, both DQF-COSY (Fig. 25A) and TOCSY for $\tau_m = 25$ ms (Fig. 25B) maps yield the expected primary connectivities for the hyperfine-shifted Leu 89 in the outer edges of Zone III, with the TOCSY spectra yielding some of the remote connectivities. Since pairs of coupled protons in a paramagnetic system likely have greater spectral dispersion than in diamagnetic systems, stronger spin-lock fields are required, which may not always be attainable on a given instrument. Successful published TOCSY studies of paramagnetic molcules with large spectral dispersion have been carried out on paramagnetic model compounds with long T_2's and hence narrow lines (Luchinat *et al.*, 1990; Bertini *et al.*,

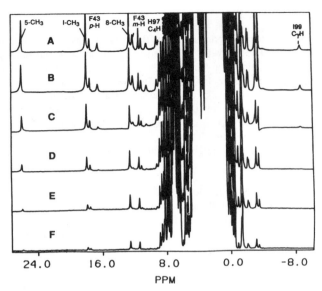

Figure 23. Effect of the length of the spin-lock mixing time on peak intensities in the first t_1 block of a TOCSY spectrum of sperm whale metMbCN in ^2H_2O, pH = 8.6 at 30 °C. (A) 500-MHz ^1H NMR reference trace collected with the f_1 channel with $t_{90} = 30\ \mu s$; selected peaks are labeled. (B)–(F) were collected with the f_2 decoupler channel with $t_{90} = 27\ \mu s$ with the MLEV-17 mixing scheme. (B) $\tau_m = 0$. (C) $\tau_m = 10$ ms. (D) $\tau_m = 20$ ms. (E) $\tau_m = 40$ ms. (F) $\tau_m = 60$ ms. All spectra consisted of 256 scans with a recycle time of 1.6 s (Yu and La Mar, unpublished results).

1991c; Ming *et al.*, 1992). Also, studies on an iron sulfur protein (*Clostridium pasteurianum* oxidized ferredoxin) with hyperfine shifted Cys proton signals from Zone II reveal that TOCSY cross peaks can be more intense than MCOSY cross peaks (Sadek *et al.*, 1993), indicating that the decay of the in-plane magnetization during the spin-lock mixing time is more than compensated by the absorptive TOCSY as opposed to antiphase COSY cross peak response. More TOCSY studies of paramagnetic proteins with Zone II and III protons are needed, but the outlook appears promising.

One of the problems with detection of coherent magnetization transfer (TOCSY) in the rotating frame is that it is difficult to completely suppress incoherent magnetization transfer (i.e., ROESY or rotating-frame NOESY) (Griesinger *et al.*, 1988; Cavanagh and Rance, 1992), in part because the efficacy of the experiment depends so critically on precise pulse phase and flip angles. Thus many TOCSY experiments rely on 90° composite pulses that optimize the pulse-power profile for a diamagnetic window, but must do so at the expense of the total spectral width since the composite pulse reduces the total spectral width of the pulse power. Preliminary experiments in our lab reveal that for a given τ_m, several of the offset broad resonances give significantly stronger TOCSY cross peaks if the composite pulse is replaced by a

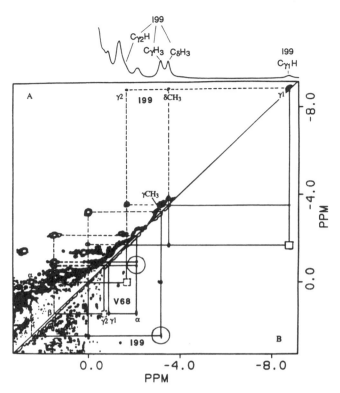

Figure 24. Comparison of the phase-sensitive TOCSY/HOHAHA spectrum with the magnitude COSY spectrum of metMbCN in ^2H_2O, pH = 8.6 at 35 °C. (A) The magnitude COSY spectrum. (B) The phase-sensitive TOCSY spectrum, collected with a 25-ms mixing time and a trim-pulse length of 100 μs. Spin system connectivities are drawn for residues of Ile 99(FG5) and Val 68(E11). The small solid-line squares indicate where cross peaks are too weak with $\tau_m = 25$ ms but are observed with $\tau_m = 10$ ms. The small dotted-line squares indicate the cross peaks which are observed only at a lower contour level. The small circles indicate remote connectivities. The 2D data were collected in 512 t_1 blocks of 128 scans over a 15,504 Hz bandwidth using 1024 complex t_2 points and a pulse repetition time of 0.6 s^{-1}. The t_{90} was 13 μs for the magnitude COSY and 27 μs for the TOCSY spectra. Processing involved 0°-shifted and 30°-shifted sine–bell windows over the collected data for the COSY and TOCSY, respectively, prior to zero-filling to 2048 × 2048 data points. Adapted from Yu *et al.* (1990).

single pulse; it is also clear that such a substitution introduces some ROESY cross peaks (Qin and La Mar, unpublished results). The response to a TOCSY experiment depends more critically than does a COSY experiment on the resonance offsets of the peaks, the nature and power of the pulse, and the strength of the spin-lock field (Cavanagh *et al.*, 1990) and hence may differ substantially for different systems with identical linewidth and *J*, or on the same system as detected by different spectrometers.

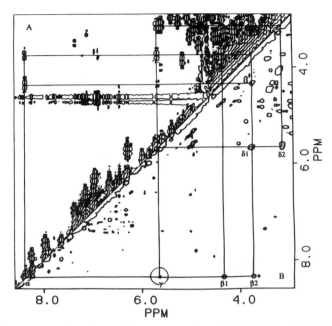

Figure 25. Comparison of portions of the phase-sensitive DQF-COSY and TOCSY/HOHAHA spectra of metMbCN in ^2H_2O, ph = 8.6, at 35 °C. (A) The phase-sensitive DQF-COSY spectrum. (B) The phase-sensitive TOCSY spectrum, collected with a 25-ms mixing time and a trim-pulse length of 100 μs. Spin-system connectivities are drawn for the residue of Leu 89(F4). The small circle indicates the relayed cross peak between C_aH and $C_\gamma H$. The TOCSY data were collected and processed as described in Fig. 24. The DQF-COSY data were collected as for the magnitude COSY data in Fig. 24 except that the window function applied is a 15°-shifted sine–bell (Yu *et al.*, 1990).

4.2.5. Alternate Determination of Remote Connectivities

All of the conventional 2D experiments designed to detect remote connectivities in diamagnetic biopolymers will be considerably less effective, if not useless, for significantly relaxed and hyperfine-shifted lines. For resolved lines this does not constitute a problem. However, there are generally numerous broadened lines whose hyperfine shifts are either negligible or sufficiently small so as to leave them obscured by the intense diamagnetic envelope of Zone IV protons. The nature of hyperfine shifted peaks, however, is that they are temperature dependent, and hence the collection of a scalar correlation map (COSY or TOCSY) with only primary connectivities at two or more temperatures leads to the unique identification of multiple spin connectivities for a set of unresolved resonances, as illustrated in Fig. 26 for *Aplysia* metMbCN (Qin and La Mar, 1992).

4.2.6. Selected Case Studies

Too few systematic studies have been carried out to date to determine the full scope and limitations of the scalar correlation experiments. On the other hand, a sufficient

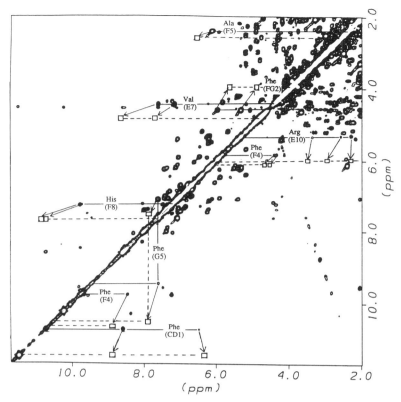

Figure 26. Portion of the 500-MHz ^{1}H NMR magnitude COSY spectrum of *Aplysia* metMbCN in ^{2}H$_2$O, pH = 8.9, 45 °C, illustrating the dramatic temperature sensitivity of cross peaks arising from hyperfine-shifted resonances. The solid lines represent the connectivities observed at 45°; the dashed lines (with an arrow connecting solid and dashed lines) represent the cross-peak connectivities observed at 25°. The spectrum was collected with 256 t_1 blocks of 196 scans per block, each with 1024 t_2 points, over a bandwidth of 15,500 Hz; the pulse sequence repetition rate is 2 s^{-1}. The data were processed by a 0°-shifted sine–bell window function and zero-filled to 2048 × 2048 data points (Qin, 1992).

range of systems has been investigated with at least partial success so that a vague picture is emerging as to what can be expected. The systems studied most extensively and with the largest success are the low-spin ferric hemoproteins, for which there are two protons in Zone I, only seven or eight protons in Zone II, and ~50 protons in Zone III. For the small metMbCN case it has been possible to identify all hyperfine-shifted spin systems, primarily by variable-temperature MCOSY experiments, with the exception of the axially bound ligand (Emerson and La Mar, 1990a; Yu *et al.*, 1990; Rajar-athnam *et al.*, 1992; Qin *et al.*, 1992; Qin and La Mar, 1992); the spin systems for some of the most strongly paramagnetically influenced residues are marked in Figs. 24 and 25. The broadest resonances exhibiting cross peaks are ~100 Hz. More recent studies on another metMbCN variant show that such studies can be extended to the backbone, taking advantage of the hyperfine shifts of backbone $C_\alpha H$, NH resonances (Qin et al., 1992; Qin and La Mar, 1992). For this study, variable-temperature MCOSY was able to unambiguously locate and identify the spin systems for all active site residues except one. This has led to sequence-specific assignment of 95% of the hyperfine-shifted resonances in a manner similar to that used for diamagnetic systems.

Limited extensions to larger isoelectronic hemoproteins have also met with surprising success. Methyl T_1's are very largely unchanged on going from the 17-kDa Mb to the 34–42 kDa heme peroxidases. The increased size leads to a proportional increase in Δ due solely to diamagnetic contributions and an effective increase in total Δ of about a factor of 2 (see Table 1). In spite of the increase in Δ, however, the COSY cross peaks from the heme side chains and the resolved nonheme resonances have allowed the establishment of the spin connectivity patterns for the $NH-C_\alpha H-C_\beta H_2$ of the axial His, the heme 2-, 4-, 6- and 7-side chains, as well as portions of some active-site side chains for HRP–CN (de Ropp et al., 1991b). The MCOSY map of 42-kDa HRP–CN illustrating these connectivities is shown in Fig. 27A; the cross peaks represent J values in the range 5–15 Hz with Δ of 50 to 120 Hz. Similarly, useful MCOSY data has been reported for the cyanide complex of lignin peroxidase (de Ropp et al., 1991a; Banci et al., 1991d) and cytochrome c peroxidase-cyanide (Satterlee et al., 1991; Banci et al., 1991c). A preliminary MCOSY map of 78-kDa LPO-CN has yielded some COSY cross peaks (Sette, de Ropp and La Mar, unpublished results). What remains to be determined for such a large protein is how completely the spin connectivities of a residue can be established considering both the increased linewidth and spectral congestion in the diamagnetic envelope due to the large increase in the number of protons in Zone IV.

The COSY connectivities for broader lines are more likely detectable for larger J values (Fig. 20). For the varied class of iron–sulfur cluster proteins (active sites in Fig. 1B, C), most of the resolved resonances arise from the contact-shifted β-methylene protons of the coordinated Cys. Thus, in spite of linewidths in the range 100–170 Hz, the expected ∼14-Hz geminal coupling allows the identification via the intense MCOSY cross peaks of all eight $C_\beta H_2$ sets in a bacterial oxidized ferredoxin, FD^{ox}, which contains two 4Fe clusters, as illustrated in Fig. 28 (Busse et al., 1991; Bertini et al., 1991b). A number of weaker cross peaks identify some of the $C_\beta H-C_\alpha H$ spin connections. The excellent sensitivity is achieved because the short T_1's (2–15 ms) allow the MCOSY map to be collected at a very rapid rate of $20\ s^{-1}$. However, the expected cross peaks for geminal protons are not always observed even for comparably broad lines; this may result from variable J coupling among the various Cys $C_\beta H$'s (Busse et al., 1991). It also appears that TOCSY with $\tau_m = 10$ ms yields not only all of the cross peaks shown in Fig. 28, but some of the missing $H_\alpha–H_\beta$ cross peaks in the diamagnetic envelope, and cross peaks appear to be more intense than for magnitude COSY (Sadek et al., 1993). Both COSY and TOCSY experiments can be expected to provide crucial information for unambiguously identifying the three protons of a given Cys whose contact shift pattern should provide information on both the orientation of the residues and the magnetic and electronic properties of the coordinated iron.

Based on the success in observing all MCOSY peaks for $J > 6$ Hz in both low-spin and high-spin small ferric heme model complexes (Keating et al., 1991), MCOSY experiments were extended to a "worst case" scenario, the high-spin ferric form of 42-kDa resting-state HRP with linewidths in the 400–500 Hz range and $T_1 \sim 2$–10 ms

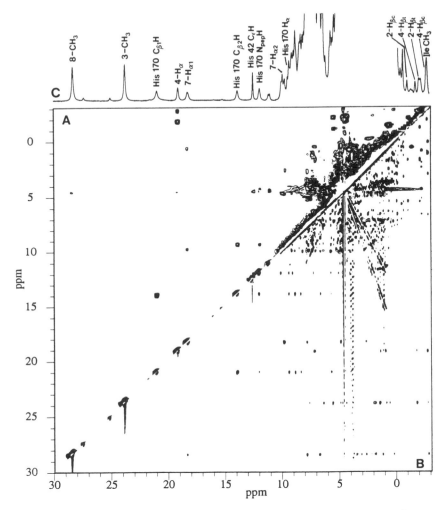

Figure 27. Split-diagonal MCOSY/NOESY spectra of HRP-CN at 500 MHz, pH = 7.0, 50 °C in ^2H_2O, with the 1D reference trace shown along the upper f_2 axis. (A) MCOSY spectrum. (B) Phase-sensitive NOESY spectrum with mixing time of 20 ms. (C) 1D reference spectrum with selected hyperfine-shifted peaks labeled. MCOSY data were collected with a repetition rate of 5 s^{-1}, using 1024 t_2 points and 256 t_1 blocks with a sweep width of 31 KHz and t_1, t_2 acquisition times of 8 ms and 32 ms, respectively. 384 scans/block yielded at total time of *circa* 5 h in a 10-mm tube. The data were processed with 0°-shifted sine–bell-squared apodization over 16 ms (t_2) and 8 ms (t_1) and symmetrized. The NOESY data were collected at a repetition rate of 5 s^{-1}, using 1024 t_2 points and 512 t_1 blocks with a sweepwidth of 31 KHz and t_1, t_2 acquisition times of 8 ms and 32 ms respectively. 160 scans/block yielded a total time of *circa* 5 h. The data set was processed with 30°-shifted sine–bell-squared apodization over 32 ms (t_2) and 8 ms (t_1). Data were phase corrected in each dimension and a polynomial baseline correction employed in f_2. Complete details and analysis of assignments from MCOSY and NOESY data of HRP-CN is given in de Ropp *et al.* (1991b). From the combination of COSY and NOESY data assignments of all heme protons as well as numerous protons of catalytically relevant amino acids were made possible. Most resolved signals here are Zone III type, with Δ = 60–120 Hz. Adapted from de Ropp *et al.* (1991b).

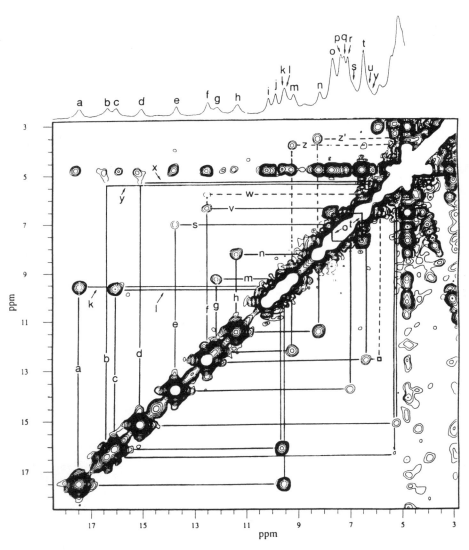

Figure 28. 500-MHz ^{1}H NMR reference spectrum (collected at 10 s^{-1}) and magnitude COSY map of *Clostridium pasteuriunum* oxidized ferredoxin in ^{2}H$_2$O, pH = 7.0 at 30 °C; the protein contains two four-iron clusters, as shown in Fig. 1c. Cross peaks are shown for the eight Cys C$_\beta$H:C$_\beta$H geminal (solid lines) and three of the eight Cys C$_\beta$H:C$_\alpha$H vicinal (dashed lines) couplings. The 2D data were collected over a 20-KHz bandwidth with 256 t_1 blocks of 256 t_2 points with 3200 scans per block and a total repetition rate of 20 s^{-1}; processing involved a 0°-shifted sine–bell–squared window and zero-filling to 1024 × 1024 points prior to Fourier transformation. The region upfield of the solvent line in the f_2 dimension is obscured by t_1 ridges and artifacts due to the rapid pulse sequence repetition rate (Busse *et al.*, 1991).

(de Ropp and La Mar, 1991). The reference trace is illustrated in Fig. 29A. Previous isotope labeling and 1D NOEs had provided the heme assignments (La Mar *et al.*, 1980b; Thanabal *et al.*, 1988b) key to testing the extreme limits of COSY cross peak detection. On the positive side, it was possible to detect the weak MCOSY cross peaks for the two propionate α-methylene groups for which $J \sim 14$ Hz is expected (Fig. 30B). In contrast to this success, however, was the failure to detect the vinyl $H_\alpha : H_{\beta t}$ cross peak for which the larger $J \sim 16$ Hz and narrower lines (400 Hz, 100 Hz) should have yielded a stronger cross peak (not shown). More recently, cross peaks for geminal protons with linewidths to 900 Hz have been reported for a high-spin ferric cytochrome c' (Banci *et al.*, 1992a).

Besides the work on several inorganic complexes, extension of scalar connectivity experiments to other classes of paramagnetic proteins have been limited. Initial studies on a high-spin ($S = \frac{3}{2}$) Co(II)-substituted carbonic anhydrase complex (~ 30 kDa) has allowed the mapping of some residues in the outer region of Zone III and the identification of several individual scalar connectivities in the inner pocket of Zone III (Banci *et al.*, 1992b). While assignments are incomplete, the initial results are promising, and promise fruitful future studies on other Co(II)-substituted Zn enzymes.

The available information indicates that scalar connectivities can be effectively mapped by MCOSY (particularly in conjunction with variable-temperature studies and TOCSY) in Zone III for small (6–18 kDa) metalloproteins. This class of proteins includes the important and numerous low-spin ferric forms of the hemoproteins, cytochromes, myoglobins and monomeric hemoglobins, and likely the oxidized four-iron cluster bacterial ferredoxins. The extension to Zone III protons of larger proteins such as the heme peroxidases is likely more problematical, although the initial results are promising (de Ropp *et al.*, 1991a, b; Banci *et al.*, 1991c, d) and are being actively pursued. Certainly the scalar correlation experiments will see decreased utility for a given zone as the size of the protein increases because of the increase in Δ relative to $\rho(T)$. The detection of MCOSY peaks involving broad peaks in high-spin ferric hemoproteins is a basis for optimism (de Ropp and La Mar, 1991; Banci *et al.*, 1992a). However, the failure to detect other cross peaks in the same system with comparable J and narrower lines reflects the crucial limits of our understanding of all of the factors that influence the detectability of such cross peaks. The utility of TOCSY for broad lines needs to be explored further. The conclusions we draw at this time are that:

1. scalar cross peaks can be detected from remarkably broad resonances,
2. it is not yet predictable when such a cross peak is observable, and
3. it will be very difficult if not impossible to map a complete spin system in Zone II.

4.3. Dipolar Correlation

4.3.1. NOESY

The equations for the 2D NOE experiment are very similar to those for the transient NOE (Section 3.2.5), with the time evolution of the intensity of the diagonal and cross

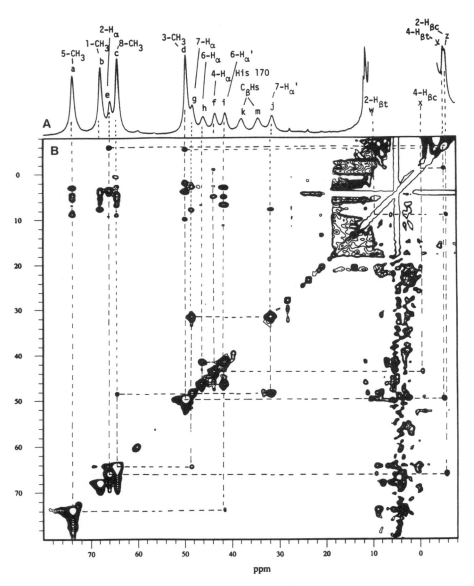

Figure 29. (A) Reference trace of resting-state HRP in ^2H_2O, pH = 7.0 at 55 °C. The peak assignments given refer to the heme structure shown in Fig. 1a. (B) Complete 300-MHz ^1H NMR NOESY map collected with a mixing time of 10 ms. The phase-sensitive data were collected on a GE Ω-300 spectrometer using 512 t_1 blocks of 1280 scans over a 60-kHz bandwidth using 1024 complex t_2 points at a pulse-sequence repetition rate of 11 s^{-1}. The data were processed by a 60°-shifted sine–bell-squared window over 256 t_1 and 256 t_2 points and zero-filled to 1024 × 1024 prior to Fourier transformation (de Ropp and La Mar, 1991).

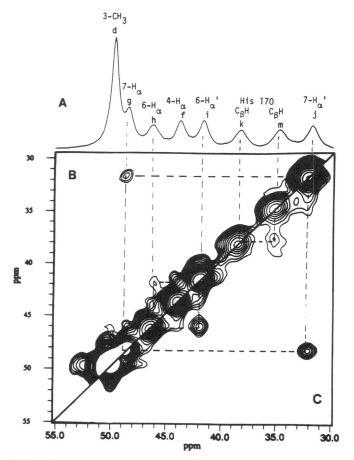

Figure 30. (A) Portion of the low-field 300-MHz ^{1}H spectrum of resting-state HRP in ^{2}H$_2$O, pH = 7.0, 55 °C. (B) Magnitude COSY map of resting-state horseradish peroxidase illustrating the only detected cross peaks involving hyperfine-shifted resonances, the 6- and 7-propionate $C_\alpha H_2$s with T_1's 3–5 ms and $\Delta \sim 400$ Hz. The data were collected using 256 t_1 blocks of 4480 scans over a 60-KHz bandwidth with 1024 t_2 points at a pulse repetition rate of 23 s^{-1}; optimal processing involved a 0°-shifted sine–bell-squared window over 256 t_1, t_2 points zero-filled to 1024 × 1024 and symmetrized. (C) NOESY map of the same sample with a mixing time of 3 ms. Note the weak cross peak between the $T_1 \sim 2$ ms, $\Delta \sim 500$ Hz, His 170 C_βHs. The data collection and processing are identical to that described in Fig. 29B (de Ropp and La Mar, 1991).

peaks given by (Neuhaus and Williamson, 1989):

$$a_{\text{dia}}(\tau_m) = \frac{M_0}{2} \frac{|\sigma|}{D} \left\{ \exp[-(\bar{\rho} - D)\tau_m] + \exp[-(\bar{\rho} + D)\tau_m] \right\} \tag{41}$$

$$a_{\text{cross}}(\tau_m) = \frac{-M_0}{2} \frac{\sigma}{D} \left\{ \exp[-(\bar{\rho} - D)\tau_m] - \exp[-(\bar{\rho} + D)\tau_m] \right\} \tag{42}$$

for a pair of spins with $\rho_i \neq \rho_j$, where M_0 is the intensity of the diagonal at $\tau_m = 0$ and $\bar{\rho}$, D were defined in Eqs. (27) and (28). Equation (42) differs from that of the transient NOE [Eq. (26)] only by a factor of 2 in both the initial buildup rate and maximum value; the maximum cross peak, however, occurs for the same value of τ_{max} as for the transient NOE [Eq. (30)]. The pulse sequence for the phase-sensitive NOESY experiment, the more conventional 2D dipolar correlation method, is given in Fig. 16 (Kumar *et al.*, 1980). The dipolar correlation develops solely in the mixing time τ_m, where direct observation is not possible. The NOESY experiment, like the COSY case, can be carried out in both magnitude- and phase-sensitive modes (with similar sensitivity improvement or TT saving and resolution loss in the magnitude experiment noted for the COSY case). Again, the resolution loss in the magnitude experiment may be irrelevant for strongly shifted and relaxed lines that are well resolved. However, recent reports suggest that the magnitude NOESY experiment may adversely influence zero-quantum magnetization transfer (Olejniczak *et al.*, 1985; Wang *et al.*, 1992) which should make the magnitude experiment much less appealing for macromolecules. The major advantage of 2D NOESY versus 1D NOEs are the same as in a diamagnetic system: uniform excitation of all resonances whether resolved or not, simultaneous sampling of all dipolar couplings, and significant improvement in resolution in crowded spectral regions. In consideration of the shortcomings in implementing the 1D NOE experiments for paramagnetic systems (Lecomte *et al.*, 1991), the elimination of off-resonance effects provide crucial advantages in the 2D experiment.

Selection of τ_m. Diamagnetic and paramagnetic systems differ significantly in that the complete range of relaxation times in the former is rather limited, while that in the latter can range over several orders of magnitude in even the most weakly paramagnetic system. For any cross peak for which the two-spin approximation is approximately valid, the optimal cross peak intensity is given by the τ_{max} expressed in Eq. (30). Analysis of this expression reveals that τ_{max} will be between ρ_i^{-1} and ρ_j^{-1}, but much closer to the shorter ρ^{-1} (or smaller *selective* T_1). Thus a reasonable guide is that the detection of a weak NOESY peak requires $\tau_m \sim T_1^{sel}$ (the shorter of the pair). Consequently, a range of τ_m may be dictated (by the wide range of T_1^{sel}) in a paramagnetic system solely to allow the detection of all of the cross peaks.

A case in point is illustrated for the strongly paramagnetic, high-spin ferric 42-kDa HRP complex. The ^1H NMR spectra resolve three pairs of methylene proton pairs ($\sigma_{ij} \sim -37 \, s^{-1}$) with T_1's 2–10 ms, which yield a NOESY map using $\tau_m = 10$ ms with reasonably intense cross peaks for the more slowly relaxing propionate α-CH_2 sets (T_1's ~ 6–8 ms), but none for the axial His β-CH_2 (T_1's ~ 2 ms) as shown in Fig. 29B (de Ropp and La Mar, 1991). Reducing τ_m to 3 ms weakens the propionate cross peak, but allows detection of the weak axial His $C_\beta H_2$ peak (Fig. 30C). It is not improbable that a very strongly coupled (large σ) pair of spins such as a methylene group in Zone II may fail to give a cross peak, while a much more weakly coupled pair in Zone III may yield a significant cross peak if τ_m is selected to be too long. Thus a knowledge of the T_1's expected is crucial to the successful setup of the NOESY experiment. This range of T_1's can be estimated by a nonselective inversion–recovery experiment or by simply collecting and processing the first t_1 block for a range of τ_m for a NOESY experiment and inspecting whether peaks of interest are still present when acquisition starts (similar to the procedure for TOCSY illustrated in Fig. 23). The two trial experiments require comparable time; however, neither can anticipate the individual relaxation rates of signals unresolved from the diamagnetic envelope.

If the intrinsic relaxivity of a metal ion in a particular protein is known, the range of T_1's can be estimated by consideration of the approximate distance from the metal of likely functional groups. Particularly useful for this are X-ray coordinates on a structural homolog of the protein of interest (Kuriyan *et al.*, 1986), which directly yield r_{ij}'s and R_{M_i}, R_{M_j} for use in Eqs. (7) and (1), respectively. The simulated intensity–time profiles for some cross peaks considered crucial to the sequence-specific identification of a Phe, Ile, and Leu, as well as some heme peaks, in metMbCN (Emerson and La Mar, 1990a) are illustrated in Fig. 31. Clearly, no single τ_m will simultaneously sample all of the necessary dipolar correlations optimally; ideally three τ_m and minimally two

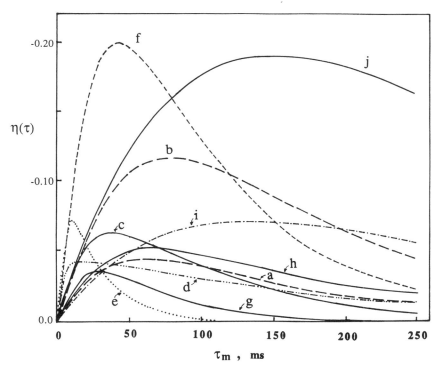

Figure 31. Plot of cross-peak intensity versus mixing time for ten pairs of proton sets in the active site of low-spin ferric metMbCN. The paramagnetic relaxation rate was calculated based on Eq. (3a) with $\rho = 40$ s^{-1} for the His 93–ring N$_\delta$H with $R_{\mathrm{Fe}} = 5.1$ Å from the iron. The cross relaxation rates were obtained from r_{ij} *via* Eq. (7). Both $R_{\mathrm{Fe}-i}$ and r_{ij} were obtained from the MbCO crystal coordinates (Kuriyan *et al.*, 1986). The vertical expansion of individual plots vary as follows: ×1 (———); ×20 (– – –), ×10 (· · · ·), ×10^2 (— · · · —), ×5 (— · —), and ×0.5 (- - - - -). The contacts represented by the lines are (*a*) heme 5-CH$_3$: Phe 43 *m*-H; (*b*) heme 5-CH$_3$: Ile 99 C$_\delta$H$_3$; (*c*) His 93 C$_\alpha$H: His 97 C$_4$H; (*d*) His 93 C$_4$H: Ile 99 C$_\delta$H$_3$; (*e*) His 93 C$_2$H: His 93 C$_\delta$H; (*f*) Ile 99 C$_{\gamma1}$H: C$_{\gamma2}$H; (*g*) Phe 43 *p*-H: *o*-H; (*h*) heme 1-CH$_3$: δ-*meso* H; (*i*) heme 1-CH$_3$: 8-CH$_3$; (*j*) 2-vinyl H$_\alpha$: H$_{\beta c}$.

values are needed to observe all cross peaks at 50% of optimal intensity (note the vertical scale for various cross peaks differ by a factor of 200). If the shortest T_1 peaks are ignored, a single optimal τ_m is at 50 ms; all but two of the connectivities in Fig. 31 are indeed readily detected at this mixing time, as well as those for another thirty hyperfine-shifted and relaxed protons in metMbCN (see Fig. 32) (Emerson and La Mar, 1990a). The two weakest cross peaks involving His 93 are missing in the NOESY map but can be identified by 1D NOEs (see Section 4.3.4). The low-field portion of the NOESY map illustrating the cross peaks from the resolved hyperfine-shifted signals is found in Fig. 32.

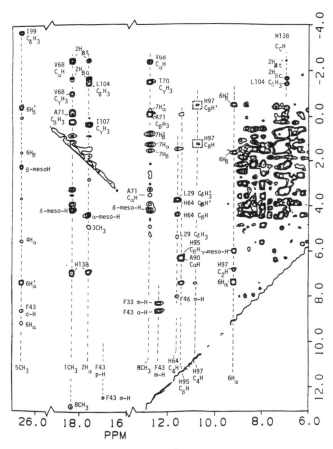

Figure 32. Low-field portion of the 500-MHz ^1H NMR NOESY spectrum of sperm whale metMbCN in ^2H_2O, pH = 8.6 at 35 °C, illustrating cross peaks from the resolved hyperfine-shifted resonances. Cross peaks in dashed boxes are detected at a lower contour level. The data were collected with 512 t_1 blocks of 256 scans over a 32-kHz bandwidth using 1024 complex t_2 points and a pulse repetition rate of 1 s^{-1}. The data were processed with a 30°-shifted sine–bell-squared window prior to zero-filling to 1024 × 1024 points (Emerson and La Mar, 1990a).

Estimation of Distances. The simulated profile for NOESY cross peaks of metMbCN in Fig. 31 shows how difficult it is to determine the initial buildup curve for any proton pair, since the initial slope portion is very short and occurs over a wide range of τ_m for different proton pairs. For many of the weakly dipolar-coupled and strongly relaxed lines, there is simply no "linear" portion near $t = 0$ over which the cross peak is likely to be detectable. Hence extracting distances from NOESY for a paramagnetic protein will be more difficult than in a diamagnetic analog if the protein is relatively small; this is particularly true if the protons are closer to the metal than the middle of Zone III. Thus realistic estimates of distance from NOESY maps are unlikely for a high-spin Mb (16 kDa, Zone II–III), but likely addressable is the low-spin form (Zone III–IV).

In a series of isoelectronic proteins of variable molecular size with strong structural homology in the active site, the ρ's are inconsequentially altered, since $S(S + 1)T_{1e}$ remains invariant, but σ for all proton pairs increases ($\sim$linearly in size). The maximum cross peak intensity occurs at the same $\tau_{\max}$, but increases with protein size. With these stronger cross peaks, the prospects improve significantly for determining their cross-peak buildup curves. One recent study presented results on NOESY cross peak intensity with mixing time for a seven-iron ferredoxin (Cheng *et al.*, 1992), but did not quantify the rise curves. However, work in our laboratory indicates that reasonable rise curves may be obtainable for the Zone III protons (the majority of the active site) of a low-spin heme peroxidase such as HRP-CN (42 kDa) (Sette, de Ropp, Hernandez, La Mar, unpublished results). Figure 33 shows that the strong cross peaks of geminal groups observed with $\tau_m \sim 20$ ms are readily detected and quantifiable for $\tau_m = 3$ ms (de Ropp *et al.*, 1991b). Indeed, it is possible to readily detect geminal cross peaks at mixing times

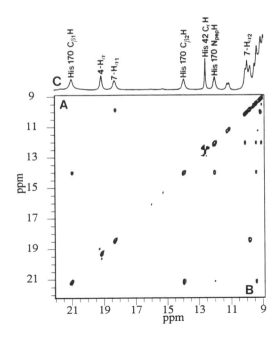

Figure 33. Split diagonal portions of the NOESY spectra of HRP-CN obtained at 500 MHz, pH = 7.0, and 50 °C in ^2H_2O. (*A*) Spectrum collected with a mixing time of 3 ms. (*B*) Spectrum collected with a mixing time of 20 ms. (*C*) 1D reference spectrum with assignments. Spectrum *A* reveals cross peaks solely from the geminal proton pairs His-170 β-CH$_2$ and heme 7-H$_\alpha$s. Spectrum *B* reveals numerous additional NOEs. Data were collected and processed as described for the NOESY data in Fig. 27. Adapted from de Ropp *et al.* (1991b).

of 0.5 to 1 ms. By collecting NOESY maps at short mixing times (0.5 ms, 1 ms, 1.5 ms, 2.2 ms, 3 ms, 5 ms, etc.), it is possible to construct well-behaved NOE rise curves for paramagnetic resonances whose initial linear portion allows calculation of σ (from the relationship $a_{\text{cross}}(\tau_m) = M_0 \sigma \tau_m$ obtained from Eq. (42) for short τ_m). Since $\sigma_{AB}/\sigma_{CD} = r_{AB}^{-6}/r_{CD}^{-6}$ from Eq. (7), knowledge of one σ and corresponding interatomic distance $A-B$ allows determination of any other distance $C-D$ if its σ can be measured (assuming equal $\tau_r S$). Thus, for low-spin HRP-CN it is possible to quantify σ from NOE build-up curves and determine distances for a large number of paramagnetically influenced protons from the heme and adjacent amino acids (not just geminal pairs). Prospects for distance determination in similar low-spin proteins with large σ are judged to be good. On the other hand, the lack of change in σ while ρ increases by a factor ~ 10 relative to HRP-CN in the native resting-state high-spin ferric HRP make the prospects for extracting distances very slim. For cases where $\rho \gg |\sigma|$, NOESY cross peaks will be difficult to detect even under optimal conditions and will not yield distance estimates; this is a case where there will be an advantage to returning to the 1D NOE (see below).

Cross Peak Asymmetry. The desire to optimize sensitivity by increasing the number of scans per block through minimizing the pulse sequence recycle time can lead to asymmetry in the NOESY map. Although NOESY cross peaks are symmetrical about the diagonal for protons which are not saturated by the rapid repetition rate, a pair of coupled spins with $\rho_i \neq \rho_j$ are differentially saturated by rapid repetition, with the result that the cross peak intensity become asymmetrical, with that of the f_1 shift for the slower-relaxing proton exhibiting reduced intensity (Anderson *et al.*, 1987). Thus both sides of the diagonal have to be inspected to establish the presence or absence of a dipolar connectivity. Conversely, the differential decrease of the two cross peaks with repetition rate can be used to determine, at least qualitatively, which one is more strongly relaxed, and hence closer to the metal, if one of the two (or both) signals are unresolved.

Asymmetrical Resolution. A problem not infrequently encountered is how to establish a unique dipolar correlation between a strongly shifted and strongly relaxed broad proton signal in Zone II (i.e., heme methyl in resting state HRP, a coordinated Cys $C_\beta H$ in any of the iron–sulfur cluster proteins) and an unshifted, very weakly relaxed amino acid whose narrow signals lies in the intense, unresolved diamagnetic envelope. In order to make acquisition time reasonable for the broad spectral window, the allowable t_1 blocks or t_2 points over the 50–100 ppm range will not provide sufficient spectral resolution to define uniquely the shift of the "diamagnetic" residue. A solution to this problem is to use different dwell times in t_1 and t_2, with the t_1 dwell time defined by the minimal necessary t_1 blocks over the hyperfine-shifted bandwidth of 50 ppm and the t_2 dwell time determined solely by the diamagnetic window of 10 ppm (resolution improves by a factor 5 and is now comparable to that in a diamagnetic system). The restricted bandwidth must be in f_2 in order to filter out fold-in peaks. The asymmetric NOESY data set should map cross peaks from broad lines with low resolution (in the f_1 dimension) with narrow peaks with high resolution (in the f_2 dimension). It has been reported by Nettesheim *et al.* (1992) that such an experiment successfully mapped the strongly relaxed, hyperfine-shifted Cys resonances to the narrow resonances of assigned residues of the "diamagnetic" envelope; an experimental map, however, was not shown.

4.3.2. ROESY

This is the rotating frame version of the NOESY experiment (Bothner-By *et al.*, 1984; Bax and Davis, 1985a), where the mixing time corresponds to a spin-lock field provided by MLEV-17 or some comparable pulse train. The advantage of ROESY over NOESY is the removal of the dependence of the NOE on $\omega \tau_r$ (all molecules have $\sigma > 0$), minimization of spin diffusion effects in large biopolymers, and allowing distinction between dipolar (NOESY) and exchange responses (EXSY, see below) in the cross peaks (opposite absorptive phases). The advantage of ROESY in diamagnetic systems occurs largely for molecules near the $\omega \tau_r \sim 1.12$ region. The selection of mixing times for optimally detecting dipolar and exchange cross peaks are the same as for NOESY and EXSY. The ROESY pulse sequence is the same as that for TOCSY (Fig. 16), except that the strength of the spin-lock field during the mixing time is lower in the former experiment (Bax and Davis, 1985a). The need to maximize the spin-lock field to cover the increased spectral dispersion in a paramagnetic system will make suppression of TOCSY cross peaks difficult (see also Kessler *et al.*, 1987; Griesinger and Ernst, 1987) and the interpretation of cross-peak intensities highly problematical. It is likely that the weak NOEs to be expected for a small paramagnetic molecule near $\omega \tau_r \sim 1.12$ are better addressed with a NOESY map of the complex in a viscous solvent, as reported for the bis-cyano complex of hemin (Yu *et al.*, 1986). The only report utilizing ROESY of which we are aware relied on this experiment to differentiate exchange and dipolar connectivities in a model heme complex (Simonis *et al.*, 1992).

4.3.3. Selected Case Studies

The close analogy between 1D NOE and 2D NOESY, together with the much more extensive experience with 1D NOEs on paramagnetic molecules since 1982, has led to 2D dipolar studies more extensive and more successful to date than scalar correlation studies. The type of system most extensively and successfully investigated by NOESY is the same that has yielded the most informative scalar correlation data, low-spin ferric hemoproteins. The focus on these systems is due to their favorable spectral properties with respect to 2D NMR—relatively narrow, well-resolved lines overwhelmingly in Zones III and IV—and to the extraordinarily high information content in both the contact shifts of the prosthetic group (Shulman *et al.*, 1971; La Mar, 1979) and the sizable dipolar shifts of nearby amino acid residues (Emerson and La Mar, 1990b; Rajarathnam *et al.*, 1992). Among the smaller proteins, this includes 12–13-kDa ferricytochromes *c* (Keller and Wüthrich, 1981; Senn and Wüthrich, 1985; Busse *et al.*, 1990) and *b* (McLachlan *et al.*, 1988; Wu *et al.*, 1991) and the cyano-met myoglobins (Rajarathnam *et al.*, 1992; Emerson and La Mar, 1990a; Qin *et al.*, 1992) and monomeric hemoglobins (16–18 kDa; Peyton *et al.*, 1991). The bis-cyano complex of the extracted ferric heme yielded the first complete 2D NOESY map for a paramagnetic molecule, although immobilization in a viscous solvent to increase σ was necessary (Yu *et al.*, 1986). The most extensive studies have been done on sperm whale metMbCN (Emerson and La Mar, 1990a) and some of the point mutants (Rajarathnam *et al.*, 1992), for which the NOESY map permits the identification in the sequence of all residues by their unique scalar connectivities (Yu *et al.*, 1990). A representative section of a map is shown in Fig. 32. The hyperfine-shifted residues exhibit essentially all cross peaks to nearby residues observed in the analogous diamagnetic MbCO complex (Dalvit and Wright, 1987). The only resonances which could not be identified in the 2D map are the $\sim$400-Hz broad axial His ring CH's, whose proximity to the iron places them as the only protons in Zone I. However, even these could be uniquely assigned by 1D NOEs (see Section 4.3.4).

For another metMbCN from an invertebrate, a strategy was explored for sequence specifically assigning only the residues with hyperfine-shifted signals using the conventional diamagnetic protein 2D methodology, taking advantage of the fact that any hyperfine-shifted proton (labile or nonlabile) yields strongly temperature-dependent COSY/NOESY cross peaks (Qin et al., 1992; Qin and La Mar, 1992). A series of variable-temperature 2D maps yielded 53 hyperfine-shifted resonances, which have been shown to account for all residues in the active site except for a portion of a single side chain closest to the metal. The hyperfine shifts imparted to the backbone $C_\alpha H$, NH, moreover, greatly simplified the sequence-sqecific assignment by providing significantly improved resolution. The backbone $C_\alpha H$, NH of ferricytochrome c near the iron center have been also assigned sequence specifically, although these protons are generally in Zone IV or close to the Zone III–IV boundary (Feng et al., 1989).

Based on the observed increased magnitude of the 1D steady-state NOEs with larger isostructural proteins (Thanabal et al., 1987a, b; Thanabal and La Mar, 1989; Dugad et al., 1990c), NOESY has been extended to the hyperfine-shifted resonances of several larger (34–42 kDa) low-spin, ferric heme peroxidase complexes: horseradish peroxidase-cyanide (de Ropp et al., 1991b), cytochrome c peroxidase-cyanide (Satterlee and Erman, 1991; Banci et al., 1991c), and lignin peroxidase-cyanide (de Ropp et al., 1991a; Banci et al., 1991d). The number of assignments in HRP-CN is more extensive because of its stability at elevated temperatures where lines are narrower; however, all NOESY maps have yielded assignments and valuable structural data. A portion of the NOESY map of HRP-CN for $\tau_m = 20$ ms is shown in Fig. 27B. Preliminary NOESY data on the low-spin cyano complex of 78-kDa lactoperoxidase, LPO-CN, indicate that the cross peaks necessary to assign the hyperfine-shifted peaks are readily detected at remarkably short mixing times (Sette, de Ropp, and La Mar, unpublished results). The larger size that results in more intense cross peaks, together with the remarkable structural homology among the heme peroxidases of diverse origin and function, can be expected to lead to highly productive NOESY studies that should provide quantitative structural information.

With the surprisingly sensitive NOESY map of low-spin 42-kDa HRP-CN, the experiment was directly extended to another "near worst-case scenario," the high-spin ferric resting-state HRP with a 100-ppm bandwidth, Δs 100–500 Hz, and T_1's 2–10 ms; all assignments for these heme resonances were known from earlier isotope labeling and 1D NOE studies (La Mar et al., 1980b; Thanabal et al., 1988b). The resulting 300-MHz NOESY map with $\tau_m \sim 10$ ms collected at a rate of 11 s^{-1} is illustrated in Fig. 29 (de Ropp and La Mar, 1991); all of the necessary dipolar connectivities are observed among the heme substituent resonances that would have allowed complete assignment. The two low-field peaks with $T_1 \sim 2$ ms which do not give cross peaks with $\tau_m = 10$ ms and must arise from the axial His $C_\beta H$, exhibit the expected cross peaks with $\tau_m = 3$ ms (Fig. 30C). More recently, several of the heme side-chain signals in a high-spin ferric dimeric ferricytochrome c' have been partially assigned based on NOESY data (Banci et al., 1992a).

The other class of proteins for which several NOESY experiments have been reported are the iron–sulfur proteins (Busse *et al.*, 1991; Skjeldal *et al.*, 1991; Banci *et al.*, 1991b; Bertini *et al.*, 1992). For the 2 × 4 Fe oxidized bacterial ferredoxin whose MCOSY map is shown in Fig. 28, NOESY confirmed the expected dipolar coupling within the Cys methylene group and helped identify other Cys C_αHs (Busse *et al.*, 1991). The cross peaks are relatively weak because the protons are near the Zone II–III interface, and the small size (~6 kDa) results in a small σ even for a CH_2 (Table 2). NOESY maps revealed several more dipolar connectivities than earlier 1D NOE studies (Bertini *et al.*, 1990), even though the steady-state NOEs should have been larger than the transient NOE or NOESY peaks; a possible reason for the loss of some of the 1D NOEs is dynamic range limitations. A related small 6-kDa four-iron high-potential iron–sulfur protein HiPiP in its reduced state yields hyperfine-shifted Cys protons in Zone II (T_1's 2–15 ms) spread over ~15 ppm. A detailed 1D NOE study had provided the pairing of the four Cys (Bertini *et al.*, 1991a). The oxidized HiPiP protein with a larger hyperfine-shifted range and similar relaxation properties yielded the C_βH signal pairing for the four Cys based on 1D NOEs (Cowan and Sola, 1990). The 10-ms τ_m NOESY map of oxidized HiPiP covering a ±40 ppm window reproduced in Fig. 34 exhibits most of the expected intra-Cys cross peaks for the hyperfine-shifted lines, as well as a sufficient number of cross peaks to the sequence-specifically assigned Zone IV envelope to allow the assignment of the individual coordinated Cys in the sequence (Nettesheim *et al.*, 1992). Similar data have been reported for another oxidized HiPiP (Banci *et al.*, 1991a). The assignment of individual Cys methylene protons by 1D NOEs in a larger 14-kDa two-Fe plant-type reduced ferredoxin (Fig. 1B) with one Fe(II) and one Fe(III) was difficult because of several near-degeneracies, rapid relaxation, and large off-resonance effects [Fe(II) protons in Zone II, Fe(III) protons in Zone I] (Dugad *et al.*, 1990b). However, the NOESY map of a closely related two-Fe reduced ferredoxin from *Anabena* (spectrum in Fig. 35A, B) reveals not only all the necessary inter- and intra-Cys dipolar contacts, as shown in Fig. 35C, but also sufficient contacts with the previously assigned "diamagnetic" Zone IV protons so as to lead to sequence-specific assignment of the Cys (Skjeldal *et al.*, 1991).

The only other class of paramagnetic metalloproteins for which 2D data are available at this time is a 30-kDa Co(II)-substituted carbon anhydrase. Preliminary data reveal a wealth of NOESY connectivities, for both the strongly contact-shifted coordinated His signals and many less strongly dipolar-shifted amino acids in the active site; the detected cross peaks involve protons in Zones II–IV (Banci *et al.*, 1992b).

Since informative 1D NOEs have now been reported for various derivatives of metal-substituted superoxide dismutase (Banci *et al.*, 1990a, b), uteroferrin (Scarrow *et al.*, 1990), hemerythrin (Maroney *et al.*, 1986), tetrameric hemoglobin (Chatfield and La Mar, 1992), and Co(II)-substituted metallothionenes (Bertini *et al.*, 1989), it is expected that NOESY on these systems will be able to duplicate the results and provide much additional information on assignment and structure. Moreover, in contrast to scalar connectivity experiments, the increased cross relaxation rates in larger proteins will invariably improve the prospects for detecting the cross peaks for a given r_{ij} for a series of isoelectronic chromophores, in spite of the increase in the diamagnetic linewidth contributions. Preliminary data indicate that the NOESY map of 78-kDa low-spin ferric lactoperoxidase yields all of the cross peaks for which 1D NOEs have been reported (Thanabal and La Mar, 1989), as well as many that cannot be observed by 1D methods

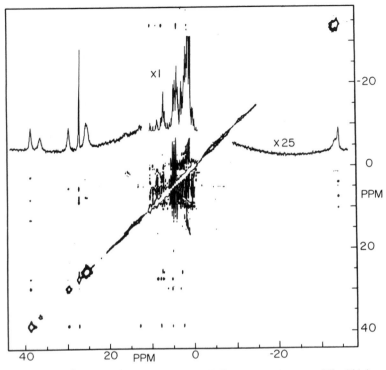

Figure 34. 500-MHz ¹H NMR NOESY spectrum of *Chromatium vinosum* oxidized high-potential iron–sulfur protein (HiPIP^ox) illustrating the cross peaks among the hyperfine-shifted protons of the coordinated Cys ligand as well as cross peaks between the Cys and noncoordinated residues, which allow the sequence-specific assignments of individual Cys. The mixing time is 10 ms and the pulse sequence repetition rate is 2 s⁻¹. The resolved protons represent primarily Zone II protons with T_1's 2–15 ms (Nettesheim *et al.*, 1992).

(Sette, de Ropp, and La Mar, unpublished). Certainly the success enjoyed by the early NOESY studies gives us a very promising outlook for extending such experiments to all paramagnetic metalloproteins which give resolved spectra in the hyperfine-shifted window.

4.3.4. NOESY Versus 1D NOE

Does the success of NOESY experiments on paramagnetic proteins make the 1D NOE experiments obsolete? Far from it! Aside from the fact that 2D experiments can not necessarily be performed on spectrometers which will allow 1D NOE studies, there are several situations where 1D NOEs are either the only practical experiment or a crucial adjunct to 2D NOESY data. Even the most informative NOESY maps of a paramagnetic metalloprotein fail to exhibit cross peaks for some of the broader lines (Emerson and La Mar, 1990a; de Ropp *et al.*, 1991a; Rajarathnam *et al.*, 1992; Skjeldal

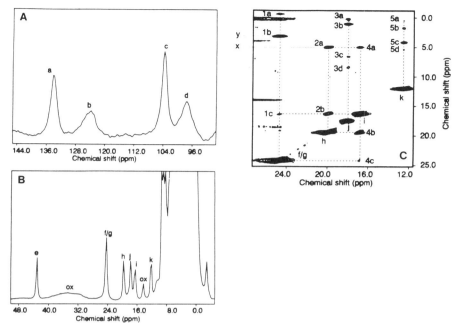

Figure 35. Reference traces and NOESY spectrum of reduced Anabaena 7120 vegetative ferredoxin at 600 MHz, pH = 8.2, 27 °C, in ^2H_2O. (A) Extreme low-field portion of the reference 1D spectrum showing the $C_\beta H$ signals of the Fe(III)-coordinated Cys. (B) Low-field portion of the reference spectrum with the $C_\beta H$ signals of the Fe(II)-coordinated Cys; note labeled resonances f to k, which appear on the diagonal of the NOESY spectrum. (C) Portion of the NOESY spectrum collected with a mixing time of 10 ms. Numerous cross peaks from the hyperfine-shifted resonances f–k both to other hyperfine peaks and to diamagnetic peaks are detected; these cross peaks are labeled 1 to 5 moving upfield along f_2 and a to d moving downfield along f_1. The NOESY data were collected with 3000 scans per t_1 block with 512 t_1 points and 4096 t_2 points and a sweepwidth of 33 KHz, with the carrier placed at 17 ppm. The repetition rate was *circa* 14 s^{-1} (Skjeldal *et al.*, 1991).

et al., 1991). This is likely only a sensitivity problem, but the necessary increase in total experiment time may be prohibitive. Collecting data in only one dimension will allow N_b times the number of scans in the 1D NOE as compared to the 2D experiment in the same TT and unchanged pulse repetition rate. Since N_b is generally ≥ 256, this leads to an improvement by a factor of 16 in sensitivity per unit time of a 1D transient NOE relative to a NOESY slice; the twofold decrease in the intrinsic NOESY response relative to the transient NOE makes the improvement closer to 32. Thus 1D steady-state NOE studies are likely the only recourse for examining dipolar contact to Zone I and many Zone II signals. Even for Zone II and III protons, the 1D experiments may provide detectable NOEs in relatively dilute samples where NOESY experiments are not practical.

The most important use of 1D NOEs will be as crucial adjuncts to successful 2D NOESY studies. Even in proteins with an advantageous disposition of most of the active-site protons in Zones II and III so favorable for 2D studies, there will be a few protons in the very unfavorable Zone I/II border. A case in point is sperm whale metMbCN (Emerson and La Mar, 1990a; Rajarathnam *et al.*, 1992), which has but two protons (resolved signals with $T_1 \sim$ 2–3 ms, $\Delta \sim$ 300–400 Hz) that do not contribute to any 2D NOESY map even for mixing time optimized for their detection. Inspection of the cross-peak simulation profile in Fig. 31 shows that the desired cross peaks would be very much weaker than those presently detected (Emerson and La Mar, 1990a). Increasing the repetition rate will cause saturation of the Zone III protons expected to exhibit NOEs; a better signal-to-noise ratio for the Zone I/II border protons would necessitate a vastly increased TT. The 1D steady-state NOE observed by saturating the broad upfield peak under rapid pulsing conditions, however, provides the crucial dipolar contacts to a 2D-assigned residue (Ile 99) that allows the unique identification of the broad peaks as shown in Fig. 36 (Emerson and La Mar, 1990a). The controls necessary to account for off-resonance effects follow the protocols outlined in Sections 3.2.3 and 3.2.4. Saturation of the low-field broad peak (under a narrow methyl) similarly gives the dipolar contact to the previously assigned His 93 N_1H, identifying the broad signal as the His 93 C_2H. Low power saturation of the narrow peak coincident with the broad His 93 C_2H peak illustrates that the NOEs to N_1H in Fig. 36 arise from His C_2H and not the 1-CH_3 (not shown). 1D NOEs were found similarly necessary and successful for identifying the 600-Hz broad imidazole ring proton of the axial His in HRP-CN (Thanabal *et al.*, 1987b), cytochrome c peroxidase-cyanide (Banci *et al.*, 1991c) and lignin peroxidase-cyanide (Banci *et al.*, 1991d).

A related advantage of 1D NOE as compared to NOESY would occur for resonances so strongly hyperfine-shifted ($>$100 ppm) as to make unreasonable demands on N_b or N_p to adequately digitize narrow resonances or on the needed pulse power to excite the strongly shifted resonance. This condition may be applicable to exploring dipolar contacts from the Fe(III)-coordinated Cys $C_\beta H$ signals in the 100–140 ppm window of the two-Fe reduced plant-type Fd with ^1H NMR spectra shown in Fig. 35A (Skjeldal *et al.*, 1991; Dugad *et al.*, 1990b). Lastly, even for NOESY-detected cross peaks, the weakness of the response for Zone II protons may not allow any estimate of the distances involved. If both the saturated and the detected peaks are resolved, however, the 1D steady-state NOE with the appropriately measured $\rho_i(T)$ will yield a reasonable estimate of σ_{ij} [via Eq. (19a)] and hence r_{ij} [via Eq. (7)].

4.4. Exchange Correlation

The basic NOESY pulse sequence (Fig. 16) registers effects arising from any incoherent transfer of magnetization; this can be due to cross relaxation and/or exchange processes (Jeener *et al.*, 1979; Macura and Ernst, 1980). When magnetization exchange between two interconverting molecular species is observed, the experiment is termed EXSY (or EXCTSY) (Boyd *et al.*, 1984) and is the 2D analog of the 1D transient saturation transfer experiment (Section 3.3). We assume here that the two interconverting molecular species give well-resolved spectra so that an EXSY cross peak is trivially distinguished from a NOESY cross peak within one or the other of the two species. The increased spectral dispersion in a paramagnetic system greatly improves the prospect

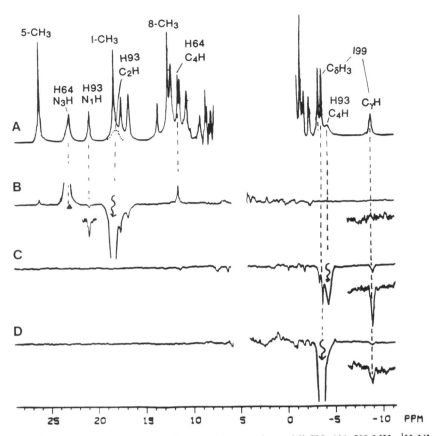

Figure 36. 1D NOEs to assign the Zone I/II protons in metMbCN. (A) 500-MHz ^{1}H NMR reference trace in ^{1}H$_2$O, pH = 8.6, 30 °C collected with a SUPERWEFT sequence. The delay between the 180° and 90° pulses is 139 ms and the recycle time was ~0.5 s. The two broad peaks under the narrow 1-CH$_3$ and upfield shoulder exhibit T_1 ~ 3 ms, which dictates that they arise from the two His 93–ring nonlabile proton. (B)–(D) NOE difference traces with the saturated peak marked by a vertical arrow; the position of the decoupler in the reference trace is marked by a triangle. (B) Saturating 1-CH$_3$ and C$_2$H at high power; note NOE to H93 N$_1$H, which assigns the broad peak to H93 C$_2$H; the NOE to N$_1$H is missing under low-power irradiation, where only 1-CH$_3$ is saturated (not shown). The positive intensity for His N$_3$H is due to saturation of this peak in the reference trace, and this gives an NOE to His 64 C$_4$H. (C) Saturate peak C$_4$H; note NOE to I99 C$_\gamma$H, which assigns it uniquely to His 93 C$_4$H. Since I99 C$_\delta$H$_3$ is off-resonance saturated in (C), we also on-resonance saturate I99 C$_\delta$H$_3$, (D), which yields a weaker NOE to I99 C$_\gamma$H than when H peak C$_4$H was saturated, dictating that the NOE to I99 C$_\gamma$H in trace (C) is from peak C$_4$H and not I99 C$_\gamma$H$_3$. Adapted from Emerson and La Mar (1990a).

for the necessary spectral resolution. (Separation of NOESY and EXSY effects is discussed by Wagner et al., 1985.) Using the notation of Section 3.3, in the simple case of equally populated two-site exchange where $k_i = k_j = k_{ex}$ in Eq. (32) and the two exchange-correlated peaks have the same T_1, the diagonal and cross-peak intensities are given by (Ernst et al., 1987)

$$a_{diag} = \frac{1}{2I_0}[1 + \exp(-2k_{ex}\tau_m)][\exp(-\tau_m/T_1)] \tag{43a}$$

$$a_{cross} = \frac{1}{2I_0}[1 - \exp(-2k_{ex}\tau_m)][\exp(-\tau_m/T_1)] \tag{43b}$$

Equation (43b) is the analog of the NOESY cross-peak intensity, Eq. (42); an exponential increase in intensity with rate $2k_{ex}$, followed by an exponential decay by T_1^{-1}. The mixing time for the maximum response is given by

$$\tau_{max} = \frac{1}{2k_{ex}}\ln(2k_{ex}T_1 + 1) \tag{44}$$

For rapid paramagnetic relaxation, i.e., $2k_{ex}T_1 < 1$, this leads to $\tau_{max} \sim T_1$, where T_1 is the selective T_1. Although this dictates that mixing times that give optimal cross-peak intensity in a paramagnetic system are much shorter than in a diamagnetic system, the intensity of the cross peak is not necessarily weaker than in an isostructural interconverting pair of diamagnetic systems. The faster decay by T_1 relaxation simply means that a faster exchange process is needed to increase the cross-peak intensity, and that can be generally provided by higher temperature.

The earliest 2D experiments involving a paramagnetic molecule were EXSY experiments that detected the cross peak between a Zone III proton of low-spin ferricytochrome c and the diamagnetic ferrocytochrome (Boyd et al., 1984; Santos et al., 1984; Turner, 1985). The weak paramagnetic relaxation effects on the former derivative allowed the use of relatively long τ_m. Several successful studies on detecting conformational processes have been reported for model compounds (Jenkins and Lauffer, 1988a, b; Luchinat et al., 1990; Walker and Simonis, 1992). This experiment has been most useful in metalloproteins for transferring the assignments between two paramagnetic states of an iron-sulfur cluster protein using the hyperfine-shifted Zone II Cys $C_\beta H_2$ protons. The EXSY map illustrating the correlation from the 11-kDa reduced two-Fe *Anabaena* Fd to oxidized Fd is shown in Fig. 37 (Skjeldal et al., 1991); the T_1's for the correlated peaks are in the range 2–10 ms, and the mixing time used was 5 ms. Similarly useful EXSY spectra have been reported for the various redox states of HiPiP (Bertini et al., 1991a) and the two four-iron clusters of the ferredoxin from *Clostridium pasteurianum* (Bertini et al., 1992). It is expected that such studies will find increasing use in the further characterization of the electronic and molecular properties of iron–sulfur redox proteins.

ACKNOWLEDGMENTS. The support of the National Institutes of Health, HL 16087 and GM 26226, and the National Science Foundation, DMB-91-0476, are gratefully acknowledged.

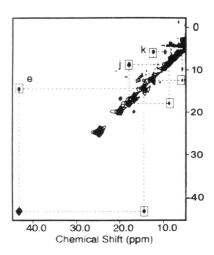

Chemical Shift (ppm)

Figure 37. EXSY spectrum of half-reduced *Anabaena* 7120 vegetative ferredoxin, obtained at 600 MHz, in ^2H_2O, pH = 8.2, and 27 °C. The boxed cross peaks, which do not appear in the NOESY spectrum of the fully reduced protein (see Fig. 35C), arise from electronic self-exchange. The dotted lines to the cross peaks link diagonal peaks in the oxidized and reduced states of the protein. (Figure 35B presents a 1D trace of the reduced protein and labels peaks *e*, *j*, *k*.) The EXSY data were collected with a 5-ms mixing time, 5600 scans per t_1 block, and a repetition rate of *circa* 16 s^{-1}. The sweep width was 71 KHz, and 4096 points in t_2 and 512 points in t_1 were utilized (Skjeldal *et al.*, 1991).

REFERENCES

Anderson, N. H., Nguyen, K. T., Hartzell, C. J., and Eaton, H. L., 1987, *J. Magn. Reson.* **74**:195.

Aue, W. P., Bartholdi, E., and Ernst, R. R., 1976, *J. Chem. Phys.* **64**:2229.

Banci, L., Bertini, I., Luchinat, C., and Scozzafava, A., 1987, *J. Am. Chem. Soc.* **109**:2328.

Banci, L., Bertini, I., Luchinat, C., Piccioli, M., Scozzafava, A., and Turano, P., 1989, *Inorg. Chem.* **28**:4650.

Banci, L., Bencini, A., Bertini, I., Luchinat, C., and Piccioli, M., 1990a, *Inorg. Chem.* **29**:4867.

Banci, L., Bertini, I., Luchinat, C., and Viezzoli, M. S., 1990b, *Inorg. Chem.* **29**:1438.

Banci, L., Bertini, I., Briganti, F., Scozzafava, A, Vicens Oliver, U., and Luchinat, C., 1991a, *Inorg. Chem. Acta* **180**:171.

Banci, L., Bertini, I., Briganti, F., Luchinat, C., Scozzafava, A., and Vicens Oliver, M., 1991b, *Inorg. Chem.* **30**:4517.

Banci, L., Bertini, I., Turano, P., Ferrer, J. C., and Mauk, A. G., 1991c, *Inorg. Chem.* **30**:4510.

Banci, L., Bertini, I., Turano, P., Tien, M., and Kirk, T. K., 1991d, *Proc. Natl. Acad. Sci. USA* **88**:6956.

Banci, L., Bertini, I., Turano, P., and Vicens Oliver, M., 1992a, *Eur. J. Biochem.* **204**:107.

Banci, L., Dugad, L. B., La Mar, G. N., Keating, K. A., Luchinat, C., and Pierattelli, R., 1992b, *Biophys. J.* **63**:530.

Barbush, M., and Dixon, D. W., 1985, *Biochem. Biophys. Res. Commun.* **129**:70.

Baumann, R., Wider, G., Ernst, R. R., and Wüthrich, K., 1981, *J. Magn. Reson.* **44**:402.

Bax, A., 1982, *Two-dimensional Nuclear Magnetic Resonance in Liquids*, Delft University Press/ Reidel, Dordrecht.

Bax, A., 1985, *J. Magn. Reson.* **65**:142.

Bax, A., and Davis, D. G., 1985a, *J. Magn. Reson.* **63**:207.

Bax, A., and Davis, D. G., 1985b, *J. Magn. Reson.* **65**:355.

Bax, A., and Drobny, G., 1985, *J. Magn. Reson.* **61**:306.

Bax, A., and Marion, D., 1988, *J. Magn. Reson.* **78**:186.

Bax, A., and Summers, M. F., 1986, *J. Am. Chem. Soc.* **108**:2093.

Bax, A., Freeman, R., and Frenkiel, J., 1981, *J. Am. Chem. Soc.* **103**:2102.

Bax, A., Ikura, M., Kay, L. E., Torchia, D. A., and Tschudin, R., 1990, *J. Magn. Reson.* **86**:304.

Bertini, I., and Luchinat, C., 1986, *NMR of Paramagnetic Molecules in Biological Systems*, Benjamin/Cummings, Menlo Park, CA.

Bertini, I., Luchinat, C., Messori, L., and Vasak, M., 1989, *J. Am. Chem. Soc.* **111**:7296.

Bertini, I., Briganti, F., Luchinat, C., and Scozzafava, A., 1990, *Inorg. Chem.* **29**:1874.

Bertini, I., Briganti, F., Luchinat, C., Scozzafava, A., and Sola, M., 1991a, *J. Am. Chem. Soc.* **113**:1237.

Bertini, I., Briganti, F., Luchinat, C., Messori, L., Monnanni, R., Scozzafava, A., and Vallini, G., 1991b, *FEBS Lett.* **289**:253.

Bertini, I., Capozzi, F., Luchinat, C., and Turano, P., 1991c, *J. Magn. Reson.* **95**:244.

Bertini, I., Briganti, F., Luchinat, C., Messori, L., Monnanni, R., Scozzafava, A., and Vallini, G., 1992, *Eur. J. Biochem.* **204**:831.

Bodenhausen, G., and Freeman, R., 1977, *J. Magn. Reson.* **28**:471.

Bodenhausen, G., Wagner, G., Rance, M., Sorensen, O. W., Wüthrich, K., and Ernst, R. R., 1984, *J. Magn. Reson.* **59**:542.

Bothner-By, A. A., Stephens, R. L., Lee, J., Warren, C. D., and Jeanloz, R. W., 1984, *J. Am. Chem. Soc.* **106**:811.

Boyd, J., Moore, G. R., and Williams, G., 1984, *J. Magn. Reson.* **58**:511.

Braunschweiler, L., and Ernst, R. R., 1983, *J. Magn. Reson.* **53**:521.

Busse, S. C., La Mar, G. N., and Howard, J. B., 1991, *J. Biol. Chem.* **266**:23714.

Busse, S. C., Moench, S. J., and Satterlee, J. D., 1990, *Biophys. J.* **58**:45.

Cavanagh, J., and Rance, M., 1992, *J. Magn. Reson.* **96**:670.

Cavanagh, J., Chazin, W. J., and Rance, M., 1990, *J. Magn. Reson.* **87**:110.

Chandrakumar, N., and Ramamoorthy, A., 1992, *J. Am. Chem. Soc.* **114**:1123.

Chatfield, M. J., and La Mar, G. N., 1992, *Arch. Biophys. Biochem.*, **295**:289.

Chatfield, M. J., La Mar, G. N., Parker, Jr., W. O., Smith, K. M., Leung, H.-K., and Morris, I. K., 1988, *J. Am. Chem. Soc.* **110**:6352.

Cheng, H., Grohmann, K., and Sweeney, W., 1990, *J. Biol. Chem.* **265**:12388.

Cheng, H., Grohmann, K., and Sweeney, W., 1992, *J. Biol. Chem.* **267**:8073.

Cowan, J. A., and Sola, M., 1990, *Biochemistry* **29**:5633.

Dalvit, C., and Wright, P. E., 1987, *J. Mol. Biol.* **194**:313.

de Ropp, J. S., and La Mar, G. N., 1991, *J. Am. Chem. Soc.* **113**:4348.

de Ropp, J. S., La Mar, G. N., Smith, K. M., and Langry, K. C., 1984, *J. Am. Chem. Soc.* **106**:4438.

de Ropp, J. S., La Mar, G. N., Wariishi, H., and Gold, M. H., 1991a, *J. Biol. Chem.* **266**:5001.

de Ropp, J. S., Yu, L. P., and La Mar, G. N., 1991b, *J. Biomolec. NMR* **1**:175.

Delikatny, E. J., Hull, W. E., and Mountford, C. E., 1991, *J. Magn. Reson.* **94**:563.

Derome, A. E., and Williamson, M. P., 1990, *J. Magn. Reson.* **88**:177.

Domaille, P. J., 1984, *J. Am. Chem. Soc.* **106**:7677.

Duben, A. G., and Hutton, W. C., 1990, *J. Magn. Reson.* **88**:60.

Dugad, L. B., La Mar, G. N., Banci, L., and Bertini, I., 1990b, *Biochemistry* **29**:2263.

Dugad, L. B., La Mar, G. N., Lee, H. C., Ikeda-Saito, M., Booth, K. S., and Caughey, W. S., 1990c, *J. Biol. Chem.* **265**:7173.

Dugad, L. B., La Mar, G. N., and Unger, S. W., 1990a, *J. Am. Chem. Soc.* **112**:1386.

Eich, G., Bodenhausen, G., and Ernst, R. R., 1982, *J. Am. Chem. Soc.* **104**:3731.

Emerson, S. D., and La Mar, G. N., 1990a, *Biochemistry* **29**:1545.

Emerson, S. D., and La Mar, G. N., 1990b, *Biochemistry* **29**:1556.

Emerson, S. D., Lecomte, J. T. J., and La Mar, G., 1988, *J. Am. Chem. Soc.* **110**:4176.

Ernst, R. R., Bodenhausen, G., and Wokaun, A., 1987, *Principles of Nuclear Magnetic Resonance in One and Two Dimensions*, Clarendon Press, Oxford, U.K.

Feng, Y., Roder, H., Englander, S. W., Wand, A. J., and Di Stefano, D. L., 1989, *Biochemistry* **28**:195.

Finzel, B. C., Poulos, T. L., and Kraut, J., 1984, *J. Biol. Chem.* **259**:13027.
Freeman, R., 1987, *A Handbook of Nuclear Magnetic Resonance*, Longman Scientific and Technical, Harlow, Essex, U.K.
Gesmar, H., and Led, J. J., 1988, *J. Magn. Reson.* **76**:183.
Granot, J., 1982, *J. Magn. Reson.* **49**:257.
Griesinger, C., and Ernst, R. R., 1987, *J. Magn. Reson.* **75**:261.
Griesinger, C., Otting, G., Wüthrich, K., and Ernst, R. R., 1988, *J. Am. Chem. Soc.* **110**:7870.
Gueron, M., 1975, *J. Magn. Reson.* **19**:58.
Gunther, H., Moskau, D., Dujardin, R., and Maercker, A., 1986, *Tetra. Lett.* **27**:2251.
Howard, J. B., and Rees, D. C., 1991, *Adv. Protein Chem.* **42**:199.
Hull, W. E., and Sykes, B. D., 1975, *J. Chem. Phys.* **63**:867.
Inubushi, T., and Becker, E. D., 1983, *J. Magn. Reson.* **51**:128.
Jeener, J., Meier, B. H., Bachmann, P., and Ernst, R. R., 1979, *J. Chem. Phys.* **71**:4546.
Jenkins, B. G., and Lauffer, R. B., 1988a, *Inorg. Chem.* **27**:4730.
Jenkins, B. G., and Lauffer, R. B., 1988b, *J. Magn. Reson.* **80**:328.
Jesson, J. P., 1973, in *NMR of Paramagnetic Molecules* (G. N. La Mar, W. D. Horrocks, Jr., and R. H. Holm, eds.), pp. 1–52, Academic Press, New York.
Kalk, A., and Berendsen, H. J. C., 1976, *J. Magn. Reson.* **24**:343.
Keating, K. A., de Ropp, J. S., La Mar, G. N., Balch, A. L., Shiau, F.-Y., and Smith, K. M., 1991, *Inorg. Chem.* **30**:3258.
Keller, R. M., and Wüthrich, K., 1978, *Biochem. Biophys. Res. Commun.* **83**:1132.
Keller, R. M., and Wüthrich, K., 1981, in *Biological Magnetic Resonance* (L. J. Berliner and J. Reuben, eds.), vol. 3, pp. 1–52, Plenum Press, New York.
Kessler, H., Griesinger, C., Kerssebaum, R., Wagner, K., and Ernst, R. R., 1987, *J. Am. Chem. Soc.* **109**:607.
Koenig, S. H., 1982, *J. Magn. Reson.* **47**:441.
Kumar, A., Ernst, R. R., and Wüthrich, K., 1980, *Biochem. Biophys. Res. Comm.* **95**:1.
Kumar, A., Hosur, R. V., and Chandrasekhar, K., 1984, *J. Magn. Reson.* **60**:143.
Kuriyan, J., Wilz, S., Karplus, M., and Petsko, G. A., 1986, *J. Mol. Biol.* **192**:133.
La Mar, G. N., 1979, in *Biological Applications of Magnetic Resonance* (R. G. Shulman, ed.), pp. 305–343, Academic Press, New York.
La Mar, G. N., Horrocks, Jr., W. D., and Holm, R. H., eds., 1973, *NMR of Paramagnetic Molecules*, Academic Press, New York.
La Mar, G. N., Budd, D. L., Viscio, D. B., Smith, K. M., and Langry, K. C., 1978, *Proc. Natl. Acad. Sci. USA* **75**:5755.
La Mar, G. N., Budd, D. L., Smith, K. M., and Langry, K. C., 1980a, *J. Am. Chem. Soc.* **102**:1822.
La Mar, G. N., de Ropp, J. S., Smith, K. M., and Langry, K. C., 1980b, *J. Biol. Chem.* **255**:6646.
La Mar, G. N., Burns, P. D., Jackson, J. T., Smith, K. M., Langry, K. C., and Strittmatter, P., 1981a, *J. Biol. Chem.* **256**:6075.
La Mar, G. N., de Ropp, J. S., Smith, K. M., and Langry, K. C., 1981b, *J. Biol. Chem.* **256**:237.
La Mar, G. N., Yamamoto, Y., Jue, T., Smith, K. M., and Pandey, R. K., 1985, *Biochemistry* **24**:3826.
La Mar, G. N., Jue, T., Nagai, N., Smith, K. M., Yamamoto, Y., Kauten, R. J., Thanabal, V., Langry, K. C., Pandey, R. K., and Leung, H.-K., 1988, *Biochim. Biophys. Acta* **952**:131.
Lecomte, J. T. J., and La Mar, G. N., 1986, *Eur. Biophys. J.* **13**:373.
Lecomte, J. T. J., Smit, J. D. G., Winterhalter, K. H., and La Mar, G. N., 1989, *J. Mol. Biol.* **209**:235.
Lecomte, J. T. J., Unger, S. W., and La Mar, G. N., 1991, *J. Magn. Reson.* **94**:112.
Lerner, L., and Bax, A., 1986, *J. Magn. Reson.* **69**:375.
Licoccia, S., Chatfield, M. J., La Mar, G. N., Smith, K. M., Mansfield, K. E., and Anderson, R. R., 1989, *J. Am. Chem. Soc.* **111**:6087.

Luchinat, C., Steuernagel, S., and Turano, P., 1990, *Inorg. Chem.* **29**:4351.

Macura, S., and Ernst, R. R., 1980, *Mol. Phys.* **41**:95.

Marion, D., and Bax, A., 1989, *J. Magn. Reson.* **83**:205.

Marion, D., and Wüthrich, K., 1983, *Biochem. Biophys. Res. Comm.* **113**:967.

Markley, J. S., Chan, T.-M., Krishnamoorthi, R., and Ulrich, E. L., 1986, in *Iron–Sulfur Protein Research* (H. Matsubara, Y. Katsube, and K. Wada, eds.), pp. 167–184, Japan Science Society Press, Tokyo.

Maroney, M. J., Kurtz, D. M., Jr., Nocek, J. M., Pearce, L. L., and Que, Jr., L., 1986, *J. Am. Chem. Soc.* **108**:6871.

Martin, G. E., and Zektzer, A. S., 1988, *Two-Dimensional NMR Methods for Establishing Molecular Connectivity*, VCH, New York.

Martin, M. L., Delpuech, J.-J., and Martin, G. J., 1980, *Practical NMR Spectroscopy*, Heyden and Son, London.

Maudsley, A. A., and Ernst, R. R., 1977, *Chem. Phys. Lett.* **50**:368.

Maudsley, A. A., Muller, L., and Ernst, R. R., 1977, *J. Magn. Reson.* **28**:463.

Mayer, A., Ogawa, S., Shulman, R. G., Yamane, T., Cavaleiro, J. A. S., Rocha Gonsalves, A. M. d'A., Kenner, G. W., and Smith, K. M., 1974, *J. Mol. Biol.* **86**:749.

McLachlan, S. J., La Mar, G. N., and Lee, K.-B., 1988, *Biochim. Biophys. Acta* **957**:430.

Mehlkopf, A. F., Korbee, D., Tiggelman, T. A., and Freeman, R., 1984, *J. Magn. Reson.* **58**:315.

Ming, L.-J., Jang, H. G., and Que, Jr., L., 1992, *Inorg. Chem.*, **31**:359.

Nettesheim, D. G., Harder, S. R., Feinberg, B. A., and Otvos, J. D., 1992, *Biochemistry* **31**:1234.

Neuhaus, D., and Williamson, M., 1989, *The Nuclear Overhauser Effect in Structural and Conformational Analysis*, VCH, New York.

Niemczura, W. P., Helms, G. L., Chesnick, A. S., Moore, R. E., and Bornemann, V., 1989, *J. Magn. Reson.* **81**:635.

Noggle, J. H., and Schirmer, R. E., 1971, *The Nuclear Overhauser Effect*, Academic Press, New York.

Oh, B.-H., and Markley, J. L., 1990a, *Biochemistry* **29**:3993.

Oh, B.-H., and Markley, J. L., 1990b, *Biochemistry* **29**:4012.

Olejniczak, E. T., Hoch, J. C., Dobson, C. M., and Poulsen, F. M., 1985, *J. Magn. Reson.* **64**:199.

Oschkinat, H., and Freeman, R., 1984, *J. Magn. Reson.* **60**:164.

Otting, G., Widmer, H., Wagner, G., and Wüthrich, K., 1986, *J. Magn. Reson.* **66**:187.

Paci, M., Disideri, A., Sette, M., Falconi, M., and Rotillio, G., 1990, *FEBS Lett.* **261**:231.

Peters, W., Fuchs, M., Sicius, H., and Kuchen, W., 1985, *Angel. Chem. Int. Ed.* **24**:231.

Peyton, D. H., La Mar, G. N., Ramaprasad, S., Unger, S. W., Sankar, S., and Gersonde, K., 1991, *J. Mol. Biol.* **221**:1015.

Phillips, W. D., and Poe, M., 1973, in *Iron Sulfur Proteins* (W. Lovenberg, ed.), New York, Academic Press, pp. 255–285.

Qin, J., 1992, Ph.D. Thesis, University of California, Davis.

Qin, J., and La Mar, G. N., 1992, *J. Biomolec. NMR* **2**:597.

Qin, J., La Mar, G. N., Ascoli, F., Bolognesi, M., and Brunori, M., 1992, *J. Mol. Biol.*, **224**:891.

Rajarathnam, K., La Mar, G. N., Chiu, M. L., and Sligar, S. G., 1992, *J. Am. Chem. Soc.*, **114**:904B.

Ramaprasad, S., Johnson, R. D., and La Mar, G. N., 1984, *J. Am. Chem. Soc.* **106**:5330.

Rance, M., Sorenson, O. W., Leupin, W., Kogler, H., Wüthrich, K., and Ernst, R. R., 1985, *J. Magn. Reson.* **61**:67.

Redfield, A. G., and Gupta, R. K., 1971, *Cold Spring Harbor Symp. Quant. Biol.* **36**:405.

Robertson, A. D., and Markley, J. L., 1990, in *Biological Magnetic Resonance* (L. J. Berliner and J. Reuben, eds.), vol. 9, pp. 155–176, Plenum Press, New York.

Sadek, M., Brownlee, R. T. C., Scrofani, S. D. B., and Wedd, A. G., 1993, *J. Magn. Reson.*, in press.

Sandström, J., 1982, *Dynamic NMR Spectroscopy*, Academic Press, New York.

Sankar, S. S., La Mar, G. N., Smith, K. M., and Fujinari, E. M., 1987, *Biochim. Biophys. Acta* **912**:220.

Santos, H., Turner, D. L., and Xavier, A. V., 1984, *J. Magn. Reson.* **59**:177.

Satterlee, J. D., 1986, *Annual Reports NMR Spectrosc.* **17**:79.

Satterlee, J. D., and Erman, J. E., 1991, *Biochemistry* **30**:4398.

Satterlee, J. D., Erman, J. E., La Mar, G. N., Smith, K. M., and Langry, K. C., 1983, *Biochim. Biophys. Acta* **743**:246.

Satterlee, J. D., Russell, D. J., and Erman, J. E., 1991, *Biochemistry* **30**:9072.

Scarrow, R. C., Pyrz, J. W., and Que, Jr., L., 1990, *J. Am. Chem. Soc.* **112**:657.

Senn, H., and Wüthrich, K., 1985, *Q. Rev. Biophys.* **18**:111.

Shulman, R. C., Glarum, S. H., and Karplus, M., 1971, *J. Mol. Biol.* **57**:93.

Simonis, U., Dallas, J. L., and Walker, F. A., 1992, *Inorg. Chem.* **31**:5349.

Skjeldal, L., Westler, W. M., Oh, B.-H., Krezel, A. M., Holden, H. M., Jacobsen, B. L., Rayment, I., and Markley, J. L., 1991, *Biochemistry* **30**:7363.

Sletten, E., Jackson, J. T., Burns, P. D., and La Mar, G. N., 1983, *J. Magn. Reson.* **52**:492.

States, D. J., Haberkorn, R. A., and Ruben, D. J., 1982, *J. Magn. Reson.* **48**:286.

Sternlicht, H., 1965, *J. Chem. Phys.* **42**:2250.

Swift, T. J., 1973, in *NMR of Paramagnetic Molecules* (G. N. La Mar, W. D. Horrocks, Jr., and R. H. Holm, eds.), pp. 52–84, Academic Press, New York.

Talluri, S., and Scheraga, H. A., 1990, *J. Magn. Reson.* **86**:1.

Tang, J., and Norris, J. R., 1988, *J. Magn. Reson.* **78**:23.

Thanabal, V., and La Mar, G. N., 1989, *Biochemistry* **28**:7038.

Thanabal, V., de Ropp, J. S., and La Mar, G. N., 1987a, *J. Am. Chem. Soc.* **109**:265.

Thanabal, V., de Ropp, J. S., and La Mar, G. N., 1987b, *J. Am. Chem. Soc.* **109**:7516.

Thanabal, V., de Ropp, J. S., and La Mar, G. N., 1988a, *J. Am. Chem. Soc.* **110**:3027.

Thanabal, V., La Mar, G. N., and de Ropp, J. S., 1988b, *Biochemistry* **27**:5400.

Timkovich, R., 1991, *Inorg. Chem.* **30**:37.

Tirendi, C. F., and Martin, J. F., 1989, *J. Magn. Reson.* **81**:577.

Torrey, H. C., 1949, *Phys. Rev.* **76**:1059.

Trewhella, J., Wright, P. E., and Appleby, C. A., 1979, *Nature* (London), **280**:87.

Turner, C. J., 1992, *J. Magn. Reson.* **96**:551.

Turner, C. J., and Patt, S. L., 1989, *J. Magn. Reson.* **85**:492.

Turner, D. L., 1985, *J. Magn. Reson.* **61**:28.

Unger, S. W., Jue, T., and La Mar, G. N., 1985a, *J. Magn. Reson.* **61**:448.

Unger, S., Lecomte, J. T. J., and La Mar, G. N., 1985b, *J. Magn. Reson.* **64**:521.

Vasavada, K. V., and Nageswara Roa, B. D., 1989, *J. Magn. Reson.* **81**:275.

Vega, A. J., and Fiat, D., 1976, *Mol. Phys.* **31**:347.

Venable, T. L., Hutton, W. C., and Grimes, R. N., 1984, *J. Am. Chem. Soc.* **106**:29.

Wagner, G., and Wüthrich, K., 1979, *J. Magn. Reson.* **33**:675.

Wagner, G., Bodenhausen, G., Muller, N., Rance, M., Sorenson, O. W., Ernst, R. R., and Wüthrich, K., 1985, *J. Am. Chem. Soc.* **107**:6440.

Walker, F. A., and Simonis, U., 1991, *J. Am. Chem. Soc.* **113**:8652.

Wang, K. Y., Borer, P. N., Levy, G. C., and Pelczer, I., 1992, *J. Magn. Reson.* **96**:165.

Waysbort, D., and Navon, G., 1975, *J. Chem. Phys.* **62**:1021.

Wider, G., Macura, S., Kumar, A., Ernst, R. R., and Wüthrich, K., 1984, *J. Magn. Reson.* **56**:207.

Wu, J., La Mar, G. N., Yu, L. P., Lee, K.-B., Walker, F. A., Chiu, M. L., and Sligar, S. G., 1991, *Biochemistry* **30**:2156.

Wüthrich, K., 1986, *NMR of Proteins and Nucleic Acids*, John Wiley and Sons, New York.

Yamamoto, Y., 1987, *FEBS Lett.* **222**:115.
Yamamoto, Y., Nanai, N., Inoue, Y., and Chujo, R., 1988, *Biochem. Biophys. Res. Comm.* **151**:262.
Yamamoto, Y., Ogawa, A., Inoue, Y., Chujo, R., and Suzuki, T., 1989, *FEBS Lett.* **247**:263.
Yu, C., Unger, S. W., and La Mar, G. N., 1986, *J. Magn. Reson.* **67**:346.
Yu, L. P., La Mar, G. N., and Rajarathnam, K., 1990, *J. Am. Chem. Soc.* **112**:9527.

2

Nuclear Relaxation in Paramagnetic Metalloproteins

Lucia Banci

1. WHY DO NUCLEI RELAX FASTER IN THE PRESENCE OF UNPAIRED ELECTRONS?

The presence of unpaired electrons in a molecule, which is then called paramagnetic, induces large effects on its NMR parameters, chemical shifts, and relaxation rates (Bertini and Luchinat, 1986). This chapter will be expressly dedicated to relaxation rates in such systems.

The relaxation rate measures the rate for the return to the equilibrium condition by a spin system put out of equilibrium by any perturbation. The equilibrium is obtained through spontaneous transitions between the spin energy levels split by the presence of an external magnetic field. These spontaneous transitions are essentially induced by the coupling of the spin system within itself and with the environment, called the lattice, which includes other particles—atoms, molecules, electrons, etc. The motions of these particles can generate random fluctuating magnetic fields, which can induce the spin transitions responsible for the relaxation processes (Abragam, 1961–78; Slichter, 1989; Banci *et al.*, 1991).

The modulation of the coupling of a nuclear magnetic moment with another magnetic moment associated with the above particles affects the relaxation behavior of the observed nucleus, as the latter magnetic moment can produce local magnetic fields

Lucia Banci • Department of Chemistry, University of Florence, 50121 Florence, Italy.

Biological Magnetic Resonance, Volume 12: NMR of Paramagnetic Molecules, edited by Lawrence J. Berliner and Jacques Reuben. Plenum Press, New York, 1993.

randomly fluctuating in such a way as to produce, among others, the correct frequency to induce nuclear spin transitions.

Electrons have magnetic moments three to five orders of magnitude larger than those of nuclei (Koenig et al., 1952; Bertini and Luchinat, 1986; Banci et al., 1986b). Therefore the local fluctuating magnetic fields produced by them are orders of magnitude larger than those produced by other magnetic nuclei.

Furthermore, electron spins have relaxation rates orders of magnitude larger than those of nuclei. Therefore, electron spin systems are always at equilibrium on the nuclear relaxation time scale and represent the lattice from the nuclear point of view.

The analysis of the effect of the coupling of a magnetic nucleus with unpaired electrons on nuclear relaxation and on the structural and dynamic properties of paramagnetic molecules, obtained from the understanding of their relaxation behavior, is the scope of this chapter.

We define the rate constant for reaching equilibrium along the z direction of the external magnetic field as the longitudinal relaxation rate T_1^{-1} and that for reaching equilibrium in the xy plane as the transversal relaxation rate T_2^{-}.

We will restrict ourself, but without loss of generality, to the analysis of the relaxation phenomena in paramagnetic metalloproteins, as they represent an important class of molecules, to which this volume is dedicated. Furthermore we will devote most of the analysis to ^1H nuclei even if the theoretical survey reported is absolutely general. Other nuclei, relevant to the characterization of metalloproteins, are explicitly treated in other chapters of the present volume.

The interaction between magnetic nuclei and unpaired electrons is called hyperfine coupling. This interaction is described by a term, in the spin Hamiltonian formalism, which has the generic form

$$\hat{\mathcal{H}} = \hat{\mathbf{I}} \cdot \mathbf{A} \cdot \hat{\mathbf{S}} \tag{1}$$

where $\hat{\mathbf{I}}$ and $\hat{\mathbf{S}}$ are the nuclear and electron angular momentum operators, respectively, and $\mathbf{A}$ is the hyperfine coupling tensor.

An unpaired electron can couple with a magnetic nucleus through two mechanisms: dipolar and contact (Abragam and Bleaney, 1970).

Dipolar coupling occurs through space; the dipolar interaction between the two magnetic moments associated with the nuclear and the electron spin I and S, respectively, is described by the spin Hamiltonian term:

$$\hat{\mathcal{H}} = -\left(\frac{\mu_0}{4\pi}\right)\frac{g_e\mu_B g_N\mu_N}{r^3}[3(\hat{\mathbf{I}}\cdot\mathbf{r})(\hat{\mathbf{S}}\cdot\mathbf{r}) - \hat{\mathbf{I}}\cdot\hat{\mathbf{S}}] \tag{2}$$

where $\mathbf{r}$ is a unit vector oriented along the line of length r connecting I and S, g_e and g_N are the electron and nuclear g factors, and μ_B and μ_N are the electron and nuclear Bohr magnetons.

As can be noticed from Eq. (2), the dipolar interaction is anisotropic, i.e., orientation dependent. The dipolar interaction energy depends on the reciprocal of the third power of the dipole–dipole distance and on the angle between the vector $\mathbf{r}$ and the direction of the external magnetic field.

Magnetic nuclei and unpaired electrons can also interact through a contact, scalar interaction, which is due to the small but finite unpaired electron spin density present on the nucleus. The latter can exist because either the unpaired electron is in an s orbital which has a finite, nonzero density on the nucleus, or because a nonzero spin density on the nucleus can be produced, through chemical bonds, by a direct spin delocalization mechanism or can be induced by spin polarization effects. Due to its nature, this interaction is isotropic, i.e., does not depend on the orientation of the molecule in the magnetic field. The spin Hamiltonian term for this interaction takes the simple form

$$\hat{\mathscr{H}} = a\,\hat{\mathbf{I}} \cdot \hat{\mathbf{S}} \tag{3}$$

where a is the isotropic hyperfine coupling constant.

2. CORRELATION TIME FOR THE NUCLEUS–UNPAIRED ELECTRON COUPLING

The various contributions to the relaxation rates due to a variety of mechanisms can be generically written as

$$T_{1,2}^{-1} \propto \langle E_{\text{int}}^2 \rangle \cdot f(\tau_c, \omega) \tag{4}$$

i.e., they depend on the averaged square of the interaction energy responsible for relaxation and on a function (spectral density) of the spin transition frequency ω and of the correlation time of the actual interaction.

The reciprocal of τ_c represents the rate at which the nuclear spins "see" the lattice fluctuating with time and then the rate at which the interaction changes with time. From another point of view, τ_c depends on the values of frequencies contained in the motions of the lattice; τ_c^{-1} is related to the highest frequencies contained in the motions of the environment of the nuclear spin under investigation.

Relationship (4) holds as long as the interaction energy (expressed in frequency units $E_{\text{int}}/\hbar$) between the spin and the lattice, whose modulation is responsible for spin relaxation, is smaller than the inverse τ_c^{-1} of the correlation time of the actual interaction. This is called Redfield limit (Redfield, 1957, 1963), and it is always met in the cases of interest here for nuclear spins.

In the case of the interaction of a nucleus with unpaired electrons, the nature of τ_c is determined by the nature of the interaction which induces nuclear relaxation. Dipolar interaction can be modulated by the rotation of the molecule that contains both magnetic moments. As the molecule rotates, the angle formed by the vector $\mathbf{r}$ connecting the electron and the nuclear magnetic moments, r with the direction of the external magnetic field changes with time; therefore, the dipolar interaction energy changes with time. The correlation time for this process is the rotational correlation time τ_r. This quantity is given, using the approximation of spherical molecules, by the Stokes–Einstein equation (Abragam, 1961–1978; Banci et al., 1991):

$$\tau_r = \frac{4\pi a^3 \eta}{3kT} \tag{5}$$

where a is the radius of the molecule, η is the viscosity of the solvent, and d is the density of the molecule, which in the case of proteins is usually taken as 1. The magnitude of τ_r is directly proportional to the size of the molecule, and consequently to its molecular weight, and in macromolecules, due to their large size, τ_r becomes quite long (τ_r around 1.4×10^{-8} s for a molecule of 30,000 MW).

Electrons relax orders of magnitude faster than nuclei (Al'tshuler and Kozyrev, 1964; Geschwind, 1972) and in macromolecules, where the rotation is slowed down by the large molecular size, the electron's magnetic moment undergoes spin transitions and therefore changes its orientation many times before the molecule changes its orientation. The electron relaxation becomes the process which modulates the dipolar interaction with the nuclear magnetic moment. The electron relaxation time as seen by the nucleus is called electron correlation time, denoted τ_s.

If chemical exchange is operative between the moiety bearing the magnetic nucleus and that bearing the unpaired electrons, this process can modulate the nucleus–unpaired electrons interaction. The correlation time for such a process is called exchange correlation time and is indicated by τ_m. The overall correlation time is determined by the fastest of the above processes:

$$\tau_c^{-1} = \tau_r^{-1} + \tau_s^{-1} + \tau_m^{-1} \tag{6}$$

The contact interaction, due to its isotropic nature, cannot be modulated by rotation, as the reorientation of the molecule does not produce any change in the interaction energy. In this case the correlation time is determined by the shorter of the electron correlation time and the exchange correlation time. Therefore for the contact interaction, τ_c^{-1} ($=\tau_e^{-1}$) is given by:

$$\tau_e^{-1} = \tau_s^{-1} + \tau_m^{-1} \tag{7}$$

3. CONTRIBUTIONS TO NUCLEAR RELAXATION DUE TO COUPLING WITH UNPAIRED ELECTRONS

We are going to present here the equations for the various contributions to nuclear relaxation, as a result of the coupling with the unpaired electrons, through the mechanisms discussed in Section 1. We will discuss the relevance and the difficulties of relaxation time measurements for obtaining structural and dynamic properties of macromolecules.

3.1. Dipolar Relaxation

Modulation of the dipolar interaction between I and S, described by the Hamiltonian [Eq. (2)] induces nuclear relaxation. Solomon (1955) derived the equation for this contribution to the longitudinal relaxation rate T_1^{-1} and to the transverse relaxation rate T_2^{-1} according to a theory proposed by Bloembergen et al. (1948) and by Abragam and Pound (1953). The equations are

$$T_{1M}^{-1} = \frac{1}{10}\left(\frac{\mu_0}{4\pi}\right)^2 \frac{g_e^2\mu_B^2 g_N^2\mu_N^2}{r_{M-I}^6}\left(\frac{3\tau_c}{1 + \omega_I^2\tau_c^2} + \frac{\tau_c}{1 + (\omega_I - \omega_S)^2\tau_c^2} + \frac{6\tau_c}{1 + (\omega_I + \omega_S)^2\tau_c^2}\right) \quad (8)$$

$$T_{2M}^{-1} = \frac{1}{20}\left(\frac{\mu_0}{4\pi}\right)^2 \frac{g_e^2\mu_B^2 g_N^2\mu_N^2}{r_{M-I}^6}\left(4\tau_c + \frac{3\tau_c}{1 + \omega_I^2\tau_c^2} + \frac{\tau_c}{1 + (\omega_I - \omega_S)^2\tau_c^2}\right.$$

$$\left. + \frac{6\tau_c}{1 + (\omega_I + \omega_S)^2\tau_c^2} + \frac{6\tau_c}{1 + \omega_S^2\tau_c^2}\right) \quad (9)$$

where ω_I and ω_S are the Larmor frequencies for the nuclear and the electron spin, respectively. This contribution depends on r_{M-I}^{-6}, which directly provides structural information on the distance between the nucleus and the metal ion.

On this basis, relaxation time measurements have been used and are used for obtaining structural information on the various residues in the active cavity of metalloproteins, interacting with the metal ion. These measurements had been also used for mapping substrate, inhibitors and other molecules interacting close to the metal ion in a paramagnetic macromolecule.

T_{1M}^{-1} and T_{2M}^{-1} are field dependent, as shown in Fig. 1, according to Eqs. (8) and (9). The first dispersion occurs at the magnetic field for which $\omega_S\tau_c = 1$ ($|\omega_S + \omega_I|$ and

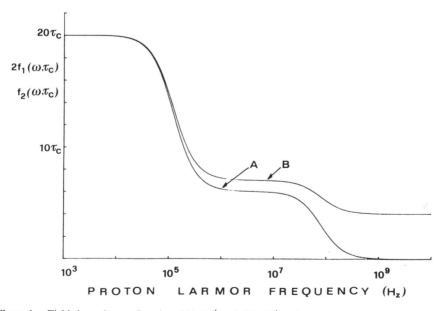

Figure 1. Field dependence of nuclear (A) T_1^{-1} and (B) T_2^{-1} in the presence of dipolar coupling with an unpaired electron. The ordinates $f_{1,2}(\omega, \tau_c)$ represent the field dependence of the spectral density functions in Eqs. (8) and (9), rspectively, and are in units of τ_c.

$|\omega_S - \omega_I| \cong \omega_S$), while the second corresponds to $\omega_I \tau_c = 1$. Also contained in T_{2M}^{-1} is a nondispersive term which keeps this contribution from vanishing even at high magnetic fields, at variance with T_{1M}^{-1}.

Equations (8) and (9) have been derived by the simplified point–dipole approximation, where the unpaired electrons are localized on the metal ion. However, some spin density can delocalize, through the chemical bonds between the paramagnetic metal ion and the donor atoms of the ligands, on the ligands themselves (Golding and Stubbs, 1979; Golding et al., 1982). Even if the fraction of unpaired electron that can delocalize is relatively small, the contribution to nuclear relaxation of the delocalized spin density may be sizable, due to the small r value.

For an exact calculation of this dipolar contribution to $T_{1,2M}^{-1}$, the unpaired electron distribution should be evaluated over all the space (Gottlieb et al., 1977; Nordenskiöld et al., 1982; Kowalewski et al., 1985). This procedure is not feasible in most of the practical cases. It is therefore customary to split the dipolar contribution into a metal-centered and a ligand-centered contribution (Gottlieb et al., 1977; Waysbort and Navon, 1978). In this approximation, all the unpaired electrons are considered as localized at a point on the metal ion, and the effect of the spin density delocalized on orbitals close to the resonating nucleus is treated parametrically.

In any case, the unpaired electrons of the metal ion are delocalized within the polyhedron coordination cage, and for nuclei far away from the metal the point dipole approximation can be safely used. This issue will be addressed in Section 4 for a case in which ligand-centered effects are dominant.

Equations (8) and (9) were derived for the case of $S = \frac{1}{2}$, but they have later been generalized to the cases with $S > \frac{1}{2}$. They become (Abragam, 1961–1978; Hertz, 1973)

$$T_{1M}^{-1} = \frac{2}{15} \left(\frac{\mu_0}{4\pi}\right)^2 \frac{g_e^2 \mu_B^2 g_N^2 \mu_N^2 S(S+1)}{r_{M-I}^6} \left(\frac{7\tau_c}{1 + \omega_S^2 \tau_c^2} + \frac{3\tau_c}{1 + \omega_I^2 \tau_c^2}\right) \tag{10}$$

$$T_{2M}^{-1} = \frac{1}{15} \left(\frac{\mu_0}{4\pi}\right)^2 \frac{g_e^2 \mu_B^2 g_N^2 \mu_N^2 S(S+1)}{r_{M-I}^6} \left(4\tau_c + \frac{13\tau_c}{1 + \omega_S^2 \tau_c^2} + \frac{3\tau_c}{1 + \omega_I^2 \tau_c^2}\right) \tag{11}$$

with the approximation $|\omega_S + \omega_I| \simeq |\omega_S - \omega_I| \simeq \omega_S$.

In all cases, these equations hold for the almost ideal case in which the electron spin functions are isotropic and not split at zero magnetic field. This means that g-anisotropy, hyperfine coupling with the metal nucleus, zero field splitting of the S manifold (for $S > \frac{1}{2}$) are all not taken into account in the above equations when computing the electron spin transition probabilities that contribute to nuclear relaxation. Therefore the above equations do not consider the possibility that the S manifold is split at zero magnetic field. The effect of these splittings is relevant for nuclear relaxation at low magnetic fields, where they can be larger than the Zeeman splitting. As the magnetic field increases, their contribution to spin level energy becomes less and less determinant. In addition, these splittings affect nuclear relaxation as long as they are larger than $\hbar \tau_c^{-1}$. Treatments which take into account a correct description of the electron spin functions, including the effect of g anisotropy, besides that of the splitting of the S manifold due to hyperfine coupling with other nuclei or to classical zero-field splitting, have been developed for a number of systems (Bertini et al., 1984, 1985a, b, c, 1990b; Banci, 1985a, b, 1986a; Sternlicht, 1965; Vasavada and Nageswara Rao, 1989). In most of the cases analytical equations cannot be derived; nevertheless, the computational procedure and the code had been derived which permit the interpretation of a large variety of experimental data.

3.2. Contact Relaxation

Unpaired electrons on the resonating nuclei can couple with the nucleus itself and modulation of this contact, hyperfine coupling can contribute to relaxation of the nuclear spin. This contribution is given by the following equations (Solomon and Bloembergen, 1956; Bloembergen, 1957):

$$T_{1M}^{-1} = \frac{2}{3} S(S+1) \frac{a^2}{\hbar^2} \frac{\tau_e}{1 + (\omega_I - \omega_S)^2 \tau_e^2} \tag{12}$$

$$T_{2M}^{-1} = \frac{1}{3} S(S+1) \frac{a^2}{\hbar^2} \left(\frac{\tau_e}{1 + (\omega_I - \omega_S)^2 \tau_e^2} + \tau_e \right). \tag{13}$$

As mentioned in the preceding section, here the correlation time is determined by the electron relaxation time and, if operative, by the exchange correlation time. Both mechanisms provide fluctuating magnetic fields.

This contribution shows only one dispersion, as it depends only on $\omega_I - \omega_S$ [Eqs. (12) and (13) and Fig. 2]. This means that only the simultaneous transitions of the

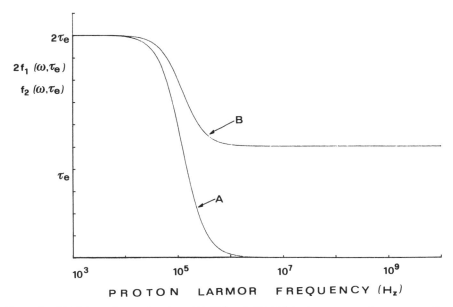

Figure 2. Field dependence of nuclear (A) T_1^{-1} and (B) T_2^{-1} in the presence of contact coupling with an unpaired electron. The ordinates $f_{1,2}(\omega, \tau_e)$ represent the field dependence of the spectral density functions in Eqs. (12) and (13), respectively, and are in units of τ_e.

nuclear and the electron spins can contribute to nuclear relaxation through scalar coupling.

From Eqs. (12) and (13), it is evident that, in the case of large τ_s values, the contact term can make a sizable contribution to T_2^{-1} and hence to the linewidth, while the contribution to T_1^{-1} is dropped to zero at the high magnetic fields currently used for high-resolution NMR spectra ($\omega_S\tau_s \gg 1$).

3.3. Curie Relaxation

Both equations for dipolar and contact contribution to nuclear relaxation were derived using the approximation that all the spin levels of the S manifold, split in the presence of an external magnetic field, are equally populated.

We should, however, take into account that the electron spins are distributed among the spin levels according to the Boltzmann law. This distribution determines a slight excess of spins in the spin level with lower energy. Such difference in population gives rise to a finite, static magnetic moment (induced magnetization), which can be expressed as a function of the expectation value $\langle \hat{S}_z \rangle$ of $\hat{S}_z$, which in turn is proportional to the external magnetic field B_0. The value of $\langle \hat{S}_z \rangle$, in the case of a single populated multiplet and of electron Zeeman splitting much smaller than kT, is given by (Griffith, 1961; Golding, 1969)

$$\langle \hat{S}_z \rangle = \frac{g_e \mu_B B_0}{3kT} S(S+1) \tag{14}$$

The nuclear spin can interact with this static electron magnetic moment, and the modulation of this interaction can induce nuclear relaxation. This interaction cannot be modulated by electron relaxation because the static magnetic moment is already time averaged over the electron spin states. Therefore this coupling can be modulated only by rotation or chemical exchange, if present.

This contribution to nuclear relaxation, called Curie relaxation, is given by (Gueron, 1975; Vega and Fiat, 1976)

$$T_{1_M}^{-1} = \frac{2}{5}\left(\frac{\mu_0}{4\pi}\right)^2 \frac{\omega_I^2 g_e^4 \mu_B^4 S^2(S+1)^2}{(3kT)^2 r^6} \frac{3\tau_c}{1+\omega_I^2\tau_c^2} \tag{15}$$

$$T_{2_M}^{-1} = \frac{1}{5}\left(\frac{\mu_0}{4\pi}\right)^2 \frac{\omega_I^2 g_e^4 \mu_B^4 S^2(S+1)^2}{(3kT)^2 r^6}\left[4\tau_c + \frac{3\tau_c}{1+\omega_I^2\tau_c^2}\right] \tag{16}$$

Note that B_0^2 has been expressed as ω_I^2/γ_I^2 in the coefficient multiplying the spectral density functions. It is interesting to note that the dispersive term in the spectral density function depends only on the nuclear Larmor frequency ω_I, as the electron spin transitions are already time averaged.

As the magnitude of the induced magnetic moment is proportional to the external magnetic field, this contribution depends on B_0^2. The dispersive term in the spectral density function increases with B_0^2 until the dispersion occurs ($\omega_I\tau_c = B_0\gamma_I\tau_c = 1$); at higher magnetic fields the field dependence is thus canceled out. On the contrary, the nondispersive term, present in T_2^{-1} but not in T_1^{-1}, does not cancel out the field dependence in T_2^{-1}, which is thus always proportional to B_0^2. From Eqs. (15) and (16), it is evident that this contribution affects only T_2^{-1} and is sizable in the case of nuclei with large γ_I (as, for example, ^{1}H), in large molecules that have long τ_r, and in the case of metal ions with high S values, i.e., large numbers of unpaired electrons and thus large magnetic moment. In addition, the contribution of Curie relaxation becomes more relevant for macromolecules containing metal ions with short τ_s, for which the dipolar contribution becomes relatively small.

3.4. Analysis of Experimental T_1 Values

Now that the equations for the contributions to T_1 and T_2 have been presented and analyzed, it is useful to try to understand the relative weight of dipolar, contact, and Curie contributions to the overall nuclear relaxation rates.

From the analysis of the equations previously reported, we can say that in the case of protons in macromolecules, the major contribution to T_{1M}^{-1} is dipolar in nature, and in absence of chemical exchange, the correlation time is τ_s. In the presence of spin delocalization over the metal ion ligand bearing the proton, the dipolar ligand-centered contribution could be sizable for protons at 4–6 Å from the metal ion and farther (Nordenskiöld et al., 1982; Golding et al., 1982; Golding and Stubbs, 1979).

As far as T_{2M}^{-1} is concerned, again the dipolar contribution is the most relevant. In small molecules (short τ_r) with a slow relaxing metal ion (long τ_s), the contact contribution could be dominant. Even in large molecules, long τ_s and large contact (hyperfine) coupling make the contact contribution sizable and dominant at large magnetic fields because of the nondispersive term. In the latter case, and again for macromolecules, Curie relaxation could be dominant.

For nuclei other than protons, the dipolar term is smaller than in protons, due to the smaller values of γ_N. On the other hand $(a/\hbar)^2$ may be much larger than for protons if we deal with directly coordinated nuclei (for example, ^{17}O, ^{15}N, ^{14}N). In such cases contact interaction may easily be the dominant mechanism for nuclear relaxation, especially for T_2^{-1}. Contributions from non-s orbitals, in ligand centered effects, cannot be negligible.

Let us consider the practical case of a proton at 5 Å from a metal ion with a $\tau_s = 10^{-12}$ s (for example, high-spin five coordinated cobalt(II), $S = \frac{3}{2}$) in a molecule of MW 30,000. In such a molecule τ_r, in the spherical approximation, is 1.4×10^{-8} s. Let us further consider that such a proton is scalarly coupled with the metal ion with a coupling constant of 0.5 MHz. This value is large and would induce a contact contribution to the paramagnetic shift of 66 ppm. Applying the appropriate equations, we calculate, at a magnetic field of 4.7 T (200 MHz for protons) the following contributions:

$$T_{1_{dip}}^{-1} = 94.3 \text{ s}^{-1}; \qquad T_{1_{cont}}^{-1} = 37.0 \text{ s}^{-1}; \qquad T_{1_{Curie}}^{-1} = 0.6 \text{ s}^{-1}$$

$$T_{2_{dip}}^{-1} = 104.5 \text{ s}^{-1}; \qquad T_{2_{cont}}^{-1} = 42.2 \text{ s}^{-1}; \qquad T_{2_{Curie}}^{-1} = 123.0 \text{ s}^{-1}$$

while they become, at 14.1 T (600 MHz for protons)

$$T_{1_{dip}}^{-1} = 74.0 \text{ s}^{-1}; \qquad T_{1_{cont}}^{-1} = 37.0 \text{ s}^{-1}; \qquad T_{1_{Curie}}^{-1} = 0.6 \text{ s}^{-1}$$

$$T_{2_{dip}}^{-1} = 85.6 \text{ s}^{-1}; \qquad T_{2_{cont}}^{-1} = 37.7 \text{ s}^{-1}; \qquad T_{2_{Curie}}^{-1} = 1104.5 \text{ s}^{-1}$$

The field dependence of the linewidth for such a proton is shown in Fig. 3.

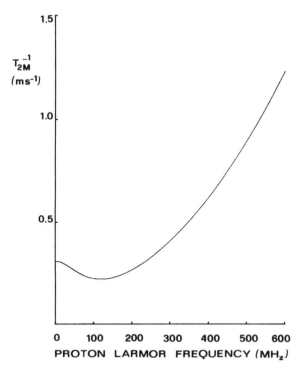

Figure 3. Field dependence of nuclear T_2^{-1} as a sum of dipolar, contact, and Curie relaxation contributions for the case of a proton at 5 Å from a cobalt(II) ion ($S = \frac{3}{2}$) with $\tau_s = 3 \times 10^{-12}$ s, $\tau_r = 1.4 \times 10^{-8}$ s, and a scalar hyperfine coupling of 0.5 MHz.

Among the several examples reported in the literature, we present here some data on derivatives of superoxide dismutase. Copper–zinc superoxide dismutase (SOD) is a dimeric enzyme of MW 32,000 containing a copper(II) ion and a zinc(II) ion linked by a histidinato residue (Fig. 4) (Tainer *et al.*, 1982). Substitution of cobalt(II) for the native zinc ion yields a derivative which has the same structural and biological properties as the native protein, but in which the copper(II) ion has much faster electron relaxation rates as a result of the magnetic coupling with the fast-relaxing cobalt(II) ion (Bertini *et al.*, 1985b), in accordance with what will be presented in Section 5.

The shortening of the τ_s of copper(II) in Cu_2Co_2SOD determines a significant decrease in T_1^{-1} and T_2^{-1} of the protons sensing the copper ion (Banci *et al.*, 1987), as discussed in Chapter 7 of this volume, thus allowing the detection of 19 reasonably sharp signals in the ^1H NMR spectrum (Fig. 5) (Bertini *et al.*, 1985b; Banci *et al.*, 1987). They are due to the protons of both the copper and the cobalt ligands, paramagnetically shifted outside the diamagnetic envelope. These signals are characterized by T_1 values ranging from 4 to 8 ms for protons of the copper-bound histidines in *meta*-like position (5 Å from the metal ion), to 2–3 ms for the *ortho*-like protons of copper-bound histidines (3 Å from the metal ion) to 2–5 ms for the *meta*-like protons of the cobalt-bound His, to 1.7 ms for the βCH_2 of the Asp residue bound to the cobalt ion. The

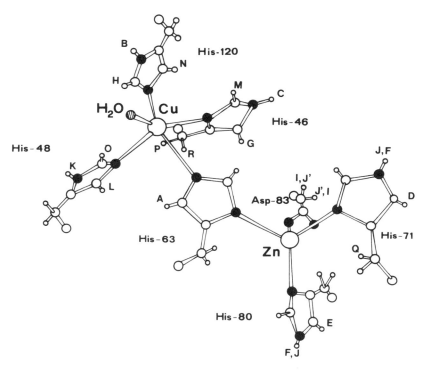

Figure 4. Schematic drawing of the active site of Cu$_2$Zn$_2$SOD as obtained from the X-ray structure of the bovine enzyme (Tainer *et al.*, 1982). The residue numbers are those of the human enzyme. The labels of the protons follow the assignment of the ^{1}H NMR spectrum of the Cu$_2$Co$_2$SOD derivative (Banci *et al.*, 1989), reported in Fig. 5.

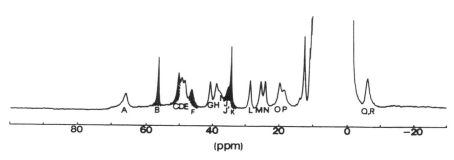

Figure 5. 300-MHz ^{1}H NMR spectrum of the Cu$_2$Co$_2$SOD derivative in H$_2$O (Bertini *et al.*, 1985b). The dashed signals disappear when the spectrum is recorded in D$_2$O. The labels indicate the resonance assignment (Banci *et al.*, 1989) to individual protons as reported in Fig. 4.

signals of *ortho*-like protons of the cobalt residues are too broad, beyond detection. The T_1 values of the signals in this system are strongly determined by the dipolar, ligand-centered contribution; indeed, the ratio between the T_1 values does not follow the ratio between the sixth power of the metal ion–proton distance, as it would if only the metal-centered dipolar contribution were operative. In particular, it was found that an *ortho*-like proton (3 Å from the metal) has a longer T_1 than a *meta*-like proton (5 Å from the metal) (Banci *et al.*, 1989). As a possible explanation, it was proposed that the unpaired electron of copper is largely delocalized onto the ligands and that the *ortho*-like proton experiences long T_1 because it has the smallest contribution from the delocalization of the unpaired electron.

4. A COMMENT ON ELECTRON RELAXATION TIMES

In macromolecules, such as proteins, molecular tumbling occurs at low rates, always slower, for molecules with MW > 10,000, than the electron relaxation rates of any paramagnetic metal ion. Therefore the correlation time in such systems is always dominated by electron relaxation. Among the transition metal ions, electron relaxation rates range from 10^8 to 10^{13} s^{-1}, depending on the nature of the metal ion itself, as reported in Table 1 (Bertini and Luchinat, 1986; Banci *et al.*, 1991).

It is important in this context, therefore, to present a brief overview of electron relaxation mechanisms (Banci *et al.*, 1986b; Bertini *et al.*, 1991d, e) and of the relaxation rates of various metal ions in order to understand the effect of the latter on the relaxation of nuclei interacting with them.

Electron spins can be relaxed, as occurs for nuclei, by fluctuating magnetic fields generated by nuclear and electron magnetic moments, electric dipoles, etc. However, electrons, unlike nuclei, have other, more efficient relaxation mechanisms related to the presence of spin–orbit coupling (Abragam and Bleaney, 1970; Geshwind, 1972).

Spin–orbit coupling arises from the coupling of the electron spin magnetic moment with the orbital angular momentum sensed by the electron, due to its motion around the electrically charged nucleus. Modulation of such coupling is due to the fluctuations of the lattice which, through spin–orbit coupling, affects the electron spin (Abragam and Bleaney, 1970).

Historically, electron relaxation mechanisms are described with different formalisms and "language" for the solid state and for solutions. However, this separation is only formal and solid state mechanisms may be operative also in solution.

The solid state mechanisms are essentially described as being due to the coupling of the vibrational transitions of the lattice and the electronic transitions through spin–orbit coupling. A quantum of vibrational energy in the lattice is called a "phonon." The two main mechanisms operative at room temperature are the Orbach (Orbach, 1961a, b, c) and the Raman (Orbach, 1961a; Van Vleck, 1939, 1940) processes. Both mechanisms involve the interaction of the electron spin with two phonons whose energy difference is equal to the energy separation of the spin states between which relaxation occurs.

In the Orbach mechanism, the electron spin system undergoes transition to an excited electronic state by absorbing energy from a vibrational transition. The transition may also involve a change of the M_s spin state. The spin can be changed, alternatively

TABLE 1
Estimated Room-Temperature Electron
Relaxation Times for Some Metal Ions at
Low Magnetic Fields (10^{-2}–10^{-3} T)

Metal ion	$\tau_{s0}(s)$
Ti^{3+}	10^{-9}–10^{-10}
V^{2+}	5×10^{-10}
V^{3+}	5×10^{-12}
VO^{2+}	10^{-8}–10^{-9}
Cr^{3+}	5×10^{-10}
Cr^{2+}	10^{-11}
Mn^{3+}	10^{-10}–10^{-11}
Mn^{2+}	10^{-9}–10^{-10}
Fe^{3+} (H.S.)	10^{-10}–10^{-11}
Fe^{3+} (L.S.)	10^{-11}–10^{-12}
Fe^{2+} (H.S.)	10^{-11}–10^{-12}
Co^{2+} (H.S.)	10^{-11}–10^{-12}
Co^{2+} (L.S.)	10^{-9}–10^{-10}
Ni^{2+}	10^{-10}–10^{-12}
Cu^{2+}	1–3×10^{-9}
Ru^{3+}	10^{-11}–10^{-12}
Re^{3+}	10^{-11}
Gd^{3+}	10^{-8}–10^{-9}
Dy^{3+}	8×10^{-13}
Ho^{3+}	8×10^{-13}
Tb^{3+}	8×10^{-13}
Tm^{3+}	8×10^{-13}
Yb^{3+}	1×10^{-12}

or again, when the electron returns to the ground state, through coupling with another vibrational transition. This mechanism requires the presence of low energy electronic excited states (i.e., within the thermal bath). It is efficient and can induce electron relaxation rates up to the order of 10^{12}–10^{13} s^{-1} (Abragam and Bleaney, 1970; Geshwind, 1972).

If no electronic excited states are available within thermal energies, the Raman mechanism is operative. In this process, the electronic system absorbs and emits two phonons differing in energy by the quantum of the electron spin transition. Even if no electronic excited state is involved, the efficiency of this process still depends on the energy gap with real electronic excited levels, as the transition probability depends on the mixing of the ground state with the excited levels. This mechanism can induce relaxation rates up to the order of 10^{10} s^{-1} (Abragam and Bleaney, 1970; Geshwind, 1972).

In solution, other relaxation mechanisms can be present that depend on fluctuations of the magnetic field around the electron spin because of the rotation and instantaneous distortions of the molecule in solution (Muus and Atkins, 1972; Kivelson and Collins, 1963). The effects on electron relaxation depend on a correlation time related to the motions that modulate the electron–lattice interaction. Such a correlation time τ_v can be determined by τ_r or by the correlation time for other types of fluctuations, such as bombardment by solvent molecules.

In the case of macromolecules, rotation is slowed down and becomes slower than the electron relaxation time. Therefore rotation in macromolecules cannot be responsible for electron relaxation, and solution-type mechanisms cannot be invoked. Still, it has been realized more and more often that efficient electron relaxation is operative, with rates comparable to those present in solids (Banci *et al.*, 1991).

Another mechanism for electron relaxation arises from bombardment of solvent molecules; they have a correlation time of the order of 10^{-11} s. Such bombardments (1) modulate the state ZFS or induce an instantaneous ZFS (Bloembergen and Morgan, 1961; McGarvey, 1957; Levanon *et al.*, 1970; Rubinstein *et al.*, 1971); and (2) modulate the hyperfine coupling with the metal nucleus (McGarvey, 1957; Kivelson, 1960; McConnell, 1956; Rogers and Pake, 1960; Bruno *et al.*, 1977). However, these processes account for electron relaxation times up to 10^{-10} s. Therefore metal ions which relax through these mechanisms (i.e., rotation or bombardment) are not suitable for NMR studies as the linewidths they induce are too broad, unless the observed nucleus is undergoing fast exchange between the bound and the free form and the free form is in large excess.

Faster electron relaxation rates can be operative for metal ions which have available low-lying excited states. In such systems, solid state-type mechanisms can also be operative in solution (Kivelson, 1966), where the molecule and, even better, the macromolecule can be "seen" as an instantaneous "microcrystal" in which lattice vibrations can be present. We can extend then the concept of phonons to solution provided that an appropriately short time scale is considered. In the case of these fast-relaxing metal ions, like cobalt(II), low-spin iron (III), and lanthanides except gadolinium(III), it has been proposed that Orbach-type mechanisms are operative also in solution (Kivelson, 1966; Banci *et al.*, 1985a, 1986b, 1991), which can account for relaxation times of the order of 10^{-11}–10^{-12} s. Therefore these metal ions are suitable for high-resolution NMR studies.

Table 1 reports the relaxation rates of some metal ions, measured as electron correlation time for nuclear relaxation at room temperature. Inspection of Table 1 is quite meaningful before attempting the characterization of metalloproteins through high-resolution NMR. Indeed, as discussed in Section 2, in macromolecules the correlation time for the dipolar interaction, which dominates T_1 and T_2 in relatively small proteins, is τ_s. Consequently, the nature of the metal ion directly determines the detectability of the NMR spectrum. Therefore, while well-resolved and relatively narrow signals are obtained for protons in cobalt(II)—or low-spin iron(III)—containing proteins of even 40,000 MW, too broad, undetectable signals are present for any molecule containing copper(II) with an MW larger than 1000.

5. ELECTRON RELAXATION IN MAGNETIC EXCHANGE–COUPLED SYSTEMS

In the preceding section, the mechanisms for elecron relaxation were presented for the case of isolated metal ions. There are, however, biological systems in which two or more metal ions are present in the molecule and the spin moments associated with them are magnetically coupled. The presence of this interaction can deeply affect the electron relaxation rates as well as the NMR parameters. While Chapter 7 of this volume is dedicated to the latter parameters, a brief overview of the effect of magnetic coupling on the relaxation rates will be presented here.

The spin moment associated with one metal ion generates fluctuating magnetic fields at the other spin moment through all the mechanisms discussed in Section 3. If the two spins are different, the fluctuations due to the relaxation of the faster metal ion provide additional relaxation pathways for the slower-relaxing metal ion. In other words, the slow-relaxing electron spin moment uses the more efficient relaxation mechanisms of the other electron spin (Banci et al., 1986b, 1987, 1990b, 1991).

The dominant term in the interaction between two metal ions is the scalar term (exchange or superexchange coupling); this interaction is essentially isotropic. The dipolar interaction is usually much smaller and is relevant for closely spaced metal ions in the absence of scalar coupling (Banci et al., 1991; Bencini and Gatteschi, 1990).

In the case of isotropic interaction an equation which relates the electron relaxation rate of the slow-relaxing metal ion M_1 with the rate of the fast one M_2 and with the extent of the coupling has been proposed. It is similar to that derived for nuclear relaxation in the case of contact interaction between nuclear and electron spins (Solomon and Bloembergen, 1956; Bloembergen, 1957). The equation is (Banci et al., 1991)

$$\tau_{sM_1}^{-1}(J) = \tau_{sM_1}^{-1}(0) + \frac{2}{3}\left(\frac{J^2}{\hbar^2}\right) S_2(S_2 + 1) \frac{\tau_{sM_2}}{1 + (\omega_{sM_1} - \omega_{sM_2})^2 \tau_{sM_2}^2} \tag{17}$$

where $\tau_{sM_1}^{-1}(J)$ is the actual value for the relaxation rate of the slow-relaxing metal ion in the coupled system, $\tau_{sM_1}^{-1}(0)$ is the relaxation rate in the absence of coupling—the difference $\tau_{sM_1}^{-1}(J) - \tau_{sM_1}^{-1}(0)$ is the enhancement due to the occurrence of magnetic coupling—and τ_{sM_2} is the relaxation time of the fast-relaxing metal ion, which is assumed not to be altered when interacting with the other metal ion.

Equation (17) shows that the enhancement of the electron relaxation rate of metal 1 is proportional to the square of the exchange coupling energy and depends on the electron relaxation time of metal 2. In the limit of vanishingly small interaction energy, the enhancement in τ_s of metal 1 is zero. Equation (17) is valid within the Redfield limit, i.e., as long as the interaction energy, in frequency units, is smaller than τ_{s2}^{-1}. Outside this limit, the enhancement in the relaxation rate of metal 1 would tend to a value close to τ_{s2}^{-1} (Bloembergen, 1949; Abragam, 1955; Banci et al., 1991).

From Eq. (17), we can estimate the magnitude of the coupling constant J capable of inducing an effect on the electron relaxation rate of a slow-relaxing metal ion. For example, if a metal ion with $\tau_{s1} = 10^{-9}$ s is coupled with a metal ion, $S_2 = \frac{3}{2}$, with $\tau_{s2} = 10^{-12}$ s, a J coupling as low as 0.1 cm^{-1} would induce a decrease in τ_{s1} of a factor of 2.

An equation has also been proposed in the case of dipolar coupling, even if this case is less common and the effect of this type of interaction drops rapidly to zero with increasing distance (Banci et al., 1991).

In the case of two identical spins, they relax at the same rate when isolated; then one spin cannot provide its own relaxation pathway to the other (Banci et al., 1991; Bertini et al., 1988).

The effect of magnetic coupling on the electron relaxation rates is of great importance for the detectability of high-resolution NMR in the case of macromolecules. In systems containing slow-relaxing metal ions, for which NMR characterization is not feasible, the presence of a second metal-ion binding site close to the first can induce magnetic coupling between the two metal ions. If one of these two ions is chosen properly, i.e., with short electron relaxation times, it will induce, through the magnetic interaction, a shortening of the relaxation rates of the slow-relaxing metal ion, thus producing sharp resonance lines for nuclei interacting with it.

When more than two metal ions are magnetically coupled, an increase of the electron relaxation rates again occurs for the slow-relaxing metal ions, toward the value of the fast-relaxing ion. In these systems, however, unlike the two-spin system, the increase in electron relaxation rates is also operative in the case of clusters of like spins. In such systems new levels, with the same S spin number, are generated as a result of the magnetic coupling (Banci et al., 1991; Bencini and Gatteschi, 1990). Therefore electron spin transitions can occur, which contribute to the electron relaxation, with Orbach-type mechanisms (Banci et al., 1991).

6. EFFECT OF FAST RELAXATION ON NOE EXPERIMENTS

Assignment of NMR spectra of paramagnetic metalloproteins has always been a difficult task because the hyperfine shift is the result of several contributions which cannot, in general, be easily predicted.

The standard techniques applied in diamagnetic systems for spectral assignment, such as NOE and 2D NMR (Wüthrich, 1986; Ernst et al., 1987; Oppenheimer and James, 1989a, b), cannot be straightforwardly transferred to paramagnetic systems. Only recently has the NOE technique been successfully applied to the latter systems (Ramaprasad et al., 1984; Lecomte and La Mar, 1986; Emerson et al., 1988; Thanabal et al., 1987a, b, 1988; see also Chapter 1 of this volume for more extensive references), a significant advance in NMR as a tool for the characterization of the metal site in metalloproteins.

The technical aspects of NOE applied to paramagnetic systems are treated in Chapter 1 of this volume; we are interested here in analyzing the effect of the nuclear relaxation times on the NOE value and in presenting in which cases NOE experiments are feasible and others in which they are not.

NOE experiments measure the fractional change in the intensity of a signal when another signal is saturated (Noggle and Schirmer, 1971; Neuhaus and Williamson, 1989; Overhauser, 1953a, b). The change is different from zero when dipolar coupling is present and then cross-relaxation is operative between the two nuclei. The NOE effect is proportional to r_{ij}^{-6}, where r_{ij} is the distance between nuclei i and j. The effect is also proportional to the selective T_1 of the observed signal. Therefore, in the case of paramagnetic molecules characterized by short relaxation times, the extent of NOE can be very small. The general equation for NOE is:

$$\eta_{I(J)} = \sigma_{IJ} \cdot T_{1Jsel} \tag{18}$$

where σ_{IJ} is the cross-relaxation rate between nucleus I and nucleus J and T_{1Isel} is the selective relaxation time of the observed nucleus I, which also includes the paramagnetic contribution.

In macromolecules we deal always with negative NOEs, as cross-relaxation is in the slow motion limit, due to the low molecular rotational rates. As a result of Eq. (18), the extent of NOE between two signals having different T_{1sel} is not symmetrical, i.e., a larger NOE is observed by saturating the faster-relaxing signal and detecting NOE on the slower-relaxing one.

As already mentioned, the presence of fast nuclear relaxation determines a reduction of the size of NOE, which can drop to a few % even for two geminal protons. The two nuclei tend to exchange energy with the unpaired electron(s) rather than through cross-relaxation (Neuhaus and Williamson, 1989; Banci et al., 1991). Furthermore, signals characterized by short T_1's experience also very short T_2's, which determine broad signals. A small fraction of such signals is very difficult to measure! In addition, signals with short T_1's and T_2's require high decoupler power levels in order to obtain sizable saturation. These high power levels could determine off-resonance saturation effects on signals close to the irradiated one.

Let us consider a generic protein of MW 30,000; it has a τ_r of $\simeq 1.4 \times 10^{-8}$ s. Two geminal protons in an aspartate or cysteine residue are approximately 1.8 Å apart and, in such a protein, they will have a cross-relaxation (in the two-spin approximation) of -23.5 s^{-1}. Two adjacent protons in a histidinate ring are 2.5 Å apart and will have a cross-relaxation of -3.3 s^{-1}. If the proton of the geminal pair, on which NOE is observed, has a T_1 value of 20 ms, an NOE of -47% will be observed, while if the T_1 value is one order of magnitude smaller, the NOE is also one order of magnitude lower, becoming as small as -4.7%. In the case of a proton of a histidine, assuming the same T_1 values, NOEs of -6.6% and of -0.7% are expected. NOEs of 0.5–1% are probably the smallest NOEs which can be safely detected among paramagnetically shifted signals.

Among the paramagnetically shifted signals, NOE has been observed and is expected to be easily detected between the protons of the metal-ion ligands that are essentially histidines, aspartate, glutamate, tyrosines, and cysteines, and the heme group in heme-containing proteins. Of course NOE has been observed between signals of the metal ligands and signals of other residues not directly coordinated to the metal ion. In such cases, the NOEs are expected to be larger that those of the first coordination sphere, as the T_1's are longer due to the larger distance from the metal ion (in most cases) and to the absence of ligand-centered effects (in all cases). However, for such protons the two-spin approximation may not hold (see below).

The application of NOE to paramagnetic molecules was first exploited by La Mar (Ramaprasad et al., 1984; Lecomte and La Mar, 1986; Emerson et al., 1988; Thanabal et al., 1987a, b, 1988); then the NMR group in Florence contributed in applying this technique to a large variety of metalloproteins (Banci et al., 1989, 1990a, b, c, 1991a, b, 1992d; Bertini et al., 1990a, 1991a, b) together with other research groups (Satterlee et al., 1987; Cowan and Sola, 1990). Studies are available on myoglobins (Emerson et al., 1988; Lecomte et al., 1986; Ramaprasad et al., 1984), peroxidases (Thanabal et al., 1987a, b, 1988; Satterlee et al., 1987; Banci et al., 1991c, d), ferredoxins (Dugad et al., 1990; Bertini et al., 1990a, 1991a), HiPIPs (Banci et al., 1991a, b; Bertini et al., 1991b, 1992), superoxide dismutase (Banci et al., 1989, 1990a, c, d), metallothioneins (Bertini et al., in press), and cobalt(II)-substituted carbonic anhydrase (Banci et al., 1992d).

TABLE 2
Some Examples of NOE Detected on Paramagnetically Shifted Signals in Metalloproteins

Group[a]	H–H distance (Å)	NOE[b] (%)	T_1 (ms)	MW	τ_r (ns)	Reference
	1.8	6	8.4	9,000	4	Banci et al., 1991a
		5	3.6	11,000	5	Dugad et al., 1990
$-C\substack{H \\ H}$		7	5.6	11,000	5	Dugad et al., 1990
		43	120	16,000	9	Lecomte and La Mar, 1986
(Asp, Glu, Cys, His)		7.7	2.4	32,000	14	Banci et al., 1989
		60	100	36,000	21	Satterlee et al., 1987
		50	80	42,000	22	Thanabal et al., 1987b
$H\!-\!C\!-\!C\substack{H \\ H}$ (Cys)	2.2–2.4	4.5	34.0	9,000	4	Banci et al., 1991a
		1.8	5.6	11,000	5	Dugad et al., 1990
(His)	2.4	6	5.4	30,000	14	Banci et al., 1992d
		0.9	1.8	32,000	14	Banci et al., 1989
		2	4.2	32,000	14	Banci et al., 1989
(His)	~3.3	0.6	3.5	32,000	14	Banci et al., 1989
		7–10	—	42,000	21	Thanabal et al., 1987b

[a] The groups bearing the two protons belong to metal ion ligands.
[b] The NOEs in the present systems are all negative; they are reported here without the minus sign.

In Table 2 some significant examples of the NOEs observed in the paramagnetic systems investigated up to now are reported; it can be seen that, as the T_1's become shorter and shorter, the NOE drops to small values, which can be difficult to "extract" from the noise. However, the NOE increases as the molecular size increases, as the cross-relaxation becomes larger due to an increase in τ_r.

The two-spin approximation in the analysis of NOE is usually correct in paramagnetic systems, which are characterized by small NOEs and short T_1 (Banci et al., 1990c, 1991; Neuhaus and Williamson, 1989; Lecomte et al., 1991). Short irradiation times are enough to reach a steady-state condition. The first-order NOE, which could be transferred to another nucleus, is small by itself and would then produce a very small effect on the other nucleus. Second-order NOE will also be very small as a result of the short relaxation times of these signals. If the third or further spins were characterized by long relaxation times, the NOE would not have time to build up, thus producing again very small, undetactable effects. In Table 3, the experimental and calculated NOEs are reported for the protons of the metal ligands in Cu_2Co_2SOD (Banci et al., 1990c). The NOEs are calculated both by the two-spin approximation and taking into account

TABLE 3
T_1 Values, Interproton Distances Inside the Active Site, and Observed and Calculated Steady-State NOE Values of Cu_2Co_2SOD[a]

Saturated signal	Observed signal	T_1[b] (ms)	X-ray distances (Å)	Experimental and calculated NOE[c] A	B	C
A (Hδ2 His-63)	K (Hδ1 His-48)	8.0	3.81	0.6 ± 0.2	0.2	(0.2)
A (Hδ2 His-63)	O (Hε1 His-48)	1.9	3.20	0.3 ± 0.1	0.1	(0.1)
A (Hδ2 His-63)	R (Hβ2 His-46)	2.4	3.42	0.1 ± 0.1	0.1	(0.1)
B (Hδ1 His-120)	H (Hε1 His-120)	1.8	2.38	0.9 ± 0.2	0.8	(0.8)
C (Hε2 His-46)	G (Hδ2 His-46)	3.5	2.47	1.3 ± 0.2	1.2	(1.2)
C (Hε2 His-46)	M (Hε1 His-46)	2.7	2.44	0.8 ± 0.2	1.0	(1.0)
G (Hδ2 His-46)	C (Hε2 His-46)	4.2	2.47	1.5 ± 0.4	1.4	(1.4)
K (Hδ1 His-48)	O (Hε1 His-48)	1.9	2.40	0.6 ± 0.2	0.8	(0.8)
L (Hδ2 His-48)	A (Hδ2 His-63)	1.5	2.73	0.3 ± 0.1	0.3	(0.3)
L (Hδ2 His-48)	P (Hβ1 His-46)	1.6	2.27	1.2 ± 0.3	0.9	(0.9)
L (Hδ2 His-48)	R (Hβ2 His-46)	2.4	3.19	0.5 ± 0.1	0.3	(0.3)
M (Hε1 His-46)	C (Hε2 His-46)	4.2	2.44	2.0 ± 0.4	1.5	(1.6)
M (Hε1 His-46)	N (Hδ2 His-120)	2.9	2.81	—	0.5	(0.5)
P (Hβ1 His-46)	L (Hδ2 His-48)	4.3	2.27	4.9 ± 1.0	2.4	(2.4)
P (Hβ1 His-46)	R (Hβ2 His-46)	2.4	1.61	7.7 ± 1.0	9.8	(9.9)
Q (Hβ1 His-71)	D (Hδ2 His-71)	3.8	2.60	1.3 ± 0.2	1.0	—
R (Hβ2 His-46)	G (Hδ2 His-46)	3.5	3.27	0.6 ± 0.2	0.2	(0.4)
R (Hβ2 His-46)	L (Hδ2 His-48)	4.3	3.19	2.0 ± 0.2	0.3	(0.5)
R (Hβ2 His-46)	P (Hβ1 His-46)	1.6	1.61	4.9 ± 0.8	6.8	(6.8)

[a] Data taken from Banci et al., 1990c.
[b] T_1 for the signal on which NOE is detected, measured at 200 MHz (Banci et al., 1989).
[c] Experimental data have been collected at 200 MHz (Banci et al., 1989). The NOE values are calculated using the reported T_1 values and $\tau_c = 1.4 \times 10^{-8}$ s. A: measured at 200 MHz; B, calculated through the two-spin system approximation; C: calculated by taking into account all the imidazole protons of the histidines in the active site; βCH_2 protons of the histidine coordinated to copper have also been considered, together with α-CH of His-46 and His-63, β-CH2 of Asp-83, γ-CH$_3$ group of Val-118, and γ-CH$_2$ of Arg-143 together with the exchangeable protons of Arg-143 ε-NH, η_1-NH$_2$ and η_2-NH$_2$. The latter are all the protons that have been observed, through a computer graphic analysis, to be at less than 3 Å distance from at least one proton of the imidazole rings.

cross-relaxation with all the nearby protons. It is shown that the largest variation in calculated NOEs for histidine ring protons is less than 20%. So spin diffusion is essentially lacking in these systems. This is a result of the very short relaxation times of the coupled protons. Only when protons with particular geometric disposition (and with long T_1) are considered, spin diffusion can be sizable. This holds for NOE involving geminal Hβ protons, as for signals Q and R, where up to 60% changes in calculated NOEs with and without spin-diffusion effects are present. It is worthwhile to note that a factor of 2 in NOE gives 10% difference in distance. Therefore, in the case of fast-relaxing systems, the short T_1 values (i.e., ≤20 ms) allow us to safely overcome the spin-diffusion problem and to estimate nucleus–nucleus distances using the two-spin approximation.

Transient NOE experiments, which had been proposed as more appropriate in order to avoid spin-diffusion problems in diamagnetic systems (Gordon and Wüthrich,

1978; Neuhaus and Williamson, 1989) can be performed also on paramagnetic systems, if care is taken in properly setting the parameters (Banci et al., 1990c). In this experiment, a spin is perturbed, usually inverted, by a selective strong r.f. pulse; then, after a delay, the spectrum is recorded. The variation in intensity of the other signals as a function of the delay is measured.

In paramagnetic systems, characterized by fast relaxation rates, the excited signal quickly relaxes to equilibrium after the selective pulse, thus reducing the extent of the transfer to the other spin. Therefore the transient NOE rapidly drops to zero with the delay time. On the other hand, some delay is necessary for NOE to be transferred. These two competing phenomena again produce small NOE results. In this experiment, the size of NOE is symmetrical; i.e., the same signal intensity variation is detected by inverting each signal, even if characterized by different relaxation times. The dependence of the NOE size as a function of the delay τ after the selective inversion is given by (Neuhaus and Williamson, 1989)

$$\eta_{IJ}(\tau) = \frac{\sigma_{IJ}}{D}(e^{-(R' + D)\tau} - e^{-(R' - D)\tau})\, e^{\tau/T_2} \tag{19}$$

where $R' = \frac{1}{2}(T_{1I\text{sel}}^{-1} + T_{1J\text{sel}}^{-1})$ and $D = (\frac{1}{4}(T_{1I\text{sel}}^{-1} - T_{1J\text{sel}}^{-1})^2 + \sigma_{IJ}^2)^{1/2}$.

With the purpose of comparing these two techniques (steady-state versus transient NOE) applied to paramagnetic systems, we performed both experiments on some signals of Cu_2Co_2SOD (Banci et al., 1990c). Figure 6 shows the comparison of the steady-state

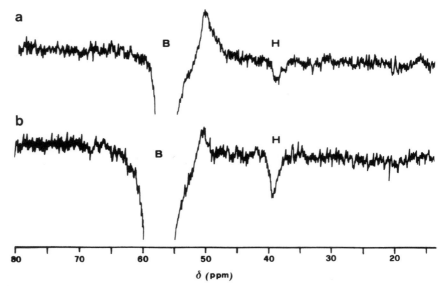

Figure 6. Comparison of the best transient NOE (a) with the steady-state NOE (b) from signal B to signal H in Cu_2Co_2SOD measured at 600 MHz. Spectrum (a) was recorded after 7 ms of delay time, for which NOE on H reaches the maximum value. The two spectra have the same signal-to-noise ratio. Adapted from Banci et al. (1990a).

NOE on signal H (Hε1 His 120) obtained by saturation of signal B (Hδ1 His 120) with the best transient NOE between the same signals, recorded after a 7-ms delay, following the selective inversion of signal B (Banci et al., 1990c). It is evident that the steady-state NOE is larger than the transient one, thus leading to greater sensitivity and smaller errors in the calculated distances. Figure 7 shows the time dependence of NOE in a transient (A) and a truncated (B) experiment with the relaxation parameters characterizing B and H. From the steady-state NOE point of view, saturation of B and detection on H is a worse experiment than the verse, as H has a much shorter T_1 than B. In such conditions, transient NOE could be larger than steady-state NOE. But this is valid, in principle, for very short delay times and for very narrow delay intervals, as fast decay of transient NOE then occurs. We said "in principle" because one of the major problems in transient NOE experiments applied to paramagnetic systems is the difficulty in obtaining the complete inversion of the signal unless the selectivity of the pulse is sacrificed. A limited selectivity of the pulse causes spillover effects, which are difficult to eliminate in the difference spectra and produce baseline distortions.

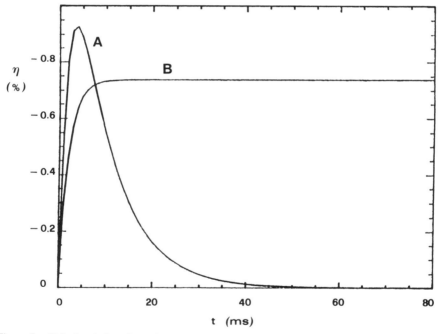

Figure 7. Calculated time dependence of transient (A) and steady-state (B) NOEs. Values are calculated at 600 MHz with $\tau_c = 1.4 \times 10^{-8}$ s, for two protons 2.4 Å apart. Relaxation rates are $T_{1I\,sel} = 8$ ms, $T_{1J\,sel} = 2$ ms. These parameters are those of peaks B and H of the spectrum of Cu$_2$Co$_2$SOD reported in Fig. 5. The time in the abscissa has different meanings in the two cases: for steady-state NOE, t is the time during which peak I is kept saturated by continuous-wave irradiation (instantaneous saturation is assumed); in the case of transient NOE, t is the delay time between the 180° selective pulse and the observation pulse (instantaneous inversion is assumed). Adapted from Banci et al. (1990a).

On the basis of the data reported in the literature and of our own experience, we can say that steady-state NOE represents the correct approach in the case of paramagnetic macromolecules, in which the NOEs under consideration are small. Application of selective pulses to fast-relaxing signals implies several technical problems. Besides the disadvantages discussed above with respect to steady-state NOE, transient experiments, if carefully performed, provide the same results as the steady-state experiments.

7. EFFECT OF FAST RELAXATION ON 2D EXPERIMENTS

In the early 80s the development of 2D NMR represented a formidable step ahead in the application of the NMR spectroscopy to the investigation of solution structure and the dynamics of proteins, DNA fragments, and other macromolecules (Wüthrich, 1986; Ernst et al., 1987; Bax, 1982, 1989; Kaptein et al., 1988; Kessler et al., 1988). With 2D NMR experiments, both scalar interactions (through bonds) and dipolar interactions (through space) can be easily determined through the detection of cross peaks between signals. COSY and TOCSY are the basic experiments for determining scalar interactions (Aue et al., 1976; Braunschweiler et al., 1983; Bax and Davis, 1985), while NOESY experiments provide information on dipolar connectivities (Macura and Ernst, 1980). These techniques allow the assignment of spectra even in crowded regions and then the obtainment of internuclear distances, in order to solve the structure of the macromolecule in solution and to understand its dynamic properties. The potential of this methodology is well settled now, with many structures of biological molecules solved by NMR spectroscopy. (For a more complete bibliography see Wüthrich, 1986; Oppenheimer and James, 1989a, b). A large number of pulse sequences have been designed for the investigation of the structure and the dynamics of diamagnetic molecules in solution. Recently, introduction of multidimensional NMR, such as 3D, 4D and even 5D NMR (Griesinger et al., 1987; Kay et al., 1990), has provided further powerful tools for the investigation of the structural character of larger and larger molecules. However, analysis of these data never implies nuclear relaxation.

Application of 2D techniques to paramagnetic systems had been delayed due to the intrinsic difficulties of these systems: large spectral widths and short relaxation times. Fast relaxation rates imply that a perturbed system returns to equilibrium before sizable or detectable magnetization transfer among nuclei has occurred. Therefore, 2D experiments in paramagnetic systems must be "tuned" to the properties of the signals investigated (Jenkins and Lauffer, 1988; Yu et al., 1986; Luchinat et al., 1990; Bertini et al., 1991c).

The fast relaxation rates determine a loss of magnetization during the various steps of the sequence (evolution t_1, mixing t_m, and detection t_2) that is not negligible and that may cause a dramatic decrease in signal intensity. Therefore, in order to obtain the maximum signal intensity, such durations should be set as short as possible. Indeed, in paramagnetic systems long acquisition times are not necessary, as the magnetization disappears from the xy plane with a constant time T_2, usually very short. After the signal has decayed, only noise is acquired. Short acquisition times allows short recycle times, if T_1's are also short, thus allowing a larger number of scans for the same exeriment time. Also, during t_1 magnetization decays with T_2; therefore, after a few experiments most of the information is lost. It is then more useful to perform fewer experiments, as resolution is not very critical in systems characterized by large linewidth, with more scans per experiment. The problem arises with the mixing time, because during this period the magnetization or the coherence transfer occurs from one set of spins to another, thus determining the appearance of cross peaks. The transfer of magnetization or coherence requires time to occur; therefore, this time cannot be reduced below certain limits, depending on the type of coupling of the two spins and the relaxation process occurring during the actual mixing time.

In NOESY experiments, which are the 2D equivalent of transient NOE experiments, during the mixing time t_m magnetization is transferred from one spin set to another dipolarly coupled spin set through cross-relaxation. This transfer occurs between the z-components of the magnetization; therefore, the process competing with this transfer is the relaxation along the z direction, which occurs with the rate constant T_1^{-1} or, more precisely, with the selective longitudinal relaxation rate. The intensity of the cross peaks as a function of the experimental time duration is (Neuhaus and Williamson, 1989)

$$I_{\text{cross peak}}(t_{1,2}) = \frac{I(0)}{2} \frac{\sigma_{IJ}}{D} \left(e^{-(R' + D)t_m} - e^{-(R' - D)t_m} \right) e^{t_{1,2}/T_2} \tag{20}$$

where $R' = \frac{1}{2}(T_{1/\text{sel}}^{-1} + T_{1J\text{sel}}^{-1})$, $D = (\frac{1}{4}(T_{1/\text{sel}}^{-1} - T_{1J\text{sel}}^{-1})^2 + \sigma_{IJ}^2)^{1/2}$. In this experiment, therefore, the best results are obtained with t_m values of the order of R' of the signals between which cross-peaks are expected.

In COSY experiments, scalar coupling between nuclei is detected. In COSY the mixing is operated by the detection pulse itself, and the development of the cross peaks occurs during times t_1 and t_2. Therefore the information is not maximal at the beginning of the FID, along t_2, and of the interferogram, along t_1, but it builds during t_1 and t_2 with a function of the coupling constant J, $\sin(\pi J t_{1,2})$. The maximum of the transfer should occur, therefore, for $t_{1,2} = (2J)^{-1}$ (Ernst et al., 1987). But competing with the building of the cross peaks, relaxation occurs both along t_1 and t_2 with time constant T_2. Thus the effect decays with a function $e^{t_{1,2}/T_2}$. In fast-relaxing systems, this function decays faster than the coherence transfer builds. Therefore, in such systems the limiting factor is T_2^{-1} and not J, and $t_{1,2}$ should be set according to the T_2 values of the signals between which COSY connectivity is expected. The maximum intensity is achieved for t_1 and t_2 of the order of T_2.

In addition, the most common COSY experiments produce cross peaks with antiphase character of the multiplet components (Ernst et al., 1987; Martin and Zektzer, 1988). As these components are usually not resolved in paramagnetic systems, they tend to cancel each other, thus drastically reducing the intensity of the cross peaks. Therefore, for systems characterized by large linewidth, sequences which give rise to in-phase cross peaks should be preferred. Alternatively, magnitude calculations on the spectra would reduce the cancellation effects of antiphase peaks, improving the signal-to-noise ratio even if there is a sizable increase in the width of the signals. Furthermore, sequences with simple phase cycles should be preferred in order to increase the sensitivity of the spectra even if more artifacts will be present. In most cases the "primitive" magnitude COSY experiment turns out to be the best for these systems.

TOCSY experiments have the advantage of having in-phase coherence transfer, which does not induce cancellation of the multiplet components of the cross peaks

(Ernst et al., 1987; Martin and Zektzer, 1988). However, in this experiment the spin lock could represent a limitation; high power levels are needed to spin-lock signals spread over a large spectral width (Luchinat et al., 1990). In most cases only a part of the spectrum can be acquired. In addition, during the time of spin lock, the signals decay with time constant $T_{1\rho}$, which is very close to T_2.

Several examples of NOESY and COSY spectra of paramagnetic systems are available now in the literature, and they tend to be relatively routine experiments, even in fast-relaxing systems. Among the many systems investigated with these techniques, elegant and significant applications are represented by heme proteins, as for example globins (Yu et al., 1990; Yamamoto et al., 1990, 1991) and peroxidases (Banci et al., 1991c, d; de Ropp and La Mar, 1991; de Ropp et al., 1991a, b; Satterlee and Erman, 1991; Satterlee et al., 1991), by iron–sulfur proteins (Banci et al., 1991b; Bertini et al., 1991a, 1992), and by cytochromes (Banci et al., 1992b).

Peroxidases are a class of heme proteins of MW 36,000–46,000 containing, in the resting state, high-spin iron(III) (Poulos, 1988). The iron ion is coordinated by the four heme pyrrole nitrogens and by the nitrogen of a histidine, acting as axial ligand (proximal histidine) (Poulos and Kraut, 1980). High-spin iron(III) has long electron relaxation times, which determine ^1H NMR signals characterized by short T_1 and large linewidth (Banci et al., 1991). The ^1H NMR spectra of such systems are also characterized by small pseudocontact shifts, because of the negligible magnetic anisotropy of high-spin iron(III), which then induce very small hyperfine shift on protons of residues not directly coordinated to the metal ion but present close to it in the active site (Bertini and Luchinat, 1986). This prevents their resolution outside the diamagnetic envelope and makes their characterization quite difficult.

On the contrary, low-spin iron(III) has a short τ_s, which determines sharp signals with relatively long T_1 and sizable pseudocontact shifts for protons of residues not directly coordinated to the metal ions as well (Bertini and Luchinat, 1986; Banci et al., 1992a). Therefore signals of protons of nonligand residues can also be resolved outside the diamagnetic envelope. For this reason the low-spin form, obtained by CN^- binding (Banci et al., 1991c, d; de Ropp and La Mar, 1991; de Ropp et al., 1991a, b; Satterlee and Erman, 1991; Satterlee et al., 1991), can provide a wealth of information and is the most studied for this class of proteins, even if 2D maps are also reported for the high-spin form (de Ropp and La Mar, 1991). A masterpiece of a study is the NMR characterization of low-spin iron(III)-containing myoglobin–CN^-: 2D experiments (Emerson and La Mar, 1990a; Yu et al., 1990), in conjunction with previous 1D NOE data (Lecomte and La Mar, 1986; Ramaprasad et al., 1984; Emerson et al., 1988), led to the assignment of about 80 signals of protons of residues in the cavity around the metal ion. This extensive assignment permitted the calculation of the dipolar shift values, correlating the magnetic properties of the metal ion with the structural features of the protein (Emerson and La Mar, 1990b).

Among peroxidases, extensive 2D NMR characterization has been reported for the low-spin form of horseradish peroxidase by La Mar and coworkers (de Ropp and La Mar, 1991; de Ropp et al., 1991b), of cytochrome c peroxidase (CcP) by us (Banci et al., 1991c) and by Satterlee and coworkers (Satterlee and Erman, 1991; Satterlee et al., 1991), and of lignine peroxidase (LIP) by us (Banci et al., 1991d) and by La Mar and coworkers (de Ropp et al., 1991a). The COSY and NOESY maps for CcP–CN^- and LIP–CN^- are reported in Figs. 8 and 9.

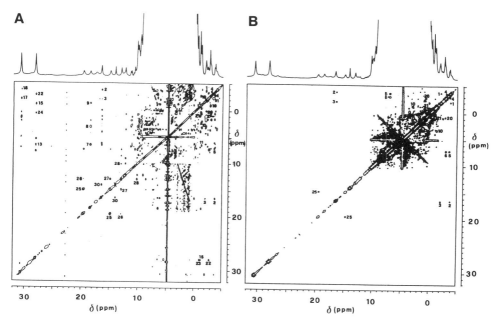

Figure 8. (A) 600-MHz 298-K NOESY spectrum of a water solution of CcP—CN⁻ obtained with 15 ms mixing time. Cross-peak assignments: 1, 4-Hβ_{cis}, 4-Hβ_{trans}; 2, 4-Hα; 4-Hβ_{cis}; 3, 4-Hα; 4-Hβ_{trans}; 4, 2-Hβ_{cis}, 2-Hβ_{trans}; 5, 2-Hα; 2-Hβ_{cis}; 6, 2- Hα; 2-Hβ_{trans}; 7, 7-Hα, 7-Hα'; 8, 7-Hα; 7-Hβ'; 9, 7-Hα, 7-Hβ; 10, 7-Hβ, 7-Hβ'; 11, 7-Hα', 7-Hβ; 12, 8-CH₃, 7-Hα; 13, 8-CH₃, 7-Hα'; 14, 8-CH₃, 7-Hβ'; 15, 8-CH₃, 7-Hβ; 16, 3-CH₃, 4-Hα; 17, 3-CH₃, 4-Hβ_{trans}; 18, 3-CH₃, 4-Hβ_{cis}; 19, δ'-CH₃ Leu-232, δ-CH₃ Leu-232; 20, Hγ Leu-232, δ-CH₃ Leu-232; 21, Hγ Leu-232, δ'-CH₃ Leu-232; 22, 8-CH₃, δ-CH₃ Leu 232; 23, 8-CH₃, δ'-CH₃ Leu-232; 24, 8-CH₃, Hγ Leu-232; 25, Hβ His-175, Hβ' His-175; 26, Hβ His-175, NH$_p$ His-175; 27, Hβ' His-175, NH$_p$ His 175; 28, Hβ' His-175, Hδ1 His-175; 29, Hβ His-175, Hδ1 His-175; 30, Hδ1 His-52, Hϵ1 His-52. Cross peaks 12, 14, 16, and 29 become detectable by decreasing the threshold. (B) 600-MHz 298-K COSY spectrum in D₂O buffer of CcP—CN⁻. The numbers indicate the same connectivities as in (A). These maps are collected on the recombinant enzyme from *E. Coli*, which contains two primary sequence differences compared to the WT CcP. Adapted from Banci *et al.* (1991c).

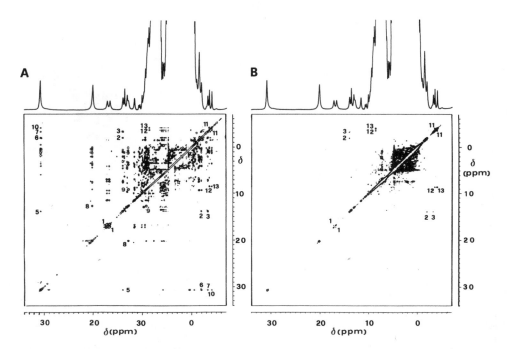

Figure 9. (A) 600-MHz 301-K NOESY spectrum of a water solution of LIP—CN⁻, obtained with 15 ms mixing time. (B) 600-MHz magnitude COSY spectrum of LIP—CN⁻ under the same experimental conditions. The corresponding cross peaks in the two spectra are labelled with the same number. Cross peaks assignment: 1, Hβ, Hβ' proximal histidine; 2, 4-Hα, 4-Hβ$_{trans}$; 3, 4-Hα, 4-Hβ$_{cis}$; 4, 4-Hβ$_{cis}$, 4-Hβ$_{trans}$; 5, 4-Hα, 3-CH₃; 6, 4-Hβ$_{trans}$, 3-CH₃; 7, 4-Hβ$_{cis}$, 3-CH₃; 8, 8-CH₃, 7-Hα; 9, 7-Hα, 7-Hα'; 10, 3-CH₃, 2-Hβ$_{trans}$; 11, 2-Hβ$_{cis}$, 2-Hβ$_{trans}$; 12, 2-Hα, 2-Hβ$_{cis}$; 13, 2-Hα, 2-Hβ$_{trans}$ (Banci *et al.*, 1991d).

NOESY cross peaks had been observed between signals with T_1 as short as 34 ms (Hβ of proximal His); all the connectivities expected between paramagnetically shifted signals had been observed, as have COSY cross peaks between signals with T_2 as short as 2.4–2.6 ms (linewidth of 120–130 Hz, again Hβ proximal His). The detection of scalar interactions between protons of some heme side chains, in conjunction with the dipolar connectivities obtained from NOESY maps, permitted the assignment of most of the heme signals and of several signals arising from residues close to the active site. (See captions to Figs. 8 and 9 for the complete cross-peak assignment.) In this respect we would like to emphasize that, while in the case of CcP-CN$^-$, some heme signals had been previously assigned with the help of selective heme deuteration (Satterlee et al., 1983), in the case of LIP-CN$^-$ complete assignment is achieved only through 2D NMR experiments (Banci et al., 1991d; de Ropp et al., 1991a). From the assignment we have learned that the 4-vinyl of the heme moiety has a different orientation in the investigated proteins. These 2D data, together with 1D NOE measurements on H$_2$O samples, provided relevant structural information on the active site. A similar characterization of the active site of horseradish peroxidase (HRP) had been reported by La Mar et al. (de Ropp et al., 1991b). It is found that the active site of these enzymes is similar but not identical. The major difference among the investigated peroxidases is the geometrical arrangement of the proximal histidine. The properties of this iron ligand are thought to modulate the redox potential of the iron ion and therefore the biochemical behavior of the enzyme. Other striking differences are also observed in the distal side, in which the enzymatic reaction occurs. This reaction involves the binding of H$_2$O$_2$ and the formation of an intermediate (Compound I) containing the Fe(IV)=O and a cation radical (Dunford, 1982; Poulos, 1988; Everse et al., 1990). The distal side of the protein contains at least two residues relevant in the catalytic process: the distal histidine, which could stabilize Compound I, and an arginine, which has been proposed to favor the binding of H$_2$O$_2$ by extracting a proton. While the structural arrangement of the distal histidine is similar in all the peroxidases investigated, significant changes in the shift and connectivity patterns of some exchangeable and nonexchangeable protons, possibly arising from the distal arginine, are observed in the various proteins, thus suggesting a different arrangement of the latter residue. This could be relevant in order to rationalize the different formation constants of Compound I in the different peroxidases. Further and more detailed analysis is available in Chapter 1 of this volume.

2D NMR techniques have been applied with success also to other proteins characterized by signals spread over a very large spectral width and with short relaxation times, such as iron–sulfur proteins, which are extensively discussed in Chapter 7 of this volume.

As mentioned previously, application of 2D techniques to paramagnetic systems is being applied to more and more extreme situations in terms of relaxation rates. In this respect, two striking examples are represented by the 2D characterization of the resting state of HRP, reported by La Mar et al. (de Ropp and La Mar, 1991) and of cytochrome c' (Cyt c') (Banci et al., 1991). The latter is a medium-sized (MW of 28,000) heme-containing protein, which is supposed to be involved in electron transfer processes. These systems contain high-spin iron(III), which, as already mentioned, is characterized by long τ_s, thus inducing fast relaxation rates on nuclei interacting with it. The downfield part of the ^{1}H NMR spectrum of Cyt c' is reported as the upper trace in Fig. 10, where the heme moiety contained in this protein is also shown.

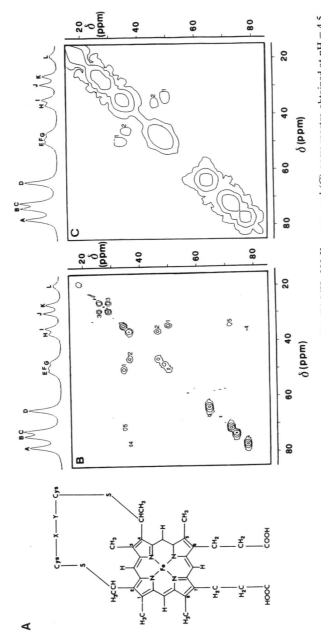

Figure 10. (A) Scheme of heme moiety present in cytochrome c'. (B) 600-MHz 300-K NOESY and (C) COSY spectra obtained at pH = 4.5. The cross-peak assignment is 1, 6/7-propionate Hα, 6/7-propionate Hα'; 2, 6/7-propionate Hα, 6/7-propionate Hα'; 3, proximal histidine Hβ, proximal histidine Hβ'; 4, 6/7-propionate Hα, 5/8-CH$_3$; 5, 6/7-propionate Hα, 5/8-CH$_3$. Adapted from Banci et al. (1992b).

The paramagnetically shifted resolved signals are due to the protons of the heme ring and of the βCH_2 of the histidine bound to the iron ion, acting as axial ligand. These signals have T_1's in the range 2–7 ms and T_2's in the range 0.3–0.4 ms. The NOESY and COSY maps of this system are reported in Fig. 10B and C, respectively; these maps show dipolar connectivities between signals with T_1 values as low as 2 ms and scalar connectivities between signals as broad as 900–1100 Hz ($T_2 \simeq 0.3$ ms).

These figures should make the reader think of how far the capabilities of NMR of paramagnetic metalloproteins have been pushed. If we take into account that the paramagnetic shift increases the resolution for the signals of the protons in the environment of the metal ions and that the fast relaxation rates are not a severe limitation nowadays, the presence of a paramagnetic metal ion, which is essentially the active site at which the biological function occurs, allows a detailed structural characterization of residues in its environment, even in proteins of medium size (MW up to 40,000–45,000) for which, in the diamagnetic case, 2D (and to some extent even 3D) techniques have not enough resolution.

8. GENERAL STRATEGY FOR NMR ANALYSIS IN PARAMAGNETIC METALLOPROTEINS

From the above analysis of the relaxation properties of paramagnetic metalloproteins, from the data reported in the literature, and from the author's own experience, the general strategy for the assignment of the NMR spectra of these systems can be outlined.

Application of 2D NMR experiments alone could be not enough to assign the spectrum and obtain structural information. Indeed, in the case of systems characterized by very large differences within the relaxation rates of the nuclei of interst, 2D experiments suitable for detecting connectivities among fast-relaxing signals, shifted over a large spectral width, could not allow the detection and resolution of cross peaks between slow-relaxing signals, experiencing smaller shift spreading, and *vice versa*. Therefore, the general approach in the case of strongly paramagnetic systems is the combined use of 1D and 2D experiments. 1D NOE experiments, which have the largest sensitivity and resolution (Neuhaus and Williamson, 1989), permit the "connection" between the fast-relaxing and the slow-relaxing signals. This approach has been successfully applied to systems with signals characterized by extreme relaxation properties (Banci *et al.*, 1992d; Bertini *et al.*, 1992a). Then 2D NMR experiments permit one to assign protons of the protein matrix and obtain the necessary structural information in order to have a complete characterization of the system.

Finally, molecular dynamics (MD) calculations on these systems, taking into account the NMR information, lead to the complete refinement of the structure in solution. Even if the papers are still a small number compared to those reported for metal-free macromolecules (for an extensive bibliography see, for example, McCammon and Harvey, 1987; Scheek *et al.*, 1989; Kaptein *et al.*, 1988), MD calculations are now also applied with success to systems containing metal ions (Merz, 1991; Merz *et al.*, 1991; Hoops *et al.*, 1991; Banci *et al.*, 1992a, c, e). Therefore another limitation in the study of metalloproteins is going to break down. Through the combination of all these techniques, expressly tailored for systems containing paramagnetic metal ions, a complete characterization of metalloproteins in solution can be achieved, sometimes at higher levels than in the case of diamagnetic systems.

Note added in proof. It has been recently reported that cross-correlation between the dipolar and Curie contributions to nuclear relaxation of a proton pair can give rise to cross peaks in COSY maps even in absence of scalar coupling between the two protons (Bertini *et al.*, 1993). This effect is more relevant for protons close in space (large dipolar cross-relaxation) and for large Curie contributions (i.e., in large molecules). It turns out that this effect could be dominant in metalloproteins.

ACKNOWLEDGMENTS.   I am grateful to Professors Ivano Bertini and Claudio Luchinat, who introduced me to the field of nuclear magnetic resonance and with whom I have had extensive, enlightening, and . . . sometimes animated discussions on several aspects of magnetic relaxation.

REFERENCES

Abragam, A., 1955, *Phys. Rev.* **98**:1729.
Abragam, A., 1961–78, *The Principles of Nuclear Magnetism*, Oxford UP, Oxford.
Abragam, A., and Bleaney, B., 1970, *Electron Paramagnetic Resonance of Transition Metal Ions*, Clarendon Press, Oxford.
Abragam, A., and Pound, R. V., 1953, *Phys. Rev.* **92**:953.
Al'tshuler, S. A., and Kozyrev, B. M., 1964, *Electron Paramagnetic Resonance*, Academic Press, New York.
Aue, W. P., Bartholdi, E., and Ernst, R. R., 1976, *J. Chem. Phys.* **64**:2225.
Banci, L., Bertini, I., and Luchinat, C., 1985a, *Inorg. Chim. Acta* **100**:173.
Banci, L., Bertini, I., and Luchinat, C., 1985b, *Chem. Phys. Lett.* **118**:345.
Banci, L., Bertini, I., Briganti, F., and Luchinat, C., 1986a, *J. Magn. Reson.* **66**:58.
Banci, L., Bertini, I., and Luchinat, C., 1986b, *Magn. Reson. Rev.* **11**:1.
Banci, L., Bertini, I., Luchinat, C., and Scozzafava, A., 1987, *J. Am. Chem. Soc.* **109**:2328.
Banci, L., Bertini, I., Luchinat, C., Piccioli, M., Scozzafava, A., and Turano, P., 1989, *Inorg. Chem.* **28**:4650.
Banci, L., Bencini, A., Bertini, I., Luchinat, C., and Piccioli, M., 1990a, *Inorg. Chem.* **29**:4867.
Banci, L., Bertini, I., and Luchinat, C., 1990b, *Struct. Bonding* **71**:113.
Banci, L., Bertini, I., Luchinat, C., and Piccioli, M., 1990c, *FEBS Lett.* **272**:175.
Banci, L., Bertini, I., Luchinat, C., and Viezzoli, M. S., 1990d, *Inorg. Chem.* **29**:1438.
Banci, L., Bertini, I., and Luchinat, C., 1991, *Nuclear and Electron Relaxation. The Electron Nucleus Hyperfine Coupling in Diluted Systems*, VCH, Weinheim.
Banci, L., Bertini, I., Briganti, F., Luchinat, C., Scozzafava, A., and Vicens Oliver, M., 1991a, *Inorg. Chim. Acta* **180**:171.
Banci, L., Bertini, I., Briganti, F., Luchinat, C., Scozzafava, A., and Vicens Oliver, M., 1991b, *Inorg. Chem.* **30**:4517.
Banci, L., Bertini, I., Turano, P., Ferrer, J. C., and Mauk, A. G., 1991c, *Inorg. Chem.* **30**:4510.
Banci, L., Bertini, I., Turano, P., Tien, M., and Kirk, T. K., 1991d, *Biochemistry* **88**:6956.
Banci, L., Bertini, I., Carloni, P., Luchinat, C., and Orioli, P. L., 1992a, *J. Am. Chem. Soc.* **114**:6994.
Banci, L., Bertini, I., Turano, P., and Vicens Oliver, M., 1992b *Eur. J. Biochem.* **204**:107.
Banci, L., Carloni, P., La Penna, G., and Orioli, P. L., 1992c, *J. Am. Chem. Soc.* **114**:6994.

Banci, L., Dugad, L. B., La Mar, G. N., Keating, K., Luchinat, C., and Pierattelli, R. 1992d, *Biophys. J.* **63**:530.

Banci, L., Schröder, S., Kollman, P. A., 1992e, *Proteins: Structure, Function and Genetics* **13**:288.

Bax, A., 1982, *Two Dimensional Nuclear Magnetic Resonance in Liquids*, Reidel, Dordrecht.

Bax, A., 1989, *Annu. Rev. Biochem.* **58**:223.

Bax, A. and Davis, D. G., 1985, *J. Magn. Reson.* **65**:355.

Bencini, A., and Gatteschi, D., 1990, *Electron Paramagnetic Resonance of Exchange-Coupled Systems*, Springer-Verlag, Berlin.

Bertini, I., and Luchinat, C., 1986, *NMR of Paramagnetic Molecules in Biological Systems*, Benjamin/Cummings, Menlo Park, CA.

Bertini, I., Luchinat, C., Mancini, M., and Spina, G., 1984, *J. Magn. Reson.* **59**:213.

Bertini, I., Briganti, F., Luchinat, C., Mancini, M., and Spina, G., 1985a, *J. Magn. Reson.* **63**:41.

Bertini, I., Lanini, G., Luchinat, C., Messori, L., Monnanni, R., and Scozzafava, A., 1985b, *J. Am. Chem. Soc.* **107**:4391.

Bertini, I., Luchinat, C., Mancini, M., and Spina, G., 1985c, in *Magneto-Structural Correlations in Exchange-Coupled Systems* (D. Gatteschi, O. Kahn, and R. D. Willett, eds.) pp. 421–461, Reidel, Dordrecht.

Bertini, I., Banci, L., Brown, R. D., Koenig, S. H., and Luchinat, C., 1988, *Inorg. Chem.* **27**:951.

Bertini, I., Briganti, F., Luchinat, C., and Scozzafava, A., 1990a, *Inorg. Chem.* **29**:1874.

Bertini, I., Luchinat, C., and Vasavada, K. V., 1990b, *J. Magn. Reson.* **89**:243.

Bertini, I., Briganti, F., Luchinat, C., Messori, L., Monnanni, R., Scozzafava, A., and Vallini, G., 1991a, *FEBS Lett.* **289**:253.

Bertini, I., Briganti, F., Luchinat, C., Scozzafava, A., and Sola, M., 1991b, *J. Am. Chem. Soc.* **113**:1237.

Bertini, I., Capozzi, F., Luchinat, C., and Turano, P., 1991c, *J. Magn. Reson.* **95**:244.

Bertini, I., Luchinat, C., and Martini, G., 1991d, in *Electron Spin Resonance Handbook* (C. P. Poole and H. Farach, eds.), CRC Press, Boca Raton, FL.

Bertini, I., Luchinat, C., and Martini, G., 1991e, in *Electron Spin Resonance Handbook* (C. P. Poole and H. Farach, eds.), CRC Press, Boca Raton, FL.

Bertini, I., Capozzi, F., Ciurli, S., Luchinat, C., Messori, L., and Piccioli, M., 1992, *J. Am. Chem. Soc.*, in press.

Bertini, I., Luchinat, C., and Tarchi, D., 1993, *Chem. Phys. Lett.*, in press.

Bertini, I., Luchinat, C., Messori, L., and Vasak, M., *Eur. J. Biochem.*, in press.

Bloembergen, N., 1949, *Physica* **15**:386.

Bloembergen, N., 1957, *J. Chem. Phys.* **27**:572.

Bloembergen, N., and Morgan, L. O., 1961, *J. Chem. Phys.* **34**:842.

Bloembergen, N., Purcell, E. M., and Pound, R. V., 1948, *Phys. Rev.* **73**:679.

Braunschweiler, L., and Ernst, R. R., 1983, *J. Magn. Reson.* **53**:521.

Bruno, G. V., Harrington, J. K., and Eastman, M. P., 1977, *J. Phys. Chem.* **81**:11.

Cowan, J. A., and Sola, M., 1990, *Biochemistry* **29**:5633.

de Ropp, J. S., and La Mar, G. N., 1991, *J. Am. Chem. Soc.* **113**:4348.

de Ropp, J. S., La Mar, G. N., Wariishi, H., and Gold, M. H., 1991a, *J. Biol. Chem.* **266**:15001.

de Ropp, J. S., Yu, L. P., and La Mar, G. N., 1991b, *J. Biomol. NMR*, **1**:175.

Dugad, L. B., La Mar, G. N., Banci, L., and Bertini, I., 1990, *Biochemistry* **29**:2263.

Dunford, H. B., 1982, *Adv. Inorg. Biochem.* **4**:41.

Emerson, S. D., and La Mar, G. N., 1990a, *Biochemistry* **29**:1545.

Emerson, S. D., and La Mar, G. N., 1990b, *Biochemistry* **29**:1556.

Emerson, S. D., Lecomte, J. T., and La Mar, G. N., 1988, *J. Am. Chem. Soc.* **110**:4176.

Ernst, R. R., Bodenhausen, G., and Wokaun, A., 1987, *Principles of Nuclear Magnetic Resonance in One and Two Dimensions*, Oxford UP, London.

Everse, J., Everse, J. K., and Grisham, M. B., eds., 1990, *Peroxidases in Chemistry and Biology*, CRC Press, Boca Raton, FL.

Geschwind, S., 1972, *Electron Paramagnetic Resonance*, Plenum Press, New York.

Golding, R. M., 1969, *Applied Wave Mechanics*, D. Van Nostrand, London.

Golding, R. M., and Stubbs, L. C., 1979, *J. Magn. Reson.* **33**:627.

Golding, R. M., Pascual, R. O., and McGarvey, B. R., 1982, *J. Magn. Reson.* **46**:30.

Gordon, S. L., and Wüthrich, K. J., 1978, *J. Am. Chem. Soc.* **100**:7094.

Gottlieb, H. P. W., Barfield, M., and Doddrell, D. M., 1977, *J. Chem. Phys.* **67**:3785.

Griesinger, C., Sorensen, O. W., and Ernst, R. R., 1987, *J. Magn. Reson.* **73**:574.

Griffith, J. S., 1961, *The Theory of Transition-Metal Ions*, Cambridge UP, Cambridge.

Gueron, M., 1975, *J. Magn. Reson.* **19**:58.

Hertz, H. G., 1973, in *Water: A Comprehensive Treatise* (F. Franks, ed.), vol. 3, Plenum Press, New York.

Hoops, S. C., Anderson, K. A., and Merz, Jr., K. M., 1991, *J. Am. Chem. Soc.* **113**:8262.

Jenkins, B. G., and Lauffer, R. B., 1988, *Inorg. Chem.* **27**:4730.

Kaptein, R., Boelens, R., Scheek, R. M., and van Gunsteren, W. F., 1988, *Biochemistry* **27**:5389.

Kay, L. E., Clore, M., Bax, A., 1990, *Science* **249**:411.

Kessler, H., Gehrke, M., and Griesinger, C., 1988, *Angew. Chem. Int. Ed. Ingl.* 490.

Kivelson, D., 1960, *J. Chem. Phys.* **33**:1094.

Kivelson, D., 1966, *J. Chem. Phys.* **45**:1324.

Kivelson, D., and Collins, G., 1963, in *Paramagnetic Resonance* (W. Low, ed.), Academic Press, New York.

Koenig, S. H., Prodell, A. G., and Kusch, P., 1952, *Phys. Rev.* **88**:191.

Kowalewski, J., Nordenskiöld, L., Benetis, N., and Westlund, P.-O., 1985, *Progr. Nucl. Magn. Reson. Spectrosc.* **17**:141.

Lecomte, J. T., and La Mar, G. N., 1986, *Eur. Biophys. J.* **13**:373.

Lecomte, J. T., Unger, S. W., and La Mar, G. N., 1991, *J. Magn. Reson.* **94**:112.

Levanon, H., Charbinsky, S., and Luz, Z., 1970, *J. Chem. Phys.* **53**:3056.

Luchinat, C., Steuernagel, S., and Turano, P., 1990, *Inorg. Chem.* **29**:4351.

Macura, S., and Ernst, R. R., 1980, *Mol. Phys.* **41**:95.

Macura, S., Huang, Y., Suter, D., and Ernst, R. R., 1981, *J. Magn. Reson.* **43**:259.

Martin, G. E., and Zektzer, A. S., 1988, *Two-Dimensional NMR Methods for Establishing Molecular Connectivity*, VCH, New York.

McCammon, J. A., and Harvey, S. C., 1987 *Dynamics of Proteins and Nucleic Acids*, Cambridge UP, Cambridge.

McConnell, H. M., 1956, *J. Chem. Phys.* **25**:709.

McGarvey, B. R., 1957, *J. Chem. Phys.* **61**:1232.

Merz, Jr., K. M., 1991, *J. Am. Chem. Soc.* **113**:406.

Merz, Jr., K. M., Murcko, M., Kollman, P. A., 1991, *J. Am. Chem. Soc.*, in press.

Muus, L. T., and Atkins, P. W., 1972, *Electronic Spin Relaxation in Liquids*, Plenum Press, New York.

Neuhaus, D., and Williamson, M., 1989, *The Nuclear Overhauser Effect in Structural and Conformational Analysis*, VCH, New York.

Noggle, J. H., and Schirmer, R. E., 1971, *The Nuclear Overhauser Effect*, Academic Press, New York.

Nordenskiöld, L., Laaksonen, A., and Kowalewski, J., 1982, *J. Am. Chem. Soc.* **104**:379.

Oppenheimer, N. J., and James, T. L., eds., 1989a, *Methods in Enzymology, Vol. 176, Nuclear Magnetic Resonance. Part A: Spectral Techniques and Dynamics,* Academic Press, San Diego, CA

Oppenheimer, N. J., and James, T. L., eds., 1989b, *Methods in Enzymology, Vol. 177, Nuclear Magnetic Resonance. Part B: Structure and Mechanism,* Academic Press, San Diego, CA

Orbach, R., 1961a, *Proc. Roy. Soc. London Ser. A* **264**:458.

Orbach, R., 1961b, *Proc. Roy. Soc. London Ser. A* **264**:485.

Orbach, R., 1961c, *Proc. Phys. Soc.* **A77**:821.

Overhauser, A. W., 1953a *Phys. Rev.* **89**:689.

Overhauser, A. W., 1953b, *Phys. Rev.* **92**:411.

Poulos, T. L., 1988, *Adv. Inorg. Biochem.* **7**:1.

Poulos, T. L., and Kraut, J., 1980, *J. Biol. Chem.* **255**:8199.

Ramaprasad, S., Johnson, R. D., and La Mar, G. N., 1984, *J. Am. Chem. Soc.* **106**:3632.

Redfield, A. G., 1957, *IBM Res. Dev.* **1**:19.

Redfield, A. G., 1963, *Phys. Rev.* **130**:589.

Rogers, R. N., and Pake, G. E., 1960, *J. Chem. Phys.* **33**:1107.

Rubinstein, M., Baram, A., and Luz, Z., 1971, *Mol. Phys.* **20**:67.

Satterlee, J. D., and Erman, J. E., 1991, *Biochemistry* **30**:4398.

Satterlee, J. D., Erman, J. E., La Mar, G. N., Smith, K. M., and Langry, K. C., 1983, *J. Am. Chem. Soc.* **105**:2099.

Satterlee, J. D., Erman, J. E., and de Ropp, J. S., 1987, *J. Biol. Chem.* **262**:11578.

Satterlee, J. D., Russell, D. J., and Erman, J. E., 1991, *Biochemistry* **30**:9072.

Scheek, R. M., van Gunsteren, W. F., Kaptein, R., 1989, in *Methods in Enzymology* (N. J. Oppenheimer and T. L. James, eds.), pp. 204–217, Academic Press, San Diego, CA.

Slichter, C. P., 1989, *Principles of Magnetic Resonance*, Springer, Berlin.

Solomon, I., 1955, *Phys. Rev.* **99**:559.

Solomon, I., and Bloembergen, N., 1956, *J. Chem. Phys.* **25**:261.

Sternlicht, H., 1965, *J. Chem. Phys.* **42**:2250.

Tainer, J. A., Getzoff, E. D., Beem, K. M., Richardson, J. S., and Richardson, D. C., 1982, *J. Mol. Biol.* **160**:181.

Thanabal, V., de Ropp, J. S., and La Mar, G. N., 1987a, *J. Am. Chem. Soc.* **109**:265.

Thanabal, V., de Ropp, J. S., and La Mar, G. N., 1987b, *J. Am. Chem. Soc.* **109**:7516.

Thanabal, V., de Ropp, J. S., and La Mar, G. N., 1988, *J. Am. Chem. Soc.* **110**:3027.

Van Vleck, J. H., 1939, *J. Chem. Phys.* **7**:72.

Van Vleck, J. H., 1940, *Phys. Rev.* **57**:426.

Vasavada, K. V., and Nageswara Rao, B. D., 1989, *J. Magn. Reson.* **81**:275.

Vega, A. J., and Fiat, D., 1976, *Mol. Phys.* **31**:347.

Waysbort, D., and Navon, G., 1978, *J. Chem. Phys.* **68**:3704.

Wüthrich, K., 1986, *NMR of Proteins and Nucleic Acids*, Wiley, New York.

Yamamoto, Y., Nanai, N., Chujo, R., and Suzuki, T., 1990, *FEBS Lett.* **264**:113.

Yamamoto, Y., Chujo, R., and Suzuki, T., 1991, *Eur. J. Biochem.* **198**:285.

Yu, C., Unger, S. W., and La Mar, G. N., 1986, *J. Magn. Reson.* **67**:346.

Yu, L. P., La Mar, G. N., and Rajarathnam, K., 1990, *J. Am. Chem. Soc.* **112**:9527.

3

Paramagnetic Relaxation of Water Protons
Effects of Nonbonded Interactions, Electron Spin Relaxation, and Rotational Immobilization

Cathy Coolbaugh Lester and Robert G. Bryant

1. INTRODUCTION

Proton relaxation by paramagnetic centers has been historically a powerful structural tool applied largely to metalloprotein systems (Dwek, 1973; Burton et al., 1979; Kushnir and Navon, 1984). The development of high-resolution multidimensional NMR has displaced some of these applications; however, the added effects of paramagnetic centers have provided new opportunities for high-resolution work in larger macromolecular systems (Emerson and La Mar, 1990a, b; Hauksson et al., 1990; Dugad et al., 1990).

The rapid development of magnetic imaging has provided a new impetus for understanding paramagnetic relaxation effects because the contrast in a magnetic image is dominated by differences in water proton relaxation times. Several very useful reviews have been presented dealing with many of the issues of the general problem (Brasch, 1983; Kang et al., 1984; Koenig and Brown, 1984; Lauffer, 1987). This discussion focuses closely on three limited but important aspects of the problems associated with

Cathy Coolbaugh Lester and Robert G. Bryant • Department of Biophysics and Department of Chemistry, University of Rochester, Rochester, New York 14642. Dr. Lester's present address is Baker Laboratory of Chemistry, Cornell University, Ithaca, New York 14853. Dr. Bryant's present address is Department of Chemistry, University of Virginia, Charlottesville, Virginia 22901.

Biological Magnetic Resonance, Volume 12: NMR of Paramagnetic Molecules, edited by Lawrence J. Berliner and Jacques Reuben. Plenum Press, New York, 1993.

making highly efficient relaxation agents, ultimately for clinical magnetic imaging:

1. the effects of outersphere relaxation;
2. the control of electron-spin relaxation rates; and
3. the effects of rotational immobilization of the paramagnetic species in a heterogeneous system such as a tissue.

2. OUTERSPHERE RELAXATION

2.1. Introduction

An understanding of water-proton relaxation induced by interactions outside the first coordination sphere of metal complexes is crucial for a quantitative understanding of relaxation as well as the design and application of contrast media for magnetic resonance imaging. The soluble contrast agents currently administered consist of metal ions complexed with multidentate ligands and a few coordinated water molecules in remaining positions. Compared with the aquo ions, these chelated systems have relatively few labile first coordination sphere protons, significantly reducing the efficiency of the metal center for promoting water-proton relaxation. Because the observed proton relaxation rate in these systems consists of contributions from both first coordination sphere interactions as well as usually weaker but important outersphere interactions, accurate quantitative characterization of paramagnetically induced proton relaxation is difficult. Furthermore, water-proton relaxation could occur through an indirect relaxation mechanism whereby relaxation of nonlabile ligand protons caused by the metal center then relaxes water protons through an intermolecular dipole–dipole coupling. This section reviews approaches to characterizing outersphere relaxation in simple metal systems.

2.2. Outersphere Relaxation Theory

When a paramagnetic center has labile ligand protons or water molecules, the solvent water-proton relaxation is predominantly caused by first-sphere relaxation mechanisms. These mechanisms include efficient relaxation of ligand protons induced by the intramolecular electron–nuclear dipole–dipole and scalar interactions transferred to the solvent nuclei through chemical exchange processes. The observed water-proton relaxation rate $T_{1\text{obs}}^{-1}$ is (McConnell, 1958; Luz and Meiboom, 1963; Dwek, 1973)

$$\frac{1}{T_{1\text{obs}}} = P_{\text{os}}\left(\frac{1}{T_{1\text{d}}} + \frac{1}{T_{1\text{os}}}\right) + \frac{P_M}{\tau_M + T_{1M}} \tag{1}$$

where τ_M is the residence time of the nuclear spin in the complex, P_M is the ratio of the number of labile protons in the first coordination sphere of the complex to the total number of protons observed, and P_{os} is $(1 - P_M)$. The relaxation rate for the unbound nucleus consists of the sum of a diamagnetic contribution $T_{1\text{d}}^{-1}$ produced by interactions with other solvent nuclei and a paramagnetic outersphere contribution $T_{1\text{os}}^{-1}$ generated by intermolecular interactions with the metal center. For spin-$\frac{1}{2}$ nuclei coordinated to a paramagnetic center, T_{1M}^{-1} is usually described by the modified Solomon–Bloembergen (MSB) equations which incorporate the derivations of Solomon, Bloembergen, and Morgan (Solomon, 1955; Solomon and Bloembergen, 1956; Bloembergen, 1957; Bloembergen and Morgan 1961) with the modifications introduced by Reuben and coworkers (Reuben et al., 1970; Benetis et al., 1983). These equations describe nuclear spin relaxation produced by intramolecular dipole–dipole and scalar interactions and work well for some aquo ions, but several assumptions incorporated into the derivation sometimes limit their applicability. These issues have been summarized in an extensive review by Kowalewski et al. (1985) and will not be presented here.

In addition to relaxation induced by first coordination sphere effects, fluctuations of the intermolecular electron–nuclear dipole–dipole interaction affects water-proton relaxation. In this case, the distance between the electron and nuclear spins is no longer fixed, and the translational motion of the solvent molecules relative to the complex produces fluctuations of the electron–nuclear dipole–dipole interaction (Abragam, 1961). Hwang and Freed (1975) have derived equations that describe relaxation produced by these time-dependent intermolecular interactions. Because forces between the spin-bearing molecules are ignored, the result is termed the "force-free" model. The derivation accounts for the exclusion of the volume accessible to the translating spins that is occupied by the spin-bearing molecules and results in a more accurate account of intermolecular relaxation than earlier treatments (Torrey, 1953; Pfeifer, 1961; Harmon and Muller, 1969). The expression for the relaxation rate of a nuclear spin I that interacts with an electron spin S is

$$\frac{1}{T_1} = \frac{32\pi}{405} \gamma_I^2 \gamma_S^2 \hbar^2 S(S+1) \left(\frac{N_a}{1000}\right) \frac{[S]}{bD} \{j_2(\omega_S - \omega_I) + 3j_1(\omega_I) + 6j_S(\omega_S + \omega_I)\} \quad (2)$$

where the spectral density function $j(\omega)$ is,

$$j(\omega) = \frac{1 + \dfrac{5z}{8} + \dfrac{z^2}{8}}{1 + z + \dfrac{z^2}{2} + \dfrac{z^3}{6} + \dfrac{4z^4}{81} + \dfrac{z^5}{81} + \dfrac{z^6}{648}} \quad (3)$$

and z is

$$z = \sqrt{\frac{2\omega b^2}{D}} \quad (4)$$

The correlation time for the translational motion τ_t is given by

$$\tau_t = \frac{b^2}{D} \quad (5)$$

Here, γ_I and γ_S are the nuclear and electron magnetogyric ratios, ω_I and ω_S are the nuclear and electron Larmor frequencies, $\hbar$ is Planck's constant divided by 2π, $[S]$ is the molar concentration of S spins, b is the distance of closest approach between the centers of the molecules on which the two spins reside, and D is the relative diffusion coefficient given by the sum of the diffusion coefficients of the I and S spin-containing molecules, $D_I + D_S$.

Equation (3) is appropriate for many stable radicals but does not apply if the electron-spin relaxation times T_{1S} and T_{2S} are comparable to the translational correlation time, τ_t. In this regime, electron-spin relaxation contributes to the time-dependence of the intermolecular dipole–dipole interaction, and the spectral density function takes on the form (Freed, 1978)

$$J_i(\omega) = \mathrm{Re}\ \frac{1 + \dfrac{s}{4}}{1 + s + \dfrac{4s^2}{9} + \dfrac{s^3}{9}} \qquad \left(s = \sqrt{\frac{b^2}{D}\left(i\omega + \frac{1}{T_{iS}}\right)}\right) \tag{6}$$

where T_{iS} is the S-spin relaxation time with $i = 1, 2$ and Re means the real part of the expression. When electron-spin relaxation times are long relative to τ_t, Eq. (6) reduces to Eq. (3).

Equations (2)–(6) assume the spins are located at the center of the molecules, which are approximated as rigid spheres. While this approximation is usually successful, Albrand et al. (1983) demonstrate that the low and high field limits for the intermolecular spectral density function may change when the spins are displaced a distance ρ from the center of the molecule (Ayant et al., 1977). In this instance, the rotational motion of the spin-containing molecules enters the relaxation equations, and the magnitude of the correction increases with increasing magnetic field strength (Albrand et al., 1983).

The electron-spin relaxation times are also a function of the magnetic field strength; however, this dependence and the underlying relaxation mechanism are not well understood for most metal systems. Fluctuations of the zero-field splitting interaction are believed to be the dominant relaxation mechanism for metal ions with $S > \frac{1}{2}$, and the origin of these fluctuations differs according to the symmetry of the metal complexes. Rubinstein et al. (1971) suggest that for octahedral aquo metal complexes with $S > \frac{1}{2}$, fluctuations of the zero-field splitting interaction produced by collisions of solvent molecules with the complex cause electron relaxation. For complexes with lower than cubic symmetry, however, Carrington and Luckhurst (1964) derive expressions based on fluctuations of the permanent zero-field splitting interaction produced by molecular rotation. Both approaches assume that the motions that produce fluctuations of the zero-field splitting interaction are fast relative to the electron-spin relaxation time; i.e., the Redfield limit is satisfied. At this limit, electron relaxation is an exponential process and the rates are equal to (Kowalewski et al., 1985)

$$\frac{1}{T_{2S}} = \left(\frac{B}{2}\right)\{3J(0) + 5J(\omega_S) + 2J(2\omega_S)\} \tag{7}$$

$$\frac{1}{T_{1S}} = B\{J(\omega_S) + 4J(2\omega_S)\} \tag{8}$$

$$J(j\omega_S) = \frac{\tau_v}{1 + (j\omega_S\tau_v)^2} \tag{9}$$

B is a constant,

$$B = \frac{\Delta^2}{25}\{4S(S + 1) - 3\} \quad \text{and} \quad \Delta^2 = \tfrac{2}{3}D^2 + 2E^2 \tag{10}$$

where Δ^2 is the trace of the square of the zero-field splitting interaction and τ_v is a correlation time that describes the fluctuations in the zero-field interactions. The constants D and E are related to the components of the diagonal zero-field splitting tensor **D** as

$$D = D_{zz} - \tfrac{1}{2}(D_{xx} + D_{yy}), \qquad E = \tfrac{1}{2}(SD_{xx} - D_{yy}) \tag{11}$$

In the presence of large zero-field splitting interactions, the Redfield limit approximations used to achieve Eqs. (7)–(11) may not be appropriate. Under these conditions electron-spin relaxation may not be characterized by a single exponential function.

The consequences of these equations are summarized graphically in Fig. 1, which shows the magnetic field strength dependence of proton spin–lattice relaxation rates

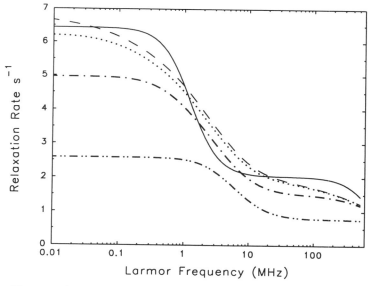

Figure 1. The magnetic field strength dependence of proton spin–lattice relaxation rates produced by dipole–dipole interactions with an $S = \tfrac{1}{2}$ electron spin. The solid line represents first-coordination-sphere relaxation and is calculated with the modified Solomon–Bloembergen equations for an intermoment distance of 2×10^{-10} m and a rotational correlation time of 2×10^{-10} s. The dashed line represents proton relaxation produced by the relative translational motions of the spin-containing molecules and was calculated with Eqs. (2)–(5) using the following parameters: $b = 2.0 \times 10^{-10}$ m, $D = 2 \times 10^{-10}$ m^2 s^{-1}. The remaining lines represent proton relaxation by translational diffusion calculated with the parameters used in computing the dashed line and the following field independent electron-spin relaxation times, where $T_{1S} = T_{2S} = T_S$: dotted line, $T_S = 10^{-8}$ s; dash–dot line, $T_S = 10^{-9}$ s; dash–dot–dot line, $T_S = 10^{-10}$ s.

produced by dipole–dipole interactions with a single electron spin. The solid line represents the magnetic field dependence of first coordination sphere relaxation calculated with the modified Solomon–Bloembergen equations, assuming that rotational diffusion dominates the effective correlation time for fluctuations of the intramolecular dipole–dipole interaction. In this case, the spectral density functions are Lorentzian. The dashed line represents relaxation produced by the translational motion of the spin-containing molecules and is calculated with Eqs. (2)–(5). The relaxation profiles are clearly different. The relaxation curve calculated for the translational diffusion case changes more slowly with increasing magnetic field strength than the Lorentzian functions characteristics of rotation. However, when electron-spin relaxation times contribute to the time dependence of the intermolecular interaction—i.e., the electron-spin relaxation times are on the order of the translational correlation time—the proton dispersion profile has a more Lorentzian shape. The effects of short electron-spin relaxation times on the relaxation profile for translational diffusion are also demonstrated in Fig. 1, which also shows that when the electron-spin relaxation time T_S becomes shorter than the translational correlation time τ_t the relaxation profile is very similar to that predicted by the solid line for relaxation produced by a rotational diffusion mechanism.

These models provide useful insight into the origins of nuclear spin relaxation in aqueous systems and can be applied to obtain information about the source of fluctuating interactions that affect relaxation in various systems. When utilized for the analysis of magnetic relaxation rate measurements over a range of magnetic field strengths, these models supply valuable information about the dynamics and interactions of water in the vicinity of paramagnetic solute molecules.

2.3. Measurements of Aqueous Metal Complexes

There are two possible relaxation mechanisms for water protons outside the first coordination sphere of paramagnetic metal complexes. The first includes the formation of an ordered second sphere of coordinated water molecules, which remain bound for a lifetime longer than the rotational correlation time of the entity. In this case, interactions of such second-sphere bound water protons with the paramagnetic center fluctuate with the same motions appropriate for first-sphere relaxation, i.e., rotational diffusion, chemical exchange between bound and free components, and electron-spin relaxation. Because water molecules are assumed to be bound to first-sphere ligands, possibly through hydrogen bonding interactions, the mechanism has been loosely referred to as second-sphere relaxation (Lauffer, 1987; Jenkins *et al.*, 1991). With these dynamical assumptions, the water proton relaxation profile is described by Eq. (1) and the dipole–dipole part of the MSB equations for first-sphere relaxation.

A second possibility consists of fluctuating intermolecular dipole–dipole interactions between water protons and the metal center that are produced by the relative translational motions of water and the complex or by electron-spin relaxation. In this case, any solvent–solute interactions are assumed to have lifetimes that are short compared to the rotational correlation times of the complex, i.e., 50 ps or longer.

Figure 2 displays the magnetic field dependence of water-proton relaxation rates for several systems, including a 5.0 mM solution of trioxalatochromate(III) ion, which contains neither labile ligand protons nor first-sphere water molecules (Lester and Bryant, 1990). The lines represent the results of fitting the data to the equations for

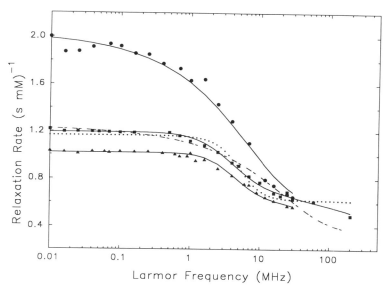

Figure 2.   The water-proton spin–lattice relaxation rates versus magnetic field strength for the following chromate(III) complexes at 282 K: (●) hexacyanochromate(II) ion, (■) trioxalatochromate(III) ion, and (▲) trimalonatochromate(III) ion. The solid lines represent fits of the data to Eqs. (2)–(6) with the parameters summarized in Table 1. The dashed line represents a fit of the data to Eqs. (2)–(5) neglecting contributions from electron-spin relaxation, with $b = 2.9 \times 10^{-10}$ m and $D = 3.8 \times 10^{-1}$ as best-fit parameters. The dotted line was calculated with the modified Solomon–Bloembergen equations, assuming three water molecular are bound to the complex for a time that is long relative to the rotational correlation time of the complex. The parameters used are $b = 3.7 \times 10^{-10}$ m, $\tau_r = 8.1 \times 10^{-11}$ s, $\tau_v = 3.1 \times 10^{-11}$ s, and $B = 6.1 \times 10^{19}$ rad^2 s^{-2}.

relaxation resulting from the formation of a bound second sphere of water, which is described by the modified Solomon–Bloembergen equations, or from the "force-free" translational diffusion model. The parameters are summarized in Table 1. The best description of the data is demonstrated for the translational diffusion model including contributions from electron-spin relaxation times, which are reported to be on the order of 10^{-9}–10^{-10} s for octahedral chromium(III) complexes (Bertini and Luchinat, 1986). The failure of the Solomon–Bloembergen formalism to account for the data demonstrates that the formation of a long-lived second coordination sphere of water molecules

TABLE 1
Parameters That Fit Data in Fig. 2

Complex ion	$b \times 10^{10}$ (m)	$D \times 10^9$ (m^2 s^{-1})	$\tau_s \times 10^{10}$ (s)	$B \times 10^{-19}$ (rad^2 s^{-2})	$\tau_v \times 10^{11}$ (s)
$[\text{Cr(CN)}_6]^{-3}$	3.7	1.8			
$[\text{Cr(mal)}_3]^{-3}$	4.5	1.12	1.4	3.0	4.8
$[\text{Cr(ox)}_3]^{-3}$	4.5	1.4	2.3	2.4	4.1

in which the water molecules rotate with the metal complex, because of a favorable hydrogen bonding arrangement with the ligand, is unimportant in this system. The translational model is more successful and provides a view of the dynamics around the metal complex. The translational diffusion coefficient derived from the fit of the data is remarkably similar to that for pure water. Thus, there is no evidence that water molecules are hydrogen-bonded to the ligand molecules for a time as long as the rotational correlation time of the complex, which is about 40 ps for similar complexes in water (Hertz, 1973).

The participation of nonlabile first coordination sphere protons in an indirect or proton mediated intermolecular mechanism for paramagnetic relaxation is addressed by examining water-proton relaxation in aqueous potassium trimalonatochromate(III), which contains six nonlabile methylene protons per metal complex. As demonstrated in Fig. 2, the relaxation efficiency of this ion is lower than that of the trioxalato-chromate(III) ion. A fit of the data to Eqs. (2)–(6) demonstrates that direct, through-space electron–nuclear magnetic couplings dominate the malonato complex relaxation. This result is expected from a consideration of the size of the interacting magnetic moments. The magnitude of the relaxation rate produced by the relative translation of two spins is proportional to the square of their magnetogyric ratios and inversely proportional to the third power of their intermoment distance. Therefore, for a single intermoment distance, the magnitude of the cross-relaxation rate produced by proton–proton dipole–dipole interactions is 4×10^5 times smaller than the rate induced by direct interactions with an unpaired electron spin. The rate induced by the relative translational motion of two protons separated by a distance of 2.4 Å is equivalent to interactions between a proton and an electron spin separated by a distance of 177 Å. Thus, for metal complexes in solution, indirect or proton-mediated relaxation pathways are expected to be unimportant.

Water-proton relaxation in the vicinity of coordinately saturated complexes of manganese(II) have also been investigated. Anionic complexes with the chelating ligands DTPA, NOTA, and DOTA are considered in addition to the pentagonal–bipyramidal cationic complex, $Mn(NH_2Et)_2$ [15]pydieneN$_5^{2+}$ (Wagnon and Jackels, 1987). The authors apply Freed's intermolecular relaxation equations, presented in Eqs. (2)–(6) to interpret the magnetic field dependence of water-proton spin–lattice relaxation in the vicinity of the anionic species and derive diffusion coefficients that are consistent with a high mobility of water surrounding the complex. The water-proton relaxation rates are lower for solutions containing the cationic species than for these containing the anionic complexes, and it is suggested that this difference reflects a larger separation between water protons and the metal center because neighboring water molecules will orient with their oxygen atoms closest to the positively charged complex.

These studies show that water-proton relaxation beyond the first coordination sphere of metal complexes is primarily controlled by fluctuations of intermolecular dipole–dipole interactions produced by the relative translational motions of the solvent and solute molecules and by electron-spin relaxation. Within the limitations of the theory used, charges in the motion of water when it passes close to the metal complexes are apparently small. In the case where the electron-spin relaxation contribution to the overall correlation time is negligible, an upper limit to the outersphere relaxation rate can be made for ions of similar size because the magnitude should scale with the square of the electron magnetic moment, which is proportional to $S(S + 1)$. Based on the water-proton relaxation rates measured at low fields for the hexacyanochromate(III) ion and the assumption that the water translational diffusion constant is 1.8×10^{-9} m^2 s^{-1}, the computed water-proton relaxation rate produced by outersphere interactions with a metal center, for three different distances of closest approach, are presented in Table 2. We emphasize that because the calculations neglect electron-spin relaxation effects, the values listed represent an upper limit to the outersphere contribution. It is also important to note that these entries depend on the validity of the point-dipole approximation; i.e., in ligand systems with significant electron delocalization, the computed outersphere rates will be in error.

The difficulty in separating first-sphere and outersphere contributions to the observed water-proton relaxation rate and those of electron delocalization is apparent in a study of water-proton relaxation in aqueous Mn(II)TPPS$_4$ (manganese(II) tetrakis[4-sulfophenyl]porphine) and Mn(III)TPPS$_4$ (Hernández and Bryant, 1991). The manganese(III) complex has been used as a magnetic contrast agent in animal experiments (Furmanski and Longley, 1988; Fiel et al., 1990). The magnetic relaxation profile, which is displayed in Fig. 3, is characteristic of the situation where the correlation time for the electron–nuclear coupling is the electron relaxation time (Koenig et al., 1987). However, when the complex is reduced to the manganese(II) form, the relaxation rate at low magnetic field strengths increases, and the dispersion curve takes on a shape characteristic of rotational reorientation, dominating the relaxation process shown in Fig. 2. The usual Solomon–Bloembergen–Morgan equations do not quantitatively account for the data in the following two ways:

1. The shape of the dispersion profile is broad and is indicative of a large contribution from translational diffusion mechanisms.

TABLE 2
Upper Limit of the Outersphere Contribution to Water-Proton Relaxation Rates (s mM)$^{-1}$, in the Vicinity of Paramagnetic Metal Ions

S	Metal ions	$S(S+1)$	Distance of closest approach ($\times 10^{-10}$ m)		
			3.7	5.6	7.4
$\frac{1}{2}$	Cu^{2+} Ti^{3+} Ni$^+$	$\frac{3}{4}$	0.28	0.19	0.14
1	Co$^+$ Ni^{2+} Cr^{4+} Cu^{3+}	2	0.75	0.70	0.52
$\frac{3}{2}$	Cr^{3+} Co^{2+} Ni^{3+}	$\frac{15}{4}$	1.41	0.93	0.69
2	Cr^{2+} Fe^{2+}	6	2.24	1.49	1.10
$\frac{5}{2}$	Mn^{2+} Fe^{3+} Cr$^+$	$\frac{35}{4}$	3.29	2.17	1.61
$\frac{7}{2}$	Gd^{3+}	$\frac{63}{4}$	5.92	3.91	2.9

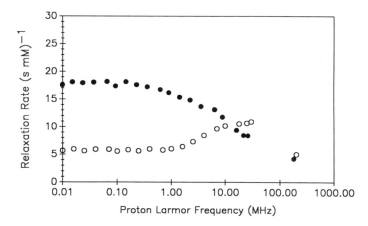

Figure 3. The water-proton nuclear spin-lattice relaxation rate per millimole of Mn(TPPS$_4$) at pH = 7 and 298 K as a function of applied magnetic field strength expressed as the proton Larmor frequency: (●) [Mn(II)TPPS$_4$]$^{4-}$, (○) [Mn(III)TPPS$_4$]$^{3-}$

2. The rate is considerably higher than predicted from first coordination sphere interactions.

If the excess relaxation is attributed to the usual outersphere interactions, the magnitude of the relaxation contribution is too large for reasonable distances of closest approach between the water protons and the metal center. This case, thus, appears to be one in which the assumptions of the point dipole approximation must be questioned, which compromises both the first coordination sphere analysis and the outer coordination sphere modeling. Nevertheless, delocalization of the electron spin in such ligand systems has been reported based on electron and nuclear magnetic resonance measurements in related complexes (Unger *et al.*, 1985). Thus, the possibility that the diffusing solvent experiences a closer proximity to the electron spin because the electron is appears to be reasonable.

2.4. Measurements in Protein Solutions

In addition to information about the state of water beyond the first coordination spheres of metal complexes, paramagnetic relaxation studies have revealed valuable information about the dynamical nature of water surrounding a protein molecule in solution. This information is provided, in part, by a study of the magnetic field dependence of water-proton spin–lattice relaxation induced by a nitroxide radical attached to the surface of a protein (Polnaszek and Bryant, 1984). By localizing the paramagnetic probe to the surface of the large-molecular-weight protein, bovine serum albumin, Polnaszek and Bryant report a diffusion coefficient that describes water molecules located within the first 10 Å of the protein. A value of $3.3 \pm 0.6 \times 10^{-10}$ m^2 s^{-1} at 286 K is reported, which is approximately five times less than the diffusion coefficient for water in the bulk, indicating that the mobility of water is perturbed only weakly by the macromolecular interface. The analysis used neglects changes induced in the protein-proton relaxation rate by the free radical that can then be transferred to the bulk water by an intermolecular cross-relaxation process. Therefore, direct electron–nuclear coupling may be overestimated; as a consequence, the magnitude of the translational diffusion coefficient deduced represents a lower limit for the diffusion coefficient of water at the protein surface.

A different interpretation of water-proton relaxation rates in aqueous protein–nitroxide solutions suggests that relaxation is controlled by first-sphere processes that involve the attachment of water to the nitroxide molecule through a hydrogen bond (Bennett *et al.*, 1987). The authors claim that these first-sphere processes are inefficient for nitroxide molecules free in solution but become important when the nitroxide is attached to a slowly rotating protein such as human serum albumin. This approach, however, is unsupported by any evidence that a nitroxide–water complex is sufficiently long-lived to meet the dynamical criteria of the model.

3. THE ELECTRON RELAXATION RATE

The water-proton relaxation efficiency of a paramagnetic metal complex such as hexaaquomanganese(II) ion often increases significantly when the rotational motion of the complex is increased by attachment of the metal to a slowly rotating macromolecule or by addition of a viscous solute such as glycerol (Burton *et al.*, 1979; Kennedy and Bryant, 1985). The observed relaxation enhancement, however, is usually less than that predicted based on the assumption that only rotational motion controls the time dependence of the dipole–dipole interaction. The implication is that other sources of magnetic field fluctuations become important. According to the MSB equations, the effective correlation time for the electron–nuclear dipole–dipole coupling includes contributions from molecular rotation, electron-spin relaxation, and chemical exchange of the coupled nucleus with the first coordination sphere of the metal center. If these events are assumed to be statistically uncorrelated, the effective correlation time is

$$\frac{1}{\tau_c} = \frac{1}{\tau_M} + \frac{1}{\tau_s} + \frac{1}{\tau_{\text{rot}}} \tag{12}$$

where τ_M is the mean residence time of the proton in the first coordination sphere of the complex, τ_s is the electron spin–spin or spin–lattice relaxation time, and τ_{rot} is the rotational correlation time for the metal complex. For small complexes, in solution, the rotational correlation time and electron-spin relaxation times are typically much shorter than the residence time for the nuclear spin in the complex, which is usually 10^{-8} s or longer (Betini and Luchinat, 1986). Equation (12) shows that increasing the rotational correlation time has only a modest effect if the electron relaxation time is short. For long rotational correlation times, the electron-spin relaxation time may limit the maximum water-proton relaxation rate that can be achieved.

Understanding how to control—i.e., lengthen—the electron-spin relaxation times is central to improving the efficiency of a metal-center-based paramagnetic relaxation agent. Few measurements of electron-spin relaxation rates, particularly for metal ions, have been reported because they have traditionally been difficult to measure. An added difficulty is that these rates are strongly dependent on magnetic field strength (Bloembergen and Morgan, 1961; Rubinstein *et al.*, 1971). Many electron relaxation rates in aqueous metal systems have been obtained indirectly by proton relaxation rate measurements (Banci *et al.*, 1986a). For current high-resolution NMR experiments conducted at high magnetic field strengths, however, the electron spin–lattice relaxation time usually falls on the high magnetic field side of the electron-spin-relaxation dispersion, and is thus quite long. As a consequence, the electron relaxation itself does not contribute to fluctuations of the magnetic interactions that control proton relaxation.

The nonmetal-based radicals have generally long electron relaxation times and correspondingly narrow resonance lines in the ESR spectrum (Berner and Kivelson, 1979; Schwartz *et al.*, 1982). However, the difficulty with these electron centers is that, unlike a metal system, there is generally no long-lived solvent complex that may impart a long correlation time like τ_{rot} to the system. Rather, the effective correlation time is limited by the solvent translational motion, which is discussed in Section 2.4, is generally rapid, and provides a limit on the effective correlation time for the electron–nuclear coupling that is in the tens of ps range (Polnaszek and Bryant, 1984). One possible way to think about solving this problem is by incorporating a radical species into a diamagnetic metal center, thus providing both the long lifetime of the nuclear spin in the metal's first coordination sphere and a long electron relaxation time. However, the radical species that seem to be sufficiently stable are also sterically bulky and place the electron center relatively far from the labile protons on the metal center (Eaton and Eaton, 1988). As a consequence, the relaxation efficiency is severely sacrificed because of the r^{-6} dependence of the electron–nuclear dipolar coupling itself. Therefore, there is considerable advantage in learning how to control electron-spin relaxation rates in metal centers directly.

Several studies have now appeared that attempt to address the issue of electron-spin relaxation times and how they depend on the dynamical characteristics of the system such as the viscosity or the rotational correlation time of the metal center (Banci *et al.*, 1985, 1986a, b). The viscosity dependence is an efficient approach to untangling the several contributions to the relaxation rates because if the viscosity-altering agent is a poor ligand, then the study may be done at constant structure of the metal center. The crucial aspect of the solution to change is the microdynamic viscosity, not the bulk viscosity. Therefore, some standard viscosity controllers that are macromolecular may provide misleading results. The more straightforward approach is to use low-molecular-weight cosolutes such as glycerol, sucrose, or low-molecular-weight polyethylene glycols.

Earlier work by Kennedy and Bryant on aqueous glycerol solutions of manganese(II) ion demonstrates that the magnetic field dependence of the water-proton relaxation rate changes smoothly from that well described by the modified Solomon–Bloembergen equations with inflections from both rotational motion and electron relaxation rates to that nearly like the relaxation profile observed for manganese bound on a protein as the glycerol concentration is raised from 0 to 75% (Koenig and Brown, 1984; Kennedy and Bryant, 1985). The electron-spin relaxation time dominates the effective correlation time for the high-viscosity solutions, as demonstrated by the increase in the nuclear relaxation rate with increasing magnetic field strength for field strengths corresponding to a proton Lamor frequency of 8 MHz. The analysis of these data based on fits to the MSB equations may suffer from interplay between the variables; however, this procedure suggests that for this highly symmetrical hexaaquo complex, the electron relaxation rate is approximately linear in the solution viscosity. More recent approaches using relaxation data obtained at much higher magnetic fields may provide a more accurate assessment of the electron relaxation rate (Banci *et al.*, 1985; Bertini *et al.*, 1991). Nevertheless, the reported manganese(II) results are consistent with the standard ideas that, for symmetrical complexes, the relaxation may be driven by distortions of the cubic symmetry induced by solvent collisions that are in turn a function of the translational and rotational mobility of both the metal complex and the solvent, hence the viscosity.

When other complexes are studied in this way, a more complex viscosity dependence is observed that suggests that competing relaxation mechanisms for the electron may be important. Hernández studied the $Gd(DOTA)^-$ and $Gd(DTPA)^{2-}$ complexes in water as a function of glycerol concentration (Hernández *et al.*, 1990). By utilizing high magnetic field relaxation rates, where the electron relaxation has dispersed and makes no contribution to the effective correlation time and observed proton relaxation rates, the authors deduce the low-magnetic-field electron relaxation rate as a function of viscosity. The results are shown in Fig. 4 for the $Gd(DTPA)^{2-}$ ion, where the independent variable is the measured rotational correlation time of the metal complex obtained from the relaxation rate at 200 MHz. The use of the 200-MHz relaxation rate to report the rotational correlation time has been confirmed using carbon-13 relaxation times for the lanthanum analog (Shukla *et al.*, 1992). The behavior of the electron relaxation rate as a function of the rotational correlation time is clearly not that expected for the standard models of nuclear or electron-spin relaxation that predict generally a linear dependence of the relaxation rate on the solution viscosity. At low glycerol concentrations, the electron relaxation rate changes with viscosity as these models predict; however, at modestly increased viscosity, the relaxation rates become nearly independent of the rotational correlation time of the metal complex and the viscosity.

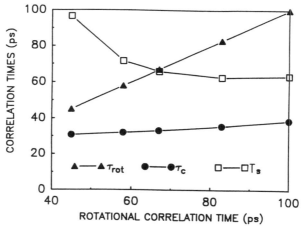

Figure 4. Correlation times for the $Gd(DTPA)^{2-}$ ion. The rotational correlation times are obtained from the water-proton spin–lattice relaxation time at 200 MHz. The effective correlation time and the electron-spin relaxation time were obtained from the water-proton relaxation data at 0.1 MHz.

If electron relaxation is assumed to be an exponential process and the system satisfies the requirements for the Redfield limit, one would expect that the electron spin-lattice relaxation time would decrease with increasing viscosity, pass through a minimum, then increase again on the slow-motion side of the minimum, which also corresponds to an inflection in the electron relaxation dispersion profile. However, Fig. 4 does not show this behavior. The electron relaxation rate becomes independent of the rotational correlation time of the complex and, therefore, the solvent–solute collision frequency as well at modest values of the viscosity. This result suggests that two processes contribute to the electron relaxation rate: one that is dependent on viscosity and one that is not. It is tempting to speculate that the viscosity-independent process may be driven by fluctuations in the electronic structure caused by intramolecular events such as low-frequency vibrational motions in the complex itself. In the solid state, these motions must dominate the electron relaxation process, and electron-spin relaxation rates are not particularly different.

A relaxation process controlled by intramolecular vibrational motions would also be expected to have a weak temperature dependence until thermal energies permitted a signficant population of excited vibrational states. Indeed, temperature-independent electron relaxation times are implicitly reported by two measurements on a manganese(III) TPPS complex where the shape of the magnetic relaxation dispersion profile provides a clear signature that the effective correlation time for the complex is the electron relaxation time (Hernández and Bryant, 1991). The consequences of these observations are very important for the potential understanding of paramagnetic relaxation agents because they suggest different strategies than usually envisioned for improving the relaxation efficiency of the complex.

If the correlation time for the electron relaxation process is limited by an intramolecular vibrational coupling of some sort, then the way to obtain longer electron relaxation times is to make the correlation time for the intramolecular process shorter. This requirement implies that the intramolecular motions of the system need to become higher in frequency, or qualitatively the metal center bonding structure needs to be made more rigid.

A complete understanding of electron-spin relaxation and its dependence on the dynamical variables of the system has not been achieved. Considerable additional data is required before any confidence in a theoretical description of the process is appropriate. The theoretical framework required becomes difficult when the symmetry of the metal center is lowered and the electron spin resonance spectrum consequently becomes more complex. The assumptions of the usual Solomon–Bloembergen equations are no longer supported, though they often provide a remarkable qualitative description of observed magnetic field dependencies. Several approaches to these situations have been made with success, but they are computationally cumbersome and do not provide the simple pictures afforded by the Solomon–Bloembergen results (Benetis et al., 1983; Westlund et al., 1984).

4. EFFECTS OF ROTATIONAL MOBILITY ON SOLVENT RELAXATION

The discussion of outer coordination sphere relaxation has shown that there is little effect of relaxation from nonlabile first coordination sphere protons in low-molecular-weight metal complexes, essentially because the magnetogyric ratio of the proton is so much smaller than that of the electron. A different situation exists, however, when a large-spin system like that associated with a protein is involved, particularly when the protein system is rotationally immobilized.

It is generally the case that solvent spins are magnetically coupled to solute spins and that the usual assumptions about proton relaxation of solute species being an intramolecular process usually fail to some extent. In the protein case, the magnetic coupling between the protein protons and the water protons is difficult to detect in terms of a nuclear Overhauser effect between the water protons and the protein protons, which is generally taken as evidence that the water correlation times at the interface are very short, on the order of 300 ps or shorter (Otting and Wüthrich, 1989; Otting et al., 1991). When the protein spin system is not free to rotate, the protein-proton intramolecular dipole–dipole couplings are no longer averaged to first order, producing two major consequences: the protein-proton resonance linewidth increases to several tens of kHz, and the magnetic communication between the protein is efficient, i.e., the spin diffusion in the rotationally immobilized protein spin system is effective, as it is in other solid systems. The importance of this second point is that a proton relaxed anywhere in the protein may communicate the effect to the remaining protons in a time quite short compared with the usual solvent-proton spin-lattice relaxation times. Therefore, a paramagnetic center may be anywhere in the protein and affect all the protons in the protein, which in turn may relax the water protons by a dipole–dipole coupling or chemical exchange mechanism between protein protons and water protons at the surface interface (Lester and Bryant, 1992). The effects are most dramatically demonstrated by examining the magnetic field dependence of water-proton relaxation in both diamagnetic and paramagnetic systems, as shown in Fig. 5.

The diamagnetic results are interesting in their own right and discussed elsewhere (Lester and Bryant, 1991). The magnetic field dependence of the water-proton relaxation in the free rotating protein solution has been the subject of some discussion for a long time. The details of the mechanism for the finding that the inflection frequency scales properly with the molecular size involves both chemical and magnetic exchange between the protein protons and the solvent protons (Hallenga and Koenig, 1976; Koenig et al., 1978; Piculell and Halle, 1986). The magnetic field dependence of the protein-proton spin-lattice relaxation rate is predicted approximately by the usual relaxation equations and has an inflection point when the rotational correlation time proton Larmor frequency product $\omega \tau_{rot}$ is approximately unity. This magnetic field dependence may then be transferred to the solvent protons by both chemical exchange effects and the intermolecular dipole–dipole cross-relaxation processes involving the solvent protons. No particular dynamical models for the water hydrodynamics are required to effect this transfer and account for the field dependence of the solvent protons by such a mechanism.

When the protein is immobilized, the correlation times for the solvent-proton–protein-proton coupling are essentially unchanged because they are dominated by the rapid solvent motion at the interface. However, the correlation time for the intramolecular proton–proton interactions in the protein are different. The magnetic field dependence of proton spin–lattice relaxation for the rotationally immobilized protein spin system is now very interesting and is given by a power law dependence first reported by Kimmich and coworkers for solid and partially hydrated protein systems (Kimmich et al., 1986; Nusser et al., 1988). The profoundly different relaxation profile for the protein protons is again coupled to the response of the solvent protons by the intermolecular

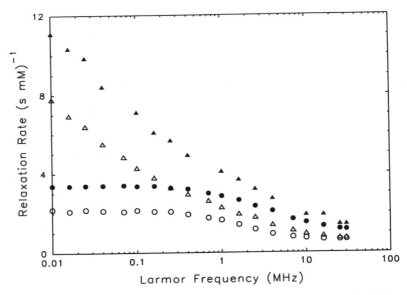

Figure 5. The water-proton spin–lattice relaxation rate as a function of magnetic field strength for aqueous copper(II)–bovine serum albumin samples at 298 K. (○) Diamagnetic bovine serum albumin solution at 1.8 mM, (●) 1.8 mM copper(II)bovine serum albumin solution with a metal–protein ratio of 1:1, (△) cross-linked diamagnetic bovine serum albumin at an initial concentration of 1.8 mM, and (▲) cross-linked copper(II)–bovine serum albumin with a metal–protein ratio of 1:1 and an initial concentration of 1.8 mM.

dipole–dipole mechanism, and the water protons reflect the same magnetic field dependence as the solid protein protons, as shown in Fig. 5. These effects explain the origin of the magnetic field dependence of the proton spin–lattice relaxation rates in heterogeneous systems like tissues, and again require nothing unique about the macromolecule or water-molecule dynamics in tissues or other heterogeneous proton-rich systems. That is, there is no need to invoke anything like bound water that has unique dynamical properties.

Addition of a paramagnetic ion to the protein system results in an increase of the water-proton relaxation rates for all field strengths, and subtraction of the relaxation rates for the diamagnetic protein yields the paramagnetic contribution to the water-proton relaxation. As demonstrated in Fig. 6, this paramagnetic contribution is much greater for the cross-linked protein system than for the solution case, particularly at low field strengths, and resembles the shape of the relaxation profile for the diamagnetic protein. Because the copper(II) ion coordinated to serum albumin contains no first coordination sphere water molecules, the paramagnetic contribution to the solvent-proton relaxation rate for the rotationally mobile protein–metal complex is predominantly outersphere in origin and is produced by fluctuations of the intermolecular electron–nuclear dipole–dipole interaction between water protons and the copper ion.

The profoundly different magnetic field dependence for the paramagnetic contribution of the cross-linked case, on the other hand, is believed to involve protons of the

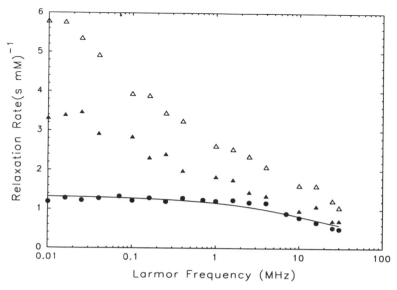

Figure 6. The paramagnetic contribution to water-proton relaxation per millimole of copper(II)–bovine serum albumin in the following systems: ($\triangle$) cross-linked at 278 K, ($\blacktriangle$) cross-linked at 298 K, and ($\bullet$) solution at 298 K. The solid line represents a fit of the data to Eqs. (2)–(5) for relaxation produced by the translational motion of water relative to the complex with $b = 2.2 \times 10^{-10}$ m and $D = 1.6 \times 10^{-9}$ m^2 s^{-1}.

immobilized macromolecule. Because the immobilized-protein protons are well coupled, except perhaps for those nearest the metal center because of frequency shifts as described by a number of authors treating spin diffusion to paramagnetic centers in solids (Lowe and Tse, 1968; Fukushima and Uehling, 1968), these protons all relax more rapidly, but still essentially as a single population. The consequence is that the protein-proton relaxation rate is augmented at all magnetic field strengths. An important feature of this system is that the protein protons appear to amplify the water-proton relaxation effects of the paramagnetic center and provide efficient proton relaxation in the absence of water in the metal ion first coordination sphere.

A larger enhancement of water-proton relaxation rates is demonstrated in the presence of a labile water molecule attached to the metal, as displayed in Fig. 7 for cross-linked manganese(II)-serum albumin. It is important to note that at low magnetic field strengths, the protein-proton coupling provides a very efficient relaxation agent. If one simply divides by the molar concentration of metal ion used and makes no careful accounting of the diamagnetic contributions, the effective millimolar relaxivity is in the range of several hundred in units of s^{-1} mM^{-1}. This result represents the situation appropriate to the tissue when a metal ion is added that binds to the nonrotating components. However, when the diamagnetic contribution from the rotationally immo-bilized protein is subtracted from the total relaxation rate, the resulting paramagnetic contribution is smaller, but still on the order of 60 s^{-1} mM^{-1} for manganese. Thus, the

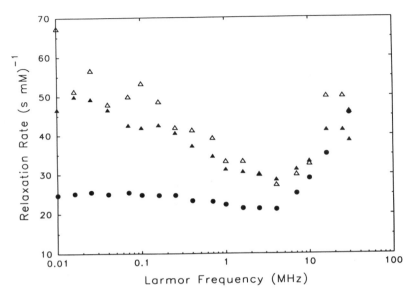

Figure 7. The paramagnetic contribution to the water-proton relaxation rate versus magnetic field strength for the following samples containing manganese(II)–bovine serum albumin: ($\bullet$) manganese(II)–bovine serum albumin solution that contains 0.6 mM bovine serum albumin and 0.12 mM manganese(II) ion; ($\triangle$) and ($\blacktriangle$) cross-linked manganese(II)–bovine serum albumin containing 1.5 mM bovine serum albumin and 0.15 mM manganese(II) ion at 278 K and 298 K, respectively.

efficiency at low magnetic field is higher than that of the present magnetic relaxation agents like Gd(DTPA)$^{2-}$ by almost an order of magnitude (Hernández *et al.* 1990).

The shape of the magnetic relaxation dispersion profile for the paramagnetic contributions is remarkable in that the field dependence is weak; i.e., the paramagnetic contribution depends approximately linearly on the log of the magnetic field strength. As of this writing, a definitive explanation for this result has not been advanced. A peak is found in the dispersion profile for the immobilized manganese(II) complex. This maximum is expected because the correlation times that drive the electron relaxation are short compared with rotational correlation times of even the dissolved protein molecule, so the electron-spin relaxation is insensitive to further immobilization of the system.

Several practical consequences of the data should be clear. First, it is possible to design a magnetic relaxation agent that may control magnetic image contrast that could be targeted to specific structures in the intra- or extrallular space that are not soluble targets, without sacrifice of paramagnetic relaxation efficiency. The amplification that results from involving all the proton spins of the protein raises the possibility that one could design metal centers without first coordination sphere water molecules that couple the paramagnetic center to the solvent by direct chemical exchange without sacrifice of sensitivity or effectiveness. Such a strategy may have considerable toxicological advantages based on both kinetic and thermodynamic stability. Nevertheless, several questions remain unexplored, including how the paramagnetic efficiency depends on the internal structure of the rotationally immobilized spin system, how the efficiency depends on the size of the immobilized spin system, and how the efficiency of the coupling to the solvent may be controlled. In addition, control of the electron relaxation rate itself must also have an impact on the efficiency just as in the solution cases discussed earlier.

REFERENCES

Abragam, A., 1961, *The Principles of Nuclear Magnetism*, Clarendon Press, Oxford.
Albrand, J. P., Taieb, M. C., Fries, P. H., and Belorizky, E., 1983, *J. Chem. Phys.* **78**:5809.
Ayant, Y., Belorizky, E., Fries, P., and Rosset, J., 1977, *J. Physique* **38**:325.
Banci, L., Bertini, I., and Luchinat, C., 1985, *Inorg. Chim. Acta* **100**:173.
Banci, L., Bertini, I., and Luchinat, C., 1986a, *Magn. Reson. Rev.* **11**:1.
Banci, L., Bertini, I., Briganti, F., and Luchinat, C., 1986b, *J. Magn. Reson.* **66**:58.
Benetis, N., Kowalewski, J., Nordenskiöd, L., Wennerström, H., and Westlund, P.-O., 1983, *Mol. Phys.* **48**:329.
Bennett, H. F., Brown III, R. D., Koenig, S. H., and Swartz, H. M., 1987, *Magn. Reson. Med.*, **4**:93.
Berner, B., and Kivelson, D., 1979, *J. Phys. Chem.* **83**:1401.
Bertini, I., and Luchinat, C., 1986, *NMR of Paramagnetic Molecules in Biological Systems*, Benjamin Cummings, Menlo Park, CA.
Bertini, I., Briganti, F., Xia, Z., and Luchinat, C., 1991, personal communication.
Bloembergen, N., 1957, *J. Chem. Phys.* **27**:572.
Bloembergen, N., and Morgan, L. O., 1961, *J. Chem. Phys.* **34**:842.
Brasch, R. C., 1983, *Radiology* **147**:781.
Burton, D. R., Forsén, S., Karlstrom, G., and Dwek, R. A., 1979, *Prog. NMR Spectrosc.* **13**:1.
Carrington, A., and Luckhurst, G. R., 1964, *Mol. Phys.* **8**:125.
Dugad, L. B., La Mar, G. N., and Unger, S. W., 1990, *J. Am. Chem. Soc.* **112**:1386.
Dwek, R. A., 1973, *Nuclear Magnetic Resonance in Biochemistry*, pp. 192–193. Clarendon Press, Oxford.
Eaton, S. S., and Eaton, G. R., 1988, *Coord. Chem. Revs.* **83**:29.
Emerson, S. D., and La Mar, G. N., 1990a, *Biochem.* **29**:1545.
Emeson, S. D., and La Mar, G. N., 1990b, *Biochem.* **29**:1556.
Fiel, R. J,. Musser, D. A., Mark, E. H., Mazurchuk, R., and Alleto, J. J., 1990, *Magn. Reson. Imag.* **8**:255.
Freed, J. H., 1978, *J. Chem. Phys.* **68**:4034.
Fukushima, E., and Uehling, E. A., 1968, *Phys. Rev.* **173**:366.
Furmanski, P., and Longley, C., 1988, *Cancer Res.* **48**:4604.
Hallenga, K., and Koenig, S. H., 1976, *Biochem.* **15**:4255.
Harmon, J. F., and Muller, B. H., 1969, *Phys. Rev.* **182**:400.
Hauksson, J. B., La Mar, G. N., Pandy, R. K., Rezzano, I. N., and Smith, K. M., 1990, *J. Am. Chem. Soc.* **112**:6198.
Hernández, G., and Bryant, R. G., 1991, *Bioconj. Chem.* **2**:394.
Hernández, G., Tweedle, M. F., and Bryant, R. G., 1990, *Inorg. Chem.* **29**:5109.
Hertz, H. G., 1973, in *Water. A Comprehensive Treatise* (F. Franks, ed.), 7, Plenum Press, New York.
Hwang, L., and Freed, J. H., 1975, *J. Chem. Phys.*, **63**:4017.
Jenkins, B. G., Armstrong, E., and Lauffer, R. B., 1991, *Magn. Reson. Med.* **17**:164.
Kang, Y. S., Gore, J. C., and Armitage, I. M., 1984, *Magn. Reson. Med.* **1**:396.
Kennedy, S. D., and Bryant, R. G., 1985, *Magn. Reson. Med.* **2**:14.

Kimmich, R., Winter, F., Nusser, W., and Spohn, K.-H., 1986, *J. Magn. Reson.* **68**:263.

Koenig, S. H., and Brown, R. D. III, 1984, *Magn. Reson. Med.* **1**:478.

Koenig, S. H., Brown, R. D. III, and Spiller, M., 1987, *Magn. Reson. Med.*, **4**:252.

Koenig, S. H., Bryant, R. G., Hallenga, K., and Jacob, G. S., 1978, *Biochem.* **17**:4348.

Kowalewski, J., Nordenskiöld, L., Benetis, N., and Westlund, P.-O., 1985, *Progr. NMR Spectrosc.* **17**:141.

Kushnir, T., and Navon, G., 1984, *J. Magn. Reson.* **56**:373.

Lauffer, R. B., 1987, *Chem. Rev.* **87**:901.

Lester, C. C., and Bryant, R. G., 1990, *J. Phys. Chem.* **94**:2843.

Lester, C. C., and Bryant, R. G., 1991, *Magn. Reson. Med.* **21**:143.

Lester, C. C., and Bryant, R. G., 1992, *Magn. Reson. Med.* **24**:236.

Lowe, I. J., and Tse, D., 1968, *Phys. Rev.* **166**:279.

Luz, Z., and Meiboom, S., 1963, *J. Chem. Phys.* **40**:2686.

McConnell, H. M., 1958, *J. Chem. Phys.* **28**:430.

Nusser, W., Kimmich, R., and Winter, F., 1988, *J. Phys. Chem.* **92**:6808.

Otting, G., and Wüthrich, K., 1989, *J. Am. Chem. Soc.* **111**:1871.

Otting, G., Liepinsh, E., and Wüthrich, K., 1991, *Science* **254**:974.

Pfeifer, H., 1961, *Ann. Phys.* **8**:1.

Piculell, L., and Halle, B., 1986, *J. Chem. Soc., Faraday Trans.* **82**:401.

Polnaszek, C. F., and Bryant, R. G., 1984, *J. Chem. Phys.* **81**:4038.

Reuben, J., Reed, G. H., and Cohn, M., 1970, *J. Chem. Phys.* **52**:1617.

Rubinstein, M., Baram, A., and Luz, Z., 1971, *Mol. Phys.* **1**:67.

Schwartz, L. J., Stillman, A. E., and Freed, J. H., 1982, *J. Chem. Phys.* **77**:5410.

Shukla, R., Ahang, X., and Tweedle, M., 1992, submitted.

Solomon, I., 1955, *Phys. Rev.* **99**:559.

Solomon, I., and Bloembergen, N., 1956, *J. Chem. Phys.* **25**:261.

Torrey, H. C., 1953, *Phys. Rev.* **92**:962.

Unger, S. W., Jui, T., and La Mar, G. N., 1985, *J. Magn. Res.* **61**:448.

Wagnon, B. K., and Jackels, S. C., 1987, *Inorg. Chem.* **28**:1923.

Westlund, P.-O., Wennerström, H., Nordenskiöld, L., Kowalewski, J., and Benetis, N., 1984, *J. Magn. Reson.* **59**:91.

4

Proton NMR Spectroscopy of Model Hemes

F. Ann Walker and Ursula Simonis

1. INTRODUCTION AND BACKGROUND

1.1. Occurrence and Properties of Iron Porphyrins (Hemes)

Hemes and heme proteins are vital components of essentially every cell of every living organism. Their roles in cells include

1. the transport of dioxygen in the red blood cells of higher animals (hemoglobin);
2. the storage of dioxygen in the muscles of higher animals (myoglobin);
3. the transport of electrons in the respiratory chains of organisms as diverse as bacteria, yeasts, algae, plants, and animals, and in photosynthetic cells from those of the simplest photosynthetic bacteria to those of higher plants (cytochromes a, b, c, d, f);
4. synthesis, modification and degradation of fatty acids, steroid and adrenal hormones, anesthetics and xenobiotics (cytochromes P-450);
5. activation and metabolism of hydrogen peroxide (peroxidases, myeloperoxidase, haloperoxidases, catalases, etc.); and
6. metabolism of the oxides of nitrogen and sulfur (nitrite reductase, sulfite oxidase, etc.).

F. Ann Walker • Department of Chemistry, University of Arizona, Tucson, Arizona 85721. Ursula Simonis • Department of Chemistry and Biochemistry, San Francisco State University, San Francisco, California 94132.

Biological Magnetic Resonance, Volume 12: NMR of Paramagnetic Molecules, edited by Lawrence J. Berliner and Jacques Reuben. Plenum Press, New York, 1993.

Hence, it is not surprising that the isolation, characterization, spectroscopic investigation, and modeling of heme proteins and both naturally occurring and synthetic hemes have been active fields of research for many years.

Among the heme proteins, a variety of axial ligations, oxidation states, and spin states of the central iron are stabilized by the protein environment created within the heme pocket. The paramagnetic states of iron that are created are particularly amenable to spectroscopic investigations by nuclear magnetic resonance (NMR), electron paramagnetic resonance (EPR), Mössbauer spectroscopy and magnetic circular dichroism (MCD). In the case of NMR spectroscopy in particular, the unpaired electrons serve to "illuminate" the heme center by causing hyperfine shifts of the protons of the heme, its axial ligand(s), and nearby protein residues. These shifts are exquisitely sensitive to heme substituents (La Mar and Walker, 1979), axial ligand effects such as histidine imidazole N–H hydrogen bonding (Satterlee, 1987; Thanabal et al., 1988), deprotonation (Satterlee, 1987), axial ligand plane orientation (Satterlee, 1987; McLachlan et al., 1988), methionine-SCH$_3$ chirality (Keller et al., 1980), and cyanide off-axis tilting (in cyanometmyoglobin) (Emerson and La Mar, 1990a, b), as well as the effect of non-bonded protein substituents (Satterlee, 1987; Lee et al., 1990) on the electronic state of the heme. However, because of the extreme sensitivity of the hyperfine shifts to all of these factors, it is often difficult to determine their relative importance. For this reason, a number of investigators, including the authors, have carried out detailed NMR investigations of appropriately designed model hemes having relatively high molecular symmetry (as compared to naturally occurring hemes) in order to probe the importance of such factors as substituent effects, electronic asymmetry due to unsymmetrical substitution, axial ligand basicity and steric effects, the effect of axial ligand plane orientation, the thermodynamics and kinetics of axial ligand binding, exchange and rotation, and hydrogen bonding effects on the NMR spectra of low-spin Fe(III) porphyrins. A number of the results of these investigations have been summarized previously in several important chapters (La Mar and Walker, 1979; La Mar 1979; Keller and Wüthrich, 1981; Goff, 1983; Satterlee, 1986; Satterlee, 1987). However, the focus in all but two of these (La Mar and Walker, 1979; Goff, 1983) has been on heme proteins, where the metal was either in the Fe(II) or Fe(III) state; hence, there has been no comprehensive summary since 1983 of the results of NMR spectroscopic investigations of model hemes. Therefore, the present chapter will focus upon the important findings since 1983 related to proton NMR spectroscopy of iron porphyrins of all known oxidation states, with only necessary reference to earlier work.

1.2. Structures and Electron Configurations of Iron Porphyrins

The general structures of iron porphyrins are shown in Fig. 1, together with the structures of closely related ring systems, iron chlorins and isobacteriochlorins, and examples of the substituents present in commonly investigated natural and synthetic hemes. The possible coordination states of synthetic and natural iron porphyrins include 4-coordinate (no axial ligands), 5-coordinate (one axial ligand), and 6-coordinate (two axial ligands) geometries. The number of axial ligands and their nature have dramatic influences on the spin state, EPR g-values, NMR isotropic shifts, Mössbauer isomer shifts, quadrupole splittings, and hyperfine coupling constants, as well as redox potentials and reactivity toward reagents such as molecular oxygen, carbon monoxide, and

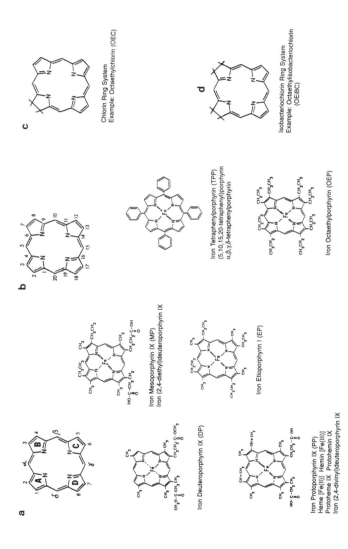

Figure 1. Structures of iron porphyrins, chlorins, and isobacteriochlorins; examples with typical nomenclature. (a) Numbering system usually used for natural porphyrins and their derivatives and examples of the most commonly studied natural porphyrin derivatives. As indicated, many natural porphyrins and their derivatives are named as 2,4-disubstituted deuteroporphyrin IX derivatives. (b) IUPAC numbering system for porphyrins and examples of commonly studied synthetic porphyrins. Tetraphenylporphyrins are named according to *either* numbering scheme. In this chapter, phenyl-substituted TPP derivatives are abbreviated as either (*p*-X)₄TPP or TpXPP by various authors, where *p*-X represents a *para* substituent. (c) The chlorin ring system. (d) The isobacteriochlorin ring system.

unsaturated hydrocarbons. The known oxidation states of iron porphyrins range from Fe(I) (d^7) to Fe(V) (d^3). Among these, the possible spin states include high-, intermediate-, and low-spin states of Fe(II) (d^6) and Fe(III) (d^5) and high- and low-spin Fe(I) (d^7) and Fe(IV) (d^4). All spin states observed to date are summarized in Fig. 2. It should be noted that the only *diamagnetic* member of this extensive series of possible oxidation and spin states is low-spin Fe(II). The NMR spectra observed for representative examples of each of the possible oxidation and spin states will be discussed in Section 2.

1.3. The Components of the Isotropic Shifts of Paramagnetic Iron Porphyrins

The observed chemical shifts of protons in paramagnetic molecules are due to a combination of diamagnetic and paramagnetic contributions. These contributions are additive:

$$\delta_{obs} = \delta_{dia} + \delta_{para} \tag{1}$$

The diamagnetic contribution δ_{dia} is that which would have been observed if the molecule contained no unpaired electrons. It is directly evaluated by recording the NMR spectrum of an appropriate diamagnetic analog of the molecule of interest. The paramagnetic contribution δ_{para} is often called the hyperfine shift δ_{hf} or the isotropic shift δ_{iso} since it is typically observed for molecules or ions in homogeneous solution, where electron spin relaxation and molecular rotation are generally rapid. The paramagnetic or hyperfine or isotropic shift includes two contributions, the contact and the dipolar or pseudocontact terms:

$$\delta_{para} = \delta_{iso} = \delta_{hf} = \delta_{obs} - \delta_{dia} = \delta_{con} + \delta_{dip} \tag{2}$$

The contact (δ_{con}) and dipolar (δ_{dip}) terms arise because spin delocalization from the unpaired electron, usually located on the metal atom, to the protons at the periphery of the molecule can occur either through chemical bonds or through space [or a combination of the two, as in the case of low-spin Fe(III) porphyrins (see below)]. The sign convention that will be used throughout this chapter is that commonly used by chemists: δ_{obs} is positive if the resonance is at higher frequency than that of tetramethylsilane (TMS). Hence, δ_{para} is positive when the resonance is at higher, and negative when it is at lower, frequency than that of the diamagnetic reference compound. This is the reverse of the sign convention used in the earlier literature, including several of the earlier reviews (La Mar and Walker, 1979; La Mar, 1979). All signs of isotropic, dipolar, and contact shifts reported in earlier publications have been changed to conform to the current sign convention.

1.3.1. The Contact Contribution

The contact contribution to the isotropic shift of protons in paramagnetic molecules originates from scalar coupling between electron spins and individual proton spins. In the most general case, the contact shift depends on the principal components of the electronic g tensor g_{ii} of the paramagnetic center, the magnetic susceptibility tensor χ_{ii}

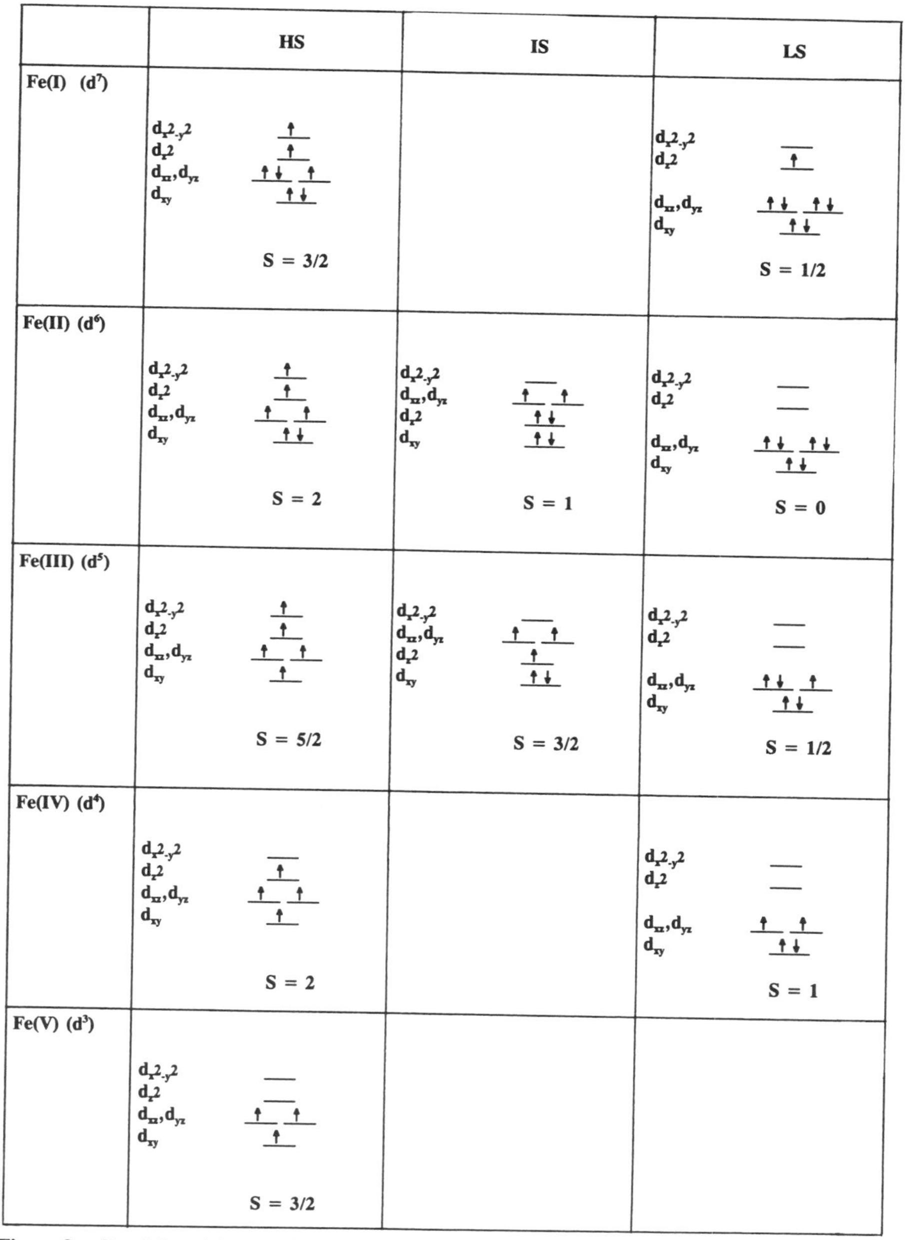

	HS	IS	LS
Fe(I) (d^7)	$d_{x^2-y^2}$ $\uparrow$ d_{z^2} $\uparrow$ d_{xz},d_{yz} $\uparrow\downarrow$ $\uparrow$ d_{xy} $\uparrow\downarrow$ S = 3/2		$d_{x^2-y^2}$ d_{z^2} $\uparrow$ d_{xz},d_{yz} $\uparrow\downarrow$ $\uparrow\downarrow$ d_{xy} $\uparrow\downarrow$ S = 1/2
Fe(II) (d^6)	$d_{x^2-y^2}$ $\uparrow$ d_{z^2} $\uparrow$ d_{xz},d_{yz} $\uparrow$ $\uparrow$ d_{xy} $\uparrow\downarrow$ S = 2	$d_{x^2-y^2}$ d_{xz},d_{yz} $\uparrow$ $\uparrow$ d_{z^2} $\uparrow\downarrow$ d_{xy} $\uparrow\downarrow$ S = 1	$d_{x^2-y^2}$ d_{z^2} d_{xz},d_{yz} $\uparrow\downarrow$ $\uparrow\downarrow$ d_{xy} $\uparrow\downarrow$ S = 0
Fe(III) (d^5)	$d_{x^2-y^2}$ $\uparrow$ d_{z^2} $\uparrow$ d_{xz},d_{yz} $\uparrow$ $\uparrow$ d_{xy} $\uparrow$ S = 5/2	$d_{x^2-y^2}$ d_{xz},d_{yz} $\uparrow$ $\uparrow$ d_{z^2} $\uparrow$ d_{xy} $\uparrow\downarrow$ S = 3/2	$d_{x^2-y^2}$ d_{z^2} d_{xz},d_{yz} $\uparrow\downarrow$ $\uparrow$ d_{xy} $\uparrow\downarrow$ S = 1/2
Fe(IV) (d^4)	$d_{x^2-y^2}$ d_{z^2} $\uparrow$ d_{xz},d_{yz} $\uparrow$ $\uparrow$ d_{xy} $\uparrow$ S = 2		$d_{x^2-y^2}$ d_{z^2} d_{xz},d_{yz} $\uparrow$ $\uparrow$ d_{xy} $\uparrow\downarrow$ S = 1
Fe(V) (d^3)	$d_{x^2-y^2}$ d_{z^2} d_{xz},d_{yz} $\uparrow$ $\uparrow$ d_{xy} $\uparrow$ S = 3/2		

Figure 2. Possible oxidation and spin states of iron porphyrins and the d-orbital configuration expected in each case. All of these states, except for high-spin d^7, have been observed for iron porphyrins and are discussed in this chapter. In addition, the spin-admixed $S = \frac{3}{2}, \frac{5}{2}$ state of Fe(III) and the alternative orbital configuration for low-spin Fe(III), with d_{xy} higher in energy than d_{xz}, d_{yz}, are also discussed.

of the molecule, and the Fermi hyperfine (contact) coupling constant A for coupling the spin of the electron to the spin of the nucleus of interest (Kurland and McGarvey, 1970):

$$\delta_{\text{con}} = \frac{A}{3\gamma_N\hbar}\left(\frac{\chi_{xx}}{g_{xx}} + \frac{\chi_{yy}}{g_{yy}} + \frac{\chi_{zz}}{g_{zz}}\right) \tag{3}$$

If a single spin state with an isotropic g tensor is populated, and to the extent that the Curie law is valid, this equation reduces to a simpler form that is usually applicable to iron porphyrins (Kurland and McGarvey, 1970; Jesson, 1973):

$$\delta_{\text{con}} = \frac{Ag\beta S(S+1)}{3\gamma_N\hbar kT} \tag{4}$$

where S is the total electron spin quantum number, γ_N is the magnetogyric ratio of the nucleus in question, and T is the absolute temperature. Thus, in simple, well-behaved systems, contact shifts are expected to vary linearly with $1/T$ and extrapolate to zero at infinite temperature ($1/T = 0$).

Interpretation of the contact contributions to the isotropic shifts of paramagnetic molecules in terms of the covalency of metal–ligand bonds relies to a large extent upon the McConnell equation (McConnell, 1956):

$$A_H = Q\rho_C/2S \tag{5}$$

where Q is a constant for a given type of proton and ρ_C is the spin density at the carbon to which the proton is attached. Spin density refers to the square of the wave function of the orbital containing the unpaired electron at a carbon position. For the case of spin delocalization to a π orbital on the carbon, $Q = -63$ MHz, and the contact shifts are negative (to low frequency). If the proton directly bound to the carbon having spin density in a π orbital is replaced by a methyl or other aliphatic group, Q is positive and somewhat variable, depending upon the rate of rotation (or preferred orientation, if rotation is slow) of the methyl or other aliphatic groups (McLachlan, 1958; Chestnut, 1958):

$$A_{\text{CH}_3} = \frac{Q_{\text{CH}_3}\rho_C}{2S} = \frac{(B_0 + B_2 \cos^2\phi)\rho_C}{2S} \tag{6}$$

where B_0 and B_2 are positive parameters and ϕ is the angle between the C—C—H plane and the p_z orbital axis on the aromatic carbon. B_2 is usually small and, since $\cos^2\phi$ is positive for all angles ϕ, Q_{CH_3} is always positive. Thus, if π spin delocalization occurs to a particular carbon atom, a proton directly bound to that carbon will have a contact shift to low frequency (δ_{con} negative), while replacement of the proton at that carbon position with a methyl or other aliphatic group will produce a contact shift to high frequency (δ_{con} positive). Such reversal in the sign of isotropic shifts for H and CH$_3$ at a given position is a clear sign of dominance of the isotropic shifts by π spin delocalization, as we will see in many parts of this chapter.

For spin delocalization to a σ-symmetry orbital on the carbon, Q is positive and depends on the number of σ bonds through which the spin is delocalized. For the electron in the $1s$ orbital of the hydrogen atom, A_H is 506.8 G or 1419 MHz, a positive quantity (Carrington and McLachlan, 1967). For σ spin delocalization to protons connected to a carbon framework, Q is also positive, and the spin density ρ_C is attenuated sharply as the number of σ bonds between the unpaired electron and the proton increases. In any case, the positive sign of Q means that the contact shifts are positive (to high frequency) when σ spin delocalization occurs to the carbon to which the proton is attached. Thus, if the contact contribution to the isotropic shifts of protons in molecules, such as iron porphyrins, can be separated from the dipolar contribution, then as directly bound protons are replaced by methyl groups at the same symmetry positions on the ring, for example, the pattern of shifts to lower and higher frequencies readily reveals the mechanism(s) of spin delocalization to each symmetry position. We will discuss this further after describing the dipolar shift and summarizing the general strategies for separation of the contact and dipolar contributions to the isotropic shift.

1.3.2. The Dipolar Contribution

The dipolar contribution to the isotropic shift has sometimes been called the pseudocontact contribution. However, this is a misnomer, since the dipolar shift contains no contribution from scalar coupling of the electron spin with the spin of the nucleus of interest, i.e., no Fermi contact term. Rather, the dipolar shift results from through-space dipole coupling of the nuclear and electron magnetic moments arising from either the magnetic anisotropy of the metal ion or from zero-field contributions in cases where the total spin of the ion is greater than $\frac{1}{2}$. For nuclei other than protons, ligand-centered dipolar shifts may also arise from unpaired spin density in p-orbitals (Kurland and McGarvey, 1970; Jesson, 1973), but this is not the case for protons. For dipolar shifts of protons by $S = \frac{1}{2}$ metal ions, it can be shown that

$$\delta_{\text{dip}} = \frac{[\chi_{zz} - \frac{1}{2}(\chi_{xx} + \chi_{yy})](3\cos^2\theta - 1)/r^3 + [\chi_{xx} - \chi_{yy}]\sin^2\theta\cos 2\Omega/i^3}{2N} \quad (7)$$

where N is Avogadro's number, χ_{ii} are the principal components of the magnetic susceptibility tensor (per mole), θ is the angle between the proton–metal vector and the z molecular axis, r is the length of this vector, and Ω is the angle between the x axis and the projection of this vector on the xy plane (Jesson, 1973; Horrocks and Greenberg, 1973, 1974). The terms $(3\cos^2\theta - 1)/r^3$ and $\sin^2\theta\cos 2\Omega/r^3$ are typically known as the axial and rhombic geometric factors, respectively, and the terms in Eq. (7) to which they belong are often called the axial and rhombic contributions to the dipolar shift, respectively. Axial geometric factors for protons at the β-pyrrole and *meso* positions (Fig. 1a, $R_1 - R_8$ and $R_\alpha - R_\delta$, respectively) of the commonly utilized model hemes tetraphenylporphinatoiron (TPPFe) and octaethylporphinato iron (OEPFe) have been calculated from X-ray crystallographic data for the low-spin Fe(III) complexes (La Mar and Walker, 1979, and references therein) and are presented in Table 1.

TABLE 1
Axial Geometric Factors for $[TPPFeL_2]^+$ and $[OEPFeL_2]^+$

	Position	$\langle(3\cos^2\theta - 1)r^{-3}\rangle$ $\times 10^{20}$ cm^{-3}	Relative G.F.
TPP	o-H	-36.0^a	1.00
	m-H	-16.7^a	0.463
	p-H	-14.8^a	0.410
	m-CH$_3$	-10.5^a	0.29
	p-CH$_3$	-11.0^a	0.31
	pyrrole-H	-70.3^a	1.95
OEP	α-CH$_2$	-42.5^a	1.18
	meso-H	-110.0^a	3.06
R-Pyridine	2,6-H	211.6^a	-5.9
	3,5-H	115^a	-3.2
	4-H	101^a	-2.8
	4-CH$_3$	68^a	-1.9
R-Imidazole	1-H	135^b	-3.75
	1-CH$_3$	80^b	-2.22
	2-H	216^b	-6.00
	2-CH$_3$	$<2^b$	~0.00
	4-H	209^b	-5.80
	5-H	134^b	-3.74
	5-CH$_3$	75^b	-2.14

[a] La Mar et al. (1977a).
[b] Data taken from Satterlee and La Mar (1976).

If (but *only* if) the complex of interest has only one thermally populated spin multiplet with effective spin S', then

$$\chi_{ii} = \frac{\beta^2 S'(S' + 1)}{3kTg_{ii}^2} \tag{8}$$

where β is the Bohr magneton, and Eq. (7) can be simplified to

$$\delta_{dip} = \frac{\beta^2 S'(S' + 1)}{18kT} \{[2g_{zz}^2 - (g_{xx}^2 + g_{yy}^2)](3\cos^2\theta - 1)/r^3 + 3(g_{xx}^2 - g_{yy}^2)\sin^2\theta\cos 2\Omega\} \tag{9}$$

The equation in this form is often used for separation of the dipolar and contact contributions to the isotropic shift. However, Horrocks and Greenberg (1973) have shown that second-order Zeeman (SOZ) contributions to the magnetic susceptibilities of heme systems range from 8% (ferricytochrome c) to 20% (metmyoglobin cyanide) of those of the first-order Zeeman (FOZ) effect, and are of opposite sign. Furthermore, excited electronic states of the metal ion that are thermally populated to a varying extent as a function of temperature can lead to non-Curie behavior (nonzero intercepts at $1/T = 0$) (Horrocks and Greenberg, 1974). As we will show in Section 2.8.4, this can lead, at the limit, to a reversal in the temperature dependence of the isotropic shifts (Safo *et al.*, 1992; Watson *et al.*, 1992). Suffice it to say that g values cannot reliably be calculated from the dipolar contributions to the isotropic shifts of protons in iron porphyrins or heme proteins, as has been reported recently for cytochrome c (Feng *et al.*, 1990); rather, magnetic susceptibilities can be calculated from Eq. (7) by fitting the structure of a protein. However, Eq. (9) is sometimes useful for estimating the sizes of the dipolar shifts in two complexes for protons at the same molecular positions when these complexes have different g values. Thus, if the size of a dipolar shift is known for one complex from some independent means, and if the g values of both complexes are known, the dipolar shift of the second complex may be estimated by scaling it by the ratio of the $g_\parallel^2$ factors for the axial and rhombic terms using Eq. (9). An example of the use of such a procedure and its shortcomings is presented in Section 2.8.4 (Watson *et al.*, 1992).

For dipolar shifts of metal complexes having $S > \frac{1}{2}$, where the g tensor is sometimes totally symmetric, magnetic anisotropy can still be present if the zero-field splitting D is large. In this case, Kurland and McGarvey (1970) have shown that the dipolar contribution to the isotropic shift exhibits a T^{-2} dependence:

$$\delta_{\text{dip}} = -\frac{28g^2\beta^2 D}{9k^2 T^2} \frac{(3\cos^2\theta - 1)}{r^3} \tag{10}$$

The classic case of high-spin Fe(III) porphyrins, where D values range from 5.6 cm^{-1} (F$^-$) to 11 cm^{-1} (Cl$^-$) to 16.4 cm^{-1} (I$^-$) (Brackett *et al.*, 1971), yields curved Curie plots due to a contact term that is linear in $1/T$ and a dipolar term due to Eq. (10) that depends on $1/T^2$ (La Mar and Walker, 1979, and references therein).

1.3.3. Separation of the Contact and Dipolar Contributions

These two terms vary in their relative and absolute contributions to the isotropic shifts of protons on the periphery of the ligands, depending upon the electron configuration and magnetic anisotropy of the metal ion involved. Some knowledge of the magnetic properties and possible electron configurations of the particular system is required in order to begin this process. An important beginning point is to obtain the temperature dependence of the isotropic shifts of the metal complex over as wide a temperature range as possible. For complexes in which axial ligands are bound to the metal, this generally means going to very low temperatures, since axial ligand dissociation may be expected to become significant at or above ambient temperatures. Hence, deuterated solvents typically used for obtaining Curie plots over wide temperature ranges include methylene chloride (mp -97 °C, bp 40 °C), chloroform (mp -64 °C, bp 61 °C), dimethylformamide (mp -61 °C, bp 153 °C), methanol (mp -98 °C, bp 65 °C) and toluene (mp -93 °C, bp 111 °C). Curie plots typically show whether there is some unusual magnetic behavior of the system. Such unusual magnetic behavior includes nonzero intercepts due to SOZ and/or excited-state contributions to the dipolar shifts, curvature due to hindered rotation of substituent groups, and curvature due to zero-field splitting contributions to the dipolar shift [Eq. (10)], as described previously (La Mar and Walker, 1979).

If the Curie plot for all proton resonances of the molecule is linear but the intercepts of the lines are nonzero, it is likely that there are significant SOZ contributions to the dipolar shifts or there is a significant difference in spin delocalization to the two $e(\pi)$ orbitals due to the effects of porphyrin substituents (see Section 2.8.2a). Further detailed approaches to the separation of the contact and dipolar contributions to the isotropic shift will be presented after we have further considered the mechanisms of spin delocalization that give rise to the contact term. An overall summary of the importance of the contact and dipolar contributions to the isotropic shifts of the known iron porphyrin spin and oxidation states is presented in Table 2. The data giving rise to this summary table are discussed in Section 2.

1.4. A Closer Look at the Contact Contribution: Mechanisms of Spin Delocalization

The unpaired electron(s) of $3d$ metalloporphyrins of varying oxidation and spin states are expected to be mainly localized either on the metal or on the porphyrin ring. If the unpaired electron is largely localized on the metal, it is expected to be located in one or more of the metal d-orbitals, while if it is localized on the porphyrin ring, it is expected to be found in either the HOMO [for porphyrin π cation radicals $(MP)^{+\cdot}$] or the LUMOs [for π anion radicals $(MP)^{-\cdot}$, or dianions $(MP)^{2-}$]. Thus, an understanding of the symmetry properties and relative energies of the metal and porphyrin orbitals will be of value in understanding the possible mechanism of spin delocalization to the protons on the periphery of the porphyrin ring.

1.4.1. The Metal Ion

Iron porphyrins range in oxidation states from Fe(I) (d^7) to Fe(IV) cation radical (d^4p^+) or possible Fe(V) (d^3). The metal may be either high-spin, intermediate-spin, or low-spin. Thus, iron porphyrins have a rich variety of potential electron configurations. The valence electrons are almost invariably located in the $3d$ orbitals of the metal, except in the rare cases of π cation radicals of Fe(II) complexes of reduced hemes bound to isonitrile ligands (Sullivan and Strauss, 1989). In 4-, 5-, and 6-coordinate metalloporphyrins, the d-orbitals consist of three general symmetry types:

1. The σ-symmetry orbitals that interact with the porphyrin nitrogens and with axial ligands (strictly speaking, in the idealized D_{4h} symmetry of 4- and 6-coordinate metalloporphyrins, these orbitals are of a_{1g} (d_{z^2}) and b_{1g} $(d_{x^2-y^2})$ symmetry, but in any case, they are engaged in σ-bonding interactions only);
2. the d_π orbitals, d_{xz} and d_{yz}, which can engage in π-bonding with both filled and empty π-symmetry orbitals of the porphyrin ring as well as the axial ligands; and
3. the d_{xy} orbital, which is nonbonding, both in relation to the porphyrin ring and the axial ligands.

The relative energies of the five d-orbitals are generally expected to be as shown in Fig. 2, although individual variations in these relative energies are observed. These are discussed in Section 2.

1.4.2. The Porphyrin Ring

Both 4- and 6-coordinate metalloporphyrins have effective D_{4h} symmetry, if we neglect the effects of planar axial ligands and potentially unsymmetrically placed

TABLE 2

Spin Delocalization Patterns of Iron Porphyrins of Various Oxidation and Spin States

Oxidation state	Number of d electrons	Spin state	Orbital of unpaired electron(s)	Dipolar shifts	Contact Shifts		Dominant type of π charge transfer
					Pyrrole-C	meso-C	
I	7	$S=\frac{1}{2}$	$(d_{z^2})^1$	Medium (+) $\alpha(g_\parallel^2-g_\perp^2)/T$	(–Shifts)	Small (σ?)	—
II	6	$S=0$		—	—	—	—
		$S=1$	$(d_{xz},d_{yz})^2$	Large (+) $\alpha(g_\parallel^2-g_\perp^2)/T$ (Small)	π	0	P→M
		$S=2$	$(d_{xz},d_{yz})^2(d_{z^2})^1(d_{x^2-y^2})^1$	None	Small π	π(?)	(?)
II Por·⁻	6 (paired)	$S=-\frac{1}{2},-\frac{1}{2}$	$(3a_{2u})^1$	Large (−) $\alpha(g_\parallel^2-g_\perp^2)/T$	π	Large π	—
III	5	$S=\frac{1}{2}$	$(d_{xz},d_{yz})^3$	Small (+) $\alpha(g_\parallel^2-g_\perp^2)/T$	π	Small (σ?)	P→M
		$S=\frac{1}{2}$	$(d_{xy})^1$	Medium (+)	0?	π	M→P
		$S=\frac{3}{2}$	$(d_{xy})^2(d_{z^2})^1(d_{xz},d_{yz})^2$	Small (+) $\alpha(g_\parallel^2-g_\perp^2)/T$	π	Small	P→M
		$S=\frac{3}{2},\frac{5}{2}$ admixed	$(d_{xy})^{2-\delta}(d_{z^2})^2(d_{yz})^2(d_{x^2-y^2})^\delta$	Small (+) mixed	π, σ mixed	$\alpha(g_\parallel^2-g_\perp^2)/T$ Mixed	P→M
		$S=\frac{5}{2}$	all 5 d	Large (+) $\alpha(D^2/T^2)$	σ	π	M→P
III Por·⁻	5	$S=3$	$(3a_{2u})^1$, plus all 5d for TPPs[a]	[a]	σ	Large π	P→M
		$S=\frac{3}{2},-\frac{1}{2}$	$(3a_{2u})^1$, $(d_{xy})^{2-\delta}(d_{z^2})^1(d_{xz},d_{yz})^2(d_{x^2-y^2})^\delta$	[a]	π	Large π	P→M
		$S=2$		[a]	σ	Large π	P→M
IV	4	$S=1$	$(3a_{2u})^1(d_{xz},d_{yz})^3$	[a](?)	π	Large π	P→M
		$S=1$	$(d_{xz},d_{yz})^2$ ferryl	[a]	π	Small	P→M
			$(d_{xz},d_{yz})^2$, but much spin delocalized to oxo	Small $\alpha(g_\parallel^2-g_\perp^2)$	Small π	0	
			phenyl $(d_{xz},d_{yz})^2$	Small	0		
IV Por·⁻	4	$S=\frac{1}{2}$	$(3a_{2u})^1(d_{xz},d_{yz})^2$ (TMP)	Small	Small π	Large π	P→M
			$(1a_{1u})^1(d_{xz},d_{yz})^2$ (TMTMP)	[a]	Large π	Small π	P→M
V	3	$S=\frac{3}{2}$	$(d_{xy})^1(d_{xy},d_{yz})^2$	[a]	~0		[a]

[a] not determined.

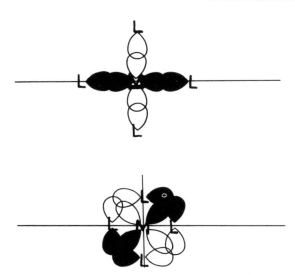

Figure 3. The symmetry-permitted interactions between metal d and porphyrin σ and π orbitals. Top: Interaction of the y-axis portion of the $d_{x^2-y^2}$ orbital with the σ-donor orbitals of two of the porphyrin nitrogens (porphyrin ring represented by the horizontal line) and the d_{z^2} orbital interacting with the σ donors of two axial ligands. Bottom: Interaction of one of the two d_π orbitals, d_{yz}, with the two porphyrin-nitrogen p_π orbitals along the y-axis, and the ligand p_π orbitals of the two axial ligands. Note the nodal plane of this group of orbitals that passes through the axial ligand and metal centers and lies perpendicular to the plane of the figure. It is this nodal plane which ensures that only e-symmetry porphyrin π orbitals can be involved in π spin delocalization to or from the metal.

porphyrin substituents. The bonding interactions between the metal atom and the porphyrin ligand are expected to be those shown in Fig. 3, where each porphyrin nitrogen supplies a σ-symmetry lone pair that points directly toward the porphyrin nitrogens along the x, $-x$, y, and $-y$ axes, and a single π-symmetry p-orbital perpendicular to the plane of the porphyrin ring. We will call this the p_z orbital of the bonding nitrogen. Not only does the coordinating nitrogen have an available π-symmetry p_z orbital, but so do each of the C_α, C_β, and C_m carbons of the porphyrin skeleton. Previous workers (Longuet-Higgins *et al.*, 1950; Zerner *et al.*, 1966; Gouterman, 1978; Antipas *et al.*, 1978; Binstead *et al.*, 1991) have developed more or less sophisticated molecular orbital treatments of the porphyrin ring and related macrocycles. In each of these treatments, π molecular orbitals are constructed based upon the symmetry properties of the idealized D_{4h}-symmetry of the porphyrin ring. The 24 p_z orbitals of the porphyrine macrocycle are thus combined to produce 24 molecular orbitals (two a_{1u}, four a_{2u}, three b_{1u}, three b_{2u}, and six e_g sets), 12 of which are filled by the 24 p_z electrons of the 24 atoms of the porphyrin nucleus. Although the theories differ in detail, all agree that the frontier and near-frontier orbitals are, in order of increasing energy, $3e(\pi) < 1a_{1u}(\pi) \sim 3a_{2u}(\pi) < 4e(\pi^*)$, using the notation of Longuet-Higgins *et al.* (1950). The $3e(\pi)$, $1a_{1u}(\pi)$, and $3a_{2u}(\pi)$ orbitals are filled, while the $4e(\pi^*)$ orbitals are empty. All theories agree further that the latter are the LUMOs of the porphyrin ring, while the HOMO(s) may be either $1a_{1u}(\pi)$, $3a_{2u}(\pi)$, or the d_{xy} or d_{π} orbitals, depending on the oxidation state of the metal, the porphyrin, and the degree of reduction of the macrocycle (chlorins and isobacteriochlorins) substituents and the axial ligand(s) present. The symmetry properties and electron distributions predicted from the simplest of these treatments are shown in Fig. 4. For electronic absorption spectra, excitations from $1a_{1u}(\pi)$ and $3a_{2u}(\pi)$

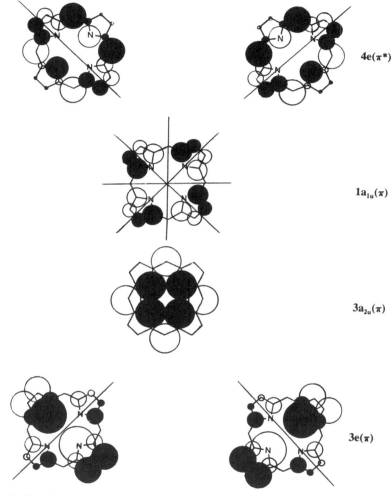

Figure 4. Nodal properties and relative energies of and electron density distributions in the frontier π orbitals of the porphyrin macrocycle, according to Longuet-Higgins *et al.* (1950). The $3e(\pi)$, $3a_{2u}(\pi)$, and $1a_{1u}(\pi)$ orbitals are filled, while the $4e(\pi^*)$ orbitals are empty and are thus the LUMOs of the porphyrin ring. Note that only the $3e(\pi)$ and $4e(\pi^*)$ orbitals have the proper symmetry (one nodal plane perpendicular to the plane of the porphyrin) for overlap with d_{xz} and d_{yz} (one each).

to $4e(\pi^*)$ give rise to the so-called α, β, and Soret bands in the visible spectra (Zerner et al., 1966; Gouterman, 1978; Antipas et al., 1978; Binstead et al., 1991); porphyrin cation radicals have either $a_{1u}(\pi)$ or $a_{2u}(\pi)$ unpaired electrons (Fajer and Davis, 1979). However, although the $1a_{1u}(\pi)$ or $3a_{2u}(\pi)$ orbital is typically the HOMO of the porphyrin ring, neither of these orbitals have the proper symmetry to overlap with the d_π orbitals of the metal. Thus, only the $3e(\pi)$ orbitals, which are lower in energy than either the $1a_{1u}(\pi)$ or the $3a_{2u}(\pi)$ orbitals, and the $4e(\pi^*)$ orbitals, which are higher in energy than either the $1a_{1u}(\pi)$ or the $3a_{2u}(\pi)$ orbitals (all shown in Fig. 4), have the proper symmetry for overlap with the metal d_π orbitals. Since the $3e(\pi)$ orbitals are filled, the type of overlap interaction that can occur between them and the metal d_π orbitals is $P \rightarrow M$ (or, more generally, $L \rightarrow M$) π bonding. Clearly, this type of bonding interaction can lead to spin delocalization from the metal to the porphyrin π system only if the metal d_π orbitals are partially, but not completely, filled. Thus, spin delocalization through interaction between the filled $3e(\pi)$ orbitals and the metal d_π orbitals can occur only if the metal-electron configuration is $(d_{xz}d_{yz})^{1-3}$. Similarly, since the $4e(\pi^*)$ orbitals of the porphyrin ring are empty, spin doclization from the metal d_π orbitals to the $4e(\pi^*)$ orbitals of the porphyrin ring can occur through $M \rightarrow P$ (or more generally, $M \rightarrow L$) π back-bonding when the metal-electron configuration is $(d_{xz}, d_{yz})^{1-3}$. (We will modify this statement slightly in Section 2.8.4., where we consider more exactly the nature of the metal orbitals as mixed by spin–orbit coupling.)

If the unpaired electron is localized in a porphyrin π-symmetry orbital, rather than in a metal d-orbital, then the system is described as a metalloporphyrin π cation or anion radical. The electron configuration of π cation radicals, $(MP)^{+\cdot}$, is typically either $(1a_{1u})^1$ or $(3a_{2u})^1$, since one or the other of these orbitals is typically the HOMO of the porphyrin π system. Anion radicals typically have the electron configuration $(4e)^1$; dianions, $(4e)^2$.

For dihydroporphyrins, also known as chlorins, the reduction of one pyrrole ring reduces the symmetry of the molecule, and thus removes the degeneracy of the $3e(\pi)$ and $4e(\pi^*)$ orbitals. Hückel molecular orbital calculations (Chatfield et al., 1988b) predict the electron density pattern and nodal properties shown in Fig. 5 for the corresponding chlorin π orbitals. An important feature of these modified orbitals is the fact that the former $1a_{1u}(\pi)$ orbital, now denoted A_5, has the proper symmetry at the nitrogens of rings A and C to allow overlap with a d_π orbital, d_{xz} in the coordinate system shown (Chatfield et al., 1988b). This has important consequences for the pattern of spin delocalization observed for both high- and low-spin Fe(III) complexes of chlorins, discussed in Sections 2.6.3 and 2.8.9, respectively. Similar modifications of the electron density distribution and symmetry properties of the A,B-tetrahydroporphyrins, better known as isobacteriochlorins, are also expected.

1.4.3. Mechanisms of Spin Delocalization through Chemical Bonds and Strategies for Separation of Contact and Dipolar Shifts

As discussed in Section 1.3.1, the McConnell equation [Eq. (5)] and the opposite sign of Q for spin density in π and σ carbon orbitals allow spin delocalization through the σ and π systems to be differentiated. Thus, all shifts of protons connected to the unpaired electron only by σ bonds should be in the same direction, whether the proton is directly attached to a conjugated π system or is separated from it by a saturated C—C bond. On the other hand, if π delocalization mechanisms are involved, the same

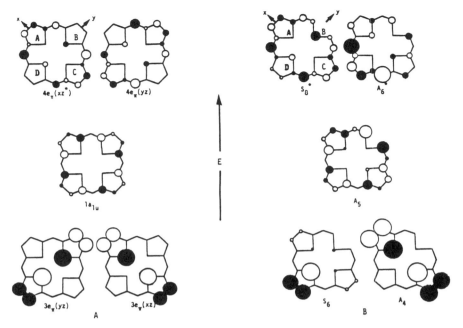

Figure 5. Comparison of the symmetry properties of the frontier π orbitals of porphine (A) and chlorin (B), excluding the $3a_{2u}(\pi)$ orbital, with appropriate symmetry and energy to interact with d_{xz}, d_{yz} iron orbitals. Because the A_5 orbital of chlorin can π bond to the iron, the noninteracting porphine counterpart MO, $1a_{1u}(\pi)$, is shown for comparison. The orientation of axes and labeling of pyrroles are shown at the top of the diagram. Phases are indicated by open and closed circles, with the size of the circle indicating relative spin density. Symmetry and energy designations are those of Gouterman (1978), with increasing number indicating increasing orbital energy and asterisks indicating antibonding orbitals. Reprinted with permission from *J. Am. Chem. Soc.* **110**:6358 (1988). © 1988 American Chemical Society.

theory predicts that the sign of the contact shift of the proton bound to the carbon atom that is part of the π system will be opposite to that of a proton that is separated from it by a saturated C—C bond (Section 1.3.1). Thus, π delocalization is characterized by a reversal in the sign of the isotropic shift when a proton directly attached to a carbon that is part of a π system is replaced by an aliphatic group such as —CH$_3$ or —CH$_2$—. Therefore, by judicious choice of model hemes, such as those shown in Fig. 1a, with $R_1 - R_8 = H$ and $R_\alpha - R_\delta = $ phenyl or alkyl, as contrasted to those with $R_1 - R_8 = $ alkyl and $R_\alpha - R_\delta = H$, it has been possible to assign the mechanism of spin delocalization to the β-pyrrole and *meso* positions of the porphyrin ring (La Mar and Walker, 1979) for a number of oxidation and spin states of the metal. These, together with recent results, are summarized in Table 2. The data giving rise to this summary table are discussed in the various subsections of Section 2.

Analysis of the pattern of isotropic shifts for a newly discovered spin or oxidation state of a metalloporphyrin often yields some generalizations as to the relative importance of the contact and dipolar contributions. For example, if the sign of the isotropic shift reverses when protons directly bound to the aromatic carbons of the macrocycle or the axial ligands are replaced by methyl or other alkyl groups, then it may be assumed that the π contact contribution dominates the isotropic shift at that position. If so, then the nodal properties of the porphyrin or axial ligand π orbital involved may allow one to evaluate the residual dipolar contribution to the isotropic shift. For example, in low-spin Fe(III) porphyrins having a $(d_{xz}, d_{yz})^3$ ground state, the isotropic shifts at the β-pyrrole positions (Fig. 1a, $R_1 - R_8$) reverse in sign when $-H$ is replaced by $-CH_2CH_3$, while those at the *meso* positions (Fig. 1a, $R_\alpha - R_\delta$) are small and shifted to lower frequency for both $-H$ and $-CH_2R$ (La Mar and Walker, 1979, and references therein). Thus, it was assumed that the contact contribution resulted from π-symmetry spin delocalization involving the $3e(\pi)$ orbitals of the porphyrin, since these orbitals have nodes at the *meso* $(R_\alpha - R_\delta)$ positions, while the $4e(\pi^*)$ orbitals have much larger coefficients for the wave functions of the *meso* carbons than for those of the β-pyrrole carbons (Longuet-Higgins *et al.*, 1950), as shown in Fig. 4. Thus, any observed isotropic shifts of *meso* substituents were assumed to be due to the dipolar contribution. In tetraphenylporphyrin complexes such as $[TPPFe(NMeIm)_2]^+$, the *ortho-*, *meta-*, and *para*-H isotropic shifts were found to be directly proportional to the calculated geometric factors for those protons, an excellent indication that the isotropic shifts at the *meso*-phenyl proton positions are overwhelmingly dipolar. Thus, scaling the *ortho-* or *meta*-H isotropic shifts by the ratio of the axial geometric factors for those protons to the axial geometric factor for the β-pyrrole protons permitted estimation of the axial dipolar contribution to the isotropic shift of the β-pyrrole protons (La Mar and Walker, 1979, and references therein). (The rhombic dipolar contribution is expected to be identically zero for totally symmetric hemes having cylindrical axial ligands such as CN^- or planar axial ligands rotating rapidly.) The same procedure was found to be less successfully applicable to other axial ligand complexes of TPPFe(III), as will be discussed in Section 2.8, because alternation in the sign of the *meso*-phenyl-H isotropic shifts (*o-*, *p*-H of opposite sign to *m*-H) indicated the existence of significant contact contribution to the isotropic shifts of those protons, in the order imidazoles $\ll$ cyanide $<$ pyridines (La Mar and Walker, 1979).

For all cases in which the $d_{x^2-y^2}$ orbital is half occupied, it has been found that there is a large σ contact contribution to the isotropic shifts of the β-pyrrole protons (La Mar and Walker, 1979). Such a σ contact contribution can be readily recognized by its attenuation as aliphatic carbons are inserted between the porphyrin ring and the protons. The presence of large σ spin density at the pyrrole positions does not rule out the possibility of large π spin density at the *meso* positions if the d_π orbitals contain at least one unpaired electron; high-spin Fe(III) porphyrins exhibit such a pattern of contact shifts, as well as a much smaller dipolar shift term that is readily separable from the contact contribution because it varies according to $1/T^2$ (La Mar and Walker, 1979), as mentioned in Section 1.3.2 [Eq. (10)].

Cases in which the dipolar contribution dominates the isotropic shifts are readily recognized by the direct proportionality of the isotropic shifts to the geometric factors of the protons of interest. An example of this situation is the isotropic shifts of the phenyl protons of $[TPPFe(R - Im)_2]^+$ complexes, as discussed in Section 2.8.2, and of the π cation radicals of Fe(III) porphyrins, as discussed in Section 2.9.

1.5. Methods of Resonance Assignment of the ^{1}H NMR Spectra of Paramagnetic Iron Porphyrins

1.5.1. Replacement of H by CH$_3$ or Other Substituents

This method was used for the original assignment of pyrrole-H, *meso*-H, and *o*-, *m*-, and *p*-phenyl-H resonances of symmetrically substituted model hemes (La Mar and Walker, 1979, and references therein), and should need no further elaboration. Balch and Renner (1986b) have used it to assign the phenyl resonances of the ferryl porphyrin complexes [TPPFe^{IV}O], and the present authors have also used it for assigning the phenyl resonances in unsymmetrically substituted derivatives of [TPPFe(NMeIm)$_2$]$^+$ (Koerner *et al.*, 1992).

1.5.2. Deuteration of Specific Groups

Smith, Goff, Dolphin, Strauss, and others have developed methods for specifically deuterating certain positions of both naturally occurring and synthetic porphyrins or metalloporphyrins. For tetraphenylporphyrin, both benzaldehyde-d_6 (and specifically deuterated benzaldehydes) and pyrrole-d_5 are commercially available, so fully, d_8-, or d_{20}-deuterated TPPH$_2$ and their metal complexes can readily be prepared. Alternatively, about 90% pyrrole deuteration can be effected during synthesis of TPPH$_2$ if propionic acid-d_1 [prepared by hydrolysis of the anhydride with D$_2$O (Shirazi and Goff, 1982)] is used as solvent in the Adler method (Adler *et al.*, 1967; Shirazi and Goff, 1982). Fully deuterated benzaldehyde can be prepared from toluene-d_8 by ceric ammonium nitrate oxidation (Fajer *et al.*, 1973). The four *meso*-H of OEP can be fully deuterated in hours if the porphyrin is fused with *p*-toluenesulfonic acid-d_1 at 120 °C (Hickman and Goff, 1984), while simultaneous ring alkyl and *meso* position deuteration can be achieved by refluxing *p*-toluenesulfonic acid-d_1 with metal-free OEPH$_2$ or etioporphyrin free base in *o*-dichlorobenzene under N$_2$ for 4–8 days. Four days of reflux results in 92% methine deuteration and 45% deuteration of the ring-adjacent positions; eight days of reflux increases the ring methyl and methylene deuteration levels to 60% (Hickman and Goff, 1984). The four *meso*-H positions of naturally occurring porphyrins deuterate at variable rates under acid conditions depending upon the nature of the electron-withdrawing/ donating substituents at the 2,4 positions (Smith *et al.*, 1979; Smith and Langry, 1983).

The 1- and 3-CH$_3$ of protoporphyrin IX can also be selectively deuterated by base-catalyzed exchange of those ring methyl protons in the presence of CH$_3$ONa/CH$_3$OD in dimethylformamide over 5 days (Evans *et al.*, 1977). Pyrrole-α-CH$_2$ protons of [OEPFeCl] and both α-CH$_2$ and methyl H of natural hemin chlorides can also be partially or totally deuterated, depending upon the length of the reaction, under base-catalyzed conditions at room temperature utilizing tetrabutylammonium hydroxide in DMSO-d_6 (Godziela *et al.*, 1986). Natural hemins are preferentially deuterated at the ring 1 and 3 methyl positions to a greater extent in the order protohemin > deuterohemin > mesohemin. No evidence of β-CH$_3$ or *meso*-H deuteration could be found, either in the ^{1}H or the ^{2}H NMR spectra of the products (Godziela *et al.*, 1986). Likewise, the two types of *meso*-H and the pyrroline-H of the OEC free base and the three types of *meso*-H and the pyrroline-H of the *ttt* and *tct* isomers of the OEiBC free bases are deuterated at different rates under acidic and basic conditions (Sullivan et al., 1991), permitting specific assignments of the resonances of each of these groups. Much more elaborate total synthesis methods are required for specific single methyl deuteration of protoporphyrin IX and related natural hemes (Kenner and Smith, 1973; Cavaleiro et al., 1974a, b; Mayer et al., 1974).

Although ^{2}H NMR investigations of deuterium-labeled diacetyldeuterohemin incorporated into sperm whale myoglobin was reported early (Oster et al., 1975), Goff and co-workers (Shirazi and Goff, 1982; Hickman and Goff, 1984; Li and Goff, 1992) have repeatedly demonstrated the utility of ^{2}H NMR spectroscopy for assignment of resonances in model hemes. As pointed out (Shirazi and Goff, 1982), the linewidths of ^{2}H resonances, in Hz, could be narrowed by as much as $(\gamma_D/\gamma_H)^2$; i.e., to 1/42 the linewidths of the ^{1}H signals although experimentally the pyrrole-H and pyrrole-D linewidths of [TPPFeCl] differ by only a factor of 10 (280 Hz and 28 Hz, respectively), as shown in Fig. 6. The o-D resonances of [d_{20}-TPPFeCl] are much sharper than the o-H of [TPPFeCl], and the m-D signal occurs as a singlet (Fig. 6) due to averaging of the two m-D environments of this 5-coordinate complex on the longer time scale (lower frequency) of deuterium NMR (Shirazi and Goff, 1982). Li and Goff (1992) have recently utilized the increased sharpness of ^{2}H resonances of the —CD$_3$ and —C$_2$D$_5$ groups, and wider spectral bandwidth (in terms of ppm) available for ^{2}H to detect the α-D of Fe(III) alkyl complexes, previously undetected (Arasasingham et al., 1989a).

La Mar, Smith and co-workers (Evans et al., 1977; La Mar et al., 1978a, b) were among the earliest to employ specifically methyl-deuterated hemins to assign the methyl resonances of model hemes and heme proteins. More recently, Strauss and co-workers (Sullivan et al., 1991) have used both ^{1}H and ^{2}H NMR spectroscopy in a complementary fashion to show that the signals that disappear from the ^{1}H spectrum upon deuteration are the only ones that appear in the ^{2}H spectrum, as shown in Fig. 7.

1.5.3. Saturation Transfer NMR Experiments

These experiments were used early by Wüthrich and co-workers to assign the heme resonances of ferricytochrome b_5 which was in chemical exchange with ferrocytochrome b_5, the heme resonances of which had been assigned by NOE techniques (Keller et al., 1973, 1976; Keller and Wüthrich, 1980). However, for hemes outside the proteins, electron exchange is sufficiently rapid that in most cases separate resonances from the Fe(II) and Fe(III) states are not observed (Weightman et al., 1971; Kimura et al., 1981; Dixon et al., 1985; Shirazi et al., 1985). However, in the case of other types of chemical exchange, such as that involving use of an ortho-phenyl substituent to hinder the rotation of an axial ligand (Simonis et al., 1992), saturation transfer provides an effective means of assigning proton resonances of groups that become equivalent by means of the chemical exchange process, as shown in Fig. 8.

1.5.4. NOE Difference Spectroscopy

This technique has been used extensively by La Mar and co-workers (Johnson et al., 1983; Ramaprasad et al., 1984a, b; Unger et al., 1985b; Yu et al., 1986) and others (Trewhella et al., 1980) to assign the resonances of the heme group in ferriheme proteins; it is much more easily applied to large molecules such as proteins, than to small ones

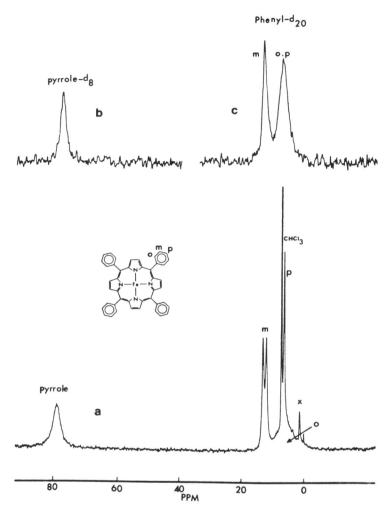

Figure 6. (a) Proton NMR spectrum of TPPFeCl, 0.01 M in CDCl₃ at 28 °C, (b) deuterium NMR spectrum of (TPP-pyrrole-d_8)FeCl, 0.01 M in CHCl at 28 °C, and (c) deuterium NMR spectrum of (TPP-phenyl-d_{20})FeCl 0.01 M in CHCl at 28 °C. Reprinted with permission from *J. Am. Chem. Soc.* **104**:6319 (1982). © 1982 American Chemical Society.

such as model hemes since the large size of the proteins ensures that the NOE will be negative. On the other hand, for the model hemes in liquids of low viscosity, the rotational correlation time τ_c is in the intermediate motion limit, such that the predicted steady-state NOE enhancement, though negative, is very small. This was earlier thought to be the reason that steady-state NOEs were sometimes not observed for model hemes (Yu *et al.*, 1986), although Barbush and Dixon (1985) showed that the dicyanoiron(III)

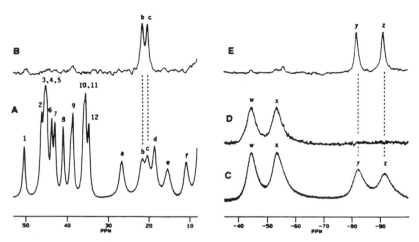

Figure 7. The 300-MHz ^{1}H NMR spectra of Fe(t-OEC)Cl (A and C) and Fe(t-OEC-5,10-d_2)Cl (D) and 46-MHz ^{2}H NMR spectra of Fe(t-OEC-7,8,15,20-d_4)Cl (B) and Fe(t-OEC-5,10-d_2)Cl (E). All of the spectra were recorded at 25 °C. Samples for ^{1}H NMR were dissolved in toluene-d_8, while samples for ^{2}H NMR were dissolved in natural-abundance toluene. Reprinted with permission from *J. Am. Chem. Soc.* **113**:5269 (1991). © 1991 American Chemical Society.

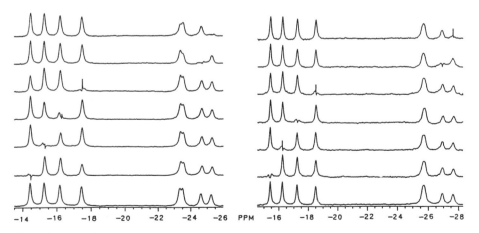

Figure 8. One-dimensional saturation transfer spectra of $[(o\text{-CONR}_2)_1\text{TPPFe(NMeIm)}_2]^+$, where the amide is derived from 3-azabicyclo-[3.2.2]nonane. Left: Spectra obtained at 0 °C; right: -15 °C. Solvent = CDCl$_3$. In the 0 °C spectra, diminution in the signals due to chemical exchange is seen for the pairs of resonances 1,4; 2,3; 5,8; 6,7 whereas at -15 °C there is no effect on the intensity of the paired resonance as its partner is irradiated.

complex of protoporphyrin IX gave small NOEs (1% negative) in DMSO-d_6 at 30 °C. Thus, the more serious limitation is that the paramagnetic relaxation undermines the NOE (Dugad et al., 1990). However, if the paramagnetic relaxation is independent of motion, as has been found to be the case, then it is possible to recover the NOE by slowing down the rate of molecular reorientation for both low-spin Fe(III) (Licoccia et al., 1989) and high-spin Fe(III) complexes (Chatfield et al., 1988b), as shown in Fig. 9 left and right, respectively. Unfortunately, few model hemes are soluble in highly viscous solvents other than ethylene glycol or supercooled DMSO; this limits the application of steady-state NOE difference spectroscopy to model hemes.

1.5.5. 2D NMR Techniques

Both scalar couplings and through-space NOEs can be detected for paramagnetic complexes under favorable circumstances. In terms of scalar-coupled systems, both COSY (homonuclear correlation spectroscopy) (Yu et al., 1986; Keating et al., 1991; Lin et al., 1992; Isaac et al., 1992) and HETCOR (heteronuclear correlation spectroscopy) (Yamamoto et al., 1990; Timkovich, 1991; Simonis et al., 1992a) of paramagnetic iron porphyrins have been successfully carried out. A two-dimensional magnitude COSY spectrum is shown in Fig. 10 (Keating et al., 1991). The success of these experiments is strongly limited by the short T_2 of most paramagnetic complexes. For bis-N-methyl-imidazole complexes of metallotetraphenylporphyrins, typical T_2^*s are 8–9 ms (Lin et al., 1992) or shorter (Walker and Simonis, 1991), while for dicyano and monocyano monoimidazole complexes of β-pyrrole substituted porphyrins, T_2^*s are usually much longer the range of [T_2^* for the ring CH_3s of (cyano)(imidazole)protohemin IX is 36–40 ms, and that of the meso-H is 21–32 ms (Bertini et al., 1991)]. Short T_2 can lead to significant cancellation of the antiphase components when J is small, preventing detection of cross peak intensity (Ernst et al., 1987; Martin and Zektzer, 1988). However, on the positive side, the fact that T_2^* is short for paramagnetic complexes means that the experiments can be run much faster than for diamagnetic compounds, since it makes no sense to acquire data over more than about $2T_2$. In fact, if the acquisition time in t_2 is much longer than $2 \times$ (longest T_2^*), significant cross peak intensity is lost (Lin et al., 1992). The need to keep the acquisition time A_t short necessarily decreases the digital resolution of 2D NMR experiments for paramagnetic compounds, since $A_t = N/F$, where N is the total number of real data points in the t_2 dimension and F is the spectral bandwidth. Likewise, acquiring many blank data blocks in the t_1 dimension only adds to the total experiment time, so the number of t_1 blocks typically acquired for model hemes with very short T_2 may be 256, 128, or even 64, yielding even poorer digital resolution in the t_1 dimension. This is not often a problem, however, because of the large dispersion of the signals of many paramagnetic complexes. The apparent digital resolution can be improved by zero-filling one or more times in either or both dimensions. We have recently found that using a 45° mixing pulse (i.e., a COSY-45 experiment) significantly increases the relative intensity of cross peaks (Lin et al., 1992). Examples of COSY spectra of high-spin and low-spin Fe(III) complexes are shown in Fig. 10 and 11, respectively.

Fewer examples of heteronuclear correlation spectra of paramagnetic iron porphyrins have been reported, but the ^{13}C–^1H heteronuclear correlation (HETCOR) map of dicyanoprotohemin IX has been reported (Yamamoto and Fujii, 1987), as has that of

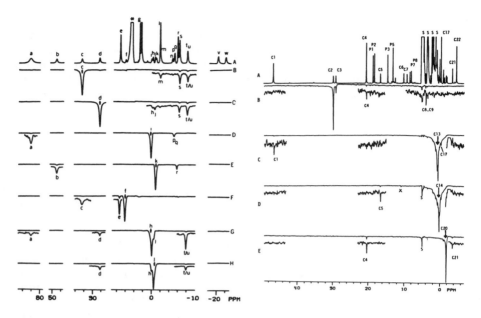

Figure 9. Resonance assignments by NOE difference spectroscopy of the dicyano complex of (pyropheophorbide methylester) iron(III) in 2:1 DMSO-d_6: D_2O at $-20\,°C$ at 360 MHz (left) and sulfhemin c in 95% ethylene glycol-d_6–5% methanol-d_4 at 30 °C at 500 MHz (right). For the spectra on the left, (A) 1D spectrum of the complex, (B) irradiation of the 8-H (peak c) yields an NOE to the 8-CH$_3$ (peak s) and to the propionate peaks t/u; hence t/u must be the 7α-CH$_2$. (C) irradiation of peak d yields NOEs to the assigned 7α-CH$_2$ (peaks t/u) and 8-CH$_3$ (peak s); hence d is due to the 7-H. (D) irradiation of peak i yields NOEs to the 5-CH$_3$ (peak a) and to the 4α-CH$_2$ (peaks p, q), thus assigning peak i as the β-*meso*-H. (E) saturation of peak k (α-*meso*-H) gives NOEs to the 2-Hα (peak b) and to peak r, which is thus identified as the 3-CH$_3$. Spectra for both traces D and E were computer corrected to eliminate off-resonance effects from peaks h and j. (F) irradiation of the δ-*meso*-H (peak f) yields an NOE to peak e, identifying it as the 1-CH$_3$, together with an expected NOE to the 8-H (peak c). (g, h) peaks h and j yield, upon saturation, reciprocal NOEs, and both show connectivity with the 7-H (peak d) and the 7α-CH$_2$ (peaks t/u); peaks h and j are thus identified as the γ-*meso*-CH$_2$. Note that for both spectra the reference frequencies were chosen to be symmetrical with respect to j and h; the NOEs thus obtained are real and indicative of geminal partners. Moreover, the spectra were computer corrected for off-resonance effects due to k and i (Licoccia *et al.*, 1989; Chatfield *et al.*, 1988). For the spectra on the right, (A) 500-MHz ^{1}D spectrum of 3-mM dicyanosulfhenmin peaks C_i. Dicyanoprotohemin (peaks P_i) and solvents (peaks S) are also present. NOE difference spectra follow. (B) Irradiation of the more downfield 6-H$_\alpha$ (peak C_2) yields NOEs to the geminal 6-H$_\alpha$ (C_3) and 6-H$_\beta$ (C_8, C_9). (C) Irradiation of the *meso*-H peak C_{13} yields an NOE to the 1-CH$_3$ (peak C_1), identifying it as the δ-*meso*-H: the 8-CH$_3$ (peak C_{17}) exhibits off-resonance effects. (D) Saturation of the *meso*-H peak C_{14} shows NOE connectivity with the 2-H$_\alpha$ (peak C_5), identifying it as the α-*meso*-H. (E) Saturation of the *meso*-H peak C_{20} produces an intensity change in the 5-CH$_3$ (peak C_4) and 4-H$_\alpha$ (peak C_{21}), assigning it as the β-*meso*-H. Reprinted with permission from *J. Am. Chem. Soc.* **110**:6356 (1988) and *J. Am. Chem. Soc.* **111**:6090 (1989). © 1988 and 1989 American Chemical Society.

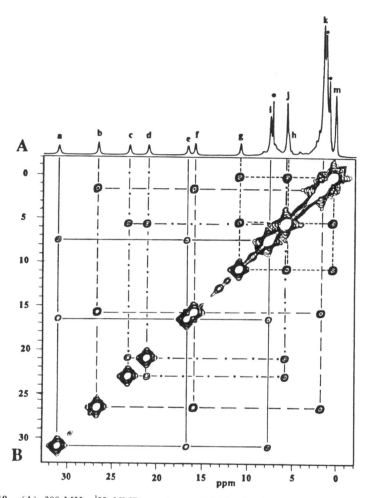

Figure 10. (A) 300-MHz ^{1}H NMR spectrum of high-spin (N-methyl-octaethylporphinato) iron(II) chloride (4) in C^{2}HCl$_3$ at 15 °C. Impurity and solvent peaks are labeled O. (B) 300-MHz MCOSY map of 4 illustrating cross peaks connecting the three signals of the four nonequivalent ethyl groups. Artifacts from solvent lines are shaded. The collected data consisted of 512 t_1 blocks of 128 scans each with 1024 t_2 points over a 50-kHz bandwidth for acquisition times in t_1 and t_2 of 10 and 20 ms, respectively. The optimal map resulted from apodizing over 512 × 512 points prior to zero-filling in t_1 for a final digitization of 49 Hz/point (Keating *et al.*, 1991). Reprinted with permission from *Inorg. Chem.* **30**:3262 (1991). © 1991 American Chemical Society.

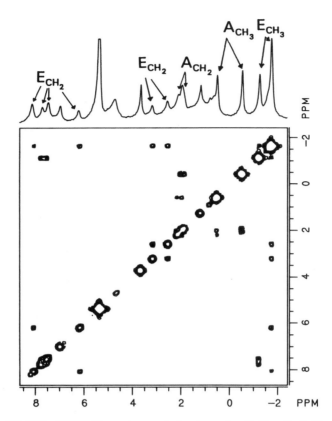

Figure 11. 300-MHz COSY-45 spectrum of the low-spin *bis*-N-methylimidazole complex (2,3,7,8,12,13-hexaethyl-17,18-diethylcarboxamido)porphinato iron(III), $[E_6A_2PFe(NMeIm)_2]^+$ in CD_2Cl_2 at 21 °C, showing the coupling pattern of methylene protons to methyl protons within the various ethyl groups (taken from Isaac *et al.*, 1992).

Geleorhinus japonicus (shark) met-cyanomyoglobin (Yamamoto *et al.*, 1990). Timkovich (1991) has recently reported the relayed $^1H-^{13}C$ scalar correlation spectrum (HMQR) and the long-range $^1H-^{13}C$ scalar correlation (heteronuclear multiple-bond correlation, HMBC) spectrum of dicyanoprotohemin IX. Parts of these 2D maps are shown in Fig. 12, where it can be seen that relayed or long-range $^1H-^{13}C$ couplings can be observed in paramagnetic complexes, at least those in which T_2 relaxation times are relatively long. We have recently obtained the $^{13}C-^1H$ HETCOR spectrum of several *bis*-N-methylimidazole complexes of unsymmetrically substituted derivatives of TPPFe(III), where T_2 are very short (Simonis *et al.*, 1992a).

Both NOESY (nuclear Overhauser enhancement and chemical exchange spectroscopy) and ROESY 2D (NOE in the rotating frame) spectra can also be valuable in obtaining dynamic and three-dimensional structural information concerning model hemes. The observation of NOESY cross peaks depends upon the mixing time τ_m being

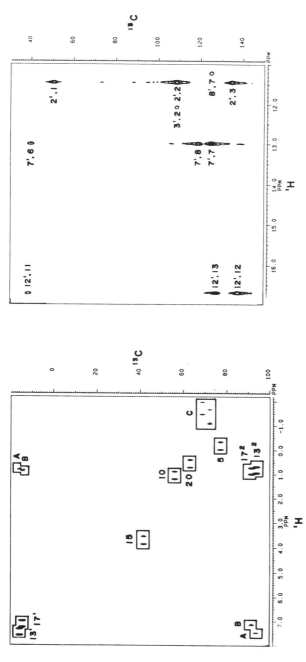

Figure 12. Examples of $^{13}C-^{1}H$ correlation spectra. Left: Portion of the HMQR spectrum of *bis*(cyano) iron(III) protoporphyrin IX. Peaks with numeric labels correspond to direct $^{1}H-^{13}C$ correlation peaks for the heme resonances given with IUB-IUPAC nomenclature. They appear as absorptive doublets with $^{1}J(^{1}H, ^{13}C)$ parallel to the proton axis. The cross peak labeled A is a relay peak between the site 13^{1} and 13^{2} spin systems. Relay peaks have dispersive shape, but only the positive contours have been drawn. B is a relay peak between sites 17^{1} and 17^{2}. C is cluster of relay peaks involving the vinyl chains at sites 3 and 8, but the direct peaks lie outside the plotted region. Peaks labeled 5, 10, 15, and 20 are direct peaks for the *meso* sites. Right: Portion of the HMBC spectrum of *bis*(cyano) iron(III) protoporphyrin IX. The peaks are labeled with two coordinates. The first is the IUB-IUPAC nomenclature for the carbon with attached proton, and the second is for the nonprotonated carbon whose chemical shift is then given along the carbon axis. For example, the peak labeled 12^{1}, 13 demonstrates the chemical shift of protons 12^{1} along the horizontal axis and the coherence transferred to the carbon 13, whose shift is given along the vertical axis. The peaks have magnitude mode line shapes. In order to show the weaker peaks, the minimum contour level was decreased and the strong peaks then show tailing and t_1 ridges. Reprinted with permission from *Inorg. Chem.* **30**:39 and 40 (1991). © 1991 American Chemical Society.

on the order of the T_1 relaxation times of the protons between which a NOE is expected (Walker and Simonis, 1991). Since the T_1 of a proton of a paramagnetic iron porphyrin can be 8–9 ms or shorter and decreases as the temperature is lowered (Walker and Simonis, 1991) [in comparison, the T_1s of ring CH_3s of dicyanoprotochemin IX are 150–190 ms (Unger et al., 1985a)], it is important to find experimental conditions that will maximize the length of T_1s. We have found that degassing the sample with argon can lengthen the T_1s by 1–3 ms (Walker and Simonis, 1991), which can make the difference between an unsuccessful and a successful experiment in some cases. We have used NOESY spectra acquired at several (low) temperatures to investigate the chemical exchange due to axial ligand rotation (at −29 and −54 °C) and to elucidate the three-dimensional structure (at −74 °C) of $[TMPFe(2\text{-}MeImH)_2]^+$ in CD_2Cl_2. Chemical exchange is evidenced by cross peaks between all protons of a given type, as shown in Fig. 13, whereas at very low temperatures, only the true NOE cross peaks are observed (Walker and Simonis, 1991), as shown in Fig. 14. From the NOE cross peaks, the low-temperature, three-dimensional structure of the complex was deduced, as discussed in Section 2.8.2c. In another example, a combination of chemical exchange and NOE cross peaks are observed for a *mono-o*-phenyl substituted derivative of $[TPPFe(NMeIm)_2]^+$ due to hindered rotation of one of the axial ligands (Simonis et al., 1992), as shown in Fig. 15B.

Rotating frame experiments appear to have mixed success with paramagnetic complexes. We have been unable to obtain HOHAHA (homonuclear Hartman–Hahn, also called TOCSY) cross peaks from any of the low-spin Fe(III) model hemes that have given COSY spectra (Simonis et al., 1992b). However, we have been able to show that ROESY spectra provide an excellent method of establishing chemical exchange in model hemes (Simonis et al., 1992b), as shown in Fig. 15A. However, the additional NOE cross peaks observed in the NOESY experiment, Fig: 15B, are not seen in the ROESY map (Fig. 15A). This is in line with predictions that the relaxation mechanism in paramagnetic molecules is mainly dominated by interaction with the spin of the unpaired electron and not by dipole–dipole interactions (Shaka, 1992; La Mar, 1992).

A combination of COSY and NOESY spectra can be utilized to assign totally the pyrrole-H resonances of unsymmetrically substituted low-spin Fe(III) porphyrins, as demonstrated in Fig. 16 for $[(o\text{-}Cl)_1(p\text{-}OCH_3)_3TPPFe(NMeIm)_2]^+$ (Simonis and Walker, 1992). Scalar couplings between resonances 1,3 and 2,4 are seen in the COSY spectrum, Fig. 16A, and these same peaks, arising from NOE between protons in close proximity, are seen in the NOESY spectrum, Fig. 16B. However, an additional set of cross peaks, between resonances 1 and 2, is also seen in the NOESY spectrum. They must arise from through-space NOEs between protons of type *b* and *c*, Fig. 16C, and they thus allow complete definition of the pattern of spin delocalization in this complex, as shown in Fig. 16D. The cross peaks due to NOEs between protons in adjacent pyrrole rings are the strongest cross peaks observed in the NOESY spectrum of this and related complexes (Simonis and Walker, 1992). We have also found that although COSY-45 spectra may be obtained at ambient temperatures, where axial ligand exchange of N-MeIm is clearly occurring, NOESY spectra of some unsymmetrically substituted low-spin Fe(III) complexes give cross peaks between all pyrrole-H resonances (Simonis and Walker, 1992), which may result from chemical exchange with the high-spin mono-N-MeIm 5-coordinate intermediate, for which the pyrrole-H resonances strongly overlap.

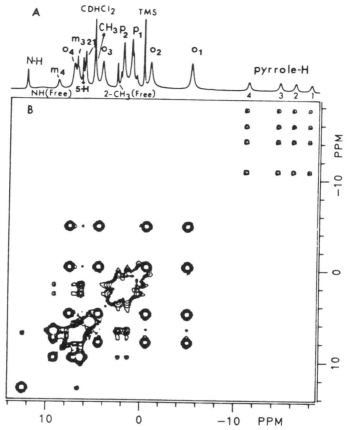

Figure 13. (A) The 300-MHz ^{1}H NMR spectrum of a 10-mM sample of [TMPFe(2-MeImH)$_2$]Cl in CD$_2$Cl$_2$ at -29 °C with assignments of all resonances. (B) NOESY map collected with optimized 80-ms mixing time showing chemical exchange (EXSY) cross peaks and 2D NOE cross peaks. The latter are observed between the NH and 5-H resonances of coordinated 2-MeImH, while a combination of NOESY and EXSY correlations are observed between *para* CH$_3$ and *meta* H resonances and between *ortho* CH$_3$ and *meta* H signals. The expected location of EXSY cross peaks between the methyl resonances of free and coordinated 2-MeImH if chemical exchange of the axial ligand were involved would be 2.85, 5.29 ppm and 5.29, 2.85 ppm. Reprinted with permission from *J. Am. Chem. Soc.* **113**:8653 (1991). © 1991 American Chemical Society.

2. PROTON NMR STUDIES OF VARIOUS IRON OXIDATION AND SPIN STATES OF PORPHYRINS AND REDUCED HEMES

2.1. Fe(I) Porphyrins

Even though iron(I) porphyrin complexes are known, only a few of them have been investigated by either proton or deuterium NMR spectroscopy. Among those

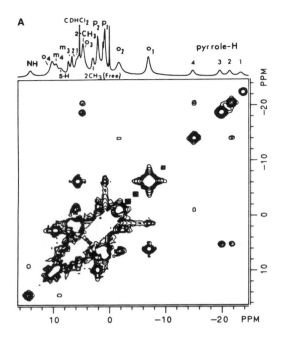

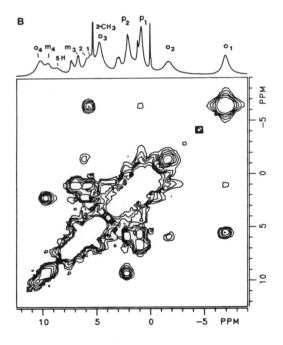

Figure 14. Proton 1D spectrum and NOESY map of [TMPFe(2-MeImH)$_2$]Cl in CD$_2$Cl$_2$ collected at $-74\ ^\circ$C with optimized mixing time of 50 ms showing 2D NOE cross peaks (A) at a level deep enough to show the cross peaks between *ortho* CH$_3$ and *meta* H and between *meta* H and *para* CH$_3$ resonances and (B) on an expanded scale at a higher level which shows the cross peaks between particular *ortho* CH$_3$ and *meta* H and between particular *meta* H and *para* CH$_3$ resonances. Reprinted with permission from *J. Am. Chem. Soc.* **113**:8655 (1991). © 1991 American Chemical Society.

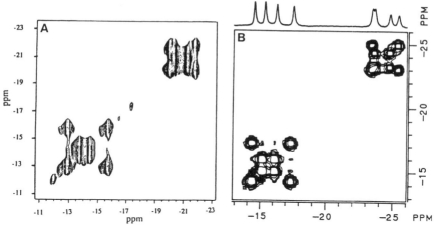

Figure 15. 2D NMR spectra of $[(o\text{-}CONR_2)_1\ TPPFe(NMeIm)_2]^+$, where the amide is derived from 3-azabio cyclo[3.2.2] nonane, in $CDCl_3$. (A) ROESY spectrum recorded at 500 MHz at room temperature; (B) NOESY specrum recorded at 300 MHz at 0 °C.

which have been studied are the *bis*-pyridine iron(I) complex of an oxyporphyrin bearing an OR group in one of the *meso* positions (Sano *et al.*, 1981) and iron(I) complexes of tetraphenylporphyrin, etioporphyrin, and octaethylporphyrin (Hickman *et al.*, 1985). Iron(I) porphyrins synthesized by two-electron reduction of the corresponding Fe(III) porphyrins using either strong reducing reagents (Cohen *et al.*, 1972; Scheidt *et al.*, 1983) or by cathodic electrolysis (Lexa *et al.*, 1974) are of synthetic value for generating (σ-alkyl)iron porphyrins (Lexa *et al.*, 1981) or may be important for dioxygen activation (Welborn *et al.*, 1981).

Proton or deuterium NMR spectroscopy of the iron(I) species are of diagnostic value in determining the ground state of the metal ion, since characteristic positive isotropic shifts for certain resonances have been observed. For tetraphenylporphinato iron(I) the deuterium pyrrole signal is observed at about +29 ppm; the deuterium methine signal of etioporphinato iron(I) and octaethylporphinato iron(I), at +14.5 ppm (Hickman *et al.*, 1985). Dipolar NMR shifts are relatively small as judged by the downfield phenyl deuterium hyperfine shifts of no more than 2 ppm. Hence the pyrrole deuterium atom experiences predominantly a contact shift. As discussed earlier, these positive isotropic shifts are associated with unpaired spin delocalization predominantly through σ-based molecular orbitals and imply that either the $d_{x^2-y^2}$ or the d_{z^2} orbital is singly populated. Based upon magnetic measurements and EPR studies of these complexes with $g_\perp = 2.3$ and $g_\parallel = 1.93$ for [TPPFe]$^-$ and $g_\perp = 2.24$ and $g_\parallel = 1.92$ for [OEPFe]$^-$ respectively, and $g_\perp > g_\parallel$ (Hickman *et al.*, 1985), a spin state of $S = \frac{1}{2}$ is widely accepted, with the unpaired electron residing in the d_{z^2} orbital, by analogy to Co(II) porphyrins (Walker, 1970). The ground state for the low-spin Fe(I) porphyrin complex is hence believed to be $(d_{xy})^2(d_{xz}, d_{yz})^4(d_{z^2})^1$. For this ground state, and again, by analogy to Co(II) porphyrins (La Mar and Walker, 1979, and references therein), it

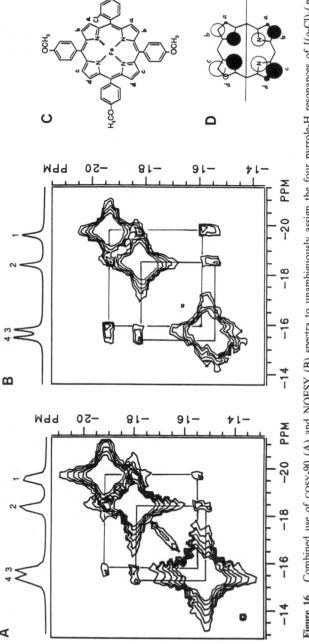

Figure 16. Combined use of COSY-90 (A) and NOESY (B) spectra to unambiguously assign the four pyrrole-H resonances of $[(o\text{-}Cl)_1(p\text{-}OCH_3)_3TPPFe(NMeIm)_2]^+$. (C) The four types of pyrrole-H are labeled a–d. The cross peak between resonances 1 and 2 in the NOESY spectrum requires that resonances 1 and 2 be due to protons b and c or vice versa. Combining this information with the 1, 3 and 2, 4 cross peaks in (A) and (B) suggests that $4 = a$, $3 = d$, $2 = b$, and $1 = c$, as summarized in (D).

is expected that a part of the positive isotropic shift of the protons at both pyrrole and *meso* positions may also be due to some dipolar shift proportional to $(g_{\parallel}^2 - g_{\perp}^2)$.

2.2. Low-Spin (Diamagnetic) Fe(II) Porphyrins

Ferrous porphyrin complexes are biologically relevant since they are active-site analogs of the oxygen transport and storage proteins, hemoglobin and myoglobin. Even though the iron(II) systems provide the closest analogy to the heme proteins, the NMR studies of these complexes are still rather limited, largely because synthetic iron(II) porphyrins are easily oxidized and proper care is needed in handling the Fe(II) compounds (Dugad and Mitra, 1984).

In order to probe structure–function relationships in the dioxygen complexes of hemoglobin and myoglobin, a variety of 5- and 6-coordinated diamagnetic iron(II) porphyrin complexes, especially dioxygen and carbon monoxide adducts, have been intensively studied (Traylor, 1981; Jones *et al.*, 1979; Suslick and Reinert, 1985; Morgan and Dolphin, 1987). We do not attempt in this chapter to give a detailed overview of all NMR studies performed for low-spin diamagnetic complexes. Only a few representative examples will be considered. The resonance shifts observed in the NMR spectra of the diamagnetic complexes can be used as a convenient reference for determining the magnitude and the direction of the isotopic shifts in structurally related paramagnetic complexes (La Mar and Walker, 1979).

2.2.1. Six-Coordinate Diamagnetic Complexes

Proton NMR spectra of 6-coordinated ferrous porphyrin complexes show proton resonances in the diamagnetic region. This can be clearly seen in Fig. 17, which depicts the proton NMR spectrum of the *bis*-piperidine ferrous complex of tetrakis(2,4,6-tri-methoxyphenylporphyrin) (Latos-Grazynzki *et al.*, 1982). Chemical shift values for this and other diamagnetic, 6-coordinated tetraphenylporphinato Fe(II) complexes are listed in Table 3. The pyrrole proton signals are shifted to high frequency by the large ring-current effect of the porphyrin macrocycle and are observed at or near 9 ppm. The phenyl-proton resonances in this and other complexes are found at 7 to 8 ppm, whereas the resonances of the coordinated piperidine ligands are shifted to low frequency as a consequence of the large ring current of the macrocycle. Similar shifts are found in the NMR spectrum of the dioxygen and carbon monoxide adducts of another Fe(II) complex containing a protected porphyrin ligand (Mispelter *et al.*, 1983). In addition to CO or O_2, a pyridine ligand which is part of a chain bridging two opposite phenyl rings is coordinated to the central metal ion. The spectra of the CO and O_2 adducts exhibit identical features and, again, reveal shifts characteristic for diamagnetic porphyrins: pyrrole proton resonances at 8.5 ppm, phenyl-H signals at 7.2 ppm, and upfield chemical shifts for the protons of the bridging chain. The pyrrole protons of the CO adduct appear at a slightly higher frequency (8.7 ppm) than those of the dioxygen complex (8.3 ppm), consistent with the π-acceptor capabilities of the axial CO ligand (Mispelter *et al.*, 1983).

Other diamagnetic Fe(II) complexes are derived from chelated protohemes, in which a coordinating imidazole ligand is provided by one of the heme substituents and CO is the sixth ligand (Traylor *et al.*, 1979). The NMR spectrum of such a protoheme

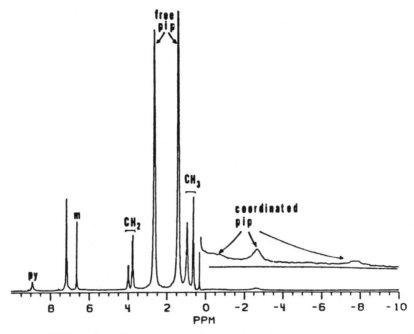

Figure 17. 360-MHz ¹H NMR spectrum of a solution of [T(2,4,6-EtO)₃PP]Fe(II)(Pip)₂ in benzene-d_6 obtained by allowing [T(2,4,6-EtO)₃PP]Fe(III)Cl to react with excess piperidine under anaerobic conditions. Reprinted with permission from *J. Am. Chem. Soc.*, **104**:5995 (1982). © 1982 American Chemical Society.

complex is shown in Fig. 18 together with the signal assignment for all protons. The porphyrin ring current affects the signals of this complex again in a characteristic way: *meso*-protons are deshielded (9.7 ppm), vinyl protons are found at 6 to 8 ppm, and side-chain protons are strongly shielded. For the *bis*-pyridine complex of protoporphyrins IX encapsulated in a cetyltrimethylammonium bromide micelle, analogous chemical shift values, listed in Table 3, are obtained (Medhi *et al.*, 1989).

2.2.2. Five-Coordinate Diamagnetic Complexes

Chemical shift values of 5-coordinated complexes are very similar to those observed for the hexacoordinated compounds. Examples for pentacoordinated species are iron(II) alkyl complexes of *meso*-tetraphenylporphyrin, such as [TPPFe(II)CH₂CH₂CH₃] and [TPPFe(II)CH₂CH₃]⁻ (Balch *et al.*, 1990d) and the mono-dioxygen or mono-(carbon monoxide) complexes of *meso*-tetrakis(trimethoxyphenylporphyrin), [T(2,4,6-(MeO)₃)PPFeO₂], [T(2,4,6-(EtO)₃)PPFeO₂], or [T(3,4,5-(MeO)₂)PPFeO₂, or of *meso*-tetrakis (α,α,α,α-pivalamidophenylporphyrin), [(TpivPP)FeO₂] (Latos-Grazynski *et al.*, 1982). The pyrrole protons of these complexes are found at 8–9 ppm, with the pyrrole protons of the iron(II) alkyl complexes less deshielded than those of the CO or

TABLE 3
¹H Chemical Shifts of 5- and 6-Coordinate Diamagnetic Low-Spin Fe(II) Complexes[a]

Compound	Temperature °C	Solvent	Chemical shifts, ppm						
			Pyrrole	Ortho	Meta	Para	Alkyl CH₂, CH₃	Ring Me	Meso-H
Five-coordinate complexes									
(T$_{piv}$PP)FeO₂	−76	toluene-d_8	8.2	0.44	9.69				
T(2,4,6-MeO)₃PPFeO₂	−75	CD₂Cl₂	9.0	3.6	6.6	4.1			
T(2,4,6-EtO)₃PPFeO₂	−75	CD₂Cl₂	9.1	3.9a, 0.7b	6.5	4.2a, 1.5b			
T(3,4,5-MeO)₃PPFeO₂	−75	toluene-d_8	9.8	7.4	3.3	4.2			
[TPPFeCH₂CH₂CH₃]⁻	23	benzene-d_6	8.1		7.0–7.5		6.0α, −0.5 −1.5β		
[TPPFeCH₂CH₃]⁻	23		8.1						
T(2,4,6-EtO)₃FeCO	25	toluene-d_8	8.61	3.87b, 0.56a	7.0–7.5	4.28b, 1.58a	−2.25α, −6.19		
Six-coordinate complexes									
T(2,4,6-EtO)₃PPFe(CO)₂	25	toluene-d_8	8.77	3.88b, 0.55a	6.55	4.29b, 1.58a			
T(2,4,6-MeO)₃PPFe(CO)₂	25	toluene-d_8	9.21	7.53	3.5	4.17			
Fe(PP)Py₂	25	CTABd						2.6–3.1c	9.4, 9.56
TPPFe-(N-MeIm)₂	−21	CD₂Cl₂	8.33	8.02B	7.64B	7.64B			
(3-MeTPP)Fe-(N-MeIm)₂	−21	CD₂Cl₂	8.41	7.83	7.53(H) 2.53(CH₃)	7.46			
(4-MeTPP)Fe-(N-MeIm)₂	−21	CD₂Cl₂	8.41	7.92	7.49C	2.61(CH₃)			
(4-OMeTPP)Fe-(N-MeIm)₂	−21	CD₂Cl₂	8.44	7.95	7.17	4.01(CH₃)			
(2,4,6-Me₃TPP)Fe-(N-MeIm)₂	−21	CD₂Cl₂	8.08	1.84(CH₃)	7.17	2.51(CH₃)			
TPPFe-(1-n-BuIm)₂	−21	CD₂Cl₂	8.40	8.01	7.64	7.64			
TPPFe-(1-t-Bu-5-MeIm)₂	−21	CD₂Cl₂	8.37	8.00A	7.67B	7.67B			
(3-MeTPP)Fe-(5-MeIm)₂	−21	CD₂Cl₂	8.32	7.77	7.58(H) 2.51(CH₃)	7.51			

[a] Data taken from Latos-Grazynski et al., (1982), Shirazi et al., (1985). Reprinted with permission from J. Am. Chem. Soc. **104**:5994 (1982); Inorg. Chem. **24**:2498 (1985). © 1982 and 1985 American Chemical Society. The following abbreviations are used: a, methyl groups, b, methylene groups, c, CH₃ masked by solvent and micelle protons; d, cetyltrimethylammonium bromide; α, α-CH₂; β, β-CH₂; A, 2-equivalent axial ligands; B, center of multiplet; C, presumably a doublet with upfield half under 2-H of free N-MeIm; T$_{piv}$PP, meso-tetrakis(α,α,α,α-pivalamidophenyl)porphyrin; TPP, meso-tetraphenylporphyrin; MeO, methoxy; EtO, ethoxy; Me, methyl; N-MeIm, N-methylimidazole; 1-n-BuIm, N-butylimidazole.

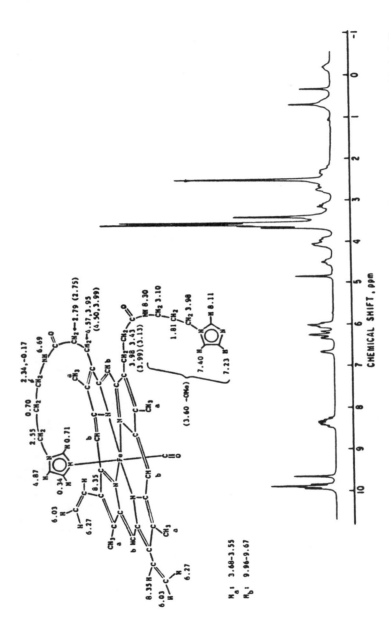

Figure 18. 220-MHz spectrum of the CO-adduct of a chelated protoheme, 0.03 M in DMSO-d_6 under 1 atm of CO, including chemical shift assignments for all protons. Reprinted with permission from *J. Am. Chem. Soc.* **101**:6719 (1979). © 1979 American Chemical Society.

O_2 adducts. Phenyl proton resonances of the TPP moiety are located at 7 to 7.5 ppm, whereas resonances for the N-alkyl hydrogens are shielded by the ring-current effect of the macrocycle. Chemical shift values of a variety of 5-coordinated ferrous porphyrin complexes are given in Table 3.

2.3. Intermediate-Spin Fe(II) Porphyrins

Four-coordinate ferrous porphyrins with an intermediate spin state ($S = 1$) have been studied by various physical methods, but assignment of the ground state for the metal ion has proved difficult. Three possible ground states, each consistent with certain physical properties, have been proposed: $^3A_{2g}(d_{xy})^2(d_{z^2})^2(d_{xz}, d_{yz})^2$, $^3Eg A(d_{xy})^2(d_{z^2})^1(d_{xz}, d_{yz})^3$, and $^3B_{2g}(d_{xy})^1(d_{z^2})^1(d_{xz}, d_{yz})^4$ (McGarvey, 1988). Most proton NMR studies performed to date for a variety of planar, 4-coordinate Fe(II) complexes containing a synthetic porphyrin macrocycle favor the $^3A_{2g}$ ground state with the electronic configuration $(d_{xy})^2(d_{z^2})^2(d_{xz}, d_{yz})^2$, as will be discussed below (Goff and La Mar, 1977; Migita and La Mar, 1980; Latos-Grazynski et al., 1982; Strauss et al., 1985; McGarvey, 1988).

2.3.1. Synthetic Fe(II) Porphyrins

The unligated complexes, synthesized in situ in benzene-d_6, toluene-d_8 (Goff and La Mar, 1977; Migita and La Mar, 1980; Strauss et al., 1985) or in dichloromethane-d_2 (Mispelter et al., 1977; Latos-Grazynski et al., 1982) by reduction of the corresponding Fe(III) porphyrins using either the chromous aqueous dithionite or the zinc amalgam reduction technique (Goff and La Mar, 1977; Mispelter et al., 1977; Migita and La Mar, 1980; Latos-Grazynski, 1982), show common features in their proton NMR spectra. The ^1H NMR spectra of [TPPFe] and the iron(II) complexes of deuteroporphyrin IX and protoporphyrin IX are shown in Figs. 19 and 20, respectively. For the synthetic complexes, large isotropic shifts are observed, as summarized in Table 4. These shifts provide information about the magnetic anisotropy and the mechanism of spin delocalization in these planar complexes.

The isotropic shifts of the Fe(II) complexes, in which the porphyrin is derived from meso-tetraphenylporphyrin (TPP), meso-substituted tetraarylporphyrins (RTPP) and octaethylporphyrin (OEP), have been analyzed in detail by Goff and La Mar (1977) and by La Mar and Walker (1979), and can be summarized as follows: With the exception of the signals for the pyrrole protons, all proton resonances are significantly shifted to higher frequencies, as seen in Figs. 19 and 20. The dipolar and contact contributions to the isotropic shifts for some of the synthetic Fe(II) complexes are listed in Table 5. Such a bias of shifts in one direction indicates that at least a major portion of the isotropic shifts originates in the dipolar interaction, a result expected, since the 4-coordinated Fe(II) molecules are highly magnetically anisotropic (Goff and La Mar, 1977).

For the synthetic tetraarylporphinato iron(II) derivatives the phenyl-H resonances are essentially totally dipolar in nature. These dominant dipolar shifts result from large magnetic anisotropies, $\chi_\parallel > \chi_\perp$. Temperature-dependent studies of these resonances show that the phenyl o-H and m-H dipolar shifts follow the Curie law with zero intercepts at $T^{-1} = 0$.

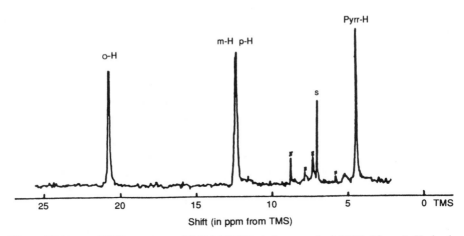

Figure 19. Proton NMR spectrum of 0.25-M TPPFe in benzene-d_6 at 25 °C. The m/p-H clearly split at low temperature to yield the resolved resonances. The o-H exhibits an 8-Hz doublet due to the m-H. S = solvent, C_6D_5H, and X = impurity. Reprinted with permission from *J. Am. Chem. Soc.* **99**:3642 (1977). © 1977 American Chemical Society.

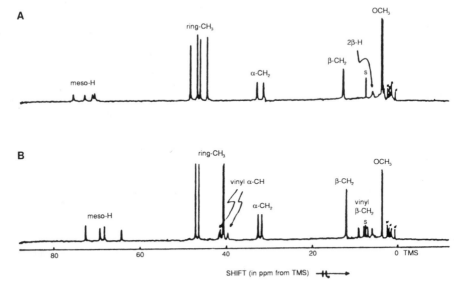

Figure 20. Proton NMR spectra of a 1.0-M benzene-d_6 solution of (A) deuteroporphyrin Fe(II), (B) (protoporphyrin IX) Fe(II) at 25 °C. S = solvent. Reprinted with permission from *J. Am. Chem. Soc.* **99**:3642 (1977). © 1977 American Chemical Society.

TABLE 4
Observed Isotropic Shifts of 4-Coordinate, Intermediate-Spin Ferrous Complexes of Synthetic and Natural Porphyrins[a]

Chemical shifts (ppm)

Synthetic porphyrins	TPP	m-CH$_3$TPP	p-CH$_3$TPP	p-OCH$_3$TPP	p-ClTPP	OEP	T-n-PrP	T$_{piv}$-TPP	basket-handle porphyrin		
									C$_{10}$	C$_{12isomer}$	C$_{12isomer}$
pyrrole-H	-4.1	-4.4	-3.9	-3.9	-3.4		-4.4	-4.8	-5.3	-3.2 / -4.2	-5.1
pyrrole-α-CH$_2$						29.7					
pyrrole-β-CH$_2$						11.0					
meso-H						65.7					
meso-α-CH$_2$							12.0				
β-CH$_2$							12.3				
γ-CH$_3$							8.4				
o-H	12.8	13.3	13.2	13.6	12.2			11.4	16.5	11.8	11.0
m-H, m^1-H	4.8	5.1	4.9	4.7	4.6			4.5, 4.6	6.6, 4.6	5.0, 4.8	4.2, 4.2
p-H	4.8	5.1						5.0	5.9	5.2	4.5
m-CH$_3$		3.5									
p-CH$_3$			3.0								
p-OCH$_3$				2.5							

Chemical shifts (ppm)

Natural porphyrins	MPFe	DPFe	PPFe	Br$_2$DPFe	Ac$_2$DPFe	Deuteroporphyrin
meso-H (average)	64.1	62.0	59.3	51.1	51.0	57.9
CH$_3$,1,3,5,8 (average)	44.3	42.5	40.2	39.0	33.4	41.6
6,7 α-CH$_2$	30.5, 30.1	29.2, 27.8	29.1, 28.3	27.5, 27.5	26.1, 23.8	27.4, 26.4
β-CH$_2$	9.8, 9.6	9.1, 9.1	8.7, 8.7	7.2, 6.9	7.2, 6.5	8.2
α-CH$_3$	0	0	0	0.7	0.4	
2,4 R$_2$	30.2, 28.4; 11.2, 10.4	3.0, 3.0	33.1, 31.5; 2.5, 1.3; 0.5, 0.5		9.2, 11.2	

[a] Data taken from Goff et al. (1977); Mispelter et al. (1977, 1980). Reprinted with permission from J. Am. Chem. Soc. **99**:3643 (1977); J. Chem. Phys. **72**:1005 (1980). © 1977 American Chemical Society and © 1980 American Institute of Physics, respectively. MP, meso-porphyrin; DP, deuteroporphyrin; PP, protoporphyrin IX; Br$_2$DP, 2,4-dibromodeuteroporphyrin; Ac$_2$DP, 2,4-diacetyldeuteroporphyrin.

TABLE 5
Dipolar and Contact Contribution to the Isotropic Shifts at 25 °C of 4-Coordinate, Intermediate-Spin Ferrous Porphyrins[a]

Porphyrin	Resonance	Isotropic shift	Dipolar shift	Contact shift
RTPPFe	o-H	12.8	11.2	−1.6
	m-H	4.8	5.0	0
	p-H	4.8	4.7	0
	m-CH$_3$	3.5	3.3	0
	p-CH$_3$	3.0	3.4	0
	p-OCH$_3$	2.5	2.8	0
	pyrrole-H	−4.1	21.8	−25.9
OEPFe	pyrrole-α-CH$_2$	29.7	13.2	16.5
OEPFe	$meso$-H	65.7	34.1	31.6

[a] Data taken from Goff et al. (1977). Reprinted with permission from J. Am. Chem. Soc. **99**:3644 (1977). © 1977 American Chemical Society.

The isotropic shifts for the pyrrole protons of these complexes are found to be small and negative (Table 4), suggesting that these shifts arise from both contact and dipolar contributions having opposite signs. The negative pyrrole-H isotropic shift is associated with the absense of spin in the $d_{x^2-y^2}$ orbital and at least one unpaired electron in the d_{xz} or d_{yz} orbital (La Mar and Walker, 1979). The Curie plot of the pyrrole-H exhibits some curvature, possibly arising from the contact contribution, which increases faster than T^{-1} as the temperature is lowered (Goff and La Mar, 1977). Contact shifts for the synthetic porphyrin complexes show that protons and α-CH$_2$ groups exhibit comparable shifts, but with opposite signs at the pyrrole and $meso$-H positions, as is also clearly shown in Table 5. This indicates that the contact shifts are dominated by π spin delocalization with positive spin density at the pyrrole positions and negative spin density at the $meso$ position. Sigma spin transfer would lead to contact shifts of the same sign for protons and α-CH$_2$ groups, contrary to what is found experimentally. The contact shifts of the pyrrole substituents reflect extensive π spin delocalization through P $\rightarrow$ Fe charge transfer with no evidence for Fe $\rightarrow$ P π back-bonding. The large π contact shift requires unpaired spin in d_{xz} and d_{yz} and strongly supports the $^3A_{2g}$ ground state configuration, $(d_{xy})^2(d_{z^2})^2(d_{xz}, d_{yz})^2$ (Goff and La Mar, 1977).

Latos-Grazynski et al. (1982) have drawn similar conclusions for the NMR spectra of the unligated Fe(II) complexes [2,4,6-(OMe)$_3$]$_4$TPP]Fe(II), [2,4,6-(OEt)$_3$]$_4$TPPFe(II), [3,4,5-(OMe)$_3$]$_4$TPPFe(II), and (TpivPP)Fe(II). The pattern of isotropic resonance shifts and their temperature dependencies, especially the nonzero intercept for the pyrrole protons, observed in these complexes are consistent in all cases with the presence of a planar, 4-coordinate iron(II) atom, $S = 1$.

Mispelter et al., (1977, 1980) have also reported proton NMR studies of ferrous ($S = 1$) complexes, including the so-called picket fence and basket handle porphyrins. The contact shift pattern of each compound indicates π delocalization, but very large anomalies are found for the temperature dependence of some resonances of the complexes studied, especially for the resonances of the basket handle porphyrin complexes. The anomalous temperature dependence of the dipolar and contact contributions to the isotropic shifts were interpreted in terms of a model that assumes that the gound state of these complexes results from stong mixing by spin–orbit coupling of two triplet states, $^3A_{2g}$ and ^3E_g, which are very close in energy. Following this model, the nature of the ground state depends critically on the energy of the d_{z^2} orbital relative to that of the degenerate d_{xz} and d_{yz} orbitals. However, for most porphyrin derivatives studied, the ground state was suggested to be $^3A_{2g}$, which is perturbed by the close-lying ^3E_g excited state.

Strauss and co-workers have reported NMR studies of OEPFe(II) and the chlorin complex OECFe(II) in toluene as a function of temperature (Strauss et al., 1985). In the solid state, the chlorin macrocycle is significantly S_4 ruffled. Large isotropic shifts with significant dipolar contributions were observed for OEPFe(II), which were analyzed as described by Goff and La Mar (1977). NMR investigations of the complex as a function of temperature revealed strict Curie behavior with intercepts near zero, suggesting that this complex is not in thermal equilibrium with other spin states, in sharp contrast to the data obtained by Mispelter et al. (1977). However, the isotropic shifts of OECFe(II), which has C_2 symmetry, could be fit to a straight line only with large nonzero intercepts, suggesting that the removal of the fourfold symmetry produces non-Curie behavior (Strauss et al., 1985). The pattern of observed positive and negative isotropic shifts strongly suggests that protons on symmetry-equivalent pyrrole rings experience isotropic shifts that move to higher frequency with decreasing temperature and have $1/T = 0$ intercepts to lower frequency of the diamagnetic value, and that protons on the unique pyrrole ring and on the pyrroline ring have isotropic shifts that move to lower frequency with decreasing temperature and have $1/T = 0$ intercepts to higher frequency than the diamagnetic value. This is consistent with in-plane magnetic anisotropy and a resultant rhombic component to the isotropic shifts. The unusual intercepts observed for the OECFe(II) complex are believed to result from a large temperature-independent paramagnetism (Strauss et al., 1985).

These results are nicely confirmed by McGarvey (1988), who developed a theory to explain the dipolar and contact shifts for 4-coordinated ferrous complexes for the general case of nonaxial symmetry. When the theory was applied to temperature studies of TTPFe(II), OEPFe(II), and OECFe(II), it was found that NMR and susceptibility data clearly support the $^3A_{2g}$ ground state for all intermediate-spin Fe(II) porphyrins, which is spin–orbit coupled with the ^3E_g state at a particular energy (600 cm^{-1} for [OECFe] and [OEPFe] and 1000 cm^{-1} for [TPPFe]). The theory successfully explains the temperature dependence of the dipolar shifts, but is less satisfactory for the interpretation of the contact shifts, particularly for the β-pyrrole positions, when the macrocycle is ruffled, as in the chlorin complex (McGarvey, 1988).

2.3.2. Natural Porphyrins

Studies of iron(II) complexes of natural porphyrins in aqueous solution face several difficulties. First, like the synthetic hemes, the natural hemes are very rapidly oxidized in air. Second, the hemes are ordinarily difficult to solubilize in aqueous media under conditions of pH and ionic strength similar to those in heme proteins. Third, the hemes are known to undergo aggregation in aqueous solutions (Medhi et al., 1989).

Like their synthetic counterparts, the iron(II) complexes of the natural porphyrins, such as the diisopropyl esters of protoporphyrin IX iron(II) [PPDiPEFe], deuteroporphyrin IX iron(II) [DPDiPEFe], and diacetyldeuteroporphyrin IX iron(II) [ADPDiP-EFe], as well as the dimethyl ester of mesoporphyrin IX iron(II) [MPDMEFe], also exhibit large isotropic shifts at low concentrations, as can be seen clearly in Table 4. Resonance assignment was possible solely on the basis of relative signal intensities and multiplet structure and by comparison with the synthetic analogs (Goff and La Mar, 1977). In contrast to the synthetic Fe(II) porphyrins, the natural porphyrins exhibit proton NMR spectra that are very sensitive to concentration changes. Substantial concentration dependence of the isotropic shifts is observed for most resonances of functional groups close to the porphyrin ring, with all signals broadened and shifted to lower frequency (Goff and La Mar, 1977; Migita and La Mar, 1980). These positive isotropic shifts and resonance broadening are believed to arise from intermolecular ring currents, dipolar shifts, and dipolar relaxation due to aggregation. A trend, more or less strongly pronounced, is found for the concentration dependence of both the *meso*-H and the ring methyl resonances of the complexes. The shifts to lower frequency level off for the *meso*-H peaks $\alpha-\gamma$ and methyls 5 and 8, whereas the *meso*-H signal δ and the CH_3 groups 1 and 3 continue to move to lower frequency. With respect to the differential line broadening induced by the intermolecular paramagnetic relaxation, it was found that at low concentration the methyl and *meso*-H linewidths increase only slightly in the region where the CH_3 and *meso*-H shifts are changing fastest. However, at higher concentration, where the shift dependence is less pronounced, the linewidth increases significantly. Hence, both the different concentration dependence of the intermolecular paramagnetic shifts and consideration of the line broadening induced by the paramagnetic relaxation provide direct evidence for the existence of two structurally different aggregates. At low concentrations, isotropic shifts and linewidths are consistent with parallel face-to-face dimers. At higher concentration, the shifts and relaxation are diagnostic of parallel stacking of the porphyrin planes (Migita and La Mar, 1980).

To overcome the difficulties associated with the studies of natural porphyrins in aqueous solution, Medhi *et al.*, (1989) have recently reported the investigation of PPFe(II) and DPFe(II) complexes inside aqueous cetyltrimethylammonium bromide (CTAB) detergent micelles. NMR studies of the planar, 4-coordinate complexes showed that the porphyrin ring methyl proton resonances have positive isotropic shifts, suggesting that the shifts are dominated by dipolar contributions. The isotropic shifts of the ring methyl protons closely obey the Curie law over the temperature range investigated (Medhi *et al.*, 1989). The ^1H NMR spectra of the Fe(II) complexes inside the micelle showed two important differences from those studied in benzene. First, the porphyrin proton resonances are much broader in the micelle due to increased rotational correlation times. Second, the position and spread of the heme methyl signals are somewhat larger in the micellar cavity than in the organic solvent. The increased methyl resonance spread may be a result of an increase in asymmetry in the porphyrin ring because of the difference in the hydrophobic interaction of the vinyl end of the ring compared to the propionic acid side chain of the heme (Medhi *et al.*, 1989).

2.4. High-Spin Fe(II) Porphyrins

2.4.1. Five-Coordinate Fe(II) Porphyrins

Five-coordinate, high-spin ferrous porphyrin complexes with imidazoles or pyridines as axial ligands are of specific interest, since they not only represent direct models of the oxygen transport and storage proteins deoxyhemoglobin and deoxymyoglobin but are also are excellent models of the reduced states of cytochrome P-450 and chloro-peroxidase. The NMR investigations of these models will be described below.

   2.4.1a. Models of Oxyhemoglobin and Oxymyoglobin. High-spin ferrous complexes can be obtained by addition of an excess of a methyl-α-substituted imidazole or pyridine to the corresponding 4-coordinated intermediate-spin Fe(II) derivatives (Goff and La Mar, 1977). A variety of models obtained in this way involve synthetic porphyrins, such as substituted and unsubstituted tetraarylporphyrins or octaethylporphyrins, as well as natural porphyrins. The addition of axial ligands affects the NMR spectra of the 4-coordinated species of the synthetic porphyrins in a distinctive way. All porphyrin resonances shift in characteristic directions; the isotropic shifts of the pyrrole-H in the TPP complexes are positive, and those of the phenyl protons are negative, with typical isotropic shifts of 43.5 ppm for the pyrrole-H and -1.0 to -1.5 ppm for the *o*-, *m*-, and *p*-phenyl protons. For 5-coordinate OEPFe(II) complexes, isotropic shifts of -7.0 ppm for the *meso*-H and 8.5 ppm for the pyrrole-α-CH$_2$ protons have been reported (Goff and La Mar, 1977). The *meso*-H protons in the natural porphyrins are shifted to lower frequency and appear in the spectral region of 0 to 6 ppm. The proton NMR spectrum of the 2-methylimidazole complex of deuteroporphyrin IX dimethyl ester iron(II) is depicted in Fig. 21. The degree of isotropic shifts in these complexes, especially the shifts for the pyrrole-H and pyrrole α-CH$_2$ protons, serve as convenient probes for both oxidation and spin state of the metal ion. The isotropic shifts of a variety of the natural

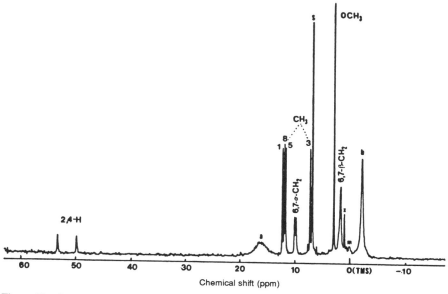

Figure 21. Proton NMR spectrum of the 2-methylimidazole complex of deuteroporphyrin IX dimethylester) iron(II) in benzene-d_6 at 25 °C. S = solvent peak; a,b-averaged axial ligand; 4,5-H and CH$_3$ peaks, respectively. Reprinted with permission from *J. Am. Chem. Soc.* **99**:6603 (1977). © 1977 American Chemical Society.

TABLE 6

Isotropic Shifts of 5-coordinate ($S = 2$) Synthetic and Natural Iron(II) Porphyrins with 2-Methylimidazole as Axial Ligand[a]

Position	TPP	m-CH$_3$TPP	p-CH$_3$TPP	OEP
Synthetic porphyrin[b]				
pyrrole-H	43.4	43.4	43.7	
pyrrole-α-CH$_2$				8.5
β-CH$_3$				−0.37
meso-H				−6.7
meso-phenyl o-H	−1.00	−1.07	−1.05	
meso-phenyl m-H	−0.70	−0.62	−0.67	
meso-phenyl p-H	−1.02	−1.03		
meso-phenyl-CH$_3$		−0.69	−0.37	

Position	MP	DP	PP	Br$_2$DP	Ac$_2$DP
Natural porphyrin[b,c]					
1,3,5,8-CH$_3$[d]	9.8	9.3	9.5	9.7	6.7
	9.6	9.0	8.8	9.7	6.7
	9.5	8.8	8.3	9.5	5.8
	5.6	4.3	3.6	3.5	−1.9
6,7 α-CH$_2$	7.3, 7.1	7.3, 7.0	7.6, 7.2	7.6, 7.4	6.2, 5.2
6,7 β-CH$_2$	−1.0	−1.1	−0.9	−0.9	−0.6
6,7 OCH$_3$	−0.4	−0.4	−0.4	−0.4	−0.3
2,4 groups	7.3, 7.1[e]	44.8, 41.3[f]	7.4, 3.6[g]		−1.2, −2.1[h]
	0, −0.6[i]		2.2, 0.9[j]		
			−1.4, −2.7[k]		

[a] Data taken from La Mar and Walker (1979).
[b] In C$_6$D$_6$ at 25 °C. Shifts in ppm, referenced to diamagnetic porphyrins.
[c] meso-H located in the region of 3 to 5 ppm from TMS not well resolved.
[d] The methyl resonances are given vertically in opposite order of their position on the porphyrin ring: The first is 8-CH$_3$, the second is 5-CH$_3$, etc. in each case.
[e] α-CH$_2$.
[f] Pyrrole-H.
[g] Vinyl-α-CH.
[h] Acetyl-CH$_3$.
[i] β-CH$_3$.
[j] Vinyl = β = CH(cis).
[k] Vinyl = β = CH($trans$).

and synthetic complexes are listed in Table 6 (Goff and La Mar, 1977; La Mar and Walker, 1979).

As mentioned earlier, meso-phenyl-H resonances yield characteristic isotropic shifts depending on whether the contact or the dipolar term dominates. The phenyl-H isotropic shifts of high-spin TTPFe(II) are small and negative (Table 6), suggesting that the high-spin complexes exhibit only very small magnetic anistropy (Goff and La Mar, 1977). Since the dipolar contribution to the isotropic shift is very small, the predominant contribution must be contact in origin. The pyrrole-H and α-CH$_2$ isotropic shifts are positive (to higher frequency) (Table 6), indicating primarily a σ spin delocalization mechanism. The small negative meso-H shifts suggest only moderate π delocalization to the meso position. This contact shift pattern, with dominance of σ spin transfer to the pyrrole positions, is characteristic of an unpaired electron in the $d_{x^2-y^2}$ orbital, leading to a ground state of $(d_{xy})^2(d_{xz}, d_{yz})^2(d_{z^2})^1(d_{x^2-y^2})^1$ electron configuration for the ferrous ion. Moderate π delocalization to the *meso* position is indicative of partially filled d_π orbitals, consistent with the ground state proposed.

Other models of deoxyhemoglobin and deoxymyoglobin involve so-called tailed picket fence porphyrins (Collman *et al.*, 1980) and pocket porphyrins (Collman *et al.*, 1983), in which the axial imidazole ligand that coordinates to the metal ion is covalently linked to the porphyrin. Based on the large positive isotropic shifts observed for the pyrrole protons, it was concluded that the pocket porphyrin complexes, which form stable and reversible dioxygen adducts, remain predominantly 5-coordinate in the absence of gaseous ligands. For the tailed picket fence porphinato-iron(II) complexes, four pyrrole-H peaks were observed in the ^{1}H NMR spectra, centered at a chemical shift of 56 ppm (isotropic shift = 47.5 ppm) (Collman *et al.*, 1980, 1983). In comparison with the isotropic shifts of other monoligated TPPFe(II) complexes (Goff and La Mar, 1977), this isotropic shift clearly indicates that the iron in this complex is 5-coordinate and high-spin. The pyrrole-H peaks shift with temperature according to the Curie law, and at $-25\,°C$ new sets of resonances between 10 and -10 ppm appear. These are attributed to the disproportionation of two molecules of the 5-coordinate, high-spin Fe(II) complex to yield, intermolecular sharing of axial ligands, one 4-coordinate, intermediate-spin Fe(II) complex and one diamagnetic, 6-coordinate, low-spin Fe(II) complex (Collman *et al.*, 1980).

Another complex, in which an axial pyridine ligand is inserted in the bridge between two opposite phenyl rings (a so-called tailed basket handle porphyrin) is described by Mispelter *et al.* (1981). NMR investigations of this complex showed that only two spin-coupled resonances could be assigned to the pyrrole protons resulting from the four equivalent pyrrole rings. In agreement with observations made for the synthetic analogs discussed by Goff and La Mar (1977), it was found that the isotropic shifts were mainly contact in nature (Mispelter *et al.*, 1981). The temperature dependencies of the pyrrole-H isotropic shifts demonstrate distinct deviations from Curie behavior. This anomalous dependence of the pyrrole shifts is explained with an electronic model based on the fact that the pyridine moiety imposes an in-plane rhombic distortion causing the splitting of the ^5E term. Each of the new terms has its own spin distribution pattern into the porphyrin ligand, and hence the pyrrole-proton shift dependencies should reflect the thermal equilibrium between the two levels (Mispelter *et al.*, 1981). An alternative explanation for the temperature dependence has been offered by Latos-Grazynski (1983) based on a purely conformational model which includes the influence of the axial ligand on the spin density distribution at the pyrrole positions.

Additional models involving natural porphyrins have been reported recently (Medhi *et al.*, 1989). The proton NMR spectra of the pentacoordinated 2-methylimidazole ferrous complexes of protoporphyrin IX and deuteroporphyrin IX in aqueous cetyltrimethylammonium bromide (CTAB) solution closely resemble that of deoxymyoglobin. The isotropic shifts of the methyl protons for the two complexes inside the micelle show linear temperature dependencies over the range studied, indicating that the shifts obey the Curie law strictly. Based on similarities found for the isotropic shift pattern of the high-spin complex in aqueous solution and that of the complex studied in benzene solution (Goff and La Mar, 1977), it is assumed that the contact and dipolar contributions to the isotropic shifts are very similar in the two solvents and that the isotropic shifts are mainly influenced by the contact term. This further implies that the dipolar contribution to the isotropic shift and hence the magnetic anistropy is also very small. The small positive shifts of the CH_2 group (21 ppm) and the large positive pyrrole-H shifts (45.8 ppm and 43.7 ppm) in the deuteroporphyrin complex are consistent with this conclusion and are characteristic of primarily σ-spin transfer (Medhi et al., 1989).

For a 5-coordinated heme c undecapeptide complex embedded in sodium dodecyl-sulfate (SDS) micelles, the heme proton resonances are observed in the spectral region of 20 to 30 ppm. By analogy to similar ferrous complexes of natural heme derivatives, the paramagnetic shifts in the heme c undecapeptide complex is expected to be predominantly contact in origin, with σ spin delocalization through the antibonding $d_{x^2-y^2}$ metal orbital (Mazumdar et al., 1991).

2.4.1b. Models for the Reduced States of Cytochrome P-450 and Chloroperoxidase. There are two different model complexes for the reduced states of cytochrome P-450 and chloroperoxidase. Paramagnetic alkyl and aryl mercaptide (thiolate) complexes of both synthetic and natural heme derivative iron(II) porphyrins are very useful models, since their coordination geometry resembles that of the enzymes. The iron atom in these models is bound to the four nitrogen atoms of the porphyrin macrocycle as well as to an axial mercaptide or thiolate ligand, which is represented by cysteinate in the enzymes. The model complexes are especially valuable, since the effect of mercaptide coordination on the electronic structure and spin delocalization mechanism can be determined by NMR spectroscopy (Lukat and Goff, 1990; Yu and Goff, 1989; Parmely and Goff, 1980). The first structural indication that pentacoordinated high-spin iron(II) mercaptide mesoporphyrin derivatives could indeed be active site analogs for the reduced state of cytochrome P-450 was given by Caron et al. (1979).

Other models for cytochrome P-450 heme alkylation involve N-substituted porphyrins, such as N-alkyl and N-aryl porphyrins, which are known to be formed as inactivation products of hepatic cytochrome P-450 by a variety of molecules, including simple alkenes such as ethylene (Balch et al., 1985; Komives et al., 1988; Battioni et al., 1987).

i. Mercaptide Complexes. The 5-coordinated mercaptide Fe(II) complexes, in which the porphyrin is derived from substituted and unsubstituted tetraphenylporphyrins or from octaethylporphyrin, deuteroporphyrin, protoporphyrin IX dimethylester and etioporphyrin I and the mercaptide moiety from *n*-butyl-, *n*-propyl-, and ethyl mercaptide (BuS^-, PrS^-, EtS^-), are obtained by chromous acetylacetonate reduction of the corresponding iron(III) chloride complexes (Lukat and Goff 1990). Proton NMR spectroscopic investigations performed for these complexes have proven essential in the assignment of the cysteine resonances in the spectra of the enzyme P-450$_{cam}$. As can be seen in Fig. 22, the proton resonances of the CH_2 groups of the axial ligand reveal strongly hyperfine-shifted signals for both the enzyme and the model complexes. The α-CH_2 shifts for the mercaptide ligands of the model compounds are very large and appear at around +250 ppm, whereas the signals for the β-CH_2 protons are located at around -2 ppm for the TPP complexes and at -5 to -7 ppm for the natural derivatives (Lukat and Goff, 1990). Based on this assignment, the methylene protons for the axial cysteinate of ferrous cytochrome P-450 and ferrous peroxidase are found to resonate at 279 ppm and 200 ppm, respectively (Lukat and Goff, 1990). As summarized in Table 7 and shown in Fig. 22, the pyrrole protons of the TPP complexes have chemical shifts of 59 to 64 ppm, whereas the phenyl-H have fairly small shifts (*o*-H, 8–10 ppm; *m*-H,

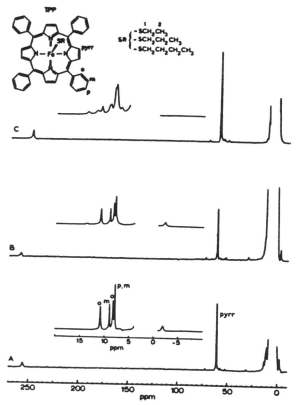

Figure 22. 300-MHz proton NMR spectrum of 2.0-mM [(TPP)FeSR]⁻ in DMSO-d_6 at 25 °C. (A) SR = n-butylmercaptide; (B) SR = n-propylmercaptide; (C) SR = ethylmercaptide. Reprinted with permission from H. M. Goff *Biochim. Biophys. Acta* **1037**:356 (1990). © 1990 Elsevier Science Publishers BV.

8–9 ppm; p-H, at 8 ppm). The ring CH₂ protons for the octaethylporphyrin and the etioporphyrin complexes resonate at around 11 and 15 ppm, whereas the *meso*-H chemical shift is −0.3 to −0.4 ppm (Lukat *et al.*, 1990; Parmely and Goff 1980).

Large positive isotropic shifts are observed for the pyrrole and ring methyl protons, as can be seen in Table 7. In order to determine what effects the coordination of the mercaptide has on the electronic structure and spin delocalization, attempts have been undertaken to analyze these isotropic shifts. Comparison of the isotropic shifts of the mercaptide complex [(TPP)Fe(BuS)]⁻ with those of the corresponding 2-methylimidazole TPP complex leads to the conclusion that the separation of contact and dipolar shifts using the method of locating a resonance insulated from contact spin density is not practical for the mercaptide complexes (Parmely and Goff, 1980). However, the comparison of the isotropic shifts of the mercaptide complexes with the corresponding 2-methylimidazole complex reveals a downfield bias in isotropic shift values for all

TABLE 7

Chemical Shifts for NMR Resonances of High-Spin Fe(II) Porphyrin Mercaptide Complexes at 25 °C in DMSO-d_6[a]

Isotropic shifts (ppm)

Porphyrin	Pyrrole-H	o-H	m-H	p-H	m-CH$_3$	p-CH$_3$	BuS: α-CH$_2$, β-CH$_2$
[(TPP)Fe-SBu]⁻	61.0	10.9, 8.2	8.9, 7.9	7.9			256[a],	−1.8
[(p-CH$_3$)TPPFe-SBu]⁻	61.2	10.7, 8.1	8.8, 8.7				NA	NA
[(m-CH$_3$)TPPFe-SBu]⁻	60.9, 60.3, 59.7	10.6, 7.9	8.8, 7.8	7.8	3.04, 2.5		NA	NA
[(3,4,5-OCH$_3$)TPPFe-SBu]⁻	66.3	10.4, 8.1			4.62, 3.91	4.40	NA	NA

Porphyrin	ring-CH$_2$	ring-CH$_3$	β-CH$_3$	meso-H			BuS: α-CH$_2$, β-CH$_2$
[(OEP)Fe-SBu]⁻	14.8, 10.8	3.6		−0.3			242, −5.9
[(EtIO)Fe-SBu]⁻	14.5, 10.6	10.2	3.4	−0.4			250, 241, −5.8, −7.2

Porphyrin	pyrrole-H	o-H	m-H	p-H	m-CH$_3$	p-CH$_3$	m-OCH$_3$	p-OCH$_3$
[R(TPP)Fe-SBu]⁻	52.3 (CH$_2$)	2.9, 0.2 (CH$_3$)	1.2, 0.2 (meso-H)	0.2	0.5, 0.0	0.8	0.7, 0.0	0.2
[(OEP)Fe-SBu]⁻	10.7, 6.7	1.7	−10.4					

[a] Data taken from Parmely and Goff (1980); Lukat and Goff (1990). Reprinted with permission of the publisher from *J. Inorg. Biochem.* **12**:272, 277 (1980); *Biochim. Biophys. Acta* **1037**:354 (1990). © 1980 and 1990 Elsevier Science Publishing Co., Inc.
[b] NA-not available.

resonances of the mercaptide species with the exception of the *meso*-H signal. These downfield shifts are believed to result from a change in sign of the dipolar shift term rather than from spin delocalization through σ-type molecular orbitals. The upfield *meso*-H isotropic shift can only be explained by π spin delocalization. Nevertheless, the isotropic shifts of at least 250 ppm for the methylene resonances of the coordinated mercaptide ligand support a very efficient spin delocalization for this ligand (Parmely and Goff, 1980).

ii. Hydroxo (fluoro) Complexes. In addition to mercaptide ligands, other anions such as hydroxide or fluoride can also bind to 4-coordinate ferrous tetraphenylporphyrin or etioporphyrin complexes. Characteristic ^{1}H NMR shifts reveal that the resulting fluoro or hydroxo complexes $[(TPP)Fe(OH)]^-$, $[(TPP)FeF]^-$, $[(Etio)Fe(OH)]^-$, and $[EtioFeF]^-$ are 5-coordinate and high-spin (Yu *et al.*, 1989; Shin *et al.*, 1987). The ^{1}H NMR spectrum of the hydroxo complex is shown in Fig. 23. The positive pyrrole-H shifts in the TPP complexes (32.8 and 30.3 ppm for the hydroxo and fluoro complexes, respectively) and the positive ring methyl and methylene shifts in the OEP complexes again indicate mainly σ spin delocalization (Yu and Goff, 1989; Shin *et al.*, 1987). The pyrrole-H resonances in $[(TPP)Fe(OH)]^-$ and in $[(TPP)FeF]^-$ are less downfield-shifted than those in other 5-coordinate complexes. This bias to smaller frequencies is not associated with a large dipolar shift contribution since the phenyl

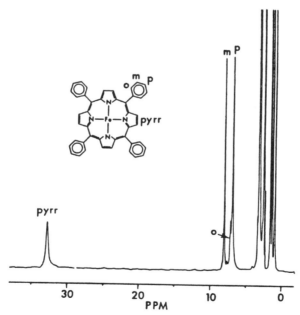

Figure 23. 360-MHz spectrum of $[(TPP)Fe(OH)]^-$ formed from reaction of 1-mM TPPFeCl in DMSO-d_6 with two equivalents of 1-M (TBA)OH in methanol under anerobic conditions. Reprinted with permission from *Inorg. Chem.* **26**:4104 (1987). © 1987 American Chemical Society.

signals show only small shifts from the diamagnetic values (Shin *et al.*, 1987), but rather reflects the importance of π spin delocalization in the porphyrin ring. Positive π spin density at the pyrrole carbon positions results in a partial cancelling of the positive isotropic shifts due to σ delocalization (Shin *et al.*, 1987).

The methyl and methylene proton signals in $[(\text{Etio})\text{Fe}(\text{OH})]^-$ and $[\text{EtioFeF}]^-$ appear at low field ($CH_3 = 18.3$ ppm, $CH_2 = 14.9$, 12.5 ppm; $CH_3 = $ masked, $CH_2 = $ 13.0, 14.0 ppm, respectively) and resemble those in other high-spin Fe(II) complexes. This shift pattern is consistent with σ spin delocalization from the singly occupied $d_{x^2-y^2}$ orbital. The *meso*-H resonances are at -12.0 ppm and -10.5 ppm, respectively, which implies that π spin delocalization to the *meso* positions is important. This suggests Fe $\rightarrow$ F π back-bonding from the partially filled d_{xz} and d_{yz} orbitals.

iii. N-alkyl (aryl) Porphyrins. As already mentioned above, N-alkylporphyrins have been identified as metabolic products of drugs by animal liver cytochromes P-450 (Battioni *et al.*, 1987). To model certain aspects of this mechanism, iron(II) monohalide complexes of N-substituted porphyrin complexes such as N-methyltetraphenylporphy-rin, N-methyloctaethylporphyrin (Balch *et al.*, 1985a) or [N-(2-phenyl-2-oxoethyl) tetra-phenylporphyrin (Komives *et al.*, 1988) and iron(II) complexes containing a metallacycle (Battioni *et al.*, 1987) have been structurally characterized.

An isotropic shift pattern characteristic of paramagnetic compounds and consistent with the C_s geometry of the complexes (Balch *et al.*, 1985a), is observed in their NMR spectra. The NMR spectrum of the halide complex of N-methyltetraphenylporphinato iron(II) is shown in Fig. 24. Representative chemical shift values are given for a variety

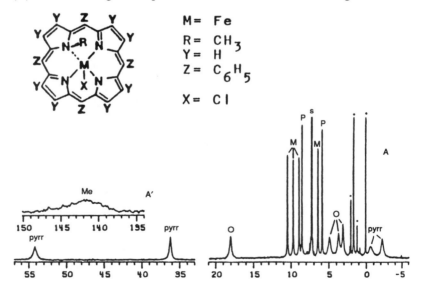

Figure 24. 360-MHz spectrum of the N-methyl iron(II) complex of tetraphenylporphyrin in CDCl$_3$ at $-50\,^{\circ}$C. Resonance assignment: Me = N-methyl; pyrr = pyrrole; o,m,p = *ortho, meta, para* phenyl-H, S = solvent, * = impurity. Reprinted with permission from *Inorg. Chem.* **24**:1438 (1985). © 1985 American Chemical Society.

of these complexes in Table 8. The N-methyl, N-aryl, or N-alkyl resonances are observed in a unique window of +60 to −160 ppm (Balch *et al.*, 1985a). Such large positive isotropic shifts are characteristic for N-alkyl groups and reflect a σ spin transfer mechanism (Balch *et al.*, 1985a). Due to the paramagnetism of the ferrous ion, significant resonance broadening is observed, which for one complex renders the N-alkyl signals undetectable (Komives *et al.*, 1988). Other characteristic porphyrin resonances are also shifted out of the normal diamagnetic region. They include two of the four pyrrole protons that occur in the 25–60 ppm region (the other two are observed at slightly lower frequency than TMS), the pyrrole CH_2 protons which appear in the spectral region of 5 to 35 ppm, and the *meso*-H, which are at −9.5 ppm. The phenyl-H resonance pattern, in which one set of phenyl groups exhibits negative *o*-H and *p*-H isotropic shifts and positive *m*-H shifts of essentially the same magnitude, is consistent with π spin delocalization at the *meso* positions and argues strongly for negligible axial anisotropy for one ring, and thus for the entire complex. In the absence of significant axial anisotropy, it can be assumed that the in-plane anisotropy must also be negligible. These conclusions, reached under the assumption that the halide complexes exhibit axial distortion, are consistent with those for complexes having a variety of other axial ligands (Goff *et al.*, 1977; Parmely and Goff, 1980).

Since the dipolar shifts are negligible, the pyrrole-H isotropic shifts are thus predominantly contact in origin and reflect the effect of lowering the symmetry on the bonding. Even though it has been assumed that the pyrrole-H of the unique N-alkylated pyrrole ring has the smallest contact shift, the pyrrole protons have not been unambiguously assigned in these complexes (Balch *et al.*, 1985a). It might be possible to achieve complete assignment of the pyrrole-H resonances using two-dimensional NMR spectroscopic techniques, as discussed in Section 1.5.

Addition of imidazole ligands to the ferrous halide complexes of N-methyltetraphenylporphyrin and N-methyloctaethylporphyrin results in the formation of the corresponding pentacoordinated imidazole complexes in which imidazole replaces the axial halide ligand. The overall porphyrin resonance pattern and the peak width in these adducts are essentially retained. The addition of various imidazole ligands affects the pyrrole-H and phenyl-H shifts only slightly, indicating that the metal–porphyrin interactions in the imidazole adducts are essentially the same as in the halide complex. Due to the similarities in shifts, it is very likely that the hyperfine shifts in the imidazole complexes also result from contact interactions. The positive hyperfine shifts of most imidazole resonances are consistent with σ spin delocalization. The small negative hyperfine shift for the imidazole 2H, which suggests that contributions from π bonding cannot be neglected, can be explained with some contributions from Fe → Im charge transfer (Balch *et al.*, 1985a).

iv. Nitrene Complexes. Since it has been proposed that the iron metabolite complexes found upon metabolic oxidation of 1,1-dialkylhydrazines by hepatic cytochrome P-450 may involve an iron–nitrene bond (Mahy *et al.*, 1984), there has been a growing interest in porphinato iron–nitrene complexes. High-spin, ferrous nitrene complexes in which the nitrene moiety is inserted into the iron–porphyrin pyrrole nitrogen bond, have been synthesized by reaction of tetraarylporphinato iron(II) complexes with the free nitrene NNC_9H_{18} at −80 °C. ^1H NMR data favor the well-defined high-spin ferrous ($S = 2$) state for this complex. The resonance pattern observed for the protons of the

TABLE 8

Chemical Shifts for N-Substituted Tetraphenylporphyrin Fe(II) Complexes in CDCl₃ at 23 °C[a]

Resonance	Porphyrin				
	$R=CH_3$ $Y=H$ $Z=C_6H_5$ $X=Cl$	$R=CH_3$ $Y=H$ $Z=p\text{-}C_6H_4CH_3$	$R=CH=C(p\text{-}C_6H_5Cl)_2$ $Y=H$ $Z=C_6H_5$	$R=CH_2COC_6H_5$ $T=H$ $Z=C_6H_5$ $X=Cl$	$R=CH_3$ $Y=CH_2CH_3$ $Z=H$ $X=Cl$
Pyrrole-H	41.9 31.7 −0.25 −0.44	42.0 31.1 0.26 0.12	48.8 27.4 7.9 0.9	51.01 31.40 −0.40 −1.10	
Ortho phenyl	14.6 6.5 5.1 4.3	14.6 6.2 5.0 4.4	20.1 3.1 0.0 −0.5	18.18 3.91 2.87 2.40	
Meta phenyl	9.4 9.3 8.5 6.8	9.2 9.1 8.2 6.6	10.7 9.5 7.5 5.3	10.40 9.40 8.91 6.16	
Para phenyl	8.3 6.6	4.4 3.0	8.4 5.6	8.73 5.70	
R-N	105	104	161 14.6 11.8 4.7 −7.8	5.31 5.11 4.81 N-CH2:[b]	121

Meso-H	6.7
	−8.7
Methylene	30.2, 26.0, 22.5
	20.5, 16.2, 15.4
	10.5, 5.4
Methyl	7.3
	5.5
	2.05
	0.21

[a] Data taken from Balch et al. (1985); Komives et al. (1988). Reprinted with permission from *Inorg. Chem.* **24**:1438, 1440 (1985); *Inorg. Chem.* **27**:3114 (1988). © 1985 and 1988 American Chemical Society.
[b] Too broad to be observed.

nitrene complex parallels that of the mercaptide complexes. As for the mercaptide complexes, the pyrrole protons give rise to the downfield signal at 66.8 ppm. Phenyl proton signals are observed between 7 and 10 ppm whereas the peaks of the axial nitrene ligand are found at 81.5, 23.5, and −15.6 ppm, verifying indeed that the complex is high-spin and pentacoordinated.

2.4.2. Six-Coordinate High-Spin Fe(II) Porphyrins

The only 6-coordinated high-spin iron(II) complex that is stable in solution has been reported recently by Mazumdar and Medhi (1990). They have shown that stable 6-coordinated high-spin Fe(II) complexes can easily be formed when weak axial ligands, such as water or tetrahydrofuran (THF) are bound to the iron porphyrins embedded in detergent micelles. The complexes are stable only inside the micelles. In the absence of detergent, 50% of the molecules consist of aggregated species. Proton NMR studies performed for the *bis*-(tetrahydrofuran) complex of protoheme encapsulated in aqueous hexadecyltrimethylammonium bromide micelles reveal four broad resonances around 30 ppm which were assigned to the *meso*-H's. This is an unusually large shift when compared to the position of the *meso*-H in 5-coordinated high-spin heme complexes (0–6 ppm). Four distinct signals with a spread of 6.5 ppm are also observed for the heme methyl groups, which are slightly shifted to lower frequency than the *meso*-H resonances. Significant differences are observed in the NMR spectrum of the 6-coordinated species when compared to those of the 5-coordinated complexes. First, the heme methyl resonance spread of 6.5 ppm in the 6-coordinated complex is much smaller than that in the 5-coordinate complexes, which may be due to the smaller heme in-plane asymmetry. Second, the average shift of the four heme resonances has a larger bias to higher frequency for the micellar complex (21 ppm) than for the 5-coordinate heme (17 ppm), and the *meso*-H resonances are shifted to higher frequency in the 6-coordinate complex than in their 5-coordinate counterparts (Mazumdar and Medhi, 1990).

2.5. Fe(II) Porphyrin Radicals

Recently, Morishima *et al.*, (1986) have reported the existence of the iron(II) oxophlorin radical on the basis of unusually large isotropic shifts observed for the *meso* protons in the proton NMR spectrum of the Fe(II) radical. The NMR spectrum of the iron oxophlorin radical, recorded either in pyridine-d_5 or in pyridine/methanol/HCl solution, showed two resonances at −154 and −111 ppm, which were assigned to the *meso* protons based on deuterium labeling. The four methylene resonances were observed at 32 ppm and in the region of 6 to −5 ppm. The shifts of the *meso*-H and methylene protons are unique among those established for paramagnetic Fe(II) and Fe(III) porphyrin complexes. An unusual temperature dependence was found for all resonances. The non-Curie behavior could plausibly be explained if the Fe(II) oxophlorin radical is thermally admixed between nearly degenerate b_2 and a_2 radical states. As the temperature is increased, the a_2 contribution of the radical state becomes greater, thereby permitting an increase in spin density at the β-pyrrole carbon atoms. This could explain the non-Curie behavior of the CH_2 resonances. The temperature dependence of the *meso* protons may also be explained in terms of the temperature dependent a_2–b_2 state mixing (Morishima *et al.*, 1986).

2.6. High-Spin Fe(III) Porphyrins

2.6.1. Five-Coordinate Monomeric Porphyrin Complexes

The ^{1}H NMR spectra of simple high-spin, 5-coordinate halide complexes of synthetic and natural porphyrins have been reported and discussed in detail (La Mar and Walker, 1979, and references therein). In these complexes, the metal is significantly out of the plane, giving rise to two o-H and m-H signals for TPP derivatives and two α-CH$_2$ resonances for OEP and $meso$-alkyl porphyrin complexes. The electron configuration of high-spin Fe(III) ($S = \frac{5}{2}$) gives rise to a ^6A ground state, for which the g tensor is isotropic. However, relatively large splittings of the zero-field Kramers' doublets $\pm\frac{1}{2}$, $\pm\frac{3}{2}$, $\pm\frac{5}{2}$ give rise to dipolar shifts that are proportional to D/T^2, according to Eq. (10) in Section 1.3.2, and D values estimated from analysis of the Curie plots are similar to those measured from far infrared magnetic resonance measurements of DPDMEFeX (Brackett $et\ al.$, 1971). Contact shifts are large at both pyrrole and $meso$ positions, but the mechanism of spin delocalization differs at these two sites. At the β-pyrrole positions, the contact shifts are to higher frequency for both pyrrole-H ($\delta_{con} \sim +61$–68 ppm at 29 °C) and pyrrole-α-CH$_2$ ($\delta_{con} = +30, 34$ ppm), indicating that spin is delocalized from high-spin Fe(III) to these positions through the σ-bond framework of the porphyrin. Such large σ spin delocalization is observed only in metalloporphyrins in which the $d_{x^2-y^2}$ orbital of the metal is half-occupied (La Mar and Walker, 1979) and is an important signature of half-occupancy of this orbital. In contrast, at the $meso$ positions, the contact shifts are to lower frequency for $meso$-H ($\delta_{con} = -80$ ppm at 29 °C) and to higher frequency for $meso$-α-CH$_2$ ($\delta_{con} = +50$ ppm) (La Mar and Walker, 1979). This reversal in sign of the contact shift is a clear indication of spin delocalization through π orbitals to the $meso$ positions. Since each d-orbital of high-spin Fe(III) is half occupied, it is clearly possible that both σ and π spin delocalization can occur. Of the π-symmetry frontier orbitals of the porphyrin having proper symmetry to overlap with the d_{xz} and d_{yz} orbitals of the metal, the $3e(\pi)$ orbitals have nodes at the $meso$ positions, while the $4e(\pi^*)$ orbitals have large wave function coefficients at the $meso$ positions (Fig. 4). Thus, the mechanism of spin delocalization is identified as Fe $\rightarrow$ P π backbonding (La Mar and Walker, 1979 and references therein).

The wide-line ^{1}H NMR spectrum of polycrystalline hemin chloride at very low temperatures has been reported (Sandreczki $et\ al.$, 1979). Because the $^6A_1$ ground state is separated from the $^6A_1(\pm\frac{3}{2})$ excited state by $2D$, only the $^6A_1(\pm\frac{1}{2})$ is significantly populated when $2D \geq kT$. The observed NMR signal from the $^6A_1(\pm\frac{1}{2})$ state decreases in intensity as that state is depopulated, and the overall signal intensity decreases in intensity markedly as the temperature is lowered below 25 K. Apparently the signal from the $^6A_1(\pm\frac{1}{2})$ is extremely broad (Sandreczki $et\ al.$, 1979).

Since 1979 the solution proton NMR spectra of a number of additional monomeric, 5-coordinate, high-spin Fe(III) porphyrins have been reported, including those having X$^-$ =

1. OH$^-$ (Cheng $et\ al.$, 1982; Miyamoto $et\ al.$, 1983; Woon $et\ al.$, 1986),
2. OR$^-$ (Goff $et\ al.$, 1984; Behere and Goff, 1984; Arasasingham $et\ al.$, 1989b, 1990),
3. OO^{2-} (Shirazi and Goff, 1982),

    4. OOR^- (Arasasingham et al., 1987, 1989b),

    5. N_3^- (Guilard et al., 1991),

    6. NCS^- (Behere et al., 1982; Dugad and Mitra, 1984),

    7. $RCOO^-$ (Oumous et al., 1984),

    8. NO_3^- (Phillipi et al., 1981a),

    9. $OTeF_5^-$ (Kellett et al., 1989),

  10. SO_4^{2-} (Phillippi et al., 1981a; Crawford and Ryan, 1991),

  11. HSO_4^- (Crawford and Ryan, 1991),

  12. RSO_3^- (Phillippi et al., 1981a; Crawford and Ryan, 1991),

  13. (triazolate)$^-$, and

  14. (tetrazolate)$^-$ (Guilard et al., 1991).

Of these, the NO_3^- complex is known to have the anion binding in a bidentate fashion to Fe(III) in the solid state (Phillippi et al., 1981a), and both it and the RSO_3^- complexes (as well as the $[(TPPFe)_2SO_4]$ dimer discussed in Section 2.6.4) are believed to have bidentate anions in solution, based upon the similarity of their pyrrole-H isotropic shifts at 299 K (72.5, 72.9 ppm, respectively), as compared to monodentate Cl^-, RO^-, and $RCOO^-$ (78–80 ppm) (Phillippi et al 1981a). [Carboxylates bind to Fe(III) porphyrins as monodentate ligands (Oumous et al., 1984).] Zero-field splitting constants D have been estimated for some of these anions: $OTeF_5^-$ appears to be similar to Cl^- (Kellett et al., 1989); $N_3^- < Cl^- \sim N_4C(CH_3)^- < N_4C(C_6H_5)^-$ (Guilard et al., 1991); NCS^-, Cl^-, Br^-, and I^- in [TPPFeX] have $D = 5.1$, 6.0, 12.5, and 13.5 cm^{-1}, respectively (Behere et al., 1982; Dugad and Mitra, 1984); and OH^- in TMPFeOH has $D = 8.9$ cm^{-1} (Cheng et al., 1982). Particularly notable among these 5-coordinate monomeric, high-spin Fe(III) complexes is the hydroxo complex, because of the difficulty of preventing its well-known hydrolysis to form the μ-oxo dimer:

$$2[PFeOH] \rightleftarrows [PFeOFeP] + H_2O \qquad (11)$$

This problem was solved by using tetraphenylporphyrin ligands having o,o'-substituents on the phenyl rings, such as $-CH_3$ (TMP), $-OCH_3$ (($2,4,6$-(OCH_3)$_3$)$_4$TPP) (Cheng et al., 1982) or halogens ($-F$ or $-Cl$) (Woon et al., 1986) or very bulky m,m'-substituents such as t-butyl (Miyamoto et al., 1983) or by use of superstructured porphyrins such as "basket-handle" (Lexa et al., 1985) or "picket fence" (Gunter et al., 1984) or related water-soluble "picket fence" derivatives (Hambright et al., 1987). The equilibrium represented by Eq. (11) has been utilized to develop a possible pH-sensitive contrast agent for MRI (Helpern et al., 1987). It and the additional equilibria between the diaqua and hydroxo complexes have been studied for the water-soluble tetraphenylporphinesulfonate complex of Fe(III) (Ivanca et al., 1991) and for hemins in aqueous detergent micelles (Mazumdar et al., 1988, 1991; Mazumdar, 1991). The mechanism of conversion of the hydroxide complex of Fe(III) tetratolylporphyrin to the corresponding μ-oxo dimer has been investigated (Fielding et al., 1985); it was found that the kinetics are consistent with the first step being dissociation of OH^- from an iron porphyrin followed by reaction of that 4-coordinate porphyrin with a second PFeOH to form the dimer.

    The peroxo complex $[TPPFeO_2]^-$ was prepared from [TPPFe] and superoxide O_2^- (Shirazi and Goff, 1982), and is an example of the facile redox reaction that takes place between Fe(II) and dioxygen and its redox-active reduced derivatives, superoxide and peroxide. As shown in Fig. 6, the NMR spectrum is clearly diagnostic of a high-spin Fe(III) porphyrin.

The formation of $PFe^{III}OO\text{-}t\text{-}Bu$ from PFe(II) and t-BuOOH requires working at very low temperatures $(-70\,^\circ C)$ (Arasasingham $et\ al.$, 1989a, b) to prevent thermal decomposition to $PFe^{IV}O$ and t-BuO$^{\cdot}$:

$$[TMPFe^{II}] + t\text{-}Bu\text{-}OOH \rightarrow [TMPFe^{III}\text{-}OO\text{-}t\text{-}Bu] + \text{side products} \qquad (12)$$

$$[TMPFe^{III}\text{-}OO\text{-}t\text{-}Bu] \rightarrow [TMP^{IV}Fe^{IV} = O] + t\text{-}BuO^{\cdot} \qquad (13)$$

followed by further reactions. Decomposition is catalyzed by bases, such as N-methylimidazole (Arasasingham $et\ al.$, 1989b). The chemical and NMR properties of the $(Fe^{IV}O)^{2+}$ unit will be discussed in Section 2.11. The pyrrole-H shifts of the 5-coordinate OH^-, t-Bu-O$^-$, and t-Bu-OO$^-$ complexes of TMPFe(III) at $-70\,^\circ C$ (116.4, 122, 114 ppm, respectively) (Arasasingham $et\ al.$, 1989b) suggest that OH^- and t-Bu-OO$^-$ have similar values of D, while that for t-Bu-O$^-$ may be larger.

N-methylporphyrin complexes of Fe(II) are readily oxidized to 5-coordinate Fe(III) N-methylporphyrin halide cations (Balch $et\ al.$, 1985d). At $-50\,^\circ C$ in $CDCl_3$, [MeTPPFeCl]$^+$ has pyrrole-H chemical shifts of 128, 92, and 70 ppm; the β-pyrrole chemical shift of the N-methylpyrrole unit was found to be at 2.4 ppm using [d_8-MeTPPFeCl]$^+$ and detection by ^{2}H NMR spectroscopy. The three remaining pyrrole rings appear to have average amounts of spin delocalization similar to those of [TPPFeCl]; only the modified pyrrole ring suffers sharply decreased spin transfer. The four m-H resonances of the N-MeTPP ligand have an average observed shift almost identical to the two of [TPPFeCl] at the same temperature (Balch $et\ al.$, 1985d). The linewidths of the proton resonances of the chloride and bromide complexes differ in relation to the zero-field splitting parameter D $(\Delta \propto D^{-2})$, with $D_{Br} > D_{Cl}$, as for the [TPPFeX] analogs (La Mar and Walker, 1973b). The chemical shifts for [MeOEP-FeCl]$^+$ and [OEPFeCl] show similar features, apart from the large asymmetry in spin delocalization caused by N-methylation (Balch $et\ al.$, 1985d).

2.6.2. Six-Coordinate Monomeric Porphyrin Complexes

In the presence of ligands of medium coordinative ability, with counterions that bind weakly, Fe(III) porphyrins from high-spin 6-coordinate complexes. Such is the case of [TPPFeI] in $1:3$ mixed $CDCl_3/CD_2Cl_2$ solutions with added DMSO (Zobrist and La Mar, 1978). Varying the concentrations of both [TPPFeI] and DMSO confirmed that the stoichiometry of the complex was 2 DMSO:1 Fe(III). The isotropic shift of the pyrrole-H of [TPPFe(DMSO)$_2$]$^+$I$^-$ in $3:1$ $CD_2Cl_2:CDCl_3$ at $-72\,^\circ C$ is ~101 ppm, as compared to ~121 ppm for the 5-coordinate [TPPFeI] parent (Zobrist and La Mar, 1978). Investigation of natural hemin derivatives substituted in the 2,4 positions in DMSO (Budd $et\ al.$, 1979) showed that the increasing spread of the methyl isotropic shifts as the porphyrin 2,4 substituents are made more electron-withdrawing is similar to that observed in low-spin ferric complexes, discussed in Sections 2.8.2. and 2.8.3. Strongly electron-withdrawing 2,4 substituents induce a ring methyl isotropic shift spread similar to that observed in metaquomyoglobins, suggesting that the in-plane asymmetry in proteins may arise from peripheral heme–apoprotein interactions

TABLE 9
Isotropic Shifts of 5- and 6-Coordinate High-Spin Fe(III) Porphyrins, Chlorins, and Isobacteriochlorins[a]

	Chemical shifts (ppm)				
Macrocycle (anion)	Pyrr-α-CH$_2$	Pyrr-CH$_3$	Pyrr-H	meso-H	meso-H position
Five-coordinate complexes					
Porphyrins					
DPDME(Cl$^-$)[b]	36, 39, 41	45	66	-67	
OEP(Cl$^-$)[d]	39.8	—	—	-55.6	
	Pyrr-α-CH$_2$ / Pyrroline-α-CH$_2$	Pyrroline-α-CH$_2$	Pyrroline-H	meso-H	
Chlorins					
t-OEC(Cl$^-$)[d]	50.4, 46.3, 45.3 (3), 43.8, 42.9, 38.6, 40.9, 38.6, 35.5 (2), 34.7	26.7, 18.6, 15.4, 10.9	21.6, 20.3	-44.3 -53.2 -82.0 -91.2	[15 or 20] [15 or 20] [5 or 10] [5 or 10]
Isobacteriochlorins					
ttt-OPEiBC(Cl$^-$)[d]	70.1, 61.8, 56.8, 51.7, 40.9, 39.9, 37.6, 34.6	29.7, 27.2, 18.2 (2), 14.1 (2), 8.0 (2)	40.7, 25.9, 20.0, 4.3	-38.4 -77.7 -116.2	[15] [10, 20] [5]
tct-OEiBC(Cl$^-$)[d]	68.2, 62.4, 57.2, 50.4, 44.6, 41.3, 35.0, 34.5	28.0, 24.0, 15.9 (2), 14.3 (2), 10.5 (2)	38.7, 33.0, 28.0, 16.3	-31.1, -48.8, -69.8, -90.2, -111.3, -126.8	[15] [10, 20] [5]
	Pyrr-α-CH$_2$	Pyrr-CH$_3$	Pyrr-H	meso-H	
Six-coordinate complexes					
Porphyrins					
PP(Cl$^-$)[c]	9.0, 38.0, 35.7 (2)	63.2, 62.2, 57.7, 54.4	—	50, 46, 48 (2)	
DP(Cl$^-$)[c]	43.7 (2). 41.0 (2)	64.3, 61.2,	0.55 (2)	~29 (4)	
OEP(CF$_3$SO$_3^-$)[c]	47.9 (16)	—	—	40.1 (4)	

	Pyrr-α-CH$_2$	Pyrroline-α-CH$_2$	Pyrroline-H	meso-H	
Chlorins					
t-OEC(CF$_3$SO$_3^-$)^c	60.6 (4), 52.8 (2)	16.5 (2), 8.6 (2)	81.3 (2)	31.7	[15, 20]
	36.4 (2), 35.5 (2)			29.6	[5, 10]
Isobacteriochlorins					
ttt-OEiBC(CF$_3$SO$_3^-$)^e	84.3 (2), 83.7 (2)	19.1 (2), 17.1 (2)	131.0 (2), 72.6 (2)	34.2	[10, 20]
	30.6 (2), 29.4 (2)			21.4	[15]
				0.7	[5]
tct-OEiBC(CF$_3$SO$_3^-$)^c	85.4 (2), 82.6 (2)	16.2 (2), 12.8 (2)	125.1 (2), 69.8 (2)	29.8	[10, 20]
	30.6 (2), 29.2 (2)			17.4	[15]
				0.7	[5]

[a] Reprinted with permission from *J. Am. Chem. Soc.* **113**:5268 (1991); *J. Am. Chem. Soc.* **101**:6094, 6095 (1979). © 1979 and 1991 American Chemical Society.
[b] Caughey and Johnson (1969); solvent = CDCl$_3$, $T=35$ °C.
[c] Budd et al., (1979); solvent = DMSO-d$_6$, $T=25$ °C.
[d] Sullivan et al., (1991), solvent = toluene-d$_8$, $T=20$ °C.
[e] Sullivan et al. (1991); solvent = 50:50 v/u CD$_2$Cl$_2$/DMSO-d$_6$, $T=20$ °C.

(Budd *et al.*, 1979). Dimethylformamide (DMF) also appears to coordinate to [TPPFeI] to form 6-coordinate complexes (Zobrist and La Mar, 1978; Budd *et al.*, 1979; Dixon *et al.*, 1985).

The most striking difference in the ^{1}H NMR spectra of 5- and 6-coordinate high-spin Fe(III) porphyrins is the *reversal in sign* of the isotropic shift for the *meso*-H, as shown in Table 9. This reversal was noted early for derivatives of natural hemins in DMSO-d_6 (Budd *et al.*, 1979), but because of the extreme breadth of the *meso*-H resonances their detection in 6-coordinate high-spin heme proteins, such as the peroxidases, was thought to be impractical, and thus little emphasis was placed on interpreting the isotropic shifts of the *meso*-H. Nevertheless, the 90–95 ppm shift of the *meso*-H resonance to higher frequency as the iron goes from 5- to 6-coordination, and presumably from out-of-plane to in-plane position, begs for an explanation. The reversal in sign of the contact shift of the *meso*-H indicates that in the 6-coordinate state, HS Fe(III) does *not* engage in Fe → P π back-bonding (or any other π bonding interaction, for that matter). This point becomes important in comparing the *meso*-H shifts of porphyrins, chlorins, and isobacteriochlorins (Sullivan *et al.*, 1991), Table 9, discussed in the next subsection.

The comparison of the size of the pyrrole-H and -CH$_3$ isotropic shifts of 5- and 6-coordinate high-spin deuterohemin provides additional details: The pyrrole-H shift is ~20% smaller and the pyrrole-CH$_3$ ~25% larger in the 6- than the 5-coordinate complexes; this opposite behavior suggests a larger π contribution to the isotropic shifts at the β-pyrrole positions of the 6- than the 5-coordinate complexes (Budd *et al.*, 1979). This observation also suggests that the interaction of HS Fe(III) with the π orbitals of the porphyrin switches from largely an interaction with the $4e(\pi^*)$, in the case of the 5-coordinate complexes, as discussed in Section 2.6.1, to weak interaction with the $3e(\pi)$ orbitals in the 6-coordinate complexes. This means that while significant Fe → P back-bonding occurs in the 5-coordinate complexes, it *does not occur* in the 6-coordinate complexes; rather, weak P → Fe π bonding occurs in this case. This conclusion differs, at least in degree, for reduced hemes, discussed in Section 2.6.3.

Comparison of 5- and 6-coordinate high-spin natural hemin derivatives suggests that upon reconstitution of heme proteins with deuterohemin, the relative pyrrole-H and -CH$_3$ shifts may serve as a useful indicator of the state of occupation of the sixth site in high-spin hemoproteins (+66, +46 ppm, respectively, 5-coordinate; +55, +64 ppm, respectively, 6-coordinate; −22, +15 ppm, respectively, low-spin) (Budd *et al.*, 1979). Morishima and coworkers (1980) have found that 6-coordinate high-spin OEPFe(III) complexes could also be formed having two alcohol or one alcohol and one aliphatic amine ligand. Behere *et al.* (1984) then showed that for TPPFe(III), the *bis*-DMSO complex is high-spin ($D = 6.0$ cm^{-1}), while the *bis*-alcohol (CD$_3$OD) complex, with $\delta_{obs} = 66$ ppm at 200 K, is probably intermediate-spin or mixed high/intermediate-spin.

A different type of 6-coordinate high-spin Fe(III) porphyrin complex is produced if [TPPFeF] is reacted with hydrated tetrabutylammonium fluoride (Bu$_4$NF · 3H$_2$O) in dichloromethane (Hickman and Goff, 1983): The *bis*-fluoride anionic complex [TPPFeF$_2$]$^-$ is stable ($K_f = 4 \times 10^3$ M^{-1} at 25 °C) and is characterized by slightly larger pyrrole-H (85.8 ppm as compared to 72 ppm for the *bis*-DMSO complex discussed above) and *o,p*-phenyl-H and slightly smaller *m*-phenyl-H isotropic shifts. Lack of splitting of the *o*-and *m*-phenyl signals of [TPPFeF$_2$]$^-$ supports the assumption of diaxial fluoride coordination with the metal in the plane of the porphyrin (Hickman and Goff, 1983; Hickman et al., 1988). Greater linewidths of the ^{1}H signals of $[TPPFeF_2]^-$ suggests that the bis-fluoride has a smaller zero-field splitting constant D than the monofluoride complex.

2.6.3. High-Spin Monomeric Fe(III) Complexes of Reduced Hemes

Several naturally occurring iron(III) chlorins have been investigated by ^{1}H NMR spectroscopy, including the bis-DMSO complex of the stable green sulfhemin prosthetic group extracted from sulfmyoglobin (Chatfield et al., 1988a, b) and that of pyropheophorbidato a iron(III) (Balch et al., 1985; Keating et al., 1991). The positive meso-H isotropic shifts suggests 6-coordinate, bis-(DMSO) complexes in each case. Three of the ring CH$_3$s of sulfheme C exhibit essentially unperturbed contact shifts relative to hemin, with only the 3-CH$_3$ (on the saturated ring) exhibiting a sharply attenuated shift, thus identifying the saturated ring as B or II (Chatfield et al., 1988b), Fig. 5. For pyropheophorbidato a iron(III), large isotropic shifts for the 1-, 3-, and 5-CH$_3$ groups and a relatively small shift for the 8-CH$_3$ are consistent with the saturated ring being D or IV (Balch et al., 1985; Keating et al., 1991), Fig. 11. The patterns observed are consistent with those observed earlier for complexes of OECFe(III) (Pawlik et al., 1988), discussed in Section 2.6.3a.

2.6.3a. Octaethylchlorin Iron(III). The idealized orbital energy diagram for the iron(III) d-orbitals and the frontier π orbitals of the porphyrin and chlorin dianion macrocycles are shown in Fig. 25 (Pawlik et al., 1988). The nonequivalence in energy and symmetry properties of the former $3e(\pi)$ and $4e(\pi^*)$ orbitals, as well as the change in symmetry of the formerly a_{1u} orbital, Fig. 5, such that it can overlap with the d_{xz} orbital of the metal (Chatfield et al., 1988b) has profound effects upon the NMR spectra

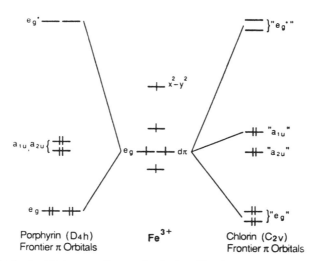

Figure 25. Idealized orbital energy diagram for the iron(III) d orbitals and the frontier π orbitals of porphyrin and chlorin macrocycle dianions. Reprinted with permission from *J. Am. Chem. Soc.* **110**:3011 (1988). © 1988 American Chemical Society.

of high-spin Fe(III) chlorins. ^{1}H NMR spectra of high-spin Fe(III) complexes of *trans*-octaethylchlorin (OEC) (X = Cl$^-$, OTeF$_5^-$, and NCS$^-$) and tetraphenylchlorin (TPC) (X = Cl$^-$, OTeF$_5^-$) have been compared to those of the corresponding OEP and TPP complexes (Pawlik *et al.*, 1988). It was found that the various types of pyrrole protons in the TPC complexes and the various types of pyrrole methylene protons in the OEC complexes exhibit a large range of isotropic shifts ($\sim$30 ppm at 300 K). For both TPC compounds, two pyrrole-H resonances are at lower frequency and one is at higher frequency than the pyrrole-H resonance for the TPP homologue, while for the OEC complexes the mirror image of this pattern is observed: Two sets of four pyrrole CH$_2$ resonances are at higher frequency and one set of four is at lower frequency than the average pyrrole CH$_2$ resonance for the OEP homologues. (The two sets arise from the fact that the out-of-plane position of Fe(III) in these 5-coordinate *trans*-OEC complexes causes each proton in the complex to be in a unique magnetic environment. The same is not true of the TPC complex, since it has a mirror plane passing through the Cl, the Fe, and the center of the pyrroline ring.) However, it should also be noted that the average pyrrole-α-CH$_2$ shift is larger for the chlorin (42 ppm) than the porphyrin (39.8 ppm), although this does not change the higher–lower frequency shift pattern by very much. This reversal in pattern on going from the TPC complexes to the OEC complexes is almost certainly due to the π component of the contact shift at the β-pyrrole position for the two types of complexes, as shown in Fig. 26. These results suggests that the amount of spin density distributed from the iron atom to the protons in the pyrrole groups of the macrocycle is nearly the same for 5-coordinate high-spin Fe(III) porphyrins and chlorins, but is distributed quite asymmetrically among the various pyrrole positions in the chlorin complexes (Pawlik *et al.*, 1988).

The differential rate of substitution of deuterium at the two chemically different types of *meso*-H positions under acidic and basic conditions (Bonnett *et al.*, 1967; Whitlock *et al.*, 1969; Sullivan *et al.*, 1991) allowed assignment of the two types of *meso*-H. For the 5-coordinate chloro iron(III) complex, the *meso*-H adjacent to the reduced ring have a much larger negative isotropic shift and hence a larger π contact shift than those farther away from it—and much more negative than the *meso*-H isotropic shift of OEPFeCl (Sullivan *et al.*, 1991)—suggesting that *neither* S_8^* *nor* A_6^*, the analogs of the $4e(\pi^*)$ orbitals of the porphyrin (Fig. 5), are involved in this π delocalization. Rather, the pattern of spin delocalization to the *meso* positions is most consistent with that expected for the filled orbital A_5 (Fig. 5). These are also the *meso* positions that exchange more rapidly with deuterium under acid conditions (Sullivan *et al.*, 1991) and are thus most electron-rich, consistent with the relative molecular orbital coefficients of A_5. Thus, for 5-coordinate chlorin complexes, the mechanism of spin delocalization appears to be Chl $\rightarrow$ Fe π donation, leading to partial cation radical character of the porphyrin π system. In contrast, for the 6-coordinate high-spin Fe(III) complex, the isotropic shift of the *meso*-H adjacent to the pyrroline ring is only slightly smaller than that of the other two. In addition, both *meso*-H have less positive isotropic shifts than the corresponding OEP complex (Table 9), suggesting a larger π spin delocalization in the 6-coordinate chlorin complex than in the corresponding porphyrin complex [OEPFe(DMSO)$_2$]$^+$ discussed above. The almost equal contact shift implied by the similarity in the isotropic shifts of the *meso*-H of [OECFe(DMSO)$_2$]$^+$ (Table 9) suggests mixed spin delocalization to both A_5 and S_8^*, or both weak Chl $\rightarrow$ Fe π donation from A_5 and Fe $\rightarrow$ Chl π back-donation to S_8^* (Fig. 5).

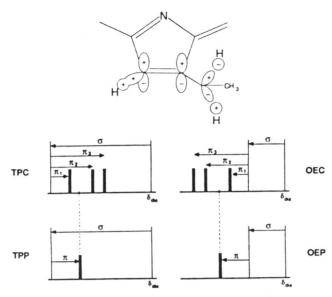

Figure 26. Top: diagram showing the relationship of pyrrole β-carbon p-π orbitals to the relevant C—H orbitals for the octaethyl macrocycles (right) and the tetraphenyl macrocycles (left), which accounts for the π contact shifts discussed in the text. Bottom: diagram showing the net ^{1}H NMR isotropic shifts for octaethyl macrocycle pyrrole-methylene protons (right) and tetraphenyl-macrocycle pyrrole protons (left) as the sum and difference, respectively, of σ and π contact shifts. Reprinted with permission from *J. Am. Chem. Soc.* **110**:3011 (1988). © 1988 American Chemical Society.

The average temperature dependence of the resonances (i.e., Curie plot slopes) of [OECFeCl] and [OECFeOTeF$_5$] is similar to the temperature dependence of the resonances for porphyrin protons in chemically similar positions (Pawlik *et al.*, 1988). For both sets of chlorin complexes, pyrroline methylene proton isotropic shifts are substantially smaller than pyrrole isotropic shifts, while pyrroline proton or methylene linewidths are larger than pyrrole proton linewidths. Large deviations in Curie plot $1/T = 0$ intercepts from diamagnetic chemical shifts are observed for many of the chlorin proton resonances, especially those of the pyrroline substituents, possibly due to temperature-independent paramagnetic (TIP) contributions (Pawlik *et al.*, 1988) or to temperature-dependent involvement of A_5 and other π MOs in spin delocalization, as we have proposed for unsymmetrically substituted [TPPFeL$_2$]$^+$ complexes (Section 2.6.8.2b.i). Whatever its origin, this behavior has been observed for all paramagnetic iron (II, III) hydroporphyrins studied to date (Strauss *et al.*, 1985; Strauss and Pawlik, 1986).

  2.6.3b. Two Octaethylisobacteriochlorin Iron(III) isomers. Five- and 6-coordinate HS Fe(III) complexes of both *trans-trans-trans-* (*ttt*) and *trans-cis-trans-*(*tct*) octaethylisobacteriochlorin, OEiBC, have also been investigated (Sullivan *et al.*, 1991). A number of resonances were assigned by means of deuterium substitution, utilizing both ^1H and ^2H NMR spectroscopy, as described in Section 1.5.2 (Fig. 7). For the 6-coordinate complexes of both isomers, the spread of the observed shifts of the pyrrole-α-CH$_2$ is larger for both isomers (ttt, 54.9; tct, 56.2 ppm) than it is for the 5-coordinate complexes (ttt, 35.5; tct, 33.7 ppm). Again, for the 5-coordinate complexes, the out-of-plane position of Fe(III) and its anion makes each pyrrole proton magnetically unique in the ttt isomer and those plus the meso-H unique in the tct isomer. The larger spread of pyrrole-α-CH$_2$ resonances for 6- versus 5-coordinate complexes is also mirrored in the pyrroline α-CH$_2$ and -H resonances (Table 9).

For the 5-coordinate complexes, the meso-H resonances are all at very low frequency, indicating large π spin delocalization to the meso positions. The large spread (ttt, 77.8; tct, 95.7 ppm) indicates π contact shifts in the order 5 ≫ 10, 20 > 15. This is the opposite order from that observed for the rate of base-catalyzed meso-H exchange for deuterium in the metal-free OEiBCH$_2$ isomers (Sullivan et al., 1991) and is thus the same as the order of decreasing electron density at the meso positions in each isomer. Thus, the π contact shift pattern at the meso positions is indicative of spin delocalization by filled π orbitals, i.e. iBC → Fe π donation, probably from the formerly $1a_{1u}(\pi)$ orbital. The meso-H isotropic shifts of the 6-coordinate DMSO adducts of the two isomers are quite different from those of the 5-coordinate complexes just discussed; as for the chlorin and porphyrin, the sign reverses, indicating a dominance of σ spin delocalization at the meso as well as the pyrrole positions. In addition, the spread of the meso-H resonances decreases to 33.5 (ttt) and 29.1 (tct) ppm. More importantly, the order of the resonances changes, so that the minor π spin delocalization that occurs at the meso positions of the 6-coordinate complexes is 5 > 15 > 10, 20. This pattern is not indicative of delocalization through any of the filled π orbitals, but rather suggests spin delocalization throgh one of the formerly $4e(\pi^*)$ orbitals of the porphyrin, i.e., Fe → iBC π back-bonding.

It is tempting to wonder whether the switch in the mechanism of π spin delocalization on going from 5- to 6-coordination in the reduced hemes (from strong L → Fe to only weak Fe → L) may play some role in the biological activity of these prosthetic groups. Although the amount of π spin transfer from either chlorin or isobacteriochlorin is greater than that for the porphyrin, in the case of the isobacteriochlorins it still amounts at most to only 6.2×10^{-6} % of an electron on average from each meso position, or a total of 2.5×10^{-5} % assuming that most of the spin density is removed from the meso positions. By comparison, 2.0×10^{-5} % of the spin density of the porphyrin meso positions is transferred to the metal, an insignificant difference.

2.6.3c. A Dioxooctaethylisobacteriochlorin Iron(III). A recent report of the ^1H NMR spectrum of a chloro iron(III) 2,7-dioxoisobacteriochlorin derived from OEP in CD$_2$Cl$_2$ (Barkigia et al., 1992) shows a very similar pattern of chemical shifts for the meso-H, in this case indicating contact shifts in the order 5 > 20 > 10 > 15, with a spread of 69 ppm. The spread of the pyrrole-α-CH$_2$ resonances is 36.7 ppm, with an average chemical shift of 49.4 ppm, and that for the pyrroline α-CH$_2$ resonances is 3.5 ppm, with an average chemical shift of 19.6 ppm. Except for the small spread of the pyrroline-α-CH$_2$ resonances, which is consistent with the structure of this reduced heme, the pattern is very similar to that of the OEiBCFeCl isomers reported by Sullivan et al. (1991), Table 9.

2.6.3d. A Dialkyloctaethylporphodimethane Iron(III). Finally, the NMR spectra of two chloro iron(III) (5,15-dialkyl)octaethylporphodimethene complexes have been reported (Botulinski *et al.*, 1988). Four pyrrole-α-CH_2 proton signals are observed, with isotropic shifts of 75.5, 66.5, 38.7, and 31.6 ppm for [(OEPMe$_2$)FeCl] and 63.8, 58.1, 39.1, and 37.6 ppm for [(OEPBu$_2$)FeCl] in $CDCl_3$ at 293 K, and *meso*-H and -CH_2 signals at 26.9 and 4.0 ppm and 21.0 and 4.9 ppm, respectively. Analysis of the Curie plot suggests that the zero-field splitting parameter D equals 15 cm^{-1} in these complexes. The isotropic shifts are shown to have substantial contact and dipolar contributions, whereas the nuclear relaxation rates $1/T_2$ and resulting linewidths appear to be dipolar in origin (Botulinski *et al.*, 1988).

2.6.4. Bridged Dimeric Complexes of High-Spin Fe(III) Porphyrins and Chlorins

A number of the properties of μ-oxo dimers of Fe(III) porphyrins, formed by hydrolysis of the hydroxo complex [Eq. (11)] or by reaction of molecular oxygen with Fe(II) porphyrins (see below), have been investigated, including their ^{1}H and ^{13}C NMR spectra, infrared and Raman spectra, Mössbauer spectra, magnetic susceptibilities, and crystal and molecular spectra. These linear PFe—O—FeP units are antiferromagnetically coupled dimers of high-spin Fe(III), although attempts to measure the coupling constant, $-2J$, from the temperature dependence of the magnetic susceptibility or NMR isotropic shifts have yielded a range of values (La Mar and Walker, 1979; Scheidt and Reed, 1981). Recent magnetic susceptibility results (Helms *et al.*, 1986) for [(TPPFe)$_2$O] in the solid state give $-2J = 271$ cm^{-1}, while values of 258 cm^{-1} (Strauss *et al.*, 1987) and 312 cm^{-1} (Boersma and Goff, 1984) have been obtained recently from ^{1}H and ^{13}C NMR results, respectively. Calculation of $-2J$ from the temperature dependence of the NMR spectra of these complexes is complicated by the possibility of dipolar, ZFS, and contact contributions to the isotropic shifts (La Mar and Walker, 1979, and references therein). In any case, the isotropic shifts of the protons of these PFe—O—FeP dimers are significantly different from those of the high-spin Fe(III) monomers (La Mar and Walker, 1979, and references therein), as shown in the lower trace of Fig. 27 for [(TPPFe)$_2$O], with all resonances occurring between 7 and 14 ppm at 320 K. Similar NMR spectra (upper trace) are observed for the reduced porphyrin analog [(TPCFe)$_2$O] with the addition of two signals from the two types of protons in the pyrroline ring (Strauss *et al.*, 1987). The tetraphenylchlorin complex has a very similar derived value of $-2J$ (265 cm^{-1}). The shift to lower frequency of the pyrroline protons suggests either

1. a sizable upfield π contact shift, possibly arising from spin delocalization via the $a_{1u}(\pi)$ counterpart of the porphyrin ring, which in the case of the chlorin ring has the proper symmetry for overlap with the d_π orbitals of the metal (discussed in Section 2.8.9 and shown in Fig. 5), or
2. a large rhombic component to the isotropic shift, as previously discussed for monomeric, 4-coordinate $S = 1$ complexes of the reduced hemes OECFe, TPCFe, and (TPiBC)Fe (Strauss *et al.*, 1985; Strauss and Pawlik, 1986).

A series of mixed-metal μ-oxo dimers of the type PFe—O—CrP' have been prepared from PFe(II) and P'Cr=O (Liston *et al.*, 1985). Their ^{1}H NMR spectra are characterized by resonances due to the PFe in the positions found for the symmetrical PFe—O—FeP dimer, and a resonance at 38 ppm due to the pyrrole-H of the chromium

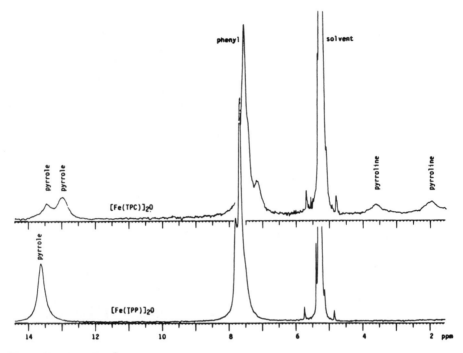

Figure 27. 200-MHz ^{1}H NMR spectra of [Fe(TPC)]$_2$O (top) and [Fe(TPP)]$_2$O in dichlorome-thane-d_2 at 320 K. Printed with permission from *Inorg. Chem.* **26**:729 (1987). © 1987 American Chemical Society.

porphyrin unit, as shown in Fig. 28. The peak at ~13 ppm was not present when OEPFe(II) was used as the iron starting material instead of TPPFe(II) (Liston *et al.*, 1985). The temperature dependence of the magnetic susceptibility of the mixed dimer leads to its formulation as antiferromagnetically coupled Fe(III) ($S = \frac{5}{2}$) and Cr(III) ($S = \frac{3}{2}$), yielding $S' = 1$, with $-2J \sim 260$–300 cm^{-1} and D_{Fe} and g_{Fe} ranging from 3.8 to 4.9 cm^{-1} and 1.98–2.05, respectively, assuming $D_{Cr} \sim -1$ cm^{-1} (Liston *et al.*, 1985). Similar binuclear μ-oxo dimers can be made in which the iron complex is the phthalocyanine.

The intermediate in the oxidation of 4-coordinate Fe(II) porphyrins to the PFe—O—FeP μ-oxo dimer, the μ-peroxo dimer C of tetra-m-tolylporphinato iron(III), has been detected by NMR specroscopy at low temperatures (Chin *et al.*, 1977), Fig. 29:

$$PFe^{II} + O_2 \rightarrow PFeO_2 \tag{14a}$$
$$\phantom{PFe^{II} + O_2 \rightarrow} A B$$

$$PFeO_2 + PFe^{II} \rightarrow PFe^{III}\!-\!OO\!-\!Fe^{III}P \rightarrow 2PFe^{IV}O \tag{14b}$$
$$\phantom{PFeO_2 + PFe^{II} \rightarrow PFe^{III}\!-\!OO} C \phantom{\!-\!Fe^{III}P \rightarrow 2PFe^{IV}} D$$

$$\downarrow$$

$$PFe^{IV}O + PFe^{II} \rightarrow PFe^{III}\!-\!O\!-\!Fe^{III}P \tag{14c}$$
$$\phantom{PFe^{IV}O + PFe^{II} \rightarrow PFe^{III}\!-\!O\!-\!F} E$$

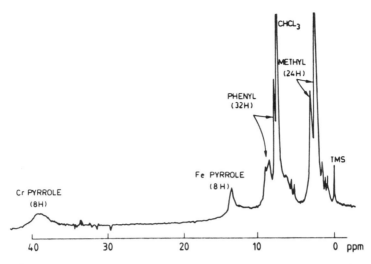

Figure 28. ^{1}H NMR spectrum of (TXP)CrOFe(TPP)CDCl$_3$, where TXP = tetrakis-(m-Xylyl)-porphyrin. Reprinted with permission from *Inorg. Chem.* **24**:1564 (1985). © 1985 American Chemical Society.

The μ-peroxo complex has a larger isotropic shift for the pyrrole-H at all temperatures and has an antiferromagnetic coupling constant $-2J$ that is about 30% smaller than that for the μ-oxo dimer E, and the magnetic moment decreases from 2.6 μ_B at $-50\,°C$ to about 2.2 μ_B at $-80\,°C$, both consistent with weaker antiferromagnetic coupling in the μ-peroxo complex (Chin *et al.*, 1977).

Similar reaction of 4-coordinate PFe(II) complexes with p-benzoquinone and its derivatives yield hydroquinone dianion-bridged Fe(III) porphyrin dimers that are characterized by very small antiferromagnetic coupling constants, $-2J \sim 7.2$–$31\ cm^{-1}$ (Kessel and Hendrickson, 1980). Accordingly, the ^{1}H NMR spectra of these complexes are fairly characteristic of high-spin Fe(III) monomeric complexes, with the pyrrole-H resonances at 62–74 ppm at 23 °C (Balch *et al.*, 1990a). However, the Curie plots show the opposite type of curvature from that of simple monomeric HS Fe(III) complexes, due to the (small) antiferromagnetic coupling between the Fe(III) centers in these dimers.

Dimeric complexes of the dinegative oxoanions SO_4^{2-} and CrO_4^{2-}, $[(PFe)_2XO_4]$ have been shown to have both ^{1}H and ^{13}C NMR resonances that are sharp and at very similar isotropic shifts to those of monomeric, high-spin Fe(III) complexes (Phillippi *et al.*, 1981a; Godziela *et al.*, 1985), suggesting that antiferromagnetic coupling is very weak or nonexistant in these dimeric complexes. Very small deviations from Curie behavior were noted, suggesting that zero-field splitting of HS Fe(III) is very small; this fact and the slightly smaller pyrrole-H isotropic shifts of $[(TPPFe)_2(SO_4)]$ suggest that the sulfate anion may be acting as a bidentate chelating ligand to each Fe(III) (Phillipi *et al.*, 1981a).

Finally, an intermolecular type of dimer formation has been observed for the unsymmetrically substituted 5-(2-hydroxyphenyl)-10,15,20-tritolylporphyrin complex of

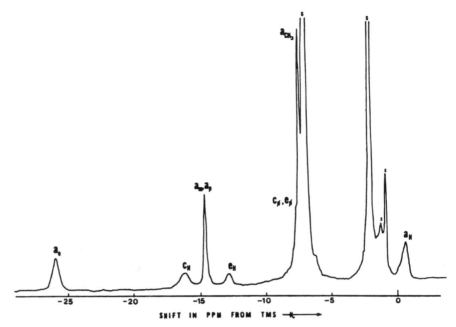

Figure 29. Proton NMR trace of a sample containing PFe, whose peaks are labeled a_i, PFe—OO—FeP, with peaks marked c_i, and PFe—O—FeP with peaks marked e_i, in toluene-d_8 at −50 °C. The subscripts o, m, and p refer to *ortho*-H, *meta*-H, and *para*-H, $\emptyset$ to the unresolved composite of *ortho*-, *meta*- and *para*-H, -CH$_3$, and H to the pyrrole-H; S = solvent peaks, and X-impurities. Reprinted with permission from *J. Am. Chem. Soc.* **99**:5487 (1977). © 1977 American Chemical Society.

Fe(III), in which the o-OH functions of one molecule are deprotonated and bind as RO⁻ ligands to the Fe(III) of a second molecule, which in turn binds its deprotonated o-OH function to the Fe(III) of the first. Despite this intramolecular dimeric structure, the proton NMR spectra reveal pyrrole-H signals in the 80 ppm region at 25 °C, diagnostic of a HS Fe(III) center, and the magnetic moments in solution ($6.0 \pm 0.2 \, \mu_B$) and solid-state ($5.91 \pm 0.05 \, \mu_B$) match the spin-only value for high-spin Fe(III), thus indicating that antiferromagnetic coupling is extremely small (Goff *et al.*, 1984).

2.7. Intermediate-Spin Fe(III) Porphyrins

In the presence of weakly coordinating anionic axial ligands such as ClO$_4^-$ (Dolphin *et al.*, 1977; Kastner *et al.*, 1978; Reed *et al.*, 1979; Goff and Shimomura, 1980; Masuda *et al.*, 1980; Goff, 1983, and references therein; Toney *et al.*, 1984a, b; Kintner and Dawson, 1991), SO$_3$CF$_3$ (Boersma and Goff, 1982), SbF$_6^-$ (Shelley *et al.*, 1985; Gupta *et al.*, 1987; Kinter and Dawson, 1991), and C(CN)$_3^-$ (Summerville *et al.*, 1978; Boersma and Goff, 1982) or two weakly basic pyridine ligands (in the case of β-pyrrole alkyl-substituted porphyrins such as OEP and etioporphyrin), such as 3-chloropyridine

(Scheidt *et al.*, 1983) and 3,5-dichloropyridine (Scheidt *et al.*, 1989; Kintner and Dawson, 1991), Fe(III) porphyrins exhibit a broad derivative-shaped EPR feature near $g = 4$ (Masuda *et al.*, 1980; Scheidt *et al.*, 1989; Boersma and Goff, 1982; Kintner and Dawson, 1991) and NMR spectra that show anti-Curie behavior (Goff and Shimomura, 1980; Goff, 1983, and references therein; Dugad and Mitra; 1984; Dugad *et al.*, 1985). Many of these spectral features are similar to those of the cytochromes c', including the EPR (Maltempo *et al.*, 1974; Maltempo, 1975; Kintner and Dawson, 1991), near-IR MCD (Rawlings *et al.*, 1977; Yoshimura *et al.*, 1985; Kintner and Dawson, 1991), Mössbauer (Moss *et al.*, 1968; Reed *et al.*, 1979; Gupta *et al.*, 1987; Scheidt *et al.*, 1989), and resonance Raman (Teraoka and Kitagawa, 1980; Hobbs *et al.*, 1990) spectra, and the NMR spectra of *Chromatium vinosum* cytochrome c', but not those of *Rhodospirillum rubrum*, *Rhodospirillum molischianum*, and *Rhodopseudomonas palustris* (La Mar *et al.*, 1990). Thus, it may be that *C. vinosum* cytochrome c' is the anomaly among this type of cytochrome c. Interestingly, it is the *C. vinosum* protein that has been studied most frequently by other spectroscopic techniques.

The cytochromes c' are known to contain 5-coordinate heme c, with no distal residues present that are capable of hydrogen bonding to endogenous ligands (Weber *et al.*, 1981; Finzel *et al.*, 1985). Moss and Maltempo have proposed the existence of an intermediate-spin ($S = \frac{3}{2}$) state with a quantum-mechanical admixture of the more common high-spin ($S = \frac{5}{2}$) state for the protein (Moss *et al.*, 1968; Maltempo, 1974; Maltempo and Moss, 1976). This differs from a simple $S = \frac{3}{2} \rightleftarrows S = \frac{5}{2}$ equilibrium in that the EPR and Mössbauer features are unique, rather than simply a superposition of the expected spectra of the two spin states, and the ground state is composed of contributions from each state (Dugad *et al.*, 1985).

The dramatic anti-Curie dependence of the pyrrole-H resonance of [TPPFeOClO$_3$] from ~ -20 ppm at -45 C to $+20$ ppm at $54\,°C$ (Goff and Shimomura, 1980) is a now-familiar hallmark of the ^{1}H NMR spectra of spin-admixed $S = \frac{3}{2}, \frac{5}{2}$ model heme species. Smaller shifts in the *meso*-H (~ 0 to ~ -9 ppm) and pyrrole-α-CH$_2$ (-40 to ~ -33 ppm from -40 to $+50\,°C$) resonances are observed for [OEPOClO$_3$] (Goff and Shimomura, 1980; Boersma and Goff, 1982; Dugad and Mitra, 1984). At low temperatures, the $S = \frac{3}{2}$ contribution to the spin-admixed state increases, leading to depopulation of the $d_{x^2-y^2}$ orbital of the $S = \frac{5}{2}$ state, thus decreasing the large σ contact shift contribution so characteristic of simple high-spin 5- and 6-coordinate Fe(III) porphyrins (Section 2.6.1, 2). The NMR spectra have been fitted to a model based upon the d-orbital splitting pattern shown in Fig. 30 (Dugad *et al.*, 1985), where $\Delta = 34,100 \text{ cm}^{-1}$, $\delta_1 = 2,500 \text{ cm}^{-1}$, $\delta_2 = 25,000 \text{ cm}^{-1}$, and $g_\parallel = 2.0$, and the best values of $\zeta = 200 \text{ cm}^{-1}$, $g_\perp = 4.77$, with the ground-state wave function composed of 61% $^4A_2$ and 39% $^6A_1$ for [TPPFeOClO$_3$], and $\zeta = 120 \text{ cm}^{-1}$, $g_\perp = 4.39$, with the ground-state wave function composed of 80% $^4A_2$ and 20% $^6A_1$ for [OEPFeOClO$_3$].

The tendency of weak-field anions to stabilize the $S = \frac{3}{2}$ ground state of the spin-admixed $\frac{3}{2}, \frac{5}{2}$ systems increases in the order CF$_3$CO$_2^- \ll$ CF$_3$SO$_3^- <$ C(CN)$_3^- \ll$ ClO$_4^-$ for TPPFe(III) (Boersma and Goff, 1982) and CF$_3$SO$_3^- <$ C(CN)$_3^- \ll$ ClO$_4^- <$ SbF$_6^-$ for OEPFe(III) (Boersma and Goff, 1982; Kintner and Dawson, 1991). The low-basicity ligand 3,5-dichloropyridine appears to stabilize the $S = \frac{3}{2}$ ground state in [OEPFe(3,5-Cl$_2$Py)$_2$]$^+$ about as well as ClO$_4^-$ or SbF$_6^-$ (Kintner and Dawson, 1991). For *para*-substituted derivatives of [TPPFeOClO$_3$], the $S = \frac{5}{2}$ contribution, as judged by the

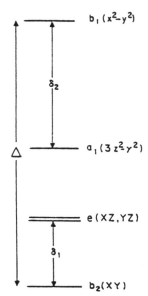

Figure 30. Single-orbital field energies of d_5 ion in tetragonal symmetry. Reprinted with permission from *Proc. Indian Acad. Sci. (Chem. Sci.)* **95**:194 (1985). © 1985 Indian Academy of Science.

pyrrole-H chemical shift at 29 °C, Fig. 31, increases in the order p-OCH$_3$ < p-H < p-Cl < p-CF$_3$ (Toney *et al.*, 1984b), so electron-donating substituents on the phenyl rings favor the $S = \frac{3}{2}$ ground state, where the $d_{x^2-y^2}$ orbital is depopulated. Consistent with this, the 2,4,6-trimethoxy derivative, [(2,4,6-(OCH$_3$)$_3$)$_4$TPPFeOClO$_3$] has the lowest reported chemical shift frequency of the pyrrole-H (Toney *et al.*, 1984a), and thus appears to be the "purest" $S = \frac{3}{2}$ complex. However, the temperature dependence of the isotropic shifts has not been reported. The shift to higher frequency (δ_{obs} positive) of the *m*-phenyl-H indicates a dipolar contribution of positive sign, consistent with the EPR parameters ($g_\perp > g_\parallel$) obtained by Dugad *et al.* (1985) for [TPPFeOClO$_3$].

2.8. Low-Spin Fe(III) Porphyrins

Low-spin Fe(III) porphyrins continue to be of major interest, both because of the richness of information that can be obtained about the nature of the orbital of the unpaired electron (this richness is aided by the good to excellent resolution of the spectra) and because of the relevance of this spin state to a variety of heme proteins (the ferricytochromes *a*, *b*, *c*, *d*, and *f*, the cyanomet forms of the oxygen-carrying proteins myoglobin and hemoglobin, the cyanide complexes of a number of peroxidases, and other related heme oxidases and monooxygenases).

It has been accepted for a number of years that low-spin Fe(III) hemes have a $(d_{xz}, d_{yz})^3$ ground state, with spin delocalization occurring through the filled $3e(\pi)$ porphyrin orbitals, i.e., P → Fe π bonding, leading to π contact shifts observed at the β-pyrrole positions, and no π contact shifts at the *meso* positions (La Mar and Walker,

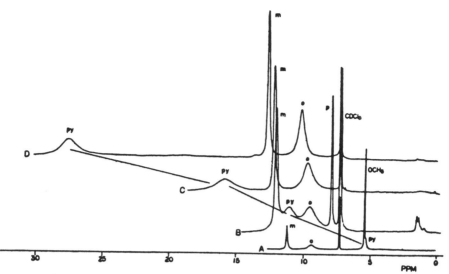

Figure 31. ^{1}H NMR spectra (250 MHz, chloroform-d, 29 °C) of 4-substituted phenyl perchlorato complexes: (A) 4-methoxyphenyl; (B) unsubstituted tetraphenyl; (C) 4-chlorophenyl; (D) 4-(trifluoromethyl)-phenyl. Proton designations: o, *ortho* H; m, *meta H*; p, *para* H; OCH$_3$, 4-methoxy H; py, pyrrole β-H. Solutions are ~2 mH in complex. Reprinted with permission from *Inorg. Chem.* **23**:4350 (1989). © 1989 American Chemical Society.

1979). In addition, it has been shown that dipolar shifts are not insignificant for this electron configuration, and a factoring method was developed, based upon the axial geometric factors $[(3\cos^2\Theta - 1)/r^3]$ of the protons at various locations around the heme and the supposition that there is no contact shift at the *meso* positions (La Mar and Walker, 1979, and references therein), as mentioned in Section 1.4.3. This method was found to work well for *bis*-(imidazole) complexes of Fe(III) porphyrins, somewhat less well for dicyano and *bis*-(pyridine) complexes, in that order (La Mar and Walker, 1979). The increasing failure of the supposition that there was no contact shift at the *meso* positions was variously ascribed to "polarization" of the σ electronic system by the π-symmetry unpaired electron and increased π-bonding interactions of the porphyrin ring with the metal due to weaker σ-bonding from axial ligands to Fe(III) (La Mar and Walker, 1979). New results (Safo *et al.*, 1992; Watson *et al.*, 1992), discussed below, clarify this situation.

In order to calculate the dipolar contribution to the isotropic shift of the protons of these complexes according to Eq. (7), it is necessary to know the principal values, χ_{ii}, of the magnetic susceptibility tensor. However, these values are usually not available. Hence, the principal values g_{ii} of the g tensor which are often readily measured by EPR spectroscopy or a combination of EPR and magnetic Mössbauer spectroscopies (Walker *et al.*, 1986), are used instead [Eq. (9)], with the cautionary notes concerning the importance of the SOZ contribution presented in Section 1.3.2. The g values of a much larger range of complexes have been measured since 1979, and a representative group of these

TABLE 10
The g Values of Selected Low-Spin Fe(III) Porphyrins

Porphyrin	L, L'	g_1^a	g_2^a	g_3^a	$(g_\parallel^2 - g_\perp^2)^b$	$(g_{xx}^2 - g_{yy}^2)^c$	Δ/λ^d	V/λ^e	a^f	b^f	c^f	$\Psi\, d_{xy}^g$	Ref.
OEP	2 NMeIm	2.986	2.273	1.506	5.20	-2.73	3.27	1.82	0.960	0.270	0.144	2.1	h
	2 4-NMe$_2$Py	2.818	2.278	1.642	4.00	-2.49	3.62	2.26	0.969	0.224	0.121	1.5	i
	2 4-CNPy	HS^j											h
	2 3-MePz	2.63	2.37	1.72	2.63^k or -3.31^l	-2.66^k or 1.30^l	2.52^k or -3.67^l	2.95^k or -1.44^l	0.976^k or 0.127^l	0.178^k or 0.178^l	0.127^k or 0.976^l	1.6^k or 95.3^l	h
	2 4-MePz	2.65	2.36	1.72	2.76^k or -3.34^l	-2.61^k or 1.45^l	3.04^k or -3.68^l	2.88^k or -1.60^l	0.977^k or 0.125^l	0.181^k or 0.181^l	0.125^k or 0.977^l	1.6k or 95.4^l	h
	2 PMe$_3$	3.54			~10.8	UK^m							h
	2 CN$^-$	3.73			~12.9	UK^m							h
TPP	2 NMeIm	2.886	2.294	1.549	4.50	-2.86	3.17	2.03	0.961	0.248	0.138	1.9	n
	2 MeImH	3.41	1.87	0.82	9.54	-2.82	2.96	0.88	0.884	0.434	0.176	3.1	o
	2 4-NMe$_2$Py	2.786	2.284	1.657	3.78	-2.47	3.60	2.34	0.970	0.216	0.120	1.4	n
	2 Py	3.70	1.12	-0.46	12.96	-1.04	2.16	0.17	0.742	0.640	0.243	5.9	p
	2 4-CNPy	2.63	2.63	0.96	-5.00^l	0.00^l	-1.75^l	0.00^l	0.285^l	0.285^l	0.897^l	80.4^l	p
	2 3-MePz	2.581	2.382	1.739	2.31^k or -3.14^l	-2.65^k or 0.99^l	2.78^k or -3.77^l	3.20^k or -1.19^l	0.977^k or 0.127^l	0.166^k or 0.166^l	0.127^k or 0.977^l	1.6^k or 95.5^l	n
	2 3-NH$_2$Pz	2.407	2.294	1.845	-1.46^k or -2.12^l	-1.86^k or 0.53^l	3.52^k or -5.11^l	4.47^k or -1.28^l	0.983^k or 0.094^l	0.118^k or 0.118^l	0.094^k or 0.983^l	0.9^k or 96.7^l	n
TMP	2 PMe$_3$	3.59			~11.3	UK^m							h
	2 t-BuNC	2.19	2.194	1.94	-1.03^l	0.00^l	-9.20^l	0.00^l	0.057^l	0.057^l	0.991^l	98.3^l	h
	2 CN$^-$	3.70	1.05	0.52	13.00	-0.83	6.87	0.44	0.817	0.547	0.091	0.8	p
	1 CN$^-$, 1 Py	3.31	1.76	0.34	9.35	-2.98	2.07	0.66	0.824	0.483	0.231	5.3	p
	2 Im$^-$	2.73	2.28	1.76	3.30	-2.10	4.25	2.70	0.983	0.190	0.102	1.0	r
	2 NMeIm	2.886	2.325	1.571	4.39	-2.94	3.09	2.07	0.966	0.244	0.140	2.0	i
	2 2-MeImH	3.17			~7.1	UK^m							r
	2 4-NMe$_2$Py	3.44	1.80	0.92	10.07	-2.39	3.64	0.89	0.891	0.429	0.150	2.2	s
	2 4-CNPy	2.53	2.53	1.56	-3.97^l	0.00^l	-2.92^l	0.00^l	0.183^l	0.183^l	0.956^l	91.4^l	i
	2 3-CH$_3$Pz	2.58	2.43	1.76	2.16^k or -3.18^l	-2.81^k or 0.75^l	2.54^k or -3.76^l	3.31^k or -0.89^l	0.983^k or 0.131^l	0.161^k or 0.161^l	0.131^k or 0.983^l	1.7^k or 96.5^l	s
	2 4-CH$_3$Pz	2.61	2.42	1.74	2.37^k or -3.31^l	-2.83^k or 0.96^l	2.62^k or -3.66^l	3.13^k or -1.06^l	0.980^k or 0.133^l	0.170^k or 0.170^l	0.133^k or 0.980^l	1.8^k or 96.1^l	h
	2 3-NH$_2$Pz	2.382	2.307	1.929	1.15^k or -1.78^l	-1.60^k or 0.35^l	3.96^k or -6.11^l	5.50^k or -1.21^l	0.998^k or 0.080^l	0.096^k or 0.096^l	0.080^k or 0.998^l	0.6^k or 99.6^l	h

		g_1	g_2	g_3		UK[m]							
PP	2 PMe$_3$	3.55	—	—	~10.9	0.00[j]	−3.31[l]	0.00[l]	0.164[l]	0.164[l]	0.979[l]	95.8[l]	h
	2 CN⁻	2.56	2.56	~1.70	−3.66[j]	−2.66	3.50	1.94	0.961	0.255	0.130	1.7	h
	2 ImH	2.92	2.25	1.55	4.79	−2.06	4.39	2.66	0.983	0.192	0.100	1.0	t
	2 Im⁻	2.74	2.27	1.76	3.38	−2.9	3.36	1.56	0.949	0.304	0.143	2.1	t
	1 CN⁻, 1 Im⁻	3.1	2.2	1.4	6.2								t
OEC	2 NMeIm	2.51	2.37	1.73	2.00[k] or −2.97[l]	−2.62[k] or 0.68[l]	2.57[k] or −3.83[l]	3.39[k] or −3.88[l]	0.972[k] or 0.127[l]	0.155[k] or 0.155[l]	0.127[k] or 0.972[l]	1.6[k] or 94.5[l]	u
OEiBC	2 NMeIm	2.49	2.37	1.71	1.93[k] or −2.98[l]	−2.69[k] or 0.58[l]	2.47[k] or −3.76[l]	3.39[k] or 0.73[l]	0.968[k] or 0.131[l]	0.155[k] or 0.155[l]	0.131[k] or 0.968[l]	1.7[k] or 93.7[l]	u
TPC	2 ImH	2.49	2.39	1.75	1.81[l] or −2.89[l]	−2.65[k] or 0.49[l]	2.46[k] or −3.92[l]	3.59[k] or 0.66[l]	0.975[k] or 0.128[l]	0.148[k] or 0.148[l]	0.128[k] or 0.975[l]	1.6[k] or 95.1[l]	v
Sulfmyoglobin cyanide		2.65	2.43	1.65	2.71[k] or −3.74[l]	−3.18[k] or 1.12[l]	2.34[k] or −3.24[l]	2.75[k] or −0.96[l]	0.970[k] or 0.149[l]	0.191[k] or 0.191[l]	0.149[k] or 0.970[l]	2.2[k] or 94.1[l]	w
Myoglobin cyanide		3.45	1.89	0.93	9.68	−2.71	3.36	0.94	0.895	0.417	0.158	2.5	x
Leghemoglobin n-butylamine		3.38	2.05	0.61[y]	9.14	−3.83	2.03	0.85	0.874	0.466	0.232	5.4	x

[a] Unless otherwise indicated, calculations using g_1, g_2, and g_3 assume $g_1 = g_{zz}$, $g_2 = g_{yy}$, $g_3 = g_{xx}$.

[b] $(g_1^2 - g_3^2) = [g_{zz}^2 - \frac{1}{2}(g_{xx}^2 + g_{yy}^2)]$, the axial magnetic anisotropy term in Eq. (9).

[c] $(g_{xx}^2 - g_{yy}^2)$ is the rhombic magnetic anisotropy term in Eq. (9).

[d] The tetragonal splitting of the d-orbitals, calculated from Eq. (17b).

[e] The rhombic splitting of the d-orbitals, calculated from Eq. (17a).

[f] Calculated from Eqs. (16a–c).

[g] $\% d_{xy} = 100c^2$.

[h] Watson and Walker (1992).

[i] Safo et al. (1991a).

[j] Only a high-spin signal observed; either complex did not form, or else it is high-spin.

[k] Calculated assuming the same assignment of g values as above.[a] Leads to $V/\Delta > 2/3$ and indicates an "improper" axis system (Taylor, 1977).

[l] Calculated assuming $g_1 = -g_{xx}$, $g_2 = g_{yy}$, $g_3 = -g_{zz}$. Gives rise to small V/Δ, indicating a "proper" axis system (Taylor, 1977).

[m] Unknown; $|g_{xx}^2 - g_{yy}^2|$ expected to be small.

[n] Walker et al. (1984).

[o] Walker et al. (1986).

[p] Inniss et al. (1988).

[q] Scheidt and Safo (1991).

[r] Quinn et al. (1982).

[s] Safo et al. (1992).

[t] Chacko and La Mar (1982).

[u] Stolzenberg et al. (1981).

[v] Peisach et al. (1973).

[w] Berzofsky et al. (1971).

[x] Thomson and Gadsby (1990).

[y] Calculated from $g_{zz}^2 + g_{yy}^2 + g_{xx}^2 = 16$ (Griffith, 1956).

is presented in Table 10, where the axial g anisotropy factor $[g_{xx}^2 - \frac{1}{2}(g_{xx}^2 + g_{yy}^2)]$ or $g_{\parallel}^2 - g_{\perp}^2$, which is proportional to the axial dipolar shift, and the rhombic g anisotropy factor, $g_{xx}^2 - g_{yy}^2$, which is proportional to the rhombic dipolar shift (see Section 1.3.2), as well as the Δ/λ and V/λ d-orbital splitting parameters and percent d_{xy} character of the orbital of the unpaired electron, discussed below, are also included. It can be seen from the data of Table 10 that the axial anisotropy increases in the order Im $<$ CN$^-$ and that the axial anisotropy of pyridine complexes depends upon the particular pyridine and the particular porphyrin. It is clear from the calculated axial anisotropies based upon EPR g values that the dipolar contribution to the isotropic shift should increase in the order Im $<$ CN$^- <$ Py for all but the least basic pyridines, but this is not usually observed (La Mar and Walker, 1979). In fact, the dipolar contribution usually decreases, while the contact contribution to the isotropic shift of *meso* substituents increases, suggesting that for dicyano and *bis*-(pyridine) complexes some delocalization occurs through the $3e(\pi)$ filled porphyrin orbitals and some through the $4e(\pi^*)$ empty porphyrin orbitals for the complexes. This point will be further emphasized below. For the time being, we note that not only have *bis*-(imidazole), dicyano, and *bis*-(pyridine) complexes of iron(III) porphyrins been studied, but also *bis*-(isonitrile), *bis*-(ammine), *bis*-(phosphine), and mixed (R)imidazole–imidazole, imidazole–imidazolate, imidazole–pyridine, pyridine–cyanide, imidazole–cyanide, and imidazole–phosphine complexes. Each of these combinations has special features that will be discussed separately. We begin with the *bis*-(isonitrile) complex, which introduces us to a "new" electron configuration for low-spin Fe(III) porphyrins.

2.8.1. *bis*-(Isonitrile) Complexes: The $(d_{xy})^1$ Ground State

It has recently been shown by Simonneaux and co-workers that when *tert*-butylisocyanide, t-BuNC, is bound to [TPPFeOClO$_3$] in CD$_2$Cl$_2$ at 298 K, the NMR spectrum is entirely contained within the window -2 to 14 ppm relative to TMS (Simonneaux *et al.*, 1989). The pyrrole-H proton resonance is at 9.73 ppm (Table 11), rather than the "expected" -17 ppm, based on previously studied *bis*-(imidazole), dicyano, and *bis*-(pyridine) complexes of (TPP)Fe(III) porphyrins (La Mar and Walker, 1979). All resonances shift in accordance with the Curie law, as shown in Fig. 32. Isotropic shifts are found to have small to negligible dipolar contributions, and there is a relatively large contact contribution to the *meso*-phenyl-H resonances. The authors recognized that the pattern of isotropic shifts observed was indicative of the $(d_{xy})^1$ ground state rather than the previously recognized, commonly observed $(d_{xz}, d_{yz})^3$ ground state, both illustrated in Fig. 33. However, the $(d_{xy})^1$ electron configuration of itself does not allow the possibility of π spin delocalization to the *meso* positions, since the d_{xy} orbital does not have the proper symmetry for π-bonding to the porphyrin π system. Thus a more sophisticated view of the orbital of the unpaired electron is required.

It has been recognized that the EPR g values of low-spin Fe(III) porphyrin complexes provide valuable information about the orbital of the unpaired electron and the degree of mixing of the three t_{2g} orbitals of octahedral symmetry, d_{xz}, d_{yz}, and d_{xy}, shown in Fig. 33, in the ground state orbital of the unpaired electron of low-spin Fe(III) porphyrins (Safo *et al.*, 1992). The Griffith theory for the EPR parameters of low-spin Fe(III) (Griffith, 1956) describes the orbital of the unpaired electron as composed of

TABLE 11
Isotropic Shifts of [TPPFe(t-BuNC)$_2$]ClO$_4$ Complexes[a]

Proton type	δ_{obs}[b]	δ_{iso}[c]	δ_{dip}[d]	δ_{con}[d]	δ_{dip}[e]	δ_{con}[e]
o-H	0.96	−7.06	0	−7.06	+0.66	−7.72
m-H	13.75	6.12	0	6.12	+0.30	5.82
m-CH$_3$	1.29	−1.22	0	−1.22	+0.18	−1.40
p-H	3.21	−4.42	0	−4.42	+0.27	−4.69
p-CH$_3$	8.95	6.31	0	6.31	+0.20	6.11
pyrrole-H	9.73	1.28	0	1.28	+1.28	0.00
t-BuNC-H	−1.87	−1.38	0	−1.38	−1.6?	~0

[a] Data taken from Simonneaux et al. (1989). Reprinted with permission from Inorg. Chem. 28:823 (1989).
© 1989 American Chemical Society.
[b] In CD$_2$Cl$_2$ at 25 °C.
[c] Diamagnetic [TPPFe(t-BuNC)$_2$] used as reference.
[d] Assuming dipolar contribution = 0 (Simonneaux et al., 1989).
[e] Assuming contact contribution at pyrrole-H position = 0 (see text).

some fraction of each of the three formerly t_{2g} d-orbitals. The highest-energy Kramer's doublet is thus expressed as

$$|\Psi\rangle^+ = a|d_{xz}\rangle^+ + ib|d_{yz}\rangle^+ + c|d_{xy}\rangle^- \tag{15a}$$

$$|\Psi\rangle^- = a|d_{xz}\rangle^- + ib|d_{yz}\rangle^- + c|d_{xy}\rangle^+ \tag{15b}$$

The coefficients a, b, and c can be calculated from the EPR g values:

$$a = (g_{zz} + g_{yy})/4K \tag{16a}$$

$$b = (g_{zz} - g_{xx})/4K \tag{16b}$$

$$c = (g_{yy} - g_{xx})/4K \tag{16c}$$

where $4K = [8(g_{zz} + g_{yy} - g_{xx})]^{1/2}$ (Taylor, 1977). The relative energies of the three formerly t_{2g} d-orbitals can also be calculated from the g values. Assuming d_{xy} to be the lowest in energy of the three, two quantities, Δ and V, in units of the spin–orbit coupling constant λ, are defined, as shown in the right-hand side of Fig. 33. The quantities Δ/λ and V/λ are calculated as follows (Taylor, 1977):

$$\frac{V}{\lambda} = E_{yz} - E_{xz} = \frac{g_{xx}}{g_{zz} + g_{yy}} + \frac{g_{yy}}{g_{zz} - g_{xx}} \tag{17a}$$

$$\frac{\Delta}{\lambda} = E_{xz} - E_{xy} - \frac{V}{2\lambda} = \frac{g_{xx}}{g_{zz} + g_{yy}} + \frac{g_{zz}}{g_{yy} - g_{xx}} - \frac{V}{2\lambda} \tag{17b}$$

Thus, the orbital of the unpaired electron in all of the cases presented in Table 10 is a mixture of contributions from the three formerly t_{2g} orbitals. In particular, for the bis-(isonitrile) complex of TPPFe(III), the ground state is largely $(d_{xy})^1$ (and thus Δ/λ is negative), but it contains small amounts (0.85% each) of d_{xz} and d_{yz} character. Simonneaux and co-workers considered the dipolar shift to be zero at all positions (Simonneaux et al., 1989), as shown in Table 11. Although the g anisotropy calculated from

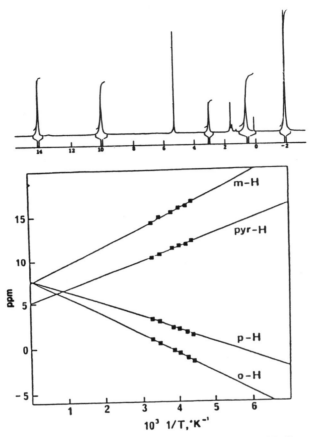

Figure 32. Top: Proton NMR spectrum of [TPPFe(*t*-BuNC)₂]ClO₄ in CD₂Cl₂ at 298 K. Bottom: Curie plot for proton resonances for [TPPFe(*t*-BuNC)₂]ClO₄ in CD₂Cl₂. Reprinted with permission from *Inorg. Chem.* **28**:824 (1989). © 1989 American Chemical Society.

the EPR *g* values is indeed small (see Table 10), it may be different in fluid solution at the temperature of the NMR experiments, as is the case of the *bis*-(pyridine) complexes of TMPFe(III), Section 2.8.4. The shift to higher frequency for the pyrrole-H resonance (+1.28 ppm, Table 11) suggests this may be the case. Thus, we have calculated the probable maximum size of the dipolar contribution, assuming the pyrrole-H isotropic shift is totally dipolar in nature. This dipolar shift (+1.28 ppm) can then be used to scale the dipolar shifts of the *meso*-phenyl-H and the *t*-BuNC-CH₃-H. The result of this procedure, shown in the far right-hand column of Table 11, is still basically the same as that presented by Simonneaux and co-workers, that the contact contribution to the *meso*-phenyl-H shifts of [TPPFe(*t*-BuNC)₂]⁺ is large for such protons (−7.72 ppm, *o*-H; +5.82 ppm, *m*-H; −4.69 ppm, *p*-H). The alternating pattern of the phenyl shifts is indicative of a contact contribution to the isotropic shifts and suggested to the authors

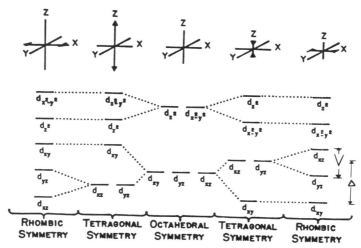

Figure 33.   Further splitting of the $3d$ orbitals of an octahedral complex due to tetragonal and rhombic distortion.

that the large value of these shifts for *meso* substituents is consistent with delocalization into the empty $4e(\pi)$ orbitals (Fig. 4) rather than the filled $3e(\pi)$ orbitals. If this were the case, we could describe the delocalization as occurring through the *nominally filled* d_{xz}, d_{yz} set to the π^* orbitals of the porphyrin ring. But since the nominally filled $(d_{xz}, d_{yz})^4$ set contains a small amount of unpaired electron character, as described by Eq. (15), delocalization of π spin density to the $4e(\pi)$ orbitals of the porphyrin is allowed, giving rise to the observed π spin delocalization to the *meso* positions, and very little to the pyrrole positions.

To test this hypothesis, we can estimate the magnitude of the expected contact shifts utilizing the predicted spin density, $\rho = 0.068$, at the *meso* carbons of the $4e(\pi^*)$ orbitals (Longuet-Higgins *et al.*, 1950), and by scaling the phenyl-H shifts to the calculated spin densities at the carbons of the benzyl radical (Carrington and McLachlan, 1967) as a model for the *meso*-phenyl odd-alternate hydrocarbon fragment. Using Eqs. (4) and (5) we can crudely estimate the magnitude of the expected contact shifts to be -64 (o), $+22(m)$, and -77 ppm (p) *if a full unpaired electron were present in the degenerate* $4e(\pi^*)$ *orbital set* (Fig. 4). For a total of 1.7% unpaired electron density in the d_{xz}, d_{yz} orbitals due to spin–orbit coupling with d_{xy}, the *maximum* contact shifts possible at the *meso*-phenyl positions are thus estimated to be approximately -1.1 (o), $+0.37$ (m), and -1.3 ppm (p) at 25 °C, clearly too small to explain the phenyl contact shifts listed in Table 11. Furthermore, preliminary NMR investigations of [OEPFe(t-BuNC)$_2$]$^+$ in CD$_2$Cl$_2$ show that the *meso*-H isotropic shift to lower frequency is extremely large ($\sim$$-48$ ppm at 25 °C) (Watson and Walker, 1992), confirming the fact that delocalization to the $4e(\pi^*)$ porphyrin orbitals by means of spin–orbit coupling of the largely $(d_{xy})^1$ ground state to the d_{xz}, d_{yz} orbitals is not sufficient to account for the observed *meso* shifts. Hence, it is possible that the largely $(d_{xy})^1$ ground state facilites thermal population of the π cation radical (i.e., Fe(II)P$^+$) configuration for these isonitrile complexes. If the d_{xy} orbital of [OEPFe(t-BuNC)$_2$]$^+$ is slightly lower in energy than the porphyrin $3a_{2u}(\pi)$ orbital (Fig. 4), it is possible to account for the *meso*-H isotropic shift by assuming 13–19% $(3a_{2u})^1$ character at 175 K and 11–17% at 310 K (Watson and Walker, 1992). These data imply an energy separation of d_{xy} and $3a_{2u}$ of only about 34 cm^{-1}. Whether this is a realistic interpretation of the NMR data remains to be verified by additional experiments. For example, measurement of the near infrared MCD spectra of these complexes could shed light on the relative energies of the formerly t_{2g} orbitals of the metal and the a_{1u}/a_{2u} π orbitals of the porphyrins at low temperatures, where only the ground state electronic configuration would be populated (Gadsby and Thomson, 1990). EPR spectroscopy at 4.2 K as compared to 77 K should show changes in the g-values, since the partial ($\sim$25%) $3a_{2u}(\pi)$ character of the unpaired electron at 77 K would not affect the g-values of the complex, because it provides no significant additional spin–orbit corrections. However, if at 4.2 K the unpaired electron is largely (99.99%) in the $3a_{2u}$ orbital, the EPR spectrum should be that of the porphyrin radical. [*Note added in proof*: The EPR spectrum of [OEPFe(t-BuNC)$_2$]$^+$ at 4.2 K is identical to that at 77 K (Watson and Walker, 1992).]

As we will see in Section 2.8.4, the $(d_{xy})^1$ electron configuration is observed not only for [TPPFe(t-BuNC)$_2$]$^+$ and [OEPFe(t-BuNC)$_2$]$^+$, but also for the complexes of increasingly low-basicity pyridines with model hemes. Thus, the $(d_{xy})^1$ configuration and its potential to mobilize partial spin delocalization to the HOMO of the porphyrin are of considerable importance in our understanding of the isotropic shifts of low-spin Fe(III) porphyrins and heme proteins.

2.8.2. *bis*-(Imidazole) Complexes

These complexes are the epitome of the opposite extreme of electronic ground states for low-spin Fe(III), as they are nearly pure $(d_{xz}, d_{yz})^3$. The $\sim$1.9% d_{xy} character of the orbital of the unpaired electron (Table 10) does not manifest itself in the pattern of spin delocalization, and there is no spin density in the $3a_{2u}(\pi)$ orbital. Thus, spin delocalization is observed to the pyrrole-H positions and not to the *meso* positions, consistent with P $\rightarrow$ Fe π delocalization through the $3e(\pi)$ filled orbitals of the porphyrin (La Mar and Walker, 1979).

La Mar and Walker (1979) have reviewed the isotropic shifts and their separation into contact and dipolar contributions) of the *bis*-(imidazole) complexes of Fe(III) with symmetrical synthetic porphyrins such as tetraphenylporphyrin (TPP), $\alpha,\beta,\gamma,\delta$-tetra-($n$-propyl)porphyrin (T-$n$-PrP), and octaethylporphyrin (OEP), as well as a number of natural porphyrins. They also reviewed nature of spin delocalization to the imidazole ligand (La Mar and Walker, 1979). Since that time there have been a number of investigations of

1. the effect of porphyrin substituents on the pattern of spin delocalization,
2. the effect of imidazole deprotonation on the pattern and extent of spin delocalization, and
3. the effect of imidazole plane orientation on the isotropic shifts of porphyrin protons.

Each of these will now be reviewed.

2.8.2a. Effect of Porphyrin Substituents on the Pattern of Spin Delocalization.

i. The Pattern of Isotropic Shifts of Unsymmetrically Phenyl-Substituted Derivatives of $[TPPFe(NMeIm)_2]^+$. Because naturally occurring hemes have no plane of symmetry perpendicular to the porphyrin plane, it is not possible to relate the orbital coefficients of the $3e(\pi)$ orbitals quantitatively to the contact shifts of protons on the periphery of the porphyrin ring. Thus, although the spread of the heme methyl isotropic shifts varies with the 2,4-substituents [4.3 ppm for $[PPFe(N-MeIm)_2]^+$, 4.8 ppm for $[DPFe(N-MeIm)_2]^+$, 4.5 ppm for $[MPFe(N-MeIm)_2]^+$) (La Mar and Walker, 1979), and to a much greater extent for the dicyano Fe(III) complexes of 2,4-disubstituted deuteroporphyrin IX discussed in Section 2.8.3 (La Mar *et al.*, 1978a), the variation does not appear to be systematic and thus does not allow a clear explanation as to the effects of electron-withdrawing or -donating substituents on the orbital of the unpaired electron. For this reason, series of *meso-* and β-pyrrole-substituted porphyrins of higher molecular symmetry have been synthesized and investigated by NMR spectroscopy (Walker, 1980; Walker *et al.*, 1982; Walker and Benson, 1982; Walker *et al.*, 1992; Lin *et al.*, 1992; Isaac *et al.*, 1992). It was found that for unsymmetrically substituted TPPs of the type $[(X)_1(Y)_3TPPFe(NMeIm)_2]^+$, where X and Y can be located in the *ortho-*, *meta-*, or *para*-position of its respective phenyl ring, the pyrrole-H resonance, centered at ~-16 ppm at 33 °C for the parent compound $[TPPFe(NMeIm)_2]^+$, is split into either three signals of intensity 1:1:2 (or four signals of equal intensity where the two signals with largest negative isotropic shift are very close together) if X is more electron-withdrawing than Y or 2:1:1 (or four signals of equal intensity having the two signals with least negative isotropic shift very close to each other) if X is more electron-donating than Y, as shown in Fig. 34. The spread of the pyrrole-H resonances was found to be linearly related to the difference in the Hammett σ-constants of the two substituents, as shown in Fig. 35 (Walker *et al.*, 1982). Since the EPR g values of all of the complexes are essentially identical (Walker *et al.*, 1984), the variation in the spread of the pyrrole-H resonances must be largely due to a change in the contact contribution as the difference in electron-donating/withdrawing properties of the phenyl substitutents increases, suggesting redistribution of the spin density in response to the difference in electronic effects of the substituents. (However, a small rhombic dipolar term, ($|\delta_{rhomb\ dip}| \leq 2$ ppm) due to the unsymmetrical substitution pattern, is also expected; see below.) The Curie plots for these unsymmetrically substituted $[TPPFe(NMeIm)_2]^+$ derivatives were found to yield average slopes and intercepts that are a function of the sum of the Hammett σ-constants of the substituents, as shown in Fig. 36, and the average slopes and intercepts of the Curie plots for complexes containing *ortho*-phenyl substituents were utilized to estimate the "apparent" σ-constants of *ortho* substituents (Walker and Benson, 1982; Walker *et al.*, 1992). It was later shown (Walker *et al.*, 1992) that the average slopes and intercepts for mono-*o*-substituted derivatives of $[TPPFe(NMeIm)_2]^+$ contain a large contribution from preferential orientation of one of the axial imidazole ligands due to steric interference with the *ortho*-substituent. Hence, the derived Hammett σ-constants for the *ortho*-substituents are not indicative simply of the electron donating/withdrawing characteristics of the *ortho*-substituents. The very small trend observed between the averge slope and the sum of the Hammett σ-constants of the *meta-* or *para-* substituents (Fig. 36) suggests that the variation in the contact shift of the pyrrole-H with the total basicity or σ-donor strength of the porphyrin nitrogens is very small (yet in the expected

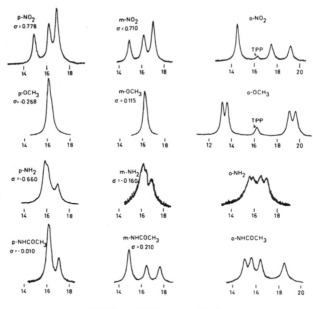

PPM upfield from TMS

Figure 34. NMR spectra of the pyrrole-H region of a series of low-spin Fe(III) complexes of [(p-, m-, or o-X)₁TPPFeCl] with N-methylimidazole, recorded at 90 MHz in CDCl₃, with Me₄Si as reference. Temperature = 33 °C. Several of the *ortho*-substituted derivatives contain a small amount of the symmetrical [TPPFe(N-MeIm)₂]⁺ (marked as "TPP"). Reprinted with permission from *J. Am. Chem. Soc.* **104**:1573 (1982). © 1982 American Chemical Society.

direction for P → Fe bonding as the mechanism of spin delocalization), and thus the spread of the pyrrole-H resonances represents the *degree of redistribution* of a constant amount of total spin density, rather than a change in the amount of spin density delocalized from Fe(III) to the porphyrin ring.

    ii. Detailed Description of the Mechanism of Spin Delocalization in Unsymmetrically Substituted Derivatives of [TPPFe(NMeIm)₂]⁺. Recently it has been possible, utilizing the COSY-45 experiment, to verify that the pyrrole-H coupling pattern observed for several of the unsymmetrically substituted [TPPFe(NMeIm)₂]⁺ complexes having one *ortho*-substituent is either peaks 1,4 and 2,3 (Fig. 37B) or 1,3 and 2,4 (Fig. 37D) (Lin *et al.*, 1992). The results have been explained by constructing linear combinations of the two $3e(\pi)$ orbitals shown in Fig. 4 so as to produce π orbitals with nodal planes passing through opposite *meso* positions, as shown in Fig. 38. The resonance and/or inductive effects of the unique substituent are then expected to modify the electron distribution in these orbitals, to either increase or decrease somewhat the π electron density at the pyrrole position nearest the unique substituent (a), and to a decreasing extent the next position (b), as well as possibly positions c and d. Depending on the distance over which the modification takes place (only a and b vs. all four positions

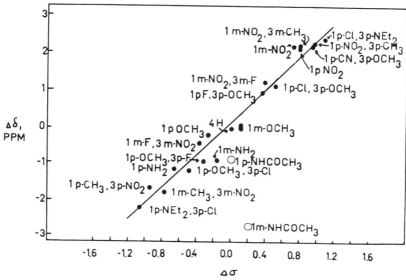

Figure 35. Plot of the separation between highest and lowest field pyrrole-H resonances versus the difference in Hammett σ-constant fdor the two types of substituents involved in the $[(X)_1(Y)_3TPPFe(N\text{-}MeIm)_2]^+Cl^-$ complexes. Reprinted with permission from *J. Am. Chem. Soc.* **104**:1574 (1982). © 1982 American Chemical Society.

a–d), the sizes of the modified MO coefficients, and hence the sizes of the circles in Fig. 38C–F, will be either $a > d > c > b$ or $a > d > b > c$. In any case, four unique electron densities are produced, yielding four different spin densities at the four pyrrole-H positions (Lin *et al.*, 1992). Hence, the correlations observed in the COSY-45 map (Fig. 37) confirm that the modified $3e(\pi)$ orbitals shown in Fig. 38C, D or E, F correctly describe the orbital into which spin density is delocalized.

A closer look at Fig. 38 and the isotropic shifts of the pyrrole-H of these unsymmetrically substituted derivatives of $[TPPFe(NMeIm)_2]^+$ provides considerably greater insight into the quantitative description of the modified π molecular orbitals of Fig. 38C–F. If we consider the case of $[(o\text{-}OEt)_1TPPFe(NMeIm)_2]^+$, the COSY-45 coupling pattern (Fig. 38B) is consistent with the modified molecular orbitals of Fig. 38C and D (Lin *et al.*, 1992). Quantitative analysis of the contributions to the contact terms for protons a–d is presented in Table 12. It should be noticed that there is an expected rhombic contribution to the dipolar shift, since this complex has in-plane asymmetry. The size of this rhombic contribution is calculated by scaling the size of the axial dipolar contribution by the relative sizes of the axial and rhombic g anisotropies, Table 10, and multiplying by $\cos 2\Omega$ for each position a–d. The angular dependence of the in-plane function, $\cos 2\Omega$, together with the fact that $g_{xx} < g_{yy}$, means that $\delta_{\text{rhomb dip}}$ is negative at positions a and d and positive at b and c. The calculated contact contributions for the orbital of Fig. 38C are consistent with small modification of the molecular orbital of Fig. 38A, while the calculated contact contributions for the orbital of Fig. 38D

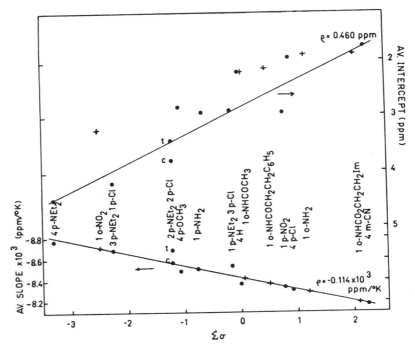

Figure 36. Plot of the average intercept and average slope of each *meta-* or *para*-substituted (circles) TPPFe(N-MeIm)$_2^+$ complex of this study against the sum of the Hammett σ-constants of the substitutents. The average slopes of *ortho*-substituted complexes (Table 1) were used to define "apparent" σ-constants for these substituents (crosses): Average intercepts of all complexes show more scatter than do the average slopes, as expected. Reprinted with permission from *J. Phys. Chem.* **88**:3496 (1982). © 1982 American Chemical Society.

are all very similar and are inconsistent with the pattern expected for slight modification of the orbital of Fig. 38B. Furthermore, the NOESY spectrum of [(*o*-OEt)$_1$TPPFe(NMeIm)$_2$]$^+$ at −30 °C shows an additional cross peak between resonances 1 and 2 of Fig. 37A (Simonis and Walker, 1992), similar to that shown in Fig. 16, that is believed to arise from an NOE between protons *b* and *c*. This result is also consistent with the pattern of spin delocalization shown in Fig. 38C. Since the supposition involved in drawing the modified orbitals of Fig. 38C and D was that only positions *a* and *b* were affected by the unique substituent (see caption to Fig. 38), the spin densities observed at positions *c* and *d* (peaks 2 and 3) should be proportional to the molecular orbital mixing coefficients c_i^2 at those positions. The predicted electron densities at positions *c* and *d* should thus be proportional to the molecular orbital coefficients c_i^2 at the two types of positions in the unmodified orbital of Fig. 38A, i.e., 0.0905 and 0.0305, a ratio of 3:1. However, the ratio of the derived contact shifts at positions *c* and *d* for the orbital of Fig. 38C (Table 12) is 1.55:1. Hence, the π molecular orbital of the porphyrin into which the spin is delocalized is actually composed of 72% of that shown

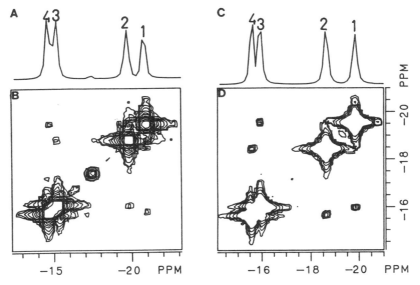

Figure 37. (A) One-dimensional 300-MHz NMR spectrum of the pyrrole-H region of [(o-OEt)$_1$TPPFe(NMeIm)$_2$]Cl, 1, in CDCl$_3$ recorded at room temperature. The small peak at-17.5 ppm is due to the presence of a small amount of [TPPFe(NMeIm)$_2$]$^+$ in the sample. (B) COSY-45 spectrum of the pyrrole-H region of recorded at room temperature, showing the 1,4 and 2,3 coupling pattern. (C) One-dimensional 300-MHz NMR spectrum of the pyrrole-H region of [(o-Cl)$_1$(p-OCH$_3$)$_3$TPPFe(NMeIm)$_2$]$^+$Cl$^-$, 2, in CD$_2$Cl$_2$ recorded at room temperature. (D) COSY-45 spectrum of the pyrrole-H region of 2 recorded at room temperature, showing the 1,3 and 2,4 coupling pattern. Reprinted with permission from *Inorg. Chem.* **31**:4217 (1992). © 1992 American Chemical Society.

in Fig. 38C and 28% of that shown in Fig. 38D. Similar analysis of the pyrrole-H isotropic shifts at −30 °C yields a combination of 75% of that shown in Fig. 38C and 25% of that shown in Fig. 38D for the orbital of the unpaired electron. Saying this another way, the π MO of Fig. 38C contains spin density 75% of the time, while that in Fig. 38D contains spin density 25% of the time at −20 °C. This is equivalent to saying that the orbital of Fig. 38C is the ground state orbital into which the unpaired electron is delocalized by P → Fe π donation, while the orbital of Fig. 38D is the excited state. Using these percentages it can be estimated that the two orbitals differ in energy by about 28 cm^{-1}. The temperature-dependent variation in the partial delocalization of the unpaired electron to these two modified $3e(\pi)$ orbitals is the probable cause of the nonzero intercepts of the Curie plots for low-spin Fe(III) porphyrins.

    iii. Unsymmetrically β-Pyrrole-substituted Porphyrins. The NMR spectra of a series of β-octasubstituted porphyrin isomers prepared from 3,4-diethylpyrrole and 3,4-*bis*-(diethylcarboxamido)pyrrole have also been investigated. The isomer having six ethyl (E) and two diethylcarboxamide (A) substituents [E$_6$A$_2$PFe(NMeIm)$_2$]$^+$ shows the multiplicity of resonances expected for a complex of idealized C_{2v} symmetry, if one

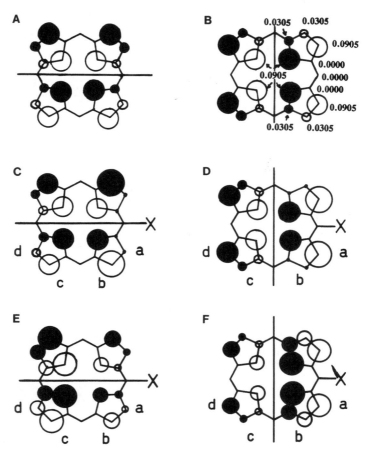

Figure 38. Symmetry properties of and electron density distribution in the filled $3e(\pi)$ porphyrin orbitals that have proper symmetry for overlap with the d_{xz} and d_{yz} orbitals of low-spin Fe(III). (A, B) The linear combinations of the $3d(\pi)$ orbitals that are appropriate for *meso*-substituted porphyrins. (C, D) Exaggerated modifications of these orbitals due to the presence of one uniquely substituted phenyl ring X if the effect of this substituent is felt only as far as the adjacent pyrrole positions a and b. (E, F) Exaggerated modifications of these orbitals if the effect of the unique substituent is felt over all four pyrrole positions a–d. If the phenyl substituent is electron donating, the modified orbital shown in Fig. 1C or 1E should be slightly lower in energy than the orbital shown in Fig. 1D or 1F, while if the substituent is electron-withdrawing, the orbital shown in Fig. 1D or 1F should be slightly lower in energy than that shown in Fig. 1C or 1E. It is not possible to predict unambiguously which (or rather, what fraction of each) of the two modified orbitals will be favored for spin delocalization, especially when *ortho*-phenyl substituents are present. However, independent of whether the substituent is electron-donating or -withdrawing, Fig. 1C–1F indicate that a different amount of spin density will be placed at each of the four symmetry-related pyrrole positions, a, b, c, and d. Since we expect the proton attached to the carbon with largest spin density to have the largest hyperfine shift, in either case the coupling pattern predicted by this simplest theory is a,b and c,d or pyrrole-H peaks 1,4 and 2,3 in the case of Fig. 1C, D or peaks 1,3 and 2,4 in the case of Fig. 1E, F. Reprinted with permission from *Inorg. Chem.* **31**:4216 (1992). © 1992 American Chemical Society.

TABLE 12

Separation of the Dipolar and Contact Contributions for $[(o\text{-OEt})_1\text{TPPFe-(NMeIm)}_2]^{+a}$

Orbital shown in Fig.	Resonance	Position	T (°C)	δ_{obs} (ppm)	δ_{iso} (ppm)	$\delta_{ax\,dip}$ (ppm)	$\delta_{rhomb\,dip}$ (ppm)	δ_{con} (ppm)	δ_{con}/T
38C	1	b	-21	-21.2	-30.2	-6.3	+2.1	-26.0	0.08
	2	c		-20.0	-29.0	-6.3	+2.1	-24.8	0.08
	3	d		-15.4	-24.4	-6.3	-2.1	-16.0	0.05
	4	a		-14.6	-23.6	-6.3	-2.1	-15.2	0.05
38D	1	a	-21	-21.2	-30.2	-6.3	-2.1	-21.8	0.07
	2	d		-20.0	-29.0	-6.3	-2.1	-20.6	0.07
	3	c		-15.4	-24.4	-6.3	+2.1	-20.2	0.06
	4	b		-14.6	-23.6	-6.3	+2.1	-19.4	0.06
38C	1	b	-30	-28.8	-37.8	-7.6	+2.5	-32.7	0.13
	2	c		-27.2	-36.2	-7.6	+2.5	-31.1	0.12
	3	d		-19.6	-28.6	-7.6	-2.5	-18.5	0.07
	4	a		-18.8	-27.8	-7.6	-2.5	-17.7	0.07
38D	1	a	-30	-28.8	-37.8	-7.6	-2.5	-27.7	0.11
	2	d		-27.8	-36.2	-7.6	-2.5	-26.1	0.10
	3	c		-19.6	-28.6	-7.6	+2.5	-23.5	0.09
	4	b		-18.8	-27.8	-7.6	+2.5	-22.7	0.09

^{a} Data taken from Walker (1992).

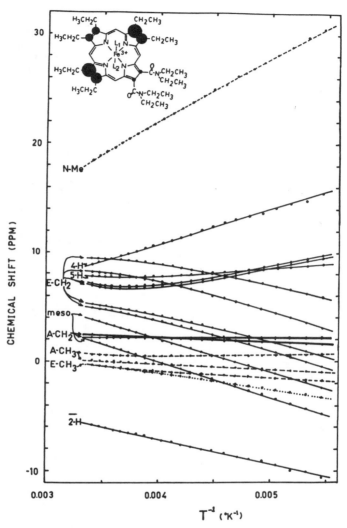

Figure 39. Curie plot for the proton resonances of the *bis*-(N-methylimidazole) complex of [(2,3,7,8,12,13-hexaethyl)-17,18-*bis*-(diethylcarboxamido)]porphinatoiron(III), [E$_6$A$_2$PFe (NHeIm)$_2$]$^+$. Note that the E-CH$_2$ resonances having similar temperature dependencies are geminal pairs, as demonstrated by the COSY-45 spectrum shown in Fig. 11. Curvature of the E-CH$_2$ plots is believed to be due to restricted rotation about the β-pyrrole-C-CH$_2$ bond. The inset shows the pattern of spin delocalization to the pyrrole positions.

also realizes that all ethyl CH_2 protons are diastereotopic, due to the fact that the carboxamide substituents are out of the porphyrin plane. Couplings of ethyl and diethyl-carboxamide protons were established by COSY techniques, as shown in Fig. 11. The Curie plot (Fig. 39) shows that the ethyl CH_2 resonances exhibit curvature (Isaac *et al.*, 1992), probably due to restricted rotation of the ethyl groups, as postulated previously for $[OEPFe(NMeIm)_2]^+$ (La Mar and Walker, 1973a, 1979). The isotropic shifts are consistent with the orbital of the unpaired electron having its nodal plane along the molecular symmetry axis, with large electron density in the pyrrole rings on either side of the pyrrole ring having the carboxamide substituents (insert to Fig. 39), and then modifying that orbital by decreasing the electron density at the positions of the electron-withdrawing carboxamide substuents and the pyrrole-ethyl groups closest to the unique pyrrole ring.

2.8.2b. The Shifts of Coordinated Imidazole Ligands and the Effect of Imidazole Deprotonation on the Pattern and Extent of Spin Delocalization.

i. Neutral Imidazole Ligands. The isotropic shifts of the protons of the imidazole ligands of $[TPPFe(R-Im)_2]^+$, and their separation into contact and dipolar contributions, have previously been reported (Satterlee and La Mar, 1976; La Mar and Walker, 1979). The dipolar shifts of all imidazole protons are positive (to high frequency), while the contact contributions are negative at all ring positions. Reversal in the sign of the contact shift when $-H$ is replaced by $-CH_3$ at the 1-, 2-, and 5- positions of the imidazole ring confirms that the mechanism of spin delocalization is largely π in nature, as discussed in Section 1.3.1 and 1.4.3. The sizes of the contact shifts at each ring position are reflective of the molecular orbital coefficients obtained by Satterlee and La Mar (1976) for the lowest-energy π^* orbital, suggesting to these authors that the axial imidazoles interacted with low-spin Fe(III) as π acceptors, and that the mechanism of spin delocalization is Fe $\rightarrow$ Im π back-bonding. However, since that time it has been found that the imidazole ligand shifts are much more consistent with delocalization into the filled π orbital that has large density on the bonding nitrogen (Chacko and La Mar, 1982). Hence, it now appears that imidazoles interact with low-spin Fe(III) as π donors and that the mechanism of spin delocalization is Im $\rightarrow$ Fe π bonding. The same conclusion has been reached concerning the interaction of strongly basic pyridines with low-spin Fe(III), while weakly basic pyridines are found to act as π acceptors toward low-spin Fe(III) (Safo *et al.*, 1992; Watson *et al.*, 1992), as discussed in Section 2.8.

In the unsymmetrically phenyl-substituted derivatives of $[TPPFe(NMeIm)_2]^+$ discussed in Section 2.8, it is found that two $N-CH_3$ and two 2-H signals of the imidazole moiety are well resolved when one phenyl ring has an *ortho* or *meta* substituent (Walker *et al.*, 1982, 1992). Thus, the N-MeIm ligands on the two sides of the porphyrin plane have different contact or dipolar shifts. These differences are undoubtedly caused by differences in the two Fe—N bond lengths, which would be expected to affect both the contact and dipolar contributions to the isotropic shifts of the protons of the two ligands: Slight lengthening of the Fe—N bond of one ligand is expected to decrease both the contact and dipolar contributions, while slight shortening will increase both contributions. Interestingly, the two $N-CH_3$ signals appear to be more or less equally offset from the position of the single $N-CH_3$ signal in unsymmetrically substituted $[TPPFe(NMeIm)_2]^+$ complexes that have no *ortho*-phenyl substituents (Walker *et al.*, 1992), suggesting that one bond is slightly shortened and one slightly lengthened when a single *ortho*-phenyl substituent is present. Although the molecular structures of none of these complexes have been determined by X-ray crystallography, it has been suggested that the *ortho* substituent may cause buckling of the porphyrin ring, so that the metal is displaced slightly from a perfectly in-plane position toward the side on which the *ortho* substituent resides, thus strengthening its bond to the ligand on that side, while weakening its bond to the ligand on the opposite side of the porphyrin ring (Walker *et al.*, 1992).

ii. Imidazolate Ligands. Deprotonation of the coordinated imidazole ligands of $[PPFe(ImH)_2]^+$ results in large shifts in the resonances of the protoporphyrin and imidazole ring protons to lower frequency (Chacko and La Mar, 1982), as shown in Fig. 40, and a concommitant decrease in the *g* anisotropy of the complex (Table 10). Thus, the dipolar contributions to the isotropic shifts of imidazolate ligand protons are decreased in magnitude as compared to those of the neutral ligands (Chacko and La Mar, 1982). The contact contribution is decreased to somewhat over half that observed for the neutral ligand at H-4' and approximately doubled at the H-5' position of the ring, suggesting that deprotonation of the imidazole leads to an increase in the strength of L → Fe π donation (Chacko and La Mar, 1982).

The isotropic shifts of the 1-, 3-, 5-, and 8-methyl protons of the protoporphyrin ring of the *bis*-(imidazole) complex decrease sharply upon deprotonation of the two imidazole ligands (Fig. 40), with the average methyl isotropic shift decreasing from 15.6 to 9.7 ppm at 25 °C (Chacko and La Mar, 1982). Only about 1.2–1.4 ppm of this decrease in isotropic shift is due to the decrease in the dipolar contribution; thus the contact contribution decreases by 5–6 ppm when the imidazole ligands are deprotonated (Chacko and La Mar, 1982). This decrease in the contact contribution is consistent with the increase, mentioned in the previous paragraph, in the strength of L → Fe π donation of the axial ligands upon deprotonation.

2.8.2c. Effect of Imidazole Plane Orientation on the Isotropic Shifts of Low-Spin Iron(III) Porphyrins. For some time it has been suggested that the orientation of planar axial ligands may be significant in explaining the large spread of the methyl resonances of the protoporphyrin ring of heme proteins (La Mar and Walker, 1979; La Mar, 1979; Walker and Benson, 1982; Satterlee, 1986, 1987, and references therein). A number of investigations of model hemes have been carried out aimed at testing this hypothesis (Traylor and Berzinis, 1980; Goff, 1980; Walker, 1980; Walker *et al.*, 1983; Zhang *et al.*, 1990, Walker *et al.*, 1992). Since the barrier to rotation of axial ligands in model hemes is very low (Nakamura and Groves, 1988; Walker and Simonis, 1991; Walker *et al.*, 1992), a number of systems that involve covalent attachment of one or both axial ligands have been investigated (Traylor and Berzinis, 1980; Goff, 1980; Walker, 1980; Walker *et al.*, 1984). However, such covalent attachment often leads to lower symmetry than desired (Traylor and Berzinis, 1980; Walker *et al.*, 1984). Nevertheless, the importance of axial ligand plane orientation was demonstrated by the fact that eight ring-CH_3 resonances were observed for the diastereomeric mixture of two coordination isomers of the dichelated protohemin monocyanides shown in Fig. 41, with a total spread of 17.1 ppm in DMSO-d_6 at ambient temperature (Traylor and Berzinis, 1980). Although two isomers are involved, this approaches the known spread of the $-CH_3$ resonances in heme proteins: 11.5 ppm for cyanoferricytochrome *c* (Wüthrich, 1969), 22.3 ppm for sperm whale cyanometmyoglobin (Emerson and La Mar, 1990a), and 27.1 ppm for

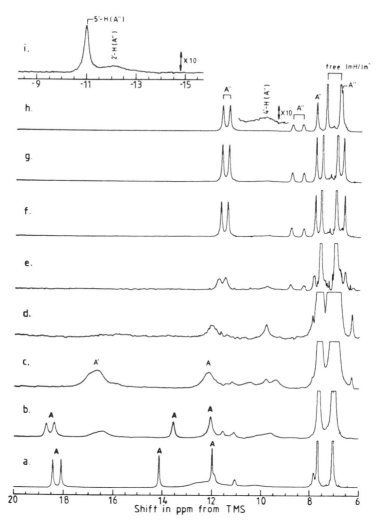

Figure 40. Portions of the 360-MHz ^{1}H NMR spectra illustrating the formation of the monoimid-azole complex A′ and *bis*(imidazolate) complex A″ by the addition of NaOD/D$_2$O to an ~0.002 M solution of the *bis*(imidazole) complex A in (C$_2$H$_3$)$_2$SO at 25 °C. (a) No base, (b–h) increasing amount of 0.2-M NaOD/D$_2$O added at each stage, and (i) upfield portion of the spectrum in (h): the total IMH concentration is ~0.006 M. Reprinted with permission from *J. Am. Chem. Soc.* **104**:7004 (1982). © 1982 American Chemical Society.

HRPCN (Thanabal *et al.*, 1987), where axial ligands are held in particular orientations by the protein.

Smaller spreads (~5 ppm) of the four pyrrole-H resonances, observed for *mono-ortho*-oxoalkylimidazole appended derivatives of [TPPFe(NPrIm)$_2$]$^+$ (Goff, 1980), were

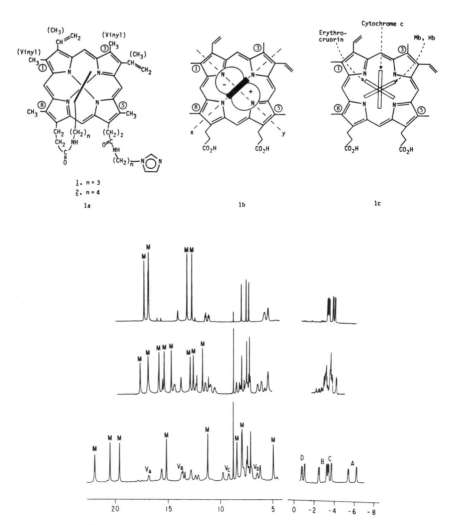

Figure 41. Top: (1a) Structure of one of the four diasteromeric isomers of the dichelated protohemin monocyanide complexes of Traylor and Berzinis (1980); (1b) interaction of the imidazole π orbital with the d_{yz} orbital of the metal; (1c) orientation of axial histidine imidazole planes in three heme proteins. Bottom: NMR spectra of cyanide complexes of the hemins I (lower trace), 2 (middle trace), and the complex with freely rotating N-methylimidazole 3 (upper trace). Methyl group resonances are identified as M, and vinyl group resonances are identified as V_A, V_B, V_C, and V_D (methines), and A, B, C, and D (methylenes). Reprinted with permission from *J. Am. Chem. Soc.* **102**:2845 and 2846 (1980). © 1980 American Chemical Society.

ascribed to a combination of "tension" and electronic asymmetry. A similar spread of the pyrrole-H resonances were observed for a series of *mono-ortho*-NHCOR derivatives of $[TPPFe(NMeIm)_2]^+$, including one in which the amide substituent provided a pendant imidazole ligand (Walker, 1980). In each of these cases of pendant imidazole derivatives of TPPFe(III), a much larger spread in the four pyrrole-H resonances was observed when the sixth ligand was 2-MeImH (Goff, 1980; Walker, 1980), suggesting that "the 2-methyl group must sterically interact with the porphyrin to yield a strained linkage which serves to magnify bonding constraints imposed by the *trans*-imidazole appendage" (Goff, 1980).

The "crowding" of the axial ligands by bulky *ortho*-phenyl substituents has also been tried (Walker *et al.*, 1983; Zhang *et al.*, 1990; Walker and Simonis, 1991; Walker *et al.*, 1992), with mixed success. The *bis*-(NMeIm) complexes of the $\alpha\beta\alpha\beta$ and $\alpha\alpha\beta\beta$ atropisomers of the Fe(III) state of the "picket fence" porphyrin (*tetra*-(o-pivalamidophenyl)porphyrin, which were expected to enforce perpendicular and parallel orientations of the axial ligands, respectively, appeared to show the expected multiplicity of pyrrole-H resonances (one and two, respectively). However, it was not possible to completely separate the two atropisomers, and the *o*-pivalamidophenyl groups underwent rotation about the *meso*-phenyl bond during iron insertion (Walker *et al.*, 1983).

A large series of *mono*-(*o*-substituted) derivatives of $[TPPFe(NMeIm)_2]^+$ have been prepared and investigated in detail by ^{1}H NMR spectroscopy (Walker *et al.*, 1992). It was found that only the *mono*-(*ortho*-carboxamido) substituent, *o*-CONR$_2$, is able to significantly hinder the rotation of one axial ligand; the other ligand is presumed to rotate rapidly over the investigated temperature range (+60 to −90 °C). Up to eight pyrrole-H resonances are observed at low temperatures, as shown in Fig. 42; as the temperature is raised, a chemical exchange process begins that leads, in the case of the dimethylcarboxamide derivative, to coalescence of the eight resonances to the four peaks expected for a *mono-ortho*-substituted derivative of $[TPPFe(NMeIm)_2]^+$ (Zhang *et al.*, 1990). ROESY and temperature-dependent NOESY (Fig. 15), as well as saturation transfer experiments (Fig. 8), confirm that resonances 1,4; 2,3; 5,8; and 6,7 are in chemical exchange (Simonis *et al.*, 1992b). At 21 °C, the pattern of pyrrole-H resonances for each of the complexes is different, as shown in Fig. 43, and the progress toward chemical exchange averaging of the resonances appears to reflect the relative sizes of the carboxamide substituents (Walker *et al.*, 1992). The dicyano complex of 1 (Fig. 42) has a total spread of the pyrrole-H resonances of ∼2 ppm at −60 °C, as compared to ∼10 ppm for the *bis*-(ImH) complex at the same temperature (Simonis and Walker, 1992). The eight-peak pattern observed at low temperatures has been explained in terms of the combined effects of

1. unsymmetrical phenyl substitution and
2. hindered axial ligand plane orientation, caused by the bulky *o*-carboxamide substituent (Walker *et al.*, 1992),

as shown schematically in Fig. 44. The nature of the axial ligand also has a dramatic effect upon the spread of the pyrrole-H resonances of these complexes, with hindered

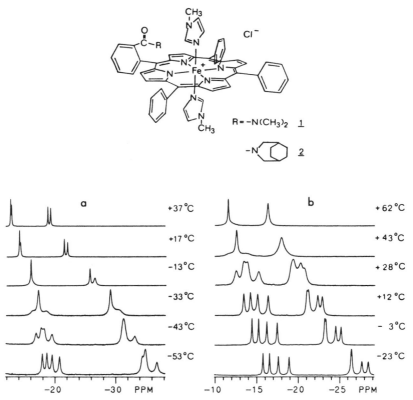

Figure 42. Top: Structures of the two $[(o\text{-}CONR_2)_1TPPFe(N\text{-}MeIm)_2]^+$ complexes. Bottom: NMR spectra (300 MHz) (using a General Electric GN-300) of the pyrrole proton resonances of (a) complex 1 and (b) complex 2 as a function of temperature, both in $CDCl_3$. In 1 the coalescence from eight to two peaks occurs at around -27 °C, in 2 at approximately 60 °C. Reprinted with permission from *J. Am. Chem. Soc.* 112:6125 (1990). © 1990 American Chemical Society.

imidazoles and basic pyridines producing a much smaller spread than nonhindered imidazoles (Simonis and Walker, 1989). This is presumably because the hindered imidazoles (2-MeImH and BzImH) and pyridines are constrained to lie over the *meso* positions of the porphyrin ring (Scheidt and Lee, 1987), thus making the symmetry planes of the unsymmetrical substituent effect and the axial ligand plane effect (Fig. 44) perpendicular to each other, thereby effectively cancelling or at least decreasing their combined effect, while nonhindered imidazoles can lie over the porphyrin nitrogens N_1 and N_3 (or N_2 and N_4), thereby maximizing their combined effect.

Nakamura and Groves (1988) showed that the NMR spectrum of the tetramesitylporphyrin complex, $[TMPFe(2\text{-}MeImH)_2]^+$, has four pyrrole-H, four *ortho*-CH$_3$, four *meta*-H, and two *para*-CH$_3$ resonances at temperatures below -30 °C, even though the porphyrin is symmetrically substituted. The multiplicity of porphyrin resonances was

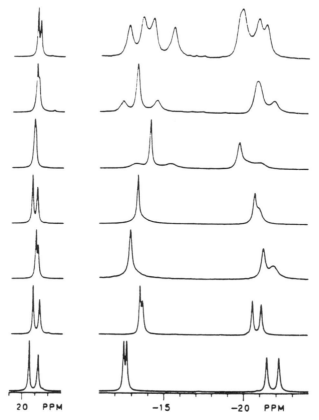

Figure 43. Pyrrole-H patterns observed at 21 °C in CDCl₃ for the seven [(o-CONR₂)₁-TPPFe(NMeIm)₂]Cl complexes of this study. Substituents from bottom to top: -CON(CH₂)₄, -CON(CH₃)₂, -CON(C₂H₅)₂, -CON(CH₂)₆, -CON(CH₂)₄O, -CON(CH₂)₅, -CON(C₈H₁₄). Reprinted with permission from *New J. Chem.* **16**:615 (1992). © 1992 Revues Scientifiques et Techniques.

explained in terms of fixed orientations of the 2-methylimidazole ligands at low temperatures. It was assumed that the projections of the axial ligand planes were perpendicular to one another and lay over the porphyrin nitrogens (Nakamura and Groves, 1988). However, it was later shown by NOESY techniques at very low temperatures (Figs. 13, 14, and 45) that the observed NOEs were consistent with the axial ligands lying over the *meso* positions of the porphyrin ring in perpendicular planes, with the 2-CH₃ group of each ligand in close proximity to one type of *o*-CH₃, and only two pyrrole-H giving rise to NOE cross peaks (Walker and Simonis, 1991), as shown schematically in Fig. 45. Such orientation of the hindered 2-MeImH ligands is expected in order that the Fe—N_ax bond lengths can be short enough to allow the complex to be low-spin, and it has been observed by X-ray crystallography for [TPPFe(2-MeImH)₂]⁺Cl⁻

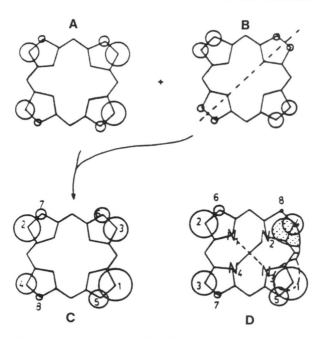

Figure 44. Combination of (A) the electronic effect of one *meso*-phenyl substituent (from Fig. 38F) and (B) the fixed orientation of one axial ligand plane (from Figs. 4 and 41) to produce (C) up to eight different π-electron densities at the β-pyrrole positions of the $[(o\text{-}CONR_2)_1TPPFe(NMeIm)_2]^+$ complexes. The β-pyrrole π-electron densities shown reflect the degree of unpaired electron densities at each β-pyrrole position, and thus the relative isotropic shift of the pyrrole-H at each position. (D) Further modification of this pattern due to the solvation effect of the alkyl groups of the amide on one pyrrole ring, necessary to explain the observed coalescence behavior of the pyrrole-H signals. Reprinted with permission from *New J. Chem.* **16**:616 (1992). © 1992 Revues Scientifiques et Techniques.

(Kirner *et al.*, 1978). Confirmation of the perpendicular orientation of the 2-MeImH ligands over the *meso* positions of the porphyrin ring was provided by the fact that only two equal-intensity, widely separated ^{13}C peaks were observed for the *meso*-^{13}C-labelled $[TMPFe(2\text{-}MeImH)_2]^+$ complex (Nakamura and Nakamura, 1991), rather than the three expected (with $1:1:2$ intensity ratio) from an over-porphyrin-nitrogen perpendicular orientation of the 2-methylimidazoles. Magnetic anisotropy appears to arise from the off-axis binding of the 2-MeImH ligands (Walker and Simonis, 1991), as suggested schematically in Fig. 45; binding two unsymmetrically substituted pyridine molecules to TMPFe(III) does *not* produce multiple resonances from the TMP ligand protons. *Thus, the porphyrin ring is not sensitive to the symmetry of the axial ligands unless their substituents require that they bind to the metal off the axis normal to the plane of the porphyrin ring.* The extrapolated intercepts of the Curie plot for this complex were used to describe the nature of the orbital of the unpaired electron (Walker and Simonis, 1991).

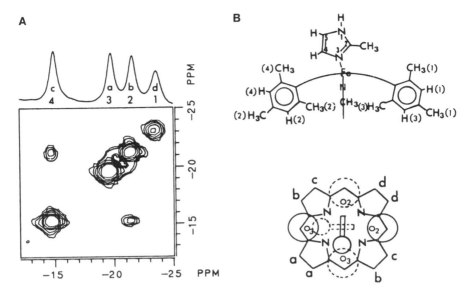

Figure 45. (A) The pyrrole H region of the NOESY map of [TMPFe(2-MeImH)$_2$]Cl at $-74\,°C$ at a low contour level, showing only one set of cross peaks, between pyrrole signals 4 and 2. (B) Proposed model of the structure of [TMPFe(2-MeImH)$_2$]Cl, showing the schematic representation of the ruffling of the porphyrin ring which bends the mesityl groups to move the *ortho*-CH$_3$ out of the way of the upper axial ligand. The opposite bend of the other two mesityl groups provides an oblong binding cavity for the lower axial ligand, at right angles to that shown. The unique pyrrole-H protons are labeled a–d in accord with the symmetry of the molecule. From information discussed in the text, H$_a$ gives rise to peak 3, H$_b$ to 2, H$_c$ to 4, and H$_d$ to 1. Assignments of phenyl resonances are given in parentheses in (B). Solid circles and lines represent ortho CH$_3$ groups and ligands above the plane, while dotted symbols represent those below the plane. Reprinted with permission from *J. Am. Chem. Soc.* **23**:8656 (1991). © 1991 American Chemical Society.

The same multiplicity of pyrrole-H, *o*-CH$_3$, *m*-H, and *p*-CH$_3$ resonances as observed for [TMPFe(2-MeImH)$_2$]$^+$ (Nakamura and Groves, 1988; Walker and Simonis, 1991) has recently been reported for the corresponding 2-methylbenzimidazole complex (Nakamura and Nakamura, 1991), although the isotropic shifts of the TMP protons, their anti-Curie behavior, and the single observed EPR feature at $g = 2.6$ are all consistent with a $(d_{xz}, d_{yz})^4(d_{xy})^1$ ground state having thermal occupancy of the $(d_{xy})^2(d_{xz}, d_{yz})^3$ excited state, as observed for low-basicity pyridine complexes of this porphyrin (Safo *et al.*, 1992; Watson *et al.*, 1992), discussed in Section 2.8.4. The 2-methylbenzimidazole ligand is the most highly sterically hindered axial ligand known to bind to iron porphyrins to produce low-spin Fe(III) complexes; it is expected to produce extreme ruffling of the porphyrin core in order to reduce steric interactions between the *ortho*-methyl groups of the TMP and the 2-methyl and benzo rings of this ligand. The fact that the complex exhibits a $(d_{xy})^1$ ground state suggests that ruffling of the porphyrin ring may be an important contributor to the stabilization of this unusual ground state.

2.8.3. Dicyano Complexes

Cyanide has been used as a ligand for both model hemes and heme proteins for many reasons:

1. It enforces the low-spin state for most Fe(III) systems.
2. It is not a planar ligand and hence does not create in-plane magnetic anisotropy.
3. It is small and can enter the heme pocket of most heme proteins.
4. It may (or may not) be a good mimic of other ligands (O_2, CO) that bind to heme proteins.
5. Since it binds more readily to Fe(III) than Fe(II), it produces a coordination geometry, in a paramagnetic state of iron, that is relevant to many active sites of heme proteins; the paramagnetism "illuminates" the heme in the NMR spectrum of the protein.
6. Finally, and most importantly, the NMR spectra of dicyanohemin complexes are well resolved, since the lines are quite narrow, due to very rapid electron spin relaxation in these complexes; both T_1 and T_2 are much longer for dicyano and monocyano monoimidazole complexes of most if not all Fe(III) porphyrins (Bertini et al., 1991; Unger et al., 1985a).

This excellent resolution, combined with the availability of high-field NMR spectrometers, has recently made it possible to identify a number of iron porphyrins in coals and to assess their degree of carbonification (Bonnett et al., 1990; Czechowski and Latos-Grazynski, 1990), to allow detection of very long-range isotope shifts in β-vinyl deuterated derivatives of dicyano-protohemin IX (Medforth et al., 1991), and to allow a variety of heteronuclear 2D NMR experiments to be carried out on dicyanoprotohemin IX (Timkovich, 1991), as shown in Fig. 12. Dicyano and (cyano)-(pyridine) complexes of natural hemins have also been studied in aqueous detergent micelles (Mazumdar, 1990; Mazumdar et al., 1990).

There have been a number of ^1H NMR investigations of Fe(III) porphyrins as either the dicyano or mono-cyano, mono-L 6-coordinate, low-spin complexes since the initial report of Wüthrich, Shulman, Wyluda, and Caughey (Wüthrich et al., 1969). In this section we will concentrate on the dicyano complexes of low-spin Fe(III) porphyrins, both synthetic and naturally occurring. First, we will consider the importance of the dipolar contribution to the isotropic shifts of these complexes, which should be related to the anisotropy in the magnetic susceptibility tensor [Eq. (7)] or, with some cautions, the g tensor [Eq. (9)]. The EPR spectra of all dicyano complexes of Fe(III) porphyrins (except those of TMP and other 2,6-phenyl substituted derivatives of TPP, Table 10) are of the "large g_{max}" type (Walker et al., 1984, 1986), with only one resolved feature at $g \geq 3.4$ that is observed at temperatures below 20 K (Table 10). Fe(III) chlorins (Berzofsky et al., 1971; Keating et al., 1992) (and presumably other reduced hemes), as well as the TMP (Watson and Walker, 1992) (and presumably other 2,6-phenyl-substituted derivatives of TPP) have near-axial (reduced hemes) or axial (TMP) EPR spectra, with $g_\perp \sim 2.6$, $g_\parallel \sim 1.5$. For the "large g_{max}" porphyrin complexes, the two unresolved g values have been obtained by single-crystal EPR investigations for $[TPPFe(CN)_2]^-$, as well as the bis-(pyridine) and (CN)(pyridine) complexes (Inniss et al., 1988), Table 10. It is interesting to note that even though the "large g_{max}" EPR signal is indicative of a relatively small rhombic splitting of the d-orbitals (i.e., the energies, calculated from Eq. (12a–c), of d_{xz} and d_{yz} are not very different), the calculated axial and rhombic anisotropies are larger than those of the imidazole complexes that give rise to rhombic EPR signals (and have a large difference in energy between d_{xz} and d_{yz}) (Table 10). Nevertheless, as we shall see, the dipolar contribution to the isotropic shifts of dicyano complexes of Fe(III) porphyrins are no larger, and are sometimes smaller than those of the corresponding *bis*-(imidazole) complexes (La Mar *et al.*, 1977b). Thus, g values measured at very low temperatures (<20 K) may not be very good indicators of the magnetic anisotropies of the dicyano complexes of Fe(III) porphyrins at the temperatures of the NMR investigations.

When the NMR spectra of dicyano complexes of natural or synthetic hemins were investigated in detail it was found that the spectra are very sensitive to solvent and concentration (La Mar and Walker, 1979, and references therein), due to

1. hydrogen bonding interactions between the coordinated cyanide ligands and the solvent (La Mar *et al.*, 1977b), and
2. aggregation, at least of the cyanide complexes of the natural hemins (La Mar and Viscio, 1974; Viscio and La Mar, 1978a, b; La Mar *et al.*, 1981) and [OEPFe(CN)$_2$]$^-$.

It was found that the isotropic shifts of [TPPFe(CN)$_2$]$^-$ derivatives and [OEPFe(CN)$_2$]$^-$ experience an overall decrease as the solvent hydrogen-bond donor strength increases (La Mar *et al.*, 1977b). Part of this decrease was shown to be due to a decrease in the dipolar contribution to the isotropic shift due to a decrease in the magnetic anisotropy, and part to a decrease in the pyrrole-H contact shift, while the *meso*-phenyl contact shifts increase as the solvent hydrogen-bond donor strength increases (La Mar *et al.*, 1977b). Furthermore, the amount of contact shift of the *meso*-phenyl protons appears to increase as the temperature decreases, while the dipolar shift decreases with decreasing temperature. Although the authors were unable to account for these observations fully at the time, it now seems possible, based on recent findings for the *bis*-(pyridine) complexes to be discussed in Section 2.8.4, that the amount of $(d_{xy})^1$ character to the Fe(III) orbital of the unpaired electron may increase as the hydrogen-bond donor strength of the solvent increases and that the amount of $(d_{xy})^1$ character is a function of temperature.

The NMR spectra of a series of the dicyano complexes of Fe(III) derivatives of 2,4-disubstituted deuteroporphyrins have provided evidence of the extreme effects of peripheral substituents on the π electronic asymmetry: The spread of the ring-methyl resonances increases in the order 0.35 ppm (diethyl), 4.99 ppm (divinyl), 5.27 ppm for deuteroporphyrin itself (H), 10.48 ppm (dibromo), 12.64 ppm (disulfonate), 17.42 ppm (diacetyl), 19.4 ppm (diformyl), and 22.54 ppm (dicyano) (La Mar *et al.*, 1978a). In all cases, the isotropic shifts decrease in the order 8-CH$_3$ > 5-CH$_3$ > 3-CH$_3$ > 1-CH$_3$, with the average isotropic shift remaining within 1 ppm of the overall average value. The differences in the spread of the resonances are believed to be due to differences in the contact contribution to the isotropic shift, while the similarity in the average isotropic shifts suggests a fairly constant dipolar and overall contact contribution. Thus, porphyrin ring substituents appear to simply rearrange spin density among the four pyrrole rings; pyrrole rings that contain more strongly electron-donating substituents receive more spin density, while those that contain more strongly electron-withdrawing substituents receive less (La Mar et al., 1978a).

La Mar and co-workers have used the excellent resolution of the ^{1}H NMR spectra of dicyano iron(III) complexes of 2,4-disubstituted deuteroporphyrins to investigate in detail proton dipolar relaxation by delocalized spin density (Unger et al., 1985a). If only the unpaired electrons on the metal contribute, nuclear relaxation of protons in low-spin Fe(III) porphyrins by T_1 and T_2 processes should be given by

$$T_1^{-1} = 6Dr^{-6}T_{1e} \tag{18}$$

$$T_2^{-1} = 7Dr^{-6}T_{1e} + B(A_h)^2 T_{1e} \tag{19}$$

where $D = (1/15)\gamma_{H}^2 g^2 \beta^2 S(S+1)$ and $B = (1/3)S(S+1)$ (Unger et al., 1985a). This should mean that the T_1s of all ring CH_3 should be the same, since r^{-6} is identical for all ring CH_3. However, there is a marked difference in the T_1s of the four methyl groups of natural hemin derivatives that increases as the spread of the methyl resonances increases; the methyl group with the largest isotropic shift has the shortest T_1 (Unger et al., 1985a). These results were explained in terms of a sum of metal-centered dipolar relaxation and the dipolar relaxation by the delocalized spin density ρ in the aromatic carbon to which the methyl group is attached, yielding a modification of Eq. (18):

$$T_1^{-1} = 6D[R_M^{-6} + \rho^2 R_L^{-6}]T_{1e} \tag{20}$$

where R_M is the distance from the metal center to the proton of interest and R_L is the distance from the aromatic carbon to the proton, as illustrated in Fig. 46. The quantity ρ is proportional to the observed contact shift, as described by Eqs. (3) and (5) or (6). Thus, in line with the predictions of Eq. (20), and as shown in Fig. 46, a linear relationship was observed between the observed T_1^{-1} and $(\delta_{con})^2$. From the intercept, assuming $R_M = 6.15$ Å, T_{1e} was determined to be 1.8×10^{-12} s and independent of the basicity of the porphyrins in these model compounds. The slope of the line in Fig. 46 indicates that $R_M^{-6} = \rho^2 R_L^{-6}$ for a contact shift of $\sim$19 ppm. Assuming $R_L \sim 1.9$ Å, this yields $\rho = 0.03$. Thus, 0.03 delocalized spin in a carbon p orbital causes paramagnetic dipolar relaxation of the appended methyl proton of a magnitude comparable to that from the iron center (Unger et al., 1985a). Thus, dipolar relaxation from both ligand-centered and metal-centered spin density must be considered in analyzing heme methyl T_1s, not only in model complexes, but also in heme proteins. It is also probable that ligand-centered dipolar relaxation is a very major contributor to the short T_1s of the pyrrole-H of TPP complexes of low-spin Fe(III) (Walker and Simonis, 1991) and the meso-H of OEP and natural porphyrin complexes of high-spin Fe(III), since $R_L \sim 1.08$ Å. This could make the ligand-centered dipolar relaxation term 33 times larger for a pyrrole-H than a pyrrole-CH_3 with a contact shift of the same magnitude. The difference in T_1s observed for low-spin Fe(III) complexes of β-pyrrole substituted porphyrins (Unger et al., 1985a) as compared to those of TPP (Walker and Simonis, 1991; Lin et al., 1992) are approximately of this magnitude.

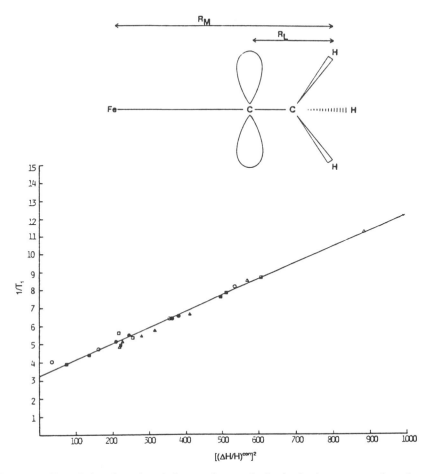

Figure 46. Top: Relaxation of methyl group by unpaired spin density on aromatic carbon to which it is attached and the paramagnetic metal center. The geometry of the interaction is characterized by distances R_M and R_L. R_M is the metal center to proton distance; R_L is the delocalized spin density to proton distance. Bottom: Plot of observed T_1^{-1} versus $[(\Delta H/H)_{con}]^2(\Delta_{con}^2)$ for the methyl groups of a variety of 2,4-R_2-deuterohemin-*bis*-cyanide complexes; ($\bullet$) $R_2 = R_4 =$ vinyl; ($\blacktriangle$) $R_2 = R_4 = H$; ($\blacksquare$) $R_2 = R_4 =$ Br; ($\bigcirc$) $R_2 = R_4 =$ acetyl; ($\triangle$) $R_2 =$ acetyl, $R_4 = H$; ($\blacksquare$) $R_2 = H$, $R_4 =$ acetyl. Reprinted with permission from S. W. Unger *et al.*, *J. Magn. Res.* **61**:455 (1985). © 1985 Academic Press, Inc.

2.8.4. *bis*-(Pyridine) Complexes

Hill and Morallee (1972) were the first to report an investigation of the proton NMR spectra of the *bis*-(pyridine) complexes of Fe(III) protoporphyrin IX. They recognized the dominance of the contact contribution to the isotropic shifts and suggested that the π orbitals involved were the $3e(\pi)$ filled orbitals and that spin delocalization occurred by P $\rightarrow$ Fe π donation. They observed a linear relationship between the isotropic shifts of the heme methyls and the basicity of the coordinated pyridine and explained the trend of decreasing isotropic shift with increasing pyridine basicity in terms of strong σ-donor pyridines effectively decreasing the positive charge on Fe(III) and thus the tendency for P $\rightarrow$ Fe π donation (Hill and Morallee, 1972). The temperature dependence of the chemical shifts of all resonances of *bis*-(pyridine) and *bis*-(R-pyridine) (where R-pyridine is a weaker base than the parent ligand, pyridine) complexes was found to depart markedly from Curie behavior; below $\sim$200 K they shift away from the diamagnetic region as the temperature is lowered, but above that temperature they reverse direction and again shift away from the diamagnetic region (Hill and Morallee, 1972). The authors suggested that a spin equilibrium was involved and postulated that it might be an $S = \frac{1}{2} \rightleftarrows \frac{3}{2}$ equilibrium (Hill and Morallee, 1972). Later work (Dugad *et al.*, 1987; Shedbalkar *et al.*, 1988) has favored an $S = \frac{1}{2} \rightleftarrows \frac{5}{2}$ spin equilibrium.

[OEPFe(R-Py)$_2$]$^+$ complexes with R-Py = pyridine and lower-basicity pyridines exhibit similar behavior (Hill *et al.*, 1979; Gregson, 1981), although the $S = \frac{3}{2}$ state is very close in energy. A model that explains the isotropic shifts and solution magnetic susceptibilities of the [OEPFe(R-Py)$_2$]$^+$ complexes has been developed that includes contributions from the ^2E, $^6A_1(\frac{5}{2})$, $^6A_1(\frac{3}{2})$, $^4A_2(\frac{3}{2})$, and $^2B_2(\frac{1}{2})$ states (Gregson, 1981). For [OEPFe(3-ClPy)$_2$]$^+$ at 320 K the percent contribution of each of these components to the E'' ground state is calculated to be 77%, 1%, 10%, 8%, and 3%, respectively. The first and second excited states, both E', are composed mainly of $^6A_1(\frac{1}{2})$ and $^2E(\frac{1}{2})$; hence, magnetic moments of these complexes may not rise above 4.5 μ_B, even though the ground state is not $S = \frac{3}{2}$ (Gregson, 1981). Interestingly, two different spin states ($S = \frac{3}{2}$ and an $S = \frac{1}{2} \rightleftarrows \frac{5}{2}$ equilibrium) have been observed for [OEPFe(3-ClPy)$_2$]$^+$ in the crystalline state, depending on the relative and absolute orientations of the 3-chloropyridine ligands: When the ϕ angles are near 45° (ligands lying over the *meso* positions) and the ligands are in perpendicular planes, the complex was found to have a spin equilibrium (Scheidt *et al.*, 1982), while when the ϕ angles are small (ligands lying nearly over the porphyrin nitrogens) and the ligands are in nearly parallel planes, the spin state of the complex was found to be $S = \frac{3}{2}$ (Scheidt *et al.*, 1983, 1987). [OEPFe(3-CNPy)$_2$]$^+$ also crystallizes as an $S = \frac{3}{2}$ ground state complex with the axial ligands in parallel planes over the porphyrin nitrogens (Safo *et al.*, 1991b). Certainly, all orientations are expected to be observed in solution, where axial ligands are expected to rotate rapidly over the temperature ranges investigated; hence the solution magnetic susceptibility and NMR results (Hill *et al.*, 1979; Gregson, 1981) and the X-ray crystallographic and solid-state magnetic susceptibility, EPR, and Mössbauer results (Scheidt *et al.*, 1982, 1983, 1987; Safo *et al.*, 1991b) are all consistent with the fact that low-basicity pyridine complexes of the type [OEPFe(R-Py)$_2$]$^+$, as well as those of the natural porphyrins (Hill and Morallee, 1972; Dugad *et al.*, 1987; Shedbalker *et al.*, 1988), are very close to the magnetic triple point of $S = \frac{1}{2}$, $\frac{3}{2}$, and $\frac{5}{2}$ (Gregson, 1981). However, in the case of the basic pyridine complex [OEPFe(4-NMe$_2$Py)$_2$]$^+$ the $S = \frac{1}{2}$ state alone is stabilized at all temperatures (La Mar *et al.*, 1977a; Safo *et al.*, 1991a), as it is for all *bis*-(imidazole) complexes.

In contrast to the complex spin state behavior of β-alkyl-substituted porphyrins discussed above, all known *bis*-(pyridine) complexes of Fe(III) tetraphenylporphyrins are low-spin (La Mar *et al.*, 1977a; La Mar and Walker, 1979; Safo *et al.*, 1992; Watson *et al.*, 1992). Nevertheless, their behavior is not as simple as that of the corresponding *bis*-(imidazole) complexes. While it has been found in all cases that the pyrrole-H resonance shifts to higher frequency as the basicity of the pyridine decreases (La Mar *et al.*, 1977a; Safo *et al.*, 1992; Watson *et al.*, 1992), all TPP complexes not having *ortho*-phenyl substituents show Curie behavior, albeit with very small temperature dependence for [TPPFe(4-CNPy)$_2$]$^+$ (La Mar *et al.*, 1977a), while TPP derivatives having *ortho*-phenyl substituents, such as [TMPFe(R-Py)$_2$]$^+$ (Safo *et al.*, 1992; Watson *et al.*, 1992) exhibit anti-Curie behavior when R-Py is a low-basicity pyridine (3-CNPy, 4-CNPy), as shown in Fig. 47. The *meta*-phenyl-H isotropic shift also increases dramatically as the basicity of the pyridine decreases, as shown in Table 13, so that for low-basicity pyridines it is no longer possible to assume that the *meso*-phenyl-H isotropic shifts are totally due to the dipolar contribution (Watson *et al.*, 1992). The shift of the

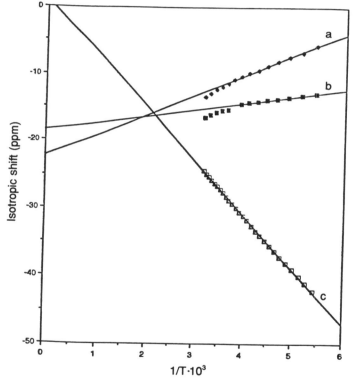

Figure 47. Curie plots of the pyrrole-H resonance of (a) [TMPFe(4-CNPy)$_2$]$^+$ in CD$_2$Cl$_2$, (b) [TMPFe(3-CNPy)$_2$]$^+$ in CD$_2$Cl$_2$, and (c) [TMPFe(N-MeIm)$_2$]$^+$ in CD$_2$Cl$_2$, showing normal (c) and anti (a and b) Curie behaviors. The fact that the pyrrole-H resonance of [TMPFe(4-CNPy)$_2$]$^+$ and [TMPFe(3-CNPy)$_2$]$^+$ shifts *away* from the diamagnetic position as the temperature is raised (a and b) is an indication that the $(d_{xy})^2(d_{xz}, d_{yz})^3$ electron configuration (which is observed for [TMPFe(N-MeIm)$_2$]$^+$, c), with its inherently large upfield shift, is increasingly populated at increasingly higher temperatures (Watson and Walker, 1992).

TABLE 13

Proton NMR Chemical Shifts of a Series of $[Fe(TMP)L_2]^+$ Complexes at $-80\ ^\circ C$ in CD_2Cl_2[a]

Ligand	$pK_a(BH^+)$[b]	$\delta(m\text{-}H)$[c]	$\delta(o,o'\text{-}H)_{Py}$ (average)[c]	$\delta(pyrr\text{-}H)$[c]	m-H δ_{iso}[c]	m-H δ_{dip}[d]	m-H δ_{con}[d]	$(o,o'\text{-}H)_{Py}$ δ_{iso}[c]	$(o,o'\text{-}H)_{Py}$ δ_{dip}[c]	$(o,o'\text{-}H)_{Py}$ δ_{con}[c]	pyrr-H δ_{iso}[d]	pyrr-H δ_{dip}[e]	pyrr-H δ_{con}[e]
4-NMe₂Py	9.70	4.85	−6.5	−30.9	−2.36	−2.36	0.00	−7.1	+30.1	−37.2	−39.7	−9.9	−29.8
2-MeHIm	7.56[f]	7.77[g]		−20.4[g]	+0.69[g]						−29.1[g]		
N-MeIm	7.33	5.01		−31.3	−2.20	−2.20	0.00				−40.1	−9.3	−30.8
3,4-Me₂Py	6.46	7.48	−25.7	−21.1	+0.27	−0.60	+0.87	−26.6	+7.7	−34.3	−29.9	−2.5	−27.4
3,5-Me₂Py	6.15	7.12	−25.6	−23.0	−0.09	−0.56	+0.47	−26.5	+7.2	−33.7	−31.8	−2.4	−29.4
4-MePy	6.02	8.22	−27.7	−18.2	+1.11	−0.37	+1.48	−28.7	+4.7	−33.4	−26.9	−1.6	−25.3
3-MePy	5.68	8.36	−29.2	−17.8	+1.15	−0.23	+1.38	−30.2	+2.9	−33.1	−26.5	−1.0	−25.5
Py	5.22	9.47	−31.5	−13.3	+2.26	0.00[h]	+2.26	−32.6	0.0	−32.6	−22.2	0.0[h]	−22.2
3-ClPy	2.84	10.85	−33.1	−10.3	+3.64	+0.33	+3.31	−34.4	−4.2	−30.2	−19.3	+1.4	−20.7
3-CNPy	1.45	13.07	−34.0	−4.4	+5.86	+0.53	+5.33	−35.5	−6.8	−28.7	−13.4	+2.2	−15.6
4-CNPy	1.1[i]	14.59	−35.2	+2.1	+7.38	+0.66	+6.72	−36.8	−8.4	−28.4	−6.9	+2.8	−9.7

[a] Data from Watson et al. (1992).
[b] pK_a values obtained from Albert (1971).
[c] Chemical shifts (ppm).
[d] Chemical shifts (ppm) $\delta_{iso} = \delta_{para} - \delta_{dia}$. The diamagnetic shifts used were those of $[TMPCoL_2]^+$ (Rafii and Walker, 1992).
[e] See text for method of separation of dipolar and contact contributions, both of which are expressed in ppm.
[f] Corrected for the presence of 2H⁺ (0.3).
[g] Average chemical shift of the multiple signals observed.
[h] Assumed value, based upon Fig. 41 (see text).
[i] Adjusted value, based on Hammett substituent constants and $pK_a(BH^+)$ values for 14 other pyridines.

pyrrole-H resonance to higher frequency and the increasing contact shift of the *meta*-phenyl-H of the TMP, the anti-Curie behavior of the weakly basic pyridine complexes, and the change in the size (and sign) of the dipolar contribution to the isotropic shift (see below) are indicative of a progressive change in the ground state of these low-spin Fe(III) complexes, as is discussed below.

We have previously shown that when axial imidazole or pyridine ligands are bound to low-spin Fe(III) porphyrins in perpendicular planes, only one resolved EPR feature, which we have called the "large g_{max}" signal, is observed (Walker *et al.*, 1986; Safo *et al.*, 1991a, 1992). The other two g values are not resolved, probably due to g strain, but they can be measured by single-crystal EPR techniques (Inniss *et al.*, 1988) or by magnetic Mössbauer spectroscopy at 4.2 K (Walker *et al.*, 1986; Safo *et al.*, 1991a). These g values, where available, are listed in Table 10. The g_{max} values of nine *bis*-(pyridine) complexes of (TMP)Fe(III) decrease in size from 3.48 to 2.53 as the pK_a of the conjugate acid of the pyridine decreases from 9.7 (4-NMe$_2$Py) to 1.1 (4-CNPy), with the type of EPR spectrum changing from "large g_{max}," with $g_{max} = g_{zz}$ in the former case to an axial signal, with $g_{max} = g_\perp = g_{xx} = g_{yy}$ in the latter (Safo *et al.*, 1992). Structures of four of these complexes have been reported (Safo *et al.*, 1991a, 1992). In all four of these structures the axial ligands are in perpendicular planes lying over the *meso* positions of the porphyrin ring, with strong ruffling of the porphyrin core in order to reduce steric crowding between the o-CH$_3$ groups and the pyridine ligands, as shown in Fig. 48. This strong ruffling produces two oblong "cavities" at right angles to each other, one above and one below the plane of the porphyrin ring, which hold the axial ligands in perpendicular planes over the *meso* positions. The EPR spectrum of [TPPFe(4-CNPy)$_2$]$^+$ is also axial in both frozen solution and the solid state, and the X-ray crystal structure shows that the axial ligands are also in perpendicular planes lying over the *meso* positions, with a strongly ruffled porphinato core, even though there are no *ortho*-phenyl substituents (Safo *et al.*, 1992). The axial EPR spectra, with $g_\perp > g_\parallel$, Table 10, are indicative of a $(d_{xy})^1$ [or at least *largely* $(d_{xy})^1$] ground state. Values of $g_\perp = 2.53$ and $g_\parallel = 1.56$, observed for [TMPFe(4-CNPy)$_2$]$^+$ (Safo *et al.*, 1992), indicate a ~93% d_{xy}, ~7% d_{xz}, d_{yz} composition of the orbital of the unpaired electron at 77 K, where the EPR spectrum was measured. Thus, low-basicity pyridines stabilize the unusual $(d_{xz}, d_{yz})^4(d_{xy})^1$ electronic ground state for low-spin Fe(III), though not as strongly as do isonitrile ligands (Simonneaux and Sodano, 1988b), as discussed in Section 2.8.1.

The change in ground state of low-spin Fe(III) from the usual $(d_{xy})^2(d_{xz}, d_{yz})^3$ to the unusual $(d_{xz}, d_{yz})^4(d_{xy})^1$ electron configuration occurs smoothly through the series of pyridine complexes of both TPPFe(III) (La Mar *et al.*, 1977a) and TMPFe(III) (Safo *et al.*, 1992; Watson *et al.*, 1992), but the trend is much more pronounced for the TMP complexes, which make it apparent that this change in the ground state was the reason for the pK_a(BH$^+$) dependence of the isotropic shifts in both cases. This pK_a(BH$^+$) dependence is due to profound effects of the electron configuration of the metal on both the dipolar and contact contributions to the isotropic shift. For the dipolar shift, a change in sign of the axial magnetic anisotropy term [Eq. (9)] occurs at some pK_a(BH$^+$) between that of 4-dimethylaminopyridine and 3-cyanopyridine (Watson *et al.*, 1992), since the former complex has $g_{zz} > g_{yy} > g_{xx}$ (Safo *et al.*, 1991a), while the latter has $g_{xx} = g_{yy} > g_{zz}$ (Safo *et al.*, 1992). The axial anisotropies calculated from the g values of these two complexes are listed in Table 14; however, as will be shown below, the

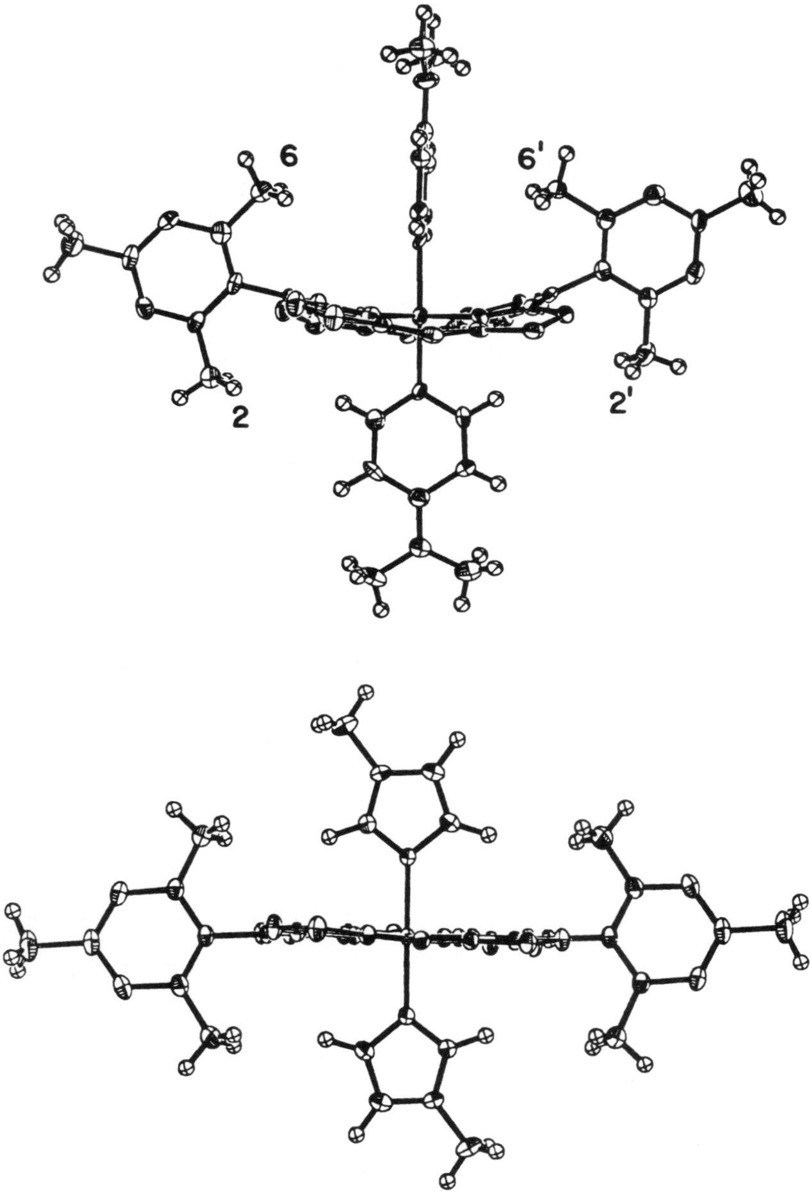

Figure 48. ORTEP diagrams showing the arrangement of the effectively coplanar mesityl rings and the axial ligands in (a) [FeTMP(4-NME$_2$Py)$_2$]ClO$_4$ and (b) molecule 2 of [FeTMP(1-MeIm)$_2$]ClO$_4$. Hydrogen atoms have been drawn artificially small to improve clarity. Reprinted with permission from *J. Am. Chem. Soc.* **113**:5508 (1991). © 1991 American Chemical Society.

TABLE 14
Comparison of Axial Magnetic Anisotropies for [TMPFeL$_2$]$^+$ Complexes Calculated from g Values and Derived from Fig. 49

Ligand	$(g_\parallel^2 - g_\perp^2)^a$	Relative $(g_\parallel^2 - g_\perp^2)^b$	Derived $(g_\parallel^2 - g_\perp^2)^c$
NMeIM	+4.39	+4.39^d	+4.39
4-NMe$_2$Py	+10.07	+4.14^d	+4.14
3,4-Me$_2$Py	$[\sim +7.8$ or $\sim -5.8]^e$	−0.07	+1.04
3,5-Me$_2$Py	$[\sim +7.5$ or $\sim -5.7]^e$	−0.20	+1.14
4-MePy	$[\sim +6.9$ or $\sim -5.6]^e$	−0.54	+0.67
3-MePy	$[\sim +6.3$ or $\sim -5.4]^e$	−1.18	+0.44
Py	$[\sim +5.6$ or $\sim -5.3]^e$	−1.44	0.00
3-ClPy	$[\sim +3.7$ or $\sim -4.7]^e$	−2.83	−0.57
3-CNPy	$[\sim +2.5$ or $\sim -4.3]^e$	−3.70	−0.94
4-CNPy	−3.97	−3.97	−1.21

^{a} Determined from EPR g values, Table 10.
^{b} Determined by factoring method using axial geometric factors from Table 1 and assuming $(g_\parallel^2 - g_\perp^2)$ is as given by g values for [TMPFe(N-MeIm)$_2$]$^+$ and [TMPFe(4-CNPy)$_2$]$^+$ (dashed line, Fig. 49).
^{c} Determined by assuming $(g_\parallel^2 - g_\perp^2)$ for [TMPFe(4-NMe$_2$Py)]$^+$ is known (see d) and $(g_\parallel^2 - g_\perp^2) = 0$ for [TMPFe(Py)$_2$]$^+$ (solid line, Fig. 49).
^{d} Calculated from the EPR g values of [TMPFe(NMeIm)$_2$]$^+$ assuming the m-H isotropic shift of the TMP ligand is totally dipolar in nature.
^{e} Calculated assuming linear change in all three g values as the basicity of the ligand changes. The two possible values reflect the two different assignments of the axes of the g tensor.

actual anisotropies in homogeneous solution at −80 °C are much smaller than those calculated from the g values [Eq. (9)].

Although only one feature (g_{max}) is resolved in the EPR spectra of the complexes of pyridines of intermediate basicities (Safo *et al.*, 1992) the *ortho*-H resonances of the pyridine ligands provide important information concerning the trend in magnetic anisotropy. As shown in Fig. 49, the dependence of the average of the isotropic shifts of the *o*- and *o'*-H of the bound pyridine ligands at −80 °C on the basicity of the pyridine changes sharply at about pK_a(BH$^+$) = 5, close to the basicity of pyridine itself (Watson *et al.*, 1992). This change in slope is suggestive that this is the point at which the axial magnetic anisotropy of the metal changes sign. If we assume that the isotropic shift of the *m*-H resonance of the TMP ligand of [TMPFe(4-NMe$_2$Py)$_2$]$^+$ is totally dipolar in nature, then this shift, together with the geometric factors for the other TMP and axial pyridine positions (Table 1), can be used to calculate the dipolar shifts at those positions. The results are $\delta_{dip} = -2.42$ ppm (*m*-H), −10.2 ppm (pyrrole-H), and +30.9 ppm (*o*-H$_{py}$). Using this *o*-H$_{Py}$ dipolar shift (Table 13), the dipolar shifts of the other complexes shown in Fig. 49 can be obtained, assuming that $\delta_{dip} = 0$ at pK_a(BH$^+$) = 5. This leads to the contact shift dependence represented by the dotted line in Fig. 49. Thus, the dipolar shift of the *o*-H of the 4-cyanopyridine ligand is predicted to be −8.4 ppm, leading to a derived dipolar shift of +2.8 ppm for the pyrrole-H of [TMPFe(4-CNPy)$_2$]$^+$ at −80 °C. The derived contact shifts at the pyrrole-H and *o*-H$_{Py}$ positions are included in Table 13. As can be seen, the contact contribution to the isotropic shift of the pyrrole-H at −80 °C decreases in magnitude from −29.8 ppm (4-NMe$_2$Py) to −9.7 ppm (4-CNPy), indicating that at that temperature, the orbital of the unpaired electron is 67% d_{xy} and 33% (d_{xz}, d_{yz}) (Watson *et al.*, 1992). At the *meso*-phenyl *m*-H position, the

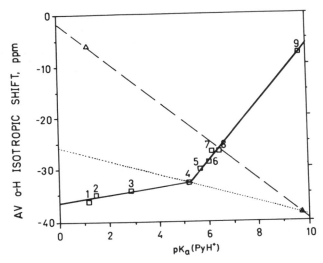

Figure 49. Plot of the average of the isotropic shifts of the *o*- and *o'*-H resonances of axial pyridine ligands of [TMPFe(R-Py)$_2$]$^+$ at -80 °C in CD$_2$Cl$_2$ versus the pK_a of the conjugate acid of the pyridine. The substituents R are 4-CN(1), 3-CN(2), 3-Cl(3), H(4), 3-Me(5), 4-Me(6), 3,5-Me$_2$(7), 3,4-Me$_2$(8), 4-NMe$_2$(9). The dashed line represents the average pyridine *o*-H contact shifts derived from the dipolar shifts calculated from the EPR *g*-values of the complexes (1) and (9), while the dotted line represents the contact shifts derived for the series of complexes if it is assumed that a change in sign of the dipolar shift occurs with basicity of the pyridine complex (4) (Watson *et al.*, 1992).

contact shift increases from 0.0 ppm (4-NMe$_2$Py) to $+6.7$ ppm (4-CNPy). This fact, combined with the alternation in sign for *o*-, *m*-, and *p*-H isotropic shifts for the corresponding [TPPFe(R-Py)$_2$]$^+$ complexes (La Mar *et al.*, 1977a), indicates significant (and increasing) π delocalization at the *meso* positions for the TMPFe(III) complexes with low-basicity pyridines. This π delocalization cannot come directly from the d_{xy} unpaired electron, since this orbital does not have the proper symmetry to engage in π-bonding with the porphyrin ring (Figs. 3, 4). Part of the π delocalization to the *meta*-phenyl-H ($\sim$1.5 ppm) could come from the small amount ($\sim$7%) of (d_{xz}, d_{yz}) character of the ground-state orbital of the unpaired electron, calculated from the *g* values, Eq. (16). This small amount of spin density in the *nominally filled* (d_{xz}, d_{yz})^4 configuration of the (d_{xy})^1 ground state can thus be transferred to the porphyrin ring by Fe $\rightarrow$ P π back-bonding, utilizing the LUMOs of the porphyrin ring, the $4e(\pi)$ orbitals as suggested by Simmoneaux *et al.* (1989) for [TPPFe (*t*-BuNC)$_2$]$^+$ (Section 2.8.1). These orbitals have large orbital coefficients at the *meso* positions, as shown in Fig. 4. However, it is also possible that the $3a_{2u}(\pi)$ orbital is close enough in energy to that of the d_{xy} orbital for there to be thermal population of the (d_{xy})^2(a_{2u})^1 state leading to large shifts of *meso* substituents, as suggested in Section 2.8.1 for the [TPPFe(*t*-BuNC)$_2$]$^+$ and [OEPFe(*t*-BuNC)$_2$]$^+$ complexes. Another alternative explanation for the large *meta*-H shifts of [TMPFe(4-CNPy)$_2$]$^+$ and the anti-Curie temperature dependence of the pyrrole-H shifts involves thermal population of low-lying quartet states, producing (d_{xz}, d_{yz})^3(d_{xy})^1(d_{z^2})^1

configurations that could allow large spin delocalization to the *meso* substituents. At the *o*-H position of the pyridine ligands the contact shift decreases in magnitude from -39.7 ppm (4-NMe$_2$Py) to -28.4 ppm (4-CNPy), a much smaller decrease than that observed for the pyrrole-H protons, suggesting a possible shift from dominant π to a combination of π and σ spin delocalization as the basicity of the pyridine decreases (Watson *et al.*, 1992). Such an increase in σ spin delocalization to weakly basic axial ligands is consistent with the possible thermal population of quartet states having an unpaired electron in d_{z^2}.

If the axial *g* anisotropies derived from the procedures presented in the previous paragraph are compared, as in Table 14, to those from Table 10, it is clear that *g* values measured at 77 K and below are not good indicators of the actual magnetic anisotropy of the complexes at the temperatures of the NMR measurements (183 K and above), and that in many cases the actual magnetic anisotropy may be of opposite sign from that predicted by the *g* values measured at very low temperatures (Table 14). This appears to be the case for the Fe(III) pyropheophorbide complex of myoglobin: The *g* values indicate (Keating *et al.*, 1992) a largely $(d_{xy})^1$ ground state, and thus $(g_\parallel^2 - g_\perp^2)$ is negative; however, the dipolar shifts of protein residues are in the same direction as (but with greater magnitude than) those of the native protein, for which $(g_\parallel^2 - g_\perp^2)$ is positive (La Mar, 1992).

2.8.5. *bis*-(Ammine) and *bis*-(Phosphine) Complexes

It has been reported that Fe(III) porphyrins undergo autoreduction in the presence of aliphatic amines (Hwang and Dixon, 1986, and references therein) and phosphines (La Mar and del Gaudio, 1977a). However, several workers have succeeded in obtaining the proton NMR spectra of several of these complexes. Kim and Goff (1990) found that although aliphatic amines caused autoreduction, the NMR spectrum of the *bis*-(ammonia) complex could be observed in the presence of some [(TPP)Fe]$_2$O. The NMR spectrum of [TPPFe(NH$_3$)$_2$]SO$_3$CF$_3$, recorded in CD$_2$Cl$_2$ at 25 °C is shown in Fig. 50. It is clearly indicative of a low-spin Fe(III) complex with the "normal" $(d_{xz}, d_{yz})^3$ ground state. An interesting feature of the spectrum is the extremely large positive isotropic shift of the ammonia protons (240.6 ppm). The Curie plots are linear for all protons of the TPP ligand, indicating a simple electronic ground state. The EPR spectrum, recorded at 5.2 K, has a "large g_{max}" signal at $g = 3.75$, nearly the limit for low-spin Fe(III), indicating that there is very little, if any, rhombic splitting of d_{xz} and d_{yz}. This is the expected situation for axial ligands that have no planes of symmetry or π-bonding orbitals.

A similar NMR spectrum is observed for the *bis*-(trimethylphosphine) complex, [TPPFe(PMe$_3$)$_2$]ClO$_4$ (Simonneaux and Sodano, 1988b); the pyrrole-H resonance is observed at -19.6 ppm in CD$_2$Cl$_2$ at 20 °C. In this case, however, the PMe$_3$ methyl-H have an isotropic shift of -2.39 ppm and a derived contact shift of -13.32 ppm. The opposite sign of the contact shift, as compared to that of the *bis*-(NH$_3$) complex discussed above, is consistent with the fact that although NH$_3$ can act only as a σ-donor, PMe$_3$ can act as both a σ-donor and a π-acceptor, due to the presence of the empty $3d$ orbitals. Clearly, this π-acceptor character has a very large effect on the observed isotropic shift of the axial ligand protons. The EPR *g* values reported in this work are $g = 2.687$, 2.088, and 1.680, from measurements at 140 K; however, at 8 K the only

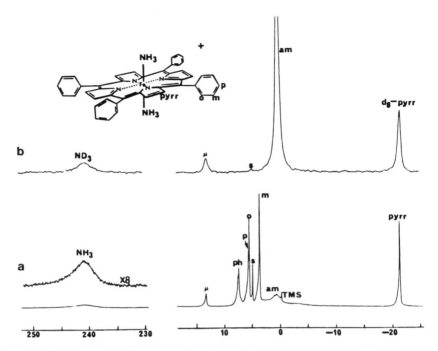

Figure 50. (a) Proton NMR spectrum of a mixture of $(TPP)Fe(NH_3)_2$-SO_3CF_3 and $[(TPP)Fe]_2O$ in CD_2Cl_2 solution at 25 °C, referenced to tetramethylsilane. Signals for $[TPPFe(NH_3)_2]^+$ are labeled as follows: pyrrole, pyrr; *ortho*, o; *meta*, m; *para*, para; coordinated ammonia, NH_3. Signals for $(TPPFe)_2O$ are labeled μ for pyrrole and ph for phenyl protons. The solvent signal is labeled s, free ammonia am, and tetramethylsilane TMS. (b) Deuterium NMR spectrum of a mixture of $(TPP\text{-}d_8)Fe(ND_3)_2SO_3CF_3$, $[(TPP\text{-}d_8)Fe]_2O$, and excess ND_3 in CH_2Cl_2 solution at 25 °C. Signals are labeled as above. Reprinted with permission from *Inorg. Chem.* **29**:3907 (1990). © 1990 American Chemical Society.

signal we find present is a "large g_{max}" signal at $g = 3.5$ (Table 1). In any case, the effective dipolar contribution to the isotropic shift at 20 °C is small at all TPP positions, and the ground state appears to be essentially pure $(d_{xz}, d_{yz})^3$. The mixed ligand PMe_3-NMeIm complex was also reported (Simonneaux and Sodano, 1988a), and is being used as a model for PMe_3-bound heme proteins, as discussed below.

2.8.6. Mixed-Ligand Complexes

The ^1H NMR spectra of a number of mixed-ligand complexes have been reported, including (N-RIm)(2-MeImH) (Walker, 1980; Goff, 1980), (NMeIm)(R-Py) (Bold, 1978), (R-Py)(R'-Py) (Bold, 1978), (ImH)(Im$^-$) (Chacko and La Mar, 1982), (ImH)(CN$^-$) (Chacko and La Mar, 1982; La Mar *et al.*, 1977b), (Im$^-$)(CN$^-$) (Chacko and La Mar, 1982), (NMeIm)(CN$^-$) (La Mar *et al.*, 1976, 1977b), (R-Py)(CN$^-$) (Bold, 1978), (ImH)(OR$^-$) (Arasasingham *et al.*, 1990), (NMeIm)(PMe$_3$) (Simonneaux and

Sodano, 1988a, b), (Py)(PMe$_3$) (Simonneaux and Sodano, 1988b), and (NMeIm)(R$^-$) and (Py)(R$^-$) (R$^-$ = aryl, alkyl, or carbanion) (Balch and Renner, 1986a). Most of these are direct models for the cyanide (Satterlee, 1986, 1987 and references therein) or phosphine (Simonneaux et al., 1987, 1990) complexes of a number of heme proteins. The mixed-ligand complexes involving R$^-$ will be discussed in Section 2.8.7.

Mixed hindered–nonhindered imidazole ligation of TPPFe(III) produces a pyrrole-H isotropic shift [-13.0 ppm for (2-MeImH)(N-MeIm) at 34 °C] that is intermediate to those of the two symmetrical bis-imidazole complexes [-11.8 ppm (2-MeImH)$_2$; -16.5 ppm (N-MeIm)$_2$] (Walker, 1980), while covalent attachment of the nonhindered imidazole to the TPP ligand produces a large spread of the pyrrole-H resonances (Walker, 1980; Goff, 1980).

Mixed pyridine complexes and pyridine-N-methylimidazole complexes of TPPFe(III) have been investigated by Bold (1978). As can be seen in Table 15, the stronger field ligand dominates the electronic properties of the low-spin Fe(III) center, making the isotropic shifts of the mixed-ligand complex more similar to those of the stronger-field ligand than to the weaker-field ligand. The same holds true for mixed (pyridine)(cyanide) and (N-methylimidazole)(cyanide) complexes of TPPFe(III) as well (Bold, 1978), as shown in Table 15.

The effect of deprotonation of N-H imidazoles on the isotropic shifts of natural porphyrins has been reported (Chacko and La Mar, 1982). For protoporphyrin and deuteroporphyrin IX complexes of low-spin Fe(III) it was found that as the axial imidazoles are sequentially deprotonated, both the average isotropic shift and the spread of the ring CH$_3$ shifts decrease, consistent with decreased contact shift resulting from decreased P → Fe π donation as the ligands became stronger π donors (Chacko and La Mar, 1982). Comparing the deuterohemin IX complexes having axial ligand combinations (ImH)$_2$, (ImH)(CN$^-$) and (CN$^-$)$_2$, the average isotropic shift of the ring CH$_3$ shifts increases in the order (CN$^-$)$_2$ < (Im)(CN$^-$) < (ImH)$_2$, while the 2,4-pyrrole-H average isotropic shifts increase in a negative sense in the order (ImH)$_2$ < (CN$^-$)$_2$ < (ImH)(CN) (Chacko and La Mar, 1982). Since the dipolar contribution is expected to shift pyrrole-H and -CH$_3$ resonances in the same direction [Eqs. (7) or (9)], this mixed trend in the isotropic shifts of -CH$_3$ and -H suggests that there are changes in both the dipolar and contact contributions to the isotropic shifts of this series of complexes. Nevertheless, the spread of both of these types of resonance changes in the order (ImH)$_2$ > (ImH)(CN) > (CN$^-$)$_2$, suggesting that the major change in the contact contribution is a decrease in its magnitude as the combined charge of the axial ligands decreases from +1 to 0 to −1. For the isocharged complexes of deuterohemin IX, where the axial ligand combinations are (Im$^-$)$_2$, (Im$^-$)(CN$^-$), and (CN$^-$)$_2$, the average shifts of the CH$_3$ protons increase in the order (Im$^-$)$_2$ < (Im$^-$)(CN$^-$) < (CN$^-$)$_2$, while those of the 2,4-pyrrole-H increase in a negative sense in the order (Im$^-$)$_2$ < (Im$^-$)(CN$^-$) < (CN$^-$)$_2$, indicating that in this series the major effect on the isotropic shifts is an increase in the contact contribution in the order (Im$^-$)$_2$ < (Im$^-$)(CN$^-$) < (CN$^-$)$_2$. Meanwhile, the spread of the CH$_3$ resonances increases in the order (Im$^-$)$_2$ ~ (Im$^-$)(CN$^-$) < (CN$^-$)$_2$ and that for the 2,4-H in the order (CN$^-$)$_2$ < (Im$^-$)(CN$^-$) < (Im$^-$)$_2$ (Chacko and La Mar, 1982). For protohemin IX having axial ligand combinations (Py)$_2$, (Py)(CN$^-$), and (CN$^-$), the (Py)(CN$^-$) and (CN$^-$)$_2$ complexes have fairly similar ring CH$_3$ isotropic shifts (~16.7, 16.2, 11.9, 10.2 ppm versus ~16.2, 15.7, 12.7, 11.2 ppm, respectively) (Mazumdar et al., 1990),

TABLE 15
Isotropic Shifts of Mixed Ligand Complexes [TPPFe(XY)][a] as Compared to Their bis-Ligand Parents

		Isotropic Shifts[a]				X-H			Y-H		
X	Y	Pyrr-H	o-H	m-H	p-H	1-Me	2,6	3,5	1-Me	2,6	3,5
CN⁻	CN⁻										
CN⁻	4-NMe₂Py	−34.6	−4.98	−1.79	−2.09		—			N.O.[b]	+9.92
CN⁻	4-MePy	−32.1	−4.50, −4.76	−1.18	−1.95		—			−8.7	+13.5
CN⁻	Py	−31.0	−4.50, −4.20	−0.90	−1.89		—			−11.6	+12.5
CN⁻	4-CNPy	−29.4	−4.40, −3.76	−0.51, −0.67	−1.77		—			−13.4	+11.2
CN⁻	NMeIm	−37.7	−5.34, −5.11	−2.35	−2.22		—		+20.2		
NMeIm	NMeIm	−36.3				+25.0			+25.0		
NMeIm	4-CNPy	−35.8				+26.3				N.O.	N.O.
NMeIm	4-MePy	−36.5				+25.6				N.O.	+14.4
NMeIm	4-NMe₂Py	−35.8				+22.3				N.O.	N.O.
4-NMe₂Py	4-NMe₂Py	−37.8	−5.54	−2.13	−2.33		−6.7	+9.5		−6.7	+9.5
4-NMe₂Py	4-MePy	−35.1					N.O.	+9.0		N.O.	+9.0
4-NMe₂Py	4-CNPy	−33.2					N.O.	+9.0		−11.0	N.O.
4-CNPy	4-CNPy	−15.9	−4.18	+2.9	−2.20		−28.6	N.O.		−28.6	N.O.
4-CNPy	4-MePy	−26.4					−21.0	N.O.		−14.0	N.O.
4-MePy	4-MePy	−31.5	−4.49	+0.64	−1.84		−20.5	+9.3		−20.5	+9.3

[a] Taken from Bold (1978). Solvent = CDCl₃, T = −60 °C.
[b] N.O. = not observed.

while the $(Py)_2$ complex has much larger isotropic shifts ($\sim$22.6, 22.0, 19.2, 17.7 ppm) (Hill and Morallee, 1972; Dugad *et al.*, 1987). Thus, the presence of the cyanide ligand in the mixed-ligand complex is able to dominate the splitting of the *d*-orbitals and stabilize the pure low-spin state, in comparison to the $S = \frac{1}{2} \rightleftarrows S = \frac{5}{2}$ spin state of the *bis*-(pyridine) complex. The same effect was noted above for mixed-ligand complexes of TPPFe(III), Table 15. Reports of the NMR spectra of the cyanide complex of the hemin octapeptide derived from proteolysis of cytochrome *c* (also called microperoxidase 8) (Smith and McLendon, 1981) and the corresponding hemin undecapeptide in aqueous detergent micelles (Mazumdar *et al.*, 1991) have also appeared.

A 6-coordinate, low-spin hemin complex having *p*-nitrophenoxide-imidazole coordination has been reported (Arasasingham *et al.*, 1990). The average isotropic shift of the four ring methyls at -60 °C is 1.1 ppm more positive than that of the *bis*-imidazole complex, while that of the four *meso*-H is 0.84 more negative, suggesting that the major effect is an increase in the contact contribution when a neutral imidazole ligand is replaced by a phenolate anion, as observed in the cyanide and imidazole mixed-ligand complexes discussed above. In agreement with this conclusion, the spread of the methyl resonances increases from 3.15 ppm to 4.96 ppm when one imidazole is replaced by the phenolate ligand. A very broad resonance at -31 ppm may be due to the *o*-H and/or *m*-H of the *p*-nitrophenolate ligand (Arasasingham *et al.*, 1990).

Mixed phosphine–axial ligand complexes of TPPFe(III) have very similar pyrrole-H shifts to those of the *bis*-L complexes: At 20 °C for (NMeIm)(PMe$_3$), $\delta_{iso} = -27.34$ ppm as compared to -27.81 ppm for (PMe$_3$)$_2$ (Simonneaux and Sodano, 1988a, b) and -26.37 ppm for (NMeIm)$_2$ (Walker and Benson, 1982). However, while for (Py)(PMe$_3$) at 20 °C, $\delta_{iso} = -28.81$ ppm (Simonneaux and Sodano, 1988b), $\delta_{iso} \sim -21$ ppm for (Py)$_2$ (La Mar *et al.*, 1977a). Thus, it would appear that the phosphine ligand stabilizes the "normal" $(d_{xz}, d_{yz})^3$ ground state of the mixed-ligand complex containing pyridine, as compared to the mixed $(d_{xy})^1$–$(d_{xz}, d_{yz})^3$ ground state of the *bis*-(pyridine) complex. (See Section 2.8.4.) The PMe$_3$ complexes of myoglobin and hemoglobin have one and two methyl-H resonances from the coordinated phosphine, respectively, suggesting some difference in the α and β heme distal environments of hemoglobin. Each of these PMe$_3$ methyl resonances has a different isotropic shift, suggesting some difference in either steric or electronic factors at the heme center of these proteins when bound to the phosphine (Simonneaux *et al.*, 1987); permutation of the ring CH$_3$ resonance order for Mb(PMe$_3$) (5 > 8 > 1 > 3) as compared to Mb(CN), Mb(N$_3$), and Mb(Im) (5 > 1 > 8 > 3) suggests a possible steric factor of the phosphine ligand (Simonneaux *et al.*, 1990).

2.8.7. Five-Coordinate Low-Spin Fe(III) Porphyrins

Reports of the NMR spectra of a number of low-spin Fe(III) complexes bound to one alkyl, aryl, or silyl axial ligand have appeared (Cocolios *et al.*, 1983; Lançon *et al.*, 1984; Balch and Renner, 1986a; Arasasingham *et al.*, 1989a; Balch *et al.*, 1990b; Kim and Goff, 1988); the chemical shifts of typical examples are presented in Table 16. The Group IV anions are expected to be extremely strong σ donors, and indeed, they are able to stabilize both low-spin Fe(II) (Balch *et al.*, 1990d) and low-spin Fe(III) in a 5-coordinate state, except when four or five electronegative fluorine atoms are present on the aryl ring, in which case the 5-coordinate complexes are high-spin (Kadish *et al.*,

TABLE 16
Chemical Shifts of 5-Coordinate Fe(III) Porphyrins with Alkyl, Aryl, and Silyl Ligands

Porphyrin	R^a	T (°C)	Solvent	δ(Rβ-H) (ppm)	δ(pyrr-H) (ppm)	δ(meso) (ppm)	δ(α-CH₂) (ppm)	Ref.
TTP	Si(CH₃)₃	24	C₆D₅CD₃	−1.2	−21.7			b
TTP	C(CH₃)₃	0	C₆D₅CD₃	−111.5	−21.0			c
TPP	CH₃	21	C₆D₆	—	−19.2			d
	C₄H₉	21	C₆D₆	−63.7	−18.4			d
	C₆H₅	21	C₆D₆	−81.0	−17.6			d
	(p-CH₃)C₆H₄	21	C₆D₆	−84.7	−16.9			d
TTP	CH₃	−70	C₆D₅CD₃	—	−34.0			e
	C₂H₅	−70	C₆D₅CD₃	154.0	−34.0			c
	n-C₃H₇	−70	C₆D₅CD₃	−27.4	−35.0			c
	CH(CH₃)₂	−70	C₆D₅CD₃	−145.6	−32.6			c
	C(CH₃)₃	−70	C₆D₅CD₃	−135.5	−33.7			c
	1-adamantane	−70	C₆D₅CD₃	−33.3	−17.9			c
	4-camphor	−70	C₆D₅CD₃	−30.8	−42.8, −48.7			c
	[NMeIm][CH₃]	−70	C₆D₅CD₃	—	−37.0			e
	[NMeIm][C₂H₅]	−70	C₆D₅CD₃	−94.0	−36.0			e
TMP	[NMeIm][C₆H₅]	−60	CDCl₃	−69.0	−34.0			f
TPP	[Py][C₆H₅]	21	C₅D₅N	−55.0	−21.2			g
(p-CF₃)₄TPPC₆H₅		21	C₆D₆	−84.27	−17.97			h

	C₆F₄H	21	C₆D₆	—	63.32			h
	C₆F₅	21	C₆D₆	—	66.67			h
	[Py][C₆H₅]	21	C₅D₅N	-58.85	-21.18			h
	[Py][C₆F₄H]	21	C₅F₅N	—	-16.63			h
	[Py][C₆F₅]	21	C₅D₅N	—	-16.43			h
OEP	CH₃	21	C₆D₆	—		3.32	2.43, -2.13	d
	n-C₄H₉	21	C₆D₆	-58.6		3.55	2.33, -1.70	d
	C₆H₅	21	C₆D₆	-79.9		5.53	4.46, -1.70	d
	C₆F₄H	21	C₆D₆	—		-48.56	39.91, 42.89	h
	C₆F₅	21	C₆D₆	—		-55.01	41.71, 42.69	g
	[Py][C₆H₅]	21	C₅D₅N	-55.7		-3.23	0.07, -2.27	g
	[Py][C₆F₄H]	21	C₅D₅N	—		1.94	3.78, 5.60	h
	[Py][C₆F₅]	21	C₅D₅N	—		2.28	4.18, 5.81	h

[a] R = alkyl, aryl, or silyl ligand.
[b] Kim and Goff (1988).
[c] Balch et al. (1990c).
[d] Cocolios et al. (1983).
[e] Arasasingham et al. (1989a).
[f] Balch and Renner (1986a).

1991). For the low-spin complexes, porphyrin β-pyrrole-H and -CH$_2$ shifts are indicative of the "normal" $(d_{xz}, d_{yz})^3$ ground state, with spin delocalization via P $\rightarrow$ Fe π bonding (Balch and Renner, 1986a). The α-H of alkyl groups had not been detected under any condition until very recently, when Li and Goff (1992) reported the ^{2}H NMR spectra of [d_8-TPPFeCD$_3$] and [d_8-TPPFeC$_2$D$_5$] in toluene at 25 °C, recorded over a very large spectral width at 55 MHz: For the CD$_3$ group, $\delta_{obs} = 532$ ppm, and for the C$_2$D$_5$ group, d_{obs} for the CD$_2$ group is 562 ppm, with the CD$_3$ signal at -117 ppm. These are the largest chemical shifts reported thus far for iron porphyrins; the low frequency of the ^{2}H nucleus makes possible the use of spectral bandwidths that correspond to such a large number of ppm. Widths of the α-CD$_3$ and -CD$_2$ resonances are reasonably narrow, 80 and 115 Hz, respectively, in both THF and toluene (Li and Goff, 1992). In light of the expected narrowing of deuterium resonances of up to a factor of 42 over those of protons at the same molecular positions (Section 1.5.2), it is not surprising that α-CH$_3$ and -CH$_2$ signals have not been detected.

It has been possible to separate the β-H and more distant H isotropic shifts of the 1-adamantane complex [TTPFe(1-ad)] into contact and dipolar contributions (Balch *et al.*, 1990b). Contact shifts within the 1-adamantane group (as well as other alkyl groups) are negative, since the unpaired electron of low-spin Fe(III) is in a π-symmetry orbital, while the spin must be delocalized to the alkyl group via σ-delocalization. It was noted, however, that whereas the β-methyl resonances of the ethyl, isopropyl, and *tert*-butyl groups have characteristic resonances at ~ -115 ppm (at 20 °C), the β-methylene groups of *n*-propyl, *n*-butyl, 1-adamantane, and 4-camphor appear at least 50 ppm to higher frequency of these because of the importance of the cos$^2 \theta$ term in Eq. (6) (Balch *et al.*, 1990b). The much smaller shift of the silyl methyl protons (Kim and Goff, 1988), Table 16, seems to suggest the possibility of competing σ and π delocalization mechanisms for this heavier coneger of the *tert*-butyl anion.

For aryl anions bound to Fe(III), the alternating sign pattern of aryl-H contact shifts (Table 16) is indicative of mainly π spin delocalization to the aryl group (Balch and Renner, 1986a). Addition of N-methylimidazole or pyridine to 5-coordinate Fe(III) porphyrins, either low-spin or high-spin, produces low-spin mixed-ligand complexes, in which the isotropic shifts of the phenyl-H or alkyl groups bound to iron are decreased (Balch and Renner, 1986a), probably due to decreased R$^-$ $\rightarrow$ Fe σ (and, in the case of R$^-$ = aryl, π as well) delocalization from the R group to the metal. Thus, the effect of adding a sixth ligand in decreasing the contact shifts of the aryl-H suggests that the aryl anion acts as a π donor rather than a π acceptor toward low-spin Fe(III). Interestingly, however, the planar nature of the aryl ligand is not sufficient to produce a resolved rhombic EPR spectrum for [TTPFe(C$_6$H$_5$)]; instead, a "large g_{max}" signal at $g = 3.54$ is observed, with a broad feature at about $g \sim 1.7$ (Arasasingham *et al.*, 1990), indicating near-degeneracy of d_{xz} and d_{yz}, probably because of the strong σ-donor nature of the aryl ligand. Addition of pyridine shifts this signal to 3.86 and the broad feature to higher field (lower g value) (Arasasingham *et al.*, 1990). Comparison of the ^{1}H chemical shifts of the phenyl and 2,3,5,6-tetrafluorophenyl complexes of both OEPFe(III) and (*p*-CF$_3$)$_4$(TPP)Fe(III) with d_5-pyridine reveals a shift to higher frequency in approximate proportion to the geometric factors for all porphyrin resonances as the electronegative fluorine atoms are substituted onto the phenyl ring; the 4-H of that ring shifts to lower frequency by about the same geometric factor proportion (Kadish *et al.*, 1991), a clear

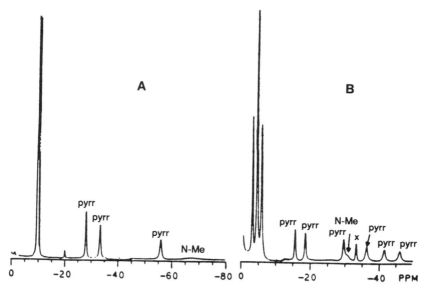

Figure 51. 300-MHz ^{1}H NMR spectra of A [(N-MeTMP)Fe^{III}(CN)$_2$] and B [(N-MeTMP)Fe^{III}(Im)$_2$]$^{2+}$ in CD$_2$Cl$_2$ at -90 °C. Resonance X$^-$ is due to the pyrrole protons of [(TMP)Fe^{III}(Im)$_2$]$^+$. In both complexes no resonances occur to low-field of 20 ppm. Reprinted with permission from *J. Am. Chem. Soc.* **112**:7555 (1990). © 1990 American Chemical Society.

indication that the major difference in the isotropic shifts is due to the dipolar contribution. Thus, it appears that the magnetic anisotropy of the low-spin 6-coordinate complexes [OEPFe(C$_6$F$_4$H)(Py)] and [(p-CF$_3$)$_4$TPPFe(C$_6$F$_4$H)(Py)], as well as the perfluorophenyl analogs, is considerably smaller than that of the phenyl complexes. The EPR spectra of these 6-coordinate, low-spin fluorophenyl complexes have not been reported.

2.8.8. Low-Spin Fe(III) Complexes of N-Alkylporphyrins

N-substituted porphyrins are formed during the enzymatic turnover of cytochrome P-450 and by the reaction of substituted hydrazines with heme proteins. Their formation results in the inactivation of the proteins involved (Ortiz de Montellano and Correia, 1983; Ortiz de Montellano and Reich, 1986). N-substituted porphyrins are also inhibitors of the enzymes ferrochelatase and heme oxygenase (Lavallee, 1987). Their Fe(II) complexes are surprisingly resistant to aerial oxidation (Anderson *et al.*, 1980), but upon oxidation with halogens or the thianthrene cation radical form the high-spin Fe(III) complexes, which readily bind axial ligands such as cyanide or imidazoles (Balch *et al.*, 1990c). The NMR spectra of both N-methyl-*tetra*-(p-tolyl)porphyrin (N-MeTTP) and N-methyl tetramesitylporphyrin (N-MeTMP) complexes having one or two cyanide and two imidazole or 5-methylimidazole ligands show the effects of the lowered symmetry of the porphyrin, as shown in Fig. 51. As can be seen for the dicyano THP complex in Fig. 51A, three of the four pyrrole-H resonances occur at very low frequencies. In

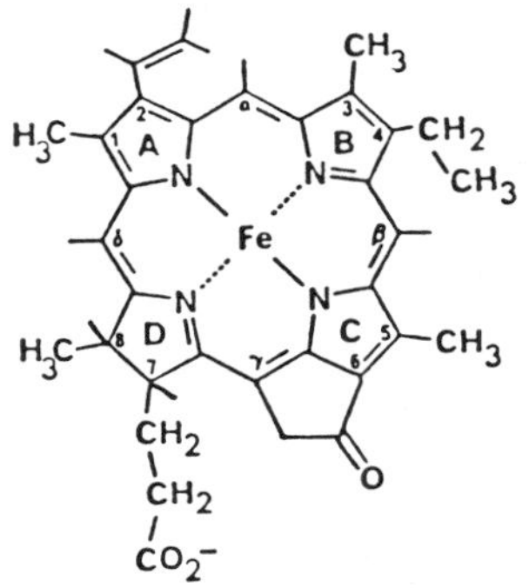

Figure 52. Structures of two "green" (reduced) hemes, both chlorins: 1, Pyropheophorbide iron; 2, sulfhemin *c*. Reprinted with permission from *J. Am. Chem. Soc.* **111**:6088 (1989). © 1989 American Chemical Society.

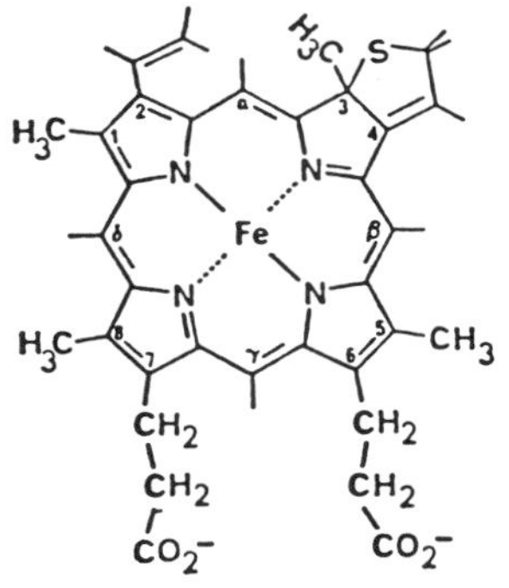

the case of the *bis*-(5-MeImH) or the *bis*(imidazole) complex, Fig. 51B, six of the eight pyrrole-H resonances are resolved, indicating the effect of the N-methyl group on hindering the rotation of at least the 5-MeImH ligand on the same side of the porphyrin plane (Balch *et al.*, 1990c). It would be interesting to see if 2D NMR methods could allow assignment of the observed pyrrole-H resonances and identification of the two that occur in the 0–7 ppm region and are obscured by the presence of excess ligand, as well as to investigate possible kinetic processes involving ligand rotation in this system.

2.8.9. Low-Spin Fe(III) Complexes of Reduced Hemes

Several investigations of the NMR spectra of the dicyano complexes of two reduced hemes have been reported (Chatfield *et al.*, 1986a, b, 1988b; Bondoc *et al.*, 1986; Licoccia *et al.*, 1989; Keating *et al.*, 1991). The structure of sulfhemin isomer *C*, extracted from sulfmyoglobin cyanide, was determined by NMR techniques (Chatfield *et al.*, 1986a, b, 1988b; Bondoc *et al.*, 1986) and shown to be reduced at pyrrole ring B or II, as shown in Fig. 52. Reduction of the porphyrin ring to the chlorin changes the symmetries and molecular orbital coefficients of the frontier orbitals (Chatfield *et al.*, 1988b), as shown in Fig. 5. The pattern of isotropic shifts of this prosthetic group is indicative of delocalization through $L \to Fe$ π donation not only from the filled A_4 orbital of the chlorin [the modification of one of the $3e(\pi)$ orbitals of the porphyrin ring], but also from the filled A_5 orbital of the chlorin, which, unlike the a_{1u} orbital of the porphyrin, has proper symmetry to overlap with the d_{xz} orbital of the metal. The NMR spectra of all three of the modified hemin complexes have been observed and classified as metmyoglobin cyanide complexes (Chatfield *et al.*, 1987, 1988a).

Similar investigation of the dicyano iron(III) complex of a chlorophyll *a* derivative, pyropheophorbide *a* methyl ester, in which it is ring D or IV that is reduced (Fig. 52), yielded total assignment of the ¹H NMR spectrum of this "green heme" complex, as well as the corresponding *bis*-(imidazole) complex (Licoccia *et al.*, 1989). More recently, the assignments have been confirmed by 2D NMR techniques (Keating *et al.*, 1991).

Again, the pattern of isotropic shifts was explained in terms of spin delocalization through $L \rightarrow Fe$ π donation from the A_4 and A_5 filled π orbitals of the chlorin (Licoccia et al., 1989) (Fig. 5).

2.8.10. Heme Apoproteins Reconstituted with Synthetic Hemins

Neya and co-workers have reconstituted apomyoglobin with *meso-tetra-(n-pro-pyl)*hemin (Neya and Funasaki, 1987) and octaethylhemin (Neya et al., 1988). In the case of the *tetra-(n-propyl)*hemin cyanide reconstituted myoglobin, the eight pyrrole-H appeared as a single resonance at -16 ppm, suggesting that this prosthetic group rotates freely in the heme pocket. This conclusion was strengthened by the extreme temperature dependence of the pyrrole-H resonances of both the deoxy and cyanide complexes. The NMR results suggested that the introduction of *meso-tetra-(n-propyl)*hemin totally disrupts the highly stereospecific hemeglobin contacts, thus making this prosthetic group mobile in the heme cavity (Neya and Funasaki, 1987). In contrast, OEPFe(III) appeared to reconstitute well; D_2O-exchangeable peaks at 27.1 and 19.9 ppm in the ^1H NMR spectrum of the cyanomet Mb complex have been assigned to the proximal and distal histidine N-H protons, respectively (Neya et al., 1988). The α-CH_2 proton peaks of the ethyl groups of the OEP were either not observed or, at least, not assigned.

2.9. Fe(III) Porphyrin Cation Radicals

One-electron oxidation of Fe(III) porphyrins may lead to either Fe(III) porphyrin cation radicals or Fe(IV) porphyrins, depending on the axial ligation of the metal and the reaction conditions. In this section, we will consider the systems that have been formulated as Fe(III) porphyrin cation radicals, and in the next section we will discuss the Fe(IV) porphyrins.

In view of the three different spin states observed for Fe(III) porphyrins, it is not surprising to find that all three have been observed for the one-electron ring-oxidized complexes. Some high-spin, monomeric Fe(III) porphyrins give rise to antiferromagnet-ically coupled $S = 2$ species (Phillippi et al., 1981b; Scholz et al., 1982; Buisson et al., 1982); others give rise to a noninteracting $S = \frac{5}{2}$, $S = \frac{1}{2}$ species (Buisson et al., 1982); spin-admixed $S = \frac{3}{2}$, $\frac{5}{2}$ Fe(III) complexes give rise to spin-admixed cation radical species of either $S = \frac{5}{2}$, $\frac{3}{2}$ (Boersma and Goff, 1984) or $S = \frac{3}{2}$, $\frac{5}{2}$ (Groves et al., 1985) ground state; and low-spin Fe(III) porphyrin complexes give rise to low-spin Fe(III) cation radical species (Goff and Phillippi, 1983).

Electrochemical oxidation of [TPPFeCl] in the presence of perchlorate or hexafluoroantimonate salts, or chemical oxidation with Fe(III) porphyrin cation radical of higher oxidation potential, produces an $S = \frac{5}{2}$, $S = \frac{1}{2}$ antiferromagnetically coupled (effectively $S_{tot} = 2$) 5-coordinate [TPPFeCl]$^+$ cation radical (Buisson et al., 1982) hav-ing its pyrrole-H resonance at 66.1 ppm (compared to 79.4 ppm for [TPPFeCl]) and its phenyl-H at 37.6, 34.4 (*ortho*), -12.4 (*meta*), and $+29.5$ ppm (*para*) (compared to ~ 6, 13.3, and 12.2, and 6.35 ppm for the parent) at 26 °C in CD_2Cl_2 (Phillippi and Goff, 1982). The large, alternating shifts of the phenyl-H are indicative of large π spin density at the *meso* positions, suggesting that the electron has been removed from the $3a_{2u}(\pi)$ orbital, Fig. 4. However, the sign of the shifts, δ_o positive, δ_m negative, is opposite to that usually seen for π cation radicals when there is large π spin delocalization to the *meso*-phenyl positions (Boersma and Goff, 1984). The X-ray crystal structure of [TTPFeCl]SbF$_6$ consists of isolated 5-coordinate [TPFeCl]$^+$ units, in which the porphyrin macrocycle is strongly S_4 ruffled (Buisson *et al.*, 1982).

In comparison, the NMR spectrum of [OEPFeCl]ClO$_4$ shows small shifts of the pyrrole-α-CH$_2$ to lower frequency (δ_{obs} = 30.5, 29.6 ppm at 20 °C compared to 43.1, 39.5 ppm at 30 °C for the parent) and a large shift of the *meso*-H to higher frequency (δ_{obs} = -18 ppm at 20 °C, compared to -54 ppm at 30 °C for [OEPFeCl]), suggesting that in this case the electron is removed from the $1a_{1u}$ orbital (Phillippi and Goff, 1982), Fig. 4. Curie plots are linear, but the *o*-H has a very large, negative intercept.

The π cation radicals of PFe—O—FeP species, the μ-oxo dimers of Fe(III) porphyrins, are convenient chemical oxidants of the chloroiron porphyrins (Phillippi and Goff, 1982), and they are interesting species in their own right. Those in which P = TPP or one of its derivatives also have larger phenyl-H isotropic shifts than the parent complexes, again with alternating signs, suggesting the same $(a_{2u})^1$ ground state for the porphyrin cation radical, this time delocalized over the two porphyrin rings; the pyrrole-H isotropic shift is affected much less (usually δ_{obs} decreases by $\sim$1 ppm). The dication of the μ-oxo dimer can also be prepared and exhibits attenuated isotropic shifts as compared to the monocation (Phillippi and Goff, 1982).

The neutrally charged μ-nitrido dimer of (TPP)Fe, [(TPPFe)$_2$N], has been studied by several groups (Summerville and Cohen, 1976; Kadish *et al.*, 1981). It is isoelectronic with the π cation radical of the μ-oxo dimers discussed above, but has rather different ^{1}H NMR spectra. Whereas [(TPPFe)$_2$O]$^+$ has resonances at 12.3 (pyrrole-H), 1.3 (*o*-H), 11.7 (*m*-H), and 3.2 ppm (*p*-H), [(TPPFe)$_2$N] has broad, overlapping resonances at 8.6, 8.3, and 7.6 ppm (Summerville and Cohen, 1976; Kadish *et al.*, 1981), suggesting that its description as a delocalized system having two Fe$^{31/2+}$ ions may be better than considering it to be a delocalized porphyrin π cation radical. One-electron oxidation with I$_2$/Ag$^+$ produces a species with a much sharper NMR spectrum, with resonances at 8.32 (pyrrole-H), 7.85 and 7.64 ppm (phenyl-H). This complex is believed to consist of Fe^{IV}—N^{-III}—Fe^{IV} (Kadish *et al.*, 1981). Very small temperature dependence over the range 26 to -55 °C suggests that the metals are strongly antiferromagnetically coupled, and the complex has a magnetic moment less than 1.8 μ_B at room temperature. Unlike [(TPPFe)$_2$O]$^+$ or [(TPPFe)$_2$O]$^{2+}$, both of which autoreduce in the presence of pyridine, [(TPPFe)$_2$N]$^+$ readily coordinates one or two pyridine ligands; stepwise addition can be observed by monitoring the pyridine *o*-H resonances at -1.26 ppm (1:1) and -1.46 ppm (2:1) (Kadish *et al.*, 1981).

In comparison to these spin-coupled Fe(III) porphyrin radical species, Nanthakumar and Goff (1992) have shown that if the 2,6-difluoro TPP complex, [(2,6-F$_2$)$_4$TPPFeCl] is oxidized electrochemically by one electron, a quite different NMR spectrum is observed: At 25 °C in CD$_3$NO$_2$ the pyrrole-H resonance moves from 80 to 85 ppm upon oxidation, while the *m*-H and *p*-H isotropic shifts reverse sign. A slightly increased rather than a 15-ppm decreased pyrrole-H shift as for [TPPFeCl]$^+$, discussed above, suggests a difference in spin coupling between high-spin Fe(III) and the porphyrin radical and is indicative of negligible spin density being placed on the β-pyrrole carbon atoms, as expected for a_{2u}-type radicals. Electrochemical removal of an additional electron appears to generate an Fe(IV) porphyrin radical (Nanthakumar and Goff, 1992), as discussed below in Section 2.11.

One-electron electrochemical oxidation of [TPPFe(ROH)₂]ClO₄, where ROH is water or ethanol, in the presence of perchlorate salts, produces [TPPFe(OClO₃)₂], a 6-coordinate complex, described as an $S = \frac{5}{2}$, $S = \frac{1}{2}$, noninteracting high-spin species because of its room-temperature magnetic moment in the solid state ($\mu_{eff} = 6.5 \pm 0.2\ \mu_B$) (Buisson et al., 1982). Its NMR spectrum at 25 °C in CD₂Cl₂ revealed isotropic shifts of 22.6 (pyrrole-H), −27.1 (o-H), 27 (m-H), −20.6 ppm (p-H) (Buisson et al., 1982), suggesting a π cation radical of a_{2u} type. However, investigation of the temperature dependence of this system in CH₂Cl₂ in the presence of electrolyte, utilizing pyrrole- and phenyl-deuterated TPP to eliminate the problem of the large amount of tetra-(n-propyl)ammonium perchlorate electrolyte present, revealed that although the pyrrole-D resonance shifts to higher frequency as the temperature is lowered, the Curie plots are not linear (Boersma and Goff, 1984). Rather, they have very steep slopes and very negative intercepts and are curved at higher temperatures (Fig. 53A), indicating that in this spin-admixed state, the ground state is largely $S = \frac{5}{2}$ (Boersma and Goff, 1984). The solution magnetic moment, $4.9 \pm 0.4\ \mu_B$, is much lower than that observed in the solid state (Buisson et al., 1982), suggesting antiferromagnetic coupling.

The complexity of the behavior of Fe(III) porphyrin radicals, and those of tetra-mesitylporphyrin in particular, is demonstrated by the properties of [TMPFe(OClO₃)₂]

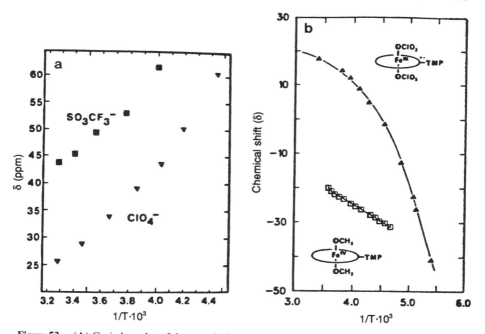

Figure 53. (A) Curie-law plot of the pyrrole deuteron signals of (TPP-d_8)Fe(SO₃CF₃)₂ and (TPP-d_8)Fe(ClO₄)₂ (iron porphyrin concentrations 2 mM, respective supporting electrolytes ∼0.25 M, CH₂Cl₂ solvent, signals referenced to (CH₃)₄Si). (B) Curie plots for (△)FeTMP(ClO₄)₂ (1) in CD₂Cl₂ and (□) FeTMP(OCH₃)₂ (2) in CDCl₃; chemical shift (δ) of the β-pyrrole proton resonance versus inverse temperature. Reprinted with permission from Inorg. Chem. 23:1676 (1984) and J. Am. Chem. Soc. 107:355 (1985). © 1984 and 1985 American Chemical Society.

in various solvents (Groves et al., 1985). This π cation radical of Fe(III) was prepared by oxidation of [TMPFeOClO$_3$] with ferric perchlorate and isolated before dissolution in CD$_2$Cl$_2$. As shown in Fig. 53B, the temperature-dependent behavior is the opposite of that observed for [TPPFe(ClO$_4$)$_2$] in CH$_2$Cl$_2$ (Fig. 53A), with the pyrrole-H resonance at very low frequency at low temperatures (δ_{iso} negative) and at high frequency at higher temperatures (δ_{iso} positive) (Groves et al., 1985). These results indicate that in the case of the TMP complex, the $S = \frac{3}{2}$ ground state is greatly favored over the spin admixed $\frac{3}{2}, \frac{5}{2}$ or $\frac{5}{2}, \frac{3}{2}$ ground state. (The properties of the TMP ligand among tetraphenylporphyrins appear to be unique in many respects, as was discussed in Section 2.8.4.) Treatment of the latter Fe(III) porphyrin cation radical complex with two equivalents of sodium methoxide produced the electron configuration isomer, the d^4 Fe(IV) porphyrin complex (Groves et al., 1985) discussed in Section 2.10.

Although complexes of the type [PFe(ImH)$_2$]$^+$ could not be electrochemically oxidized to π cation radicals, titration of the high-spin [(p-OCH$_3$)$_4$TPPFeCl]ClO$_4$ in CD$_2$Cl$_2$ with imidazole at -38 °C produced a transient complex whose pyrrole-H, o-H, and m-H resonances were at -32.7, -236.3, and $+24.8$ ppm, respectively (Goff and Phillippi, 1983). Warming to room temperature causes immediate reduction to the Fe(III) porphyrin. The temperature dependence of the pyrrole-H resonance of the transient radical was reasonably linear, and the deviation of the extrapolated shift at $1/T = 0$ from the diamagnetic position was no greater than observed for low-spin [TPPFe(ImH)$_2$]$^+$ (Goff and Phillippi, 1983). The alternating pattern of phenyl-H shifts suggests large π spin delocalization to the *meso* positions, indicative of a $3a_{2u}$ ground state for the radical; however, the signs of the shifts, δ_o negative, δ_m positive, is the reverse of that observed for the parent [TPPFeCl]ClO$_4$ complexes mentioned above, suggesting a difference in the coupling of metal and porphyrin spins. The high-spin radical complex appears to be the exception (Goff and Phillippi, 1983), as discussed above. For the corresponding low-spin etioporphyrin complex, the ring α-CH$_2$ and -CH$_3$ resonances were at 50.4 and 133.1 ppm at -51 °C; the *meso*-H signal could not be detected (Goff and Phillippi, 1983).

As an illustration of the chemical reactivity of these Fe(III) π cation radicals, [TPPFe(ClO$_4$)$_2$] reacts with the nucleophiles nitrite ion or triphenylphosphine to produce the mono-β-substituted Fe(III) porphyrins (Malek et al., 1991). The products were isolated as the 6-coordinate dichloride complexes, [(2-NO$_2$)TPPFeCl$_2$] and [(2-PPh$_3$)TPPFeCl$_2$], respectively; the latter was characterized by X-ray crystallography and ^{1}H NMR spectroscopy in CD$_2$Cl$_2$ for the *bis*-imidazole complexes (Malek et al., 1991). Curie plots for the seven pyrrole-H resonances of each complex are linear, and the average slopes and intercepts are similar to those for [TPPFe(ImH)$_2$]$^+$ (La Mar and Walker, 1973a). Phenyl-H intercepts of the two complexes deviate more from the expected diamagnetic positions, and the phenyl-H resonances of the PPh$_3^+$ moiety show extreme curvature in their temperature dependence, undoubtedly due to hindered rotation of this bulky substituent as the temperature is lowered (Malek et al., 1991).

2.10. Fe(IV) Porphyrins

The ^{1}H (and in some cases, ^{2}H) NMR spectra of three types of Fe(IV) porphyrin complexes have been reported, all of which have spin state $S = 1$: The low-spin d^4 complex [TPPFe(OCH$_3$)$_2$] mentioned in Section 2.9 above (Groves et al., 1985), the

TABLE 17
Chemical Shifts of Representative High-Valent Porphyrin Species

Compound	T (°C)	Chemical shift (ppm)				
		Pyrr-H	o-CH$_3$	m-H	p-CH$_3$	Ref.
[TMPFe(OCH$_3$)$_2$]	−78	−37.5	2.4	7.72	2.86	a
[TMPFeO]	−70	8.4	3.3	6.4, 6.0	2.6	b
[(NMeIm)TMPFeO]	−30	4.6	3.2, 1.6	7.4	2.7	b
[TMPFe(Ph)(Br)]	−50	−51	4.2, 3.8	10.6, 10.7	3.2	c
[TPPFe(m-CH$_3$C$_6$H$_4$)]ClO$_4$^{d}	−60	−72	13.4, 12.5	8.1, 7.8	9.5	c
[d_8-TPPFe(O)(F)]$^•$	−78	−1				e
[TPPFe(O)(F)]$^•$	−46	V.B.^{f}	−45 (o-H)	+47.7	−32.0 (p-H)	e
[TMPFeO]$^{+•}$	$\begin{cases} -70 \\ -77 \end{cases}$	$\begin{array}{c} -21 \\ -27 \end{array}$	$\begin{array}{c} 25, 28 \\ 24, 25 \end{array}$	$\begin{array}{c} 62 \\ 68 \end{array}$	$\begin{array}{c} 10 \\ 11.1 \end{array}$	$\begin{array}{c} g \\ h \end{array}$
[TMTMPFeO]$^{+•}$	−80	145 (CH$_3$)		56 ($meso$-H)		i

^{a} Groves *et al.* (1985).
^{b} Balch *et al.* (1984).
^{c} Balch and Renner (1986b).
^{d} Phenyl resonances: o-H, not observed; m-H, −56 ppm; p-H, −91 ppm.
^{e} Hickman *et al.* (1988).
^{f} Very broad.
^{g} Balch *et al.* (1985c).
^{h} Groves *et al.* (1981).
^{i} Fujii and Ichikawa (1992).

deep red, 5-coordinate ferryl (Fe^{IV}O)$^{2+}$ complexes (Balch *et al.*, 1984), and the 6-coordinate adducts of these ferryl complexes, the (B)(Fe^{IV}O)$^{2+}$ complexes (Chin *et al.*, 1980, 1984; La Mar *et al.*, 1983; Balch *et al.*, 1984), and the Fe(IV) phenyl complex (Balch and Renner, 1986b). The pattern of spin delocalization observed in each of these systems is unique, as summarized in Table 17, despite the similar magnetic moments. The ferryl and Fe(IV) phenyl cases have direct application to active states in heme proteins.

2.10.1. Six-Coordinate *bis*-Methoxide Fe(IV) Porphyrin

As mentioned in Section 2.9 above, Groves and co-workers (1985) have reported the valence isomerization of the Fe(III) porphyrin cation radical in the presence of methoxide:

$$[TMPFe(ClO_4)_2] + 2OCH_3^- \rightarrow [TMPFe(OCH_3)_2] + 2ClO_4^-$$
$$\text{Fe(III)P}^{-•} \qquad\qquad\qquad \text{Fe(IV)P}^{2-} \qquad\qquad (21)$$

This species is characterized by relatively large π contact shifts to the β-pyrrole positions and relatively small shifts, if any, at the *meso* positions (Groves *et al.*, 1985), thus implicating the $3e(\pi)$ orbitals in spin delocalization.

2.10.2. Five- and Six-Coordinate Ferryl (FeO)$^{2+}$ Porphyrin Complexes

Ferryl complexes have been implicated in the reaction mechanisms of peroxidases and cytochromes P-450. For horseradish peroxidase, two intermediates are spectroscopically detectable (Marnett *et al.*, 1986). Compound I, formed upon addition of peroxide to the resting Fe(III) form of the enzyme, is a green species that is formally two oxidation levels higher than the resting state and is widely believed to consist of an $(Fe^{IV}O)^{2+}$ unit complexed by a porphyrin π radical. NMR investigations of such complexes are discussed in Section 2.11. Compound II, which is red and is obtained upon one-electron reduction of compound I, also possesses a $(Fe^{IV}O)^{2+}$ unit, in this case complexed by a normal porphyrin dianion.

Model heme complexes containing the ferryl unit have been prepared and investigated by ^1H NMR spectroscopy at low temperatures by Balch and La Mar and coworkers (La Mar et al., 1983; Balch et al., 1984, 1985b; Arasasingham et al., 1989b). The ferryl species have been produced by thermal decomposition (at -30 °C) of the μ-peroxo dimer of PFe(III), formed from the reaction of 2PFe(II) with molecular oxygen at -70 °C [Eq. (14a, b) above] (Balch et al., 1984) by base-catalyzed cleavage of the μ-peroxo dimer at similar low temperatures (Chin et al., 1980; La Mar et al., 1983; Balch et al., 1984):

$$PFe{-}OO{-}FeP + 2B \rightarrow 2(B)PFeO$$
$$Fe^{III}(O_2^{2-})Fe^{III} \qquad\qquad (B)(Fe^{IV}O) \tag{22}$$

or by base-catalyzed cleavage of $PFeOOCHR_2$ complexes (Arasasingham et al., 1989b):

$$PFeOOCHR_2 + B \rightarrow (B)PFe^{IV}O + {}^\bullet OCHR_2$$
$$\rightarrow PFe^{III}OH + O{=}CR_2 + B \tag{23}$$

Both the 5- and 6-coordinate ferryl porphyrin complexes have very small observed shifts of all resonances, as summarized in Table 17. Nevertheless, the Curie plots for all resonances are strictly linear, and the observed shifts extrapolate to close to the diamagnetic positions for each type of proton, as shown in Fig. 54. Investigation of 6-coordinate imidazole or pyridine adducts of synthetic ferryl porphyrins and ferryl myoglobin led to detection of broad, weak axial histidine resonances at -11 and -16 ppm in ferryl myoglobin at 35 °C (Balch et al., 1985b). Analysis of the hyperfine shifts of the axial ligands is consistent with imidazole $\rightarrow$ Fe π donation similar to that seen in low-spin Fe(III) complexes with two imidazolate or one imidazolate and one cyanide ligand (Chacko and La Mar, 1982), although the contact shifts are much smaller in the Fe(IV) complexes (Balch et al., 1985b). For axial pyridine complexes, the dominant π bonding leads to spin polarization effects by the d_π spin, leading to contact shift patterns very similar to those reported for low-spin Fe(III) porphyrin bis-pyridine complexes (La Mar et al., 1977a) but much smaller than those reported for the Fe(IV) complexes (Balch et al., 1985b). These results indicate that the coordinated ligand is located trans to a strongly π-donating oxo ligand, where there is significant spin delocalization from the d_{xz}, d_{yz} orbitals to the p_x and p_y orbitals of oxygen in both the model compounds and ferryl myoglobin (Balch et al., 1985b). The small observed shifts are entirely consistent with theoretical calculations (Loew and Herman, 1980; Hanson et al., 1981), which indicate substantial π interaction between iron and oxygen orbitals and effective localization of unpaired spin density within the $(Fe^{IV}O)^{2+}$ unit, a significant amount of which may be localized on the oxygen, in line with the reactivity of this group (McMurry and Groves, 1986; Ortiz de Montellano, 1986).

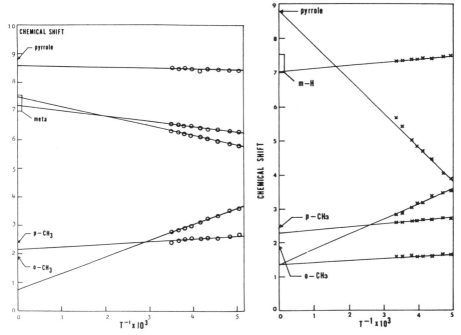

Figure 54. Left: Curie plots of the chemical shifts of [TMPFe^{IV}O]. Right: Curie plot of [(N-MeIm)TMPFe^{IV}O]. Reprinted with permission from *J. Am. Chem. Soc.* **106**:7782 and 7784 (1984). © 1984 American Chemical Society.

2.10.3. Five-Coordinate Fe^{IV} Phenyl Porphyrins

One-electron oxidation of phenyl iron(III) tetraarylporphyrin complexes with bromine in chloroform at −60 °C produces deep red solutions whose ^{1}H and ^{2}H NMR spectra indicate that they are the corresponding iron(IV) complexes (Balch and Renner, 1986b). As summarized in Table 17, the pyrrole-H are found at −60 to −70 ppm, while porphyrin aryl chemical shifts are almost exclusively dipolar. The iron phenyl resonances at ∼ −300 ppm (*o*-H), −50 to −75 ppm (*m*-H), −43 ppm (*m*-CH$_3$), ∼ −100 ppm (*p*-H), and +112 ppm (*p*-CH$_3$) (Balch and Renner, 1986b) indicate large π spin delocalization to the phenyl ring. However, the lack of alternation in sign for the *meta*-phenyl-H isotropic shift and of reversal in sign upon substitution of CH$_3$ at the *meta* position suggests some degree of σ spin delocalization to the phenyl ring. In comparison to the phenyl Fe(III) porphyrin complexes discussed in Section 2.8.7, except for the *meta*-phenyl position, the chemical shift pattern is similar for the two oxidation states, except that the shifts of the Fe(IV) phenyl complexes (Table 17) are larger than those for the Fe(III) phenyl complexes (Table 16). This is not surprising when one considers the electron configurations of the two: For the low-spin aryl Fe(III) porphyrins, the electron configuration is $(d_{xy})^2(d_{xz}, d_{yz})^3$, with one π-symmetry unpaired electron, while for the low-spin aryl Fe(IV) porphyrins, the electron configuration is $(d_{xy})^2(d_{xz}, d_{yz})^2$, with two π-symmetry unpaired electrons.

The aryl Fe(IV) porphyrins are thermally unstable and upon warming convert cleanly to N-phenylporphyrin complexes of Fe(II) by reductive elimination (Balch and Renner, 1986b). Phenyl migration, which produces a green pigment, is known to occur under aerobic conditions in protein systems such as cytochrome P-450, catalase, hemoglobin, and myoglobin. The phenyl heme is produced by the attack of aryl hydrazines in the presence of oxygen (Kunze and Ortiz de Montellano, 1983; Ortiz de Montellano and Reich, 1986) and probably goes through a phenyl Fe(IV) intermediate. These N-arylporphyrin complexes of iron are readily demetallated, leading to destruction of function of the heme protein.

2.10.4. Comparison of Fe(IV) Porphyrins and Fe(III) Porphyrin π Radicals

As was concluded by Balch and Renner (1986b), axial ligands play a major role in determining the electronic distribution within highly oxidized iron porphyrins. It has been found that an oxo ligand, a phenyl ligand, or two methoxy ligands favor the Fe(IV) porphyrin dianion structure, while halide ions, and imidazole or perchlorate ligands produce the Fe(III) porphyrin π radical monoanion electronic structure. The ability of the iron π electron to interact with the axial ligand(s) appears to make a major contribution to these differences.

2.11. Fe(IV) Porphyrin π Radicals

As mentioned in Section 2.10, the green compound I species observed in peroxidase enzymes, which contains an iron porphyrin that is two oxidation levels above PFe(III), is believed to consist of a ferryl porphyrin radical $(Fe^{IV}O)^{2+}$ $P^{-\cdot}$. The ferryl unit, complexed to a porphyrin π radical, is a transient and a very reactive species (either in the enzymes or in model hemes) and can readily insert the oxygen atom into many C-H bonds (Ortiz de Montellano, 1986; McMurry and Groves, 1986). Four groups have investigated the NMR spectra of model complexes of this reactive species (Groves *et al.*, 1981; Balch *et al.*, 1985c; Arasasingham *et al.*, 1989b; Fujii and Ichikawa, 1992; Hickman *et al.*, 1988). Nanthakumar and Goff (1992) have stabilized an Fe(IV) porphyrin radical in the absence of the oxo group by using fluoro substituents on the TPP phenyl rings. Weiss, Gold, and co-workers (Gold *et al.*, 1988; Bill *et al.*, 1990) have also generated oxoferrylporphyrin radicals by *m*-chloroperoxybenzoic acid oxidation of the chloro- and trifluoromethane–sulfonato complexes of [TMPFe(III)] and the trifluoromethane–sulfonato complex of tetra-(2,6-dichlorophenyl)porphinato iron(III), [(2,6-Cl₂)₄TPPFe(III)].

In all but one case (Nanthakumar and Goff, 1990), "hindered" or "protected" porphyrins such as tetramesitylporphyrin (TMP) (Groves *et al.*, 1981; Balch *et al.*, 1985c; Arasasingham *et al.*, 1989b), 2,7,12,17-tetramethyl-3,8,13,18-tetramesitylporphyrin (TMTMP) (Fujii and Ichikawa, 1992), or tetra-(2,6-difluorophenyl)porphyrin ((2,6-F₂)₄TPP or F₈TPP) (Nanthakumar and Goff, 1991) were used. In the case of the TMP complexes, it was found that the porphyrin radical is of the a_{2u} type, since large π spin delocalization is detected at the *meso*-phenyl-H and -CH₃ positions, as well as a smaller amount of π spin delocalization at the β-pyrrole positions (Fajer and Davis, 1979; Balch *et al.*, 1985c), as summarized in Table 17. However, the TMTMP complex has an a_{1u} unpaired electron, since the very large pyrrole-CH₃ isotropic shift indicates large spin delocalization to the pyrrole positions (Fajer and Davis, 1979; Fujii and Ichikawa, 1992). In comparison, pyrrole-CH_3 and *meso*-H shifts for the ferryl complex of protohemin are 4 and 16 ppm, respectively (La Mar *et al.*, 1983). Horseradish peroxidase compound I has been classified as an a_{2u} radical, while catalase compound I is believed to be an a_{1u} radical (Dolphin *et al.*, 1971; Dolphin and Felton, 1974).

The 6-coordinate, oxo, fluoro complex of Goff and coworkers [TPPFe(O)(F)], produced by addition of *m*-chloroperbenzoic acid to [TPPFeF$_2$]$^-$ in dichloromethane or by reaction of the μ-oxo dimer diradical [(TPPFe)$_2$O(ClO$_4$)$_2$] with excess Bu$_4$NF·3H$_2$O in CD$_2$Cl$_2$, has very small pyrrole-D shifts and large, alternating signs for the phenyl-H (Hickman *et al.*, 1988), indicative of a radical of a_{2u} type. This species has EPR signals at $g = 4.2$, 3.9, and 3.5 at 5 K but not higher, suggestive of an overall $S = \frac{3}{2}$ species composed of $S = 1$ Fe(IV) and the $S = \frac{1}{2}$ porphyrin radical (Hickman *et al.*, 1988). This is presumably the valence isomer of the Fe(V) porphyrin species obtained from the fluorinated TPP derivative (Nanthakumar and Goff, 1990) discussed in Section 2.12.

The nonferryl Fe(IV) porphyrin radical, probably [(2,6-F$_2$)$_4$TPPFeCl(ClO$_4$)$_2$], of Nanthakumar and Goff (1991), which is produced by two-electron electrochemical oxidation of [(2,6-F$_2$)$_4$TPPFeCl] in dichloromethane, has reversed phenyl-H shifts (30.5 and -2.0 ppm for *m*-H and *p*-H, respectively) from those of the one-electron reduced Fe(III) porphyrin radical [-3.5, -4.5 (*m*-H) and 16.0 ppm (*p*-H)] discussed in Section 2.9, and the pyrrole-H signal has moved from 85.5 ppm to -5 ppm upon oxidation from the Fe(III) to the Fe(IV) porphyrin radical. Although this pyrrole-H shift is similar to those observed for ferryl porphyrin radicals, Table 17, solution characterization of the two-electron oxidized product of [(2,6-F$_2$)$_4$TPPFeCl] provides no evidence of an oxo ligand (Nanthakumar and Goff, 1992). In comparison, two-electron oxidation of [(2,6-F$_2$)$_4$TPPFeCl] with *m*-chloroperbenzoic acid appears to produce the (Fe^{V}O)$^{3+}$ unit discussed below.

2.12. Fe(V) Porphyrins

The only report of a possible Fe(V) porphyrin is that of Nanthakumar and Goff (1990). By using highly electronegative fluorine substituents on the phenyl rings and the strong oxidizing agent *m*-chloroperbenzoic acid, these authors produced a red complex with Soret band at 430 nm and visible band at 550 nm, similar to those of PFe^{IV}O complexes (Gold *et al.*, 1988), but with different NMR and EPR spectra. The pyrrole-D signal was observed at +1.3 ppm at 210 K. The EPR spectrum at moderate temperatures ($g = 4.38$, 3.11, and 2.70, plus a $g = 2$ radical signal from decomposition products) is believed to be indicative of an $S = \frac{3}{2}$ species (Nanthakumar and Goff, 1990), rather than an $S = 1$ Fe(IV)/$S = \frac{1}{2}$ porphyrin radical species, which is expected to be EPR silent above $\sim$5 K (Nanthakumar and Goff, 1990).

3. THE USE OF NMR SPECTROSCOPY TO INVESTIGATE CHEMICAL REACTIONS OF MODEL HEMES

3.1. Autoreduction

Several investigators have shown that ferric porphyrins in solution can autoreduce in the presence of certain ligands. Examples of such reductions are the autoreduction of tetraphenylporphinato iron(III) chloride in the presence of piperidine (Del Gaudio and La Mar, 1976, 1978), cyanide, thiols, and phosphines (La Mar and Del Gaudio, 1977b; White-Dixon et al., 1985), or hydroxide and alkoxide ions and solid sodium hydride (Shin et al., 1987). Magnetic resonance techniques (NMR and EPR) have been successfully employed to probe the mechanism of the autoreduction and to characterize reaction intermediates and final products. It is believed that the autoreductions proceed by similar pathways for a variety of ligands and involve an intramolecular electron transfer to give Fe(II) species and ligand radicals. The mechanisms of such reactions are relevant to the understanding of both the activation of coordinated ligands in certain hemoproteins and modes of intramolecular electron transfer in cytochromes (Del Gaudio and La Mar, 1978).

Detailed NMR and EPR spectroscopic studies performed in dimethyl sulfoxide solution demonstrated that autoreduction of tetraphenylporphinato iron(III) chloride in the presence of piperidine or potassium cyanide leads to the formation of the corresponding 6-coordinate, diamagnetic dicyano or *bis*-(piperidine) ferrous complexes [TPPFe^{II}(CN)$_2$]$^{2-}$ and [TPPFe^{II}(Pip)$_2$], which are identified by their characteristic pyrrole-H shifts at 8.8–9.0 ppm. In addition to the ferrous complexes, a cyanide or a piperidine radical is formed, indicating that the iron reduction is accompanied by one-electron oxidation of the ligand molecules. The NMR spectra obtained for the autoreduction of [TPPFe^{III}(CN)$_2$]$^-$ are depicted in Fig. 55, clearly showing the conversion of the spectrum characteristic for a paramagnetic ferric species to the one typical for a diamagnetic iron(II) complex. The generation of the cyanide radical during the reduction of Fe(III) suggests that the mechanisms for the autoreduction of the dicyano complex involve homolytic bond cleavage according to the overall reaction

$$PFe^{III}(CN)_2^- \rightarrow PFe^{II}(CN)^- + \cdot CN$$

$$\downarrow \text{CN}^-$$

$$PFe^{II}(CN)_2^{2-}$$

$$(24)$$

[The ·CN radicals appear to dimerize rapidly to produce cyanogen, NCCN (La Mar and Del Gaudio, 1977b).] The reduction rate of [TPPFe^{III}(CN)$_2$]$^-$ depends on a variety of factors. Light and increased cyanide concentrations accelerate the reduction while small amounts of water inhibit the reaction. Reoxidation by molecular oxygen of the dicyano ferrous complex leads directly to the formation of the low-spin complex, as opposed to the expected μ-oxo dimer formation (Del Gaudio and La Mar, 1976; La Mar and Del Gaudio, 1977a). NMR spectroscopy of the autoreduction of TPPFe^{III}Cl in the presence of piperidine provided evidence that the reduction involves an unusual base-catalyzed intramolecular electron transfer, which yields the ferrous complex and the deprotonated piperidine radical. The electron transfer step is facilitated by the deprotonation of the coordinated piperidine by free piperidine. This conclusion is based on the observation that the reduction rate for this reaction is also critically dependent on free ligand concentration (Del Gaudio and La Mar, 1978). The observation of a large isotope effect on the reduction rate indicates that the breaking of an N—H piperidine bond is involved in the rate-determining step of the reaction. A variety of ferric and ferrous reaction intermediates has been characterized. At room temperature, addition

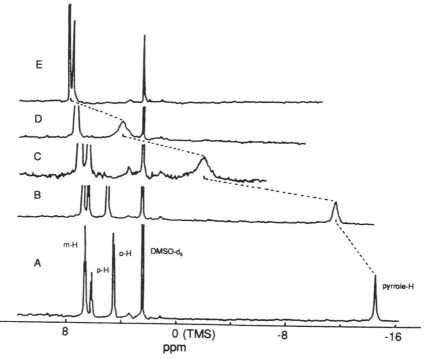

E

D

C

B

m-H

o-H   DMSO-d$_5$

A

p-H

pyrrole-H

8   0 (TMS)   -8   -16
ppm

Figure 55. Proton NMR spectra showing the autoreduction of [TPPFe^{III}(CN)$_2$]$^-$ in DMSO at 25 °C. (A) [TPPFe^{III}(CN)$_2$]$^-$; (B–D) increasing reduction; (E) final product [TPPFe^{II}(CN)$_2$]$^{2-}$. Reprinted with permission from *Adv. Chem. Ser.* **162**:212 (1977). © 1977 American Chemical Society.

of small amounts of piperidine to TPPFe^{III}Cl leads to the formation of a second high-spin complex, [TPPFe^{III}(Pip)]Cl, but complete conversion is not achieved. Addition of more piperidine results in the reduction of iron(III), and the expected high-spin [TPPFe^{II}(Pip)] and diamagnetic [TPPFe^{II}(Pip)$_2$] complexes are formed. At temperatures of −70 °C, low-spin ferric [TPPFe^{III}(Pip)$_2$]$^+$ is initially formed. The NMR spectrum of this complex is time dependent, with all paramagnetic resonances shifting from those typical of low-spin ferric species where the shifts for the two species are averaged by rapid electronic exchange (Del Gaudio and La Mar, 1978). Recently, Shin *et al.* (1987) have shown that ferric complexes of tetraphenylporphyrins and etioporphyrins undergo autoreduction in dimethyl sulfoxide solution by addition of hydroxide or alkoxide ions or solid sodium hydride. The autoreduction leads to the formation of the corresponding 5-coordinated, hydroxo ferrous ($S = 2$) complexes. This reduction can be nicely followed by NMR spectroscopy. In the case of the TPP complex, the resonances of the high-spin ferric chloride complex disappear and a new signal at 32.8 ppm, characteristic of the high-spin iron(II) species, appears in the spectrum (Shin *et al.*, 1987). The autoreduction of EtioFe^{III}Cl produces the singly coordinated hydroxo ferrous complex [(Etio)Fe^{II}(OH)]$^-$, as also evidenced by NMR spectroscopy.

Autoreduction of a series of dicyano iron(III) porphyrins studied by White-Dixon *et al.* (1985) showed that the rate of autoreduction decreases with major change in porphyrin structure. TPP complexes autoreduce faster than protoporphyrin IX species and much faster than deuteroporphyrin derivatives, for which no autoreduction can be observed.

3.2. Electron Exchange

In order to understand the pathway of electron transfer in the cytochromes, it is essential to determine the role of the protein in controlling the rate of electron transfer between two cytochrome molecules, which is currently believed to take place through the heme edges exposed to the solvent. Accurate assessment of the role of the protein demands knowledge of the factors that control electron transfer in hemes themselves; i.e., the electron self-exchange rate constants between two hemes free in solution need to be determined. Exchange rates can be measured most reliably by NMR spectroscopy using ^{1}H line-broadening techniques. White-Dixon and co-workers have employed this method to determine electron self-exchange rate constants in a variety of heme model compounds with the goal of establishing how changes in porphyrin or axial ligand structure and in macrocycle saturation affect the exchange rates (White-Dixon *et al.*, 1984, 1985, 1988; Shirazi *et al.*, 1985).

$$
\begin{array}{cccc}
L & L & & L & L \\
| & | & & | & | \\
-Fe_a^{III}- + -Fe_b^{II}- & \xrightleftharpoons{k_{exchange}} & -Fe_a^{II}- + -Fe_b^{III}- \\
| & | & & | & | \\
L & L & & L & L
\end{array}
\tag{25}
$$

The electron self-exchange rates for various N-methylimidazole complexes determined at $-21\,°C$ in CD_2Cl_2 are found to range from $8.1 \times 10^7\,M^{-1}\,s^{-1}$ for the TPP complex to $5.3 \times 10^7\,M^{-1}\,s^{-1}$ for the 3-MeTPP derivative and do not vary in a regular fashion with increase of steric crowding on the periphery of the porphyrin macrocycle. The total span is less than a factor of 2. Similar rate constants were measured for systems with sterically bulky imidazole ligands, such as N-butylimidazole. This suggests that increasing the steric bulk at either the heme site or on the axial ligand has almost no effect on the rate constants of electron exchange. This conclusion was also drawn from the analysis of the rate constants for the dicyano derivatives, $Fe^{III/II}TPP(CN)_2^{-/2-}$, for which rate constants ranging from 1.0×10^7 to $5.8 \times 10^7\,M^{-1}\,s^{-1}$ at $37\,°C$ in deuterated dimethyl sulfoxide solution were obtained (Shirazi *et al.*, 1985). However, complexes with axial imidazoles bearing NH groups have self-exchange rate constants which are a factor of 2–3 smaller than those with N-alkyl imidazole substituents. This indicates that hydrogen bonding or complete deprotonation of the axial imidazole nitrogen atom may play an important role in controlling electron transfer rates (White-Dixon *et al.*, 1985). The value of the rate constant obtained for the heme undecapeptide derived from cytochrome proteolysis ($1.3 \times 10^7\,M^{-1}\,s^{-1}$) is comparable to those of other model compounds (Kimura *et al.*, 1981) and indicates that the rate constants for model hemes are approximately a factor of 10 larger than those found in cytochromes with 80–90 amino acids. This argues further that the heme exposure to the solvent seems not to be a major factor controlling the rate constants for electron self-exchange in cytochromes.

Changes in the degree of saturation of the macrocycle affects the rate constants in a minor way. Similar rate-constant values are obtained for both the porphyrin and the octaethylisobacteriochlorin complexes, thus indicating little effect of the macrocycle on the electron transfer rate, perhaps due to counterbalancing effects in either the inner or outer sphere reorganization and orbital occupation (White-Dixon *et al.*, 1988).

3.3. Aggregation

One of the characteristic properties of porphyrins and metalloporphyrins is their tendency to dimerize or aggregate in solution. Solution studies have indicated that the structures of the aggregates vary widely with the nature of the metal ion, porphyrin substituents, and axial ligands (Snyder and La Mar, 1977; Viscio and La Mar, 1978a, b; Migita and La Mar, 1980). Although a number of concentration-dependent studies have been carried out to describe the thermodynamics of aggregation, elucidation of the solution's structural features invariably relies on NMR spectroscopy. It has been established that analysis of the concentration dependence of intermolecular paramagnetic shifts and resonance linewidths provides direct evidence of aggregation of paramagnetic metalloporphyrins in solution. Analyzing intermolecular relaxation rates (i.e., spin–lattice and spin–spin relaxation rates) is especially helpful in determining structural aspects of the porphyrin aggregates (Migita and La Mar, 1980; La Mar *et al.*, 1981). Aggregate formation is known to occur not only in high-spin and low-spin porphyrin complexes (Viscio and La Mar, 1977, 1978; Satterlee and Shelnutt, 1984; Mazumdar and Mitra, 1990; Smith and McLendon, 1981; Minch and La Mar, 1982) but also in intermediate-spin ferrous porphyrin complexes (Migita and La Mar, 1980). The latter have been discussed earlier in this chapter. It was found that 2,4-disubstituted deuteroporphyrin ferrous complexes form aggregates with either face-to-face overlap of all four pyrrole rings or parallel stacking of porphyrin planes (Migita and La Mar, 1980).

3.3.1. High-Spin Ferric Aggregates

High-spin ferric halide complexes of *p*-tolyl porphyrins also form aggregates in solution. The degree of aggregation increases with solvent dielectric constants and the size of the halide ion. In contrast to the chloro compounds, the iodo complexes show a significant tendency to aggregate. A dimer structure is proposed for these aggregates at intermediate concentrations with one-on-one pyrrole overlap; i.e., a *meso*-aryl ring of one porphyrin ring faces the iron atom of the other macrocycle (Viscio and La Mar, 1977).

Iron(III) uroporphyrin I chloride also forms a dimer in solution (Satterlee and Shelnutt, 1984). The high-frequency dimer resonances of 48.5, 32.4, and 29.1 pm of the pyrrole-CH_2 substituents are consistent with the concept of $\pi–\pi$ complex formation, with each molecule in the dimer being essentially a high-spin ferric porphyrin.

Extensive aggregation is also reported for several 5-coordinated high-spin Fe(III) derivatives of proto-, deutero-, and coproporphyrins with the dominant form of the aggregates being dimers. Based on spin–spin relaxation data, two possible four-over-four pyrrole ring overlapping structures are suggested for these aggregates (Mazumdar and Mitra, 1990).

3.3.2. Low-Spin Ferric Aggregates

Solutions of aggregates of low-spin, 6-coordinated, *bis*-cyanide ferric complexes of 2,4-disubstituted deuteroporphyrins are believed to consist of slipover-type dimers formed by π stacking between one or two pyrrole rings of two *bis*-cyanide monomers (Viscio and La Mar, 1978). The concentration-dependent linewidths and spin–lattice relaxation rates of dicyanohemin were also investigated. The results suggest that this dimer consists of stereospecific overlap of presumably parallel porphyrin planes. The 2-vinyl and 3-methyl groups of each porphyrin are in contact with the aromatic π system of the other porphyrin. These contacts and the close spacing of porphyrin planes is consistent with hydrophobic interaction involving van der Waals forces between the aromatic π system and the protons of the heme ring. This structure points out the important role of hydrophobic interactions in stabilizing the dimer (La Mar *et al.*, 1981).

Incorporation of low-spin iron(III) dicyanohemin into micelles of hexadecyltri-methylammonium bromide (CTAB) was studied by NMR spectroscopy (Minch and La Mar, 1982). It was found that dicyanohemin forms especially stable premicellar aggregates when the hemin and CTAB concentrations are roughly the same. Based on paramagnetic shift and relaxation effects, porphyrin–porphyrin interactions within the premicellar aggregate are suggested. At higher CTAB-to-dicyanohemin ratios, a more conventional dicyanohemin micelle is formed.

3.4. Dynamic Processes: Ligand Addition, Ligand Exchange, and Ligand Rotation

Early NMR investigations of dynamic processes, both the thermodynamics of lig-and binding and the kinetics of ligand exchange in iron(III) porphyrin complexes, have been reviewed in detail previously (La Mar and Walker, 1979). Since that time, additional investigations of ligand addition and exchange have appeared, and investigations of ligand rotation have also appeared.

3.4.1. Thermodynamics of Ligand Addition to Fe(III) Porphyrins

As reported previously (Satterlee and La Mar, 1976; La Mar and Walker, 1979) the equilibrium constants for ligand addition to the axial positions of iron(III) porphyrins can be readily measured by observing the relative intensity of peaks due to known proton types within the starting material, PFeX, and product, PFe(X)(L) or $PFeL_2^+X^-$ (Satterlee and La Mar, 1976; La Mar and Walker, 1979). The intermediate complex [PFe(X)(L)] can be either 6-coordinate, as implied by the parentheses and brackets, or 5-coordinate, as a tight ion pair [(PFeL)$^+$X$^-$], and is often not detected by UV–visible spectroscopy (Walker *et al.*, 1976, 1985). However, NMR spectroscopic investigations at temperatures above 25 °C indicate conclusively that it is present in measurable concentrations in many cases (La Mar and Walker, 1972; Satterlee and La Mar, 1976; Satterlee *et al.*, 1977; La Mar and Walker, 1979). Ligands for which binding constants to either tetraarylporphinato iron(III) or protohemin have been measured by NMR techniques include imidazoles (Satterlee and La Mar, 1976; Wang *et al.*, 1978; Nakamura, 1988, 1989; Nakamura and Nakamura, 1990), cyanide (Wang *et al.*, 1978), and pyridine–water (Mazumdar *et al.*, 1988). In $CDCl_3$ it was found that NMR measurement of the overall binding constants β_2 for addition of two substituted imidazole (RIm) ligands to a series of tetraphenylporphinato iron(III) halides

$$(TPP)FeX + 2RIm \rightleftarrows (TPP)Fe(RIm)_2^+X^- \quad \text{(tight ion pair)} \quad (26)$$

$$\beta_2 = \frac{[TPPFe(RIm)_2^+X^-]}{[TPPFeX][RIm]^2} \quad (27)$$

yielded values of β_2 similar to those measured by UV–visible spectroscopy (Walker *et al.*, 1976). Only the overall binding constant β_2 could be measured by NMR spectroscopy at 25 °C and below (Satterlee and La Mar, 1976). Nakamura (1988, 1989) has recently reported that the comparison of the formation constants of the TMP complexes to those of the TPP complexes showed an unexpected result. He found that the formation constant of the 2-methylimidazole complex of TMPFe(III) is a factor of 600 larger than that for the TPPFe(III) complex; i.e., 2-MeImH forms a more stable complex with TMPFeCl than with the less hindered TPPFeCl. Competitive ligation studies of 1- and 2-methylimidazole to TMPFe(III) revealed furthermore that formation of the sterically more hindered complex $[TMPFe(2\text{-MeImH})_2]^+Cl^-$ is favored over either the mixed-ligand complex or $[TMPFe(NMeIm)_2]^+Cl^-$. The increased stability of the sterically hindered porphyrin complexes was explained in terms of attractive rather than repulsive interactions between the *ortho*-methyl groups of the mesityl rings and the π system and/or the 2-alkyl substituent of the coordinated imidazoles (Nakamura, 1988). Based upon recent NOESY investigations of $[TMPFe(2\text{-MeImH})_2]^+Cl^-$ (Walker and Simonis, 1991), it is unlikely that attractive interactions between mesityl methyls and the axial ligand can stabilize the complex at ambient temperatures, where rotation of the coordinated ligands, and thus "flapping" of the mesityl groups, is rapid on the NMR time scale. An alternative explanation may be that the *bis*-2-MeImH complex is stabilized by S_4 ruffling of the porphyrin ring, even in view of the dynamic nature of this S_4 ruffling.

Nakamura and Nakamura (1990) have extended the investigations of the stability of the *bis*-(N-methylimidazole) and *bis*-(1,2-dimethylimidazole) complexes of sterically encumbered derivatives of TPPFeCl not only to TMPFeCl, as discussed above, but also the alkylated phenyl analogues of TMPFeCl where all methyl groups are replaced by either ethyl or isopropyl groups. They found that while the formation of the TPP complex suffers a large loss in bond strength when N-MeIm is replaced by 1,2-Me_2Im ($\Delta H^\circ = -26.5$ versus -20.6 Kcal/mole), the same is not true of the TMP and related complexes (ΔH° ranging from -27.2 to -28.0 (N-MeIm) and -27.8 to -29.1 Kcal/mole (1,2-Me_2Im), with error limits of 0.5–1.5 Kcal/mole). However, while the entropies of complex formation of the two TPP complexes are very similar, those for the TMP and related complexes are much more negative for the hindered than the nonhindered imidazole ($\Delta S^\circ = -69$ to -76 e.u. (N-MeIm) and -86 to -91 e.u. (1,2-Me_2Im), with error limits of 4–6 e.u.), indicating the loss of internal rotation of the 1,2-Me_2Im ligands upon binding to the TMP and related porphyrin complexes, but not TPPFe(III), at the low temperatures of these investigations (Nakamura and Nakamura, 1990). At -35 °C, the stabilities of the N-MeIm complexes of the 2,4,6-substituted TPPFe(III) derivatives increased in the order $H \ll i\text{-Pr} \sim Me < Et$, and for the 1,2-$Me_2$Im complexes $H \ll Me \sim i\text{-Pr} \sim Et$. In all cases $\beta_2(\text{N-MeIm}) > \beta_2(1,2\text{-Me}_2\text{Im})$, but by very different factors: H ($\sim 10^5$), Me (240), Et (10^3), i-Pr (160). Clearly, increasing the bulkiness of the *ortho*-phenyl substituents both increases the stability of axial ligand complexes and greatly enhances the stability of complexes of hindered axial ligands over those of TPPFe(III).

In contrast to the case of imidazoles binding to tetraphenylporphyrins, Wang *et al.* (1978) have demonstrated that binding of cyanide to high-spin protoporphyrin IX iron(III) chloride in DMSO-d_6 is a two-step process:

$$PPFe(DMSO)_2^+ + CN^- \rightleftharpoons PPFe(CN)(DMSO)_{1/0} \qquad (K_1) \qquad (28)$$

$$PPFe(CN)(DMSO)_{1/0} + CN^- \rightleftharpoons PPFe(CN)_2^- \qquad (K_2) \qquad (29)$$

with first the formation of the low-spin monocyano adduct followed by the formation of the low-spin dicyano complex. The binding constants K_1 and K_2 were found to be 1.9×10^4 and 1.1×10^3 M^{-1} at 20 °C, respectively. Imidazole binding, however, is an overall two-step process, and only the formation of the *bis*-imidazole complex can be observed. The overall binding constant is 6×10^4 M^{-2} at 20 °C. It should be noted that since the starting material already has the chloride ion dissociated, the overall β_2 ($=K_1 K_2$) is not directly comparable to the β_2 values for the addition of imidazoles to synthetic hemins in CDCl$_3$ (Satterlee and La Mar, 1976). In fact, it should be stressed that the solvent plays an important role in determining not only the coordination state of the high-spin starting material but also the degree of ion pairing of the product (Walker *et al.*, 1976).

Addition of water to a pyridine solution of protoporphyrin IX iron(III) chloride leads exclusively to the formation of the 6-coordinated aqua hemin complex [(PP)Fe(py)(H$_2$O)]Cl, when the concentration of water is high. At lower water concentration, a variety of high-spin mono- and low-spin *bis*-pyridine complexes are formed in addition to [(PP)Fe(Py)(H$_2$O)]$^+$Cl$^-$. The aqua complex shows anomalous temperature dependence of the isotropic proton shifts for the ring methyl protons, suggesting that a spin equilibrium between the high-spin and the low-spin states of the iron(III) ion in [(PP)Fe(Py)(H$_2$O)]$^+$Cl$^-$ exists in solution (Mazumdar *et al.*, 1988). Equilibrium constants were not reported.

3.4.2. Kinetics of Axial Ligand Exchange on Fe(III) Porphyrins

La Mar and Walker (1972) first reported investigations of the dynamics of axial ligand exchange in the low-spin Fe(III) complex [TPPFe(NMeIm)$_2$]$^+$Cl$^-$ in CDCl$_3$ by observing line broadening of the coordinated N-MeIm N-methyl peak as a function of temperature. It was established that even in the presence of excess ligand, ligand exchange proceeds by a dissociative mechanism involving a 5-coordinate PFeL$^+$X$^-$ transition state:

$$[PFeL_2]^+ \rightleftharpoons [[PFeL]^+]^+ + L \qquad k_1 \qquad (30)$$

$$[[PFeL]^+]^+ + L^* \rightarrow [PFeLL^*]^+ \qquad (\text{rapid}) \qquad (31)$$

Satterlee *et al.* (1977) extended these studies to characterize the axial lability of various imidazoles and pyridines in *bis*-ligated low-spin ferric complexes of synthetic porphyrins, including tetraphenyl- and tetra-n-propylporphyrins. Typical activation parameters determined in this study are $\Delta H^{\pm} = 17$–20 Kcal/mole, $\Delta S^{\pm} = 15$–19 e.u. (Satterlee et $al.$, 1977). More recently it has been shown by a combination of saturation transfer, linewidth analysis and line shape analysis in the intermediate exchange rate region that, as for the TPP complexes, ligand exchange in sterically more hindered tetramesitylporphinato iron(III) complexes is also dissociative (Nakamura, 1988, 1989). Dissociation rates of axially coordinated imidazoles measured for TPP and TMP complexes reveal that in both cases the axial lability of coordinated imidazoles increases with increasing steric bulk of the base, i.e., 2-methylimidazole complexes have larger dissociation-rate constants than the corresponding N-methylimidazole species. The increased lability is believed to be due to the steric repulsion between the alkyl substituent in the 2-position of the imidazole ring and the porphyrin macrocycle. Overall it was determined that TPP complexes are more labile than their TMP counterparts. However, the stability of the TMP complexes is invariably greater than that of their TPP counterparts, as discussed in Section 3.4.1 above. In fact, the rate constants are inversely proportional to the equilibrium constants β_2 (Nakamura, 1989), suggesting that this reaction indeed proceeds as described by Eqs. (30) and (31) and that $\beta_2 \propto K_2 = k_f/k_r$, where $k_r = k_1$.

For a number of mono-$ortho$-substituted derivatives of [TPPFe(N-MeIm)$_2$]$^+$, including a series of [(o-CONR$_2$)$_1$TPPFe(NMeIm)$_2$]$^+$ complexes having amide R groups of varying sizes, large effects are observed in the rate constants for exchange of one of the two unique N-methylimidazole ligands, with the rate constants being determined partly by the electronic effect and partly by the size of the $ortho$ substituent (Walker et $al.$, 1992). However, analysis of the NMR spectra, in combination with kinetics data obtained from analysis of the line broadening of the N-methyl peaks of the two coordinated N-MeIm ligands, indicate that for selected $ortho$-substituted complexes, the N-methylimidazole ligand that dissociates most rapidly is the one on the opposite side of the porphyrin plane from the $ortho$-substituent, probably due to buckling of the macrocycle to accommodate the substituent. Such buckling is expected to have the effect of strengthening the Fe—N bond of the ligand on the same side of the porphyrin plane as the substituent and weakening that of the ligand on the opposite side of the plane (Walker et $al.$, 1992).

3.4.3.   Kinetics of Fe(III) Porphyrin Axial Ligand Rotation

Nakamura and Groves (1988) first reported the coalescence behavior of the pyrrole-H, o-CH$_3$, p-CH$_3$, and m-H resonances of [TMPFe(2-MeImH)$_2$]$^+$ as the temperature is lowered from room temperature to -48 °C. They correctly ascribed this behavior to a slowing of axial ligand rotation as the temperature is lowered, and calculated activation parameters from complete line-shape analysis of the two p-CH$_3$ signals of the TMP ligand. For the rotation of 2-MeImH ($\Delta H^{\pm} = 12.9 \pm 0.4$ Kcal/mole; $\Delta S^{\pm} = 3.7 \pm 1.6$ e.u.) and 1,2-Me$_2$Im ($\Delta H^{\pm} = 12.6 \pm 0.8$ Kcal/mole; $\Delta S^{\pm} = 5.2 \pm 3.6$ e.u.) (Nakamura and Groves, 1988). Although their proposed structure of the low-temperature complex, in which ligand rotation is hindered, was later modified on the basis of NOESY investigations of this system (Walker and Simonis, 1991), the activation parameters clearly describe the barriers to rotation of the two 2-substituted imidazoles.

These barriers are considerably lower than those for axial ligand exchange (Satterlee *et al.*, 1977) and clearly represent a totally different kinetic process.

3.5. Bond Formation

3.5.1. Fe–C (Carbenes, CR$_2$)

Two types of carbene adducts of metalloporphyrins are known. One involves a carbene axially coordinated to the metal center; the other involves a carbene inserted into the metal–nitrogen bond. Both types, and their interconversions, have been investigated in detail by the groups of Balch and Mansuy. Carbenes are also known to add to the porphyrin itself to form homoporphyrins, N-alkylated and *meso*-substituted porphyrins (Latos-Grazynski *et al.*, 1981). An axially coordinated diamagnétic ferrous carbene complex is formed when *meso*-tetraphenylporphinato iron(II) is reacted with 2,2-*bis*(*p*-chlorophenyl)-1,1,1-trichloromethane in the presence of iron powder (Mansuy *et al.*, 1978). This diamagnetic complex can be oxidized to the paramagnetic [(*p*-ClC$_6$H$_4$)$_2$C=C:](TPP)FeX species (X = Cl, Br, I), in which the vinylidene group is inserted into the iron–nitrogen bond (Mansuy *et al.*, 1978; Latos-Grazynski *et al.*, 1981). NMR spectroscopy reveals that the latter compound is best described as an Fe(III), $S = \frac{3}{2}$, complex (Latos-Grazynski *et al.*, 1981; Balch *et al.*, 1985e). Replacing the dianion of *meso*-tetraphenylporphyrin by the dianion of protoporphyrin IX dimethyl ester results in the formation of the analogous protoporphyrin IX vinylidine derivatives, which consist of four isomers, one for insertion into each of the four unique pyrrole nitrogen–Fe bonds (Balch *et al.*, 1985e). Four pyrrole-H signals are observed in the NMR spectrum of the TPPFe(III) vinylidine complex. One single resonance is shifted to higher frequencies (20 ppm), whereas the other three "normal" pyrrole protons are shifted to lower frequencies (−18 to −45 ppm). For the OEP and protoporphyrin IX complexes, all proton methyl and methylene contact shifts are to higher frequency. Thus, the contact shift pattern at the three "normal" pyrrole protons reflects extensive spin delocalization to a π MO, as found for other $S = \frac{3}{2}$ ferric complexes (Section 2.7). The dominance of the π spin transfer mechanism can be taken as direct evidence that the $d_{x^2-y^2}$ orbital is not populated and that the ground state of the complex has the electronic configuration $(d_{xy})^2(d_{xz})^1(d_{yz})^1(d_{z^2})^1$. The vinylidine complexes exhibit axial magnetic anisotropy, with negative zero-field splitting (Balch *et al.*, 1985e). If an unsymmetrical vinylidene precursor such as Ph-CH$_2$CH-N$_2$ is used, the NMR resonance pattern of the product formed by reaction with porphyrins such as (TpClPP)Fe^{III}Cl is consistent with that expected for the vinylidene complex, which in this case is diastereotopic. The result is eight unique pyrrole-H resonances, with observed shifts again consistent with an intermediate-spin ($S = \frac{3}{2}$) complex (Artaud *et al.*, 1988, 1990). Protonation of these Fe(III) vinylidene complexes can lead to Fe(III)-N-alkyl porphyrins, while one-electron reduction produces Fe(II)-coordinated carbenes (Fe=C=CR$_2$) or Fe(II) complexes of N-vinyl porphyrins, depending on conditions (Artaud *et al.*, 1990).

3.5.2. Fe–N (Nitrenes, NNR)

Not only carbenes but also nitrenes can be inserted into the iron–pyrrole nitrogen bond of ferric and ferrous porphyrin complexes, as discussed in Section 2.4.1b(iii).

Examples of iron–nitrene complex formation have been described by Mahy *et al.* (1984, 1986, 1988). The high-spin, 5-coordinated ferrous and ferric nitrene complexes are synthesized by reacting the iron(II) or iron(III) complexes of tetraphenylporphyrin, tetratolylporphyrin, or tetra-*p*-chlorophenylporphyrin with either 1-amino-2,2,6,6-tetra-methylpiperidine in the presence of oxygen or one equivalent of iodosobenzene or with [(*p*-tolylsulfonylimino)iodo]benzene. X-ray crystal structure determination for both complexes established the insertion of the nitrene moiety into the iron–pyrrole nitrogen bond and the pentacoordination of the complexes. Proton NMR studies favor the well-established high-spin ferric ($S = \frac{5}{2}$) and high spin ferrous ($S = 2$) states for the complexes. The NMR spectrum of the Fe(II) nitrene complex has been discussed earlier in this chapter (Section 2.4.1). The ^{1}H NMR spectrum of the high-spin ferric complex in CDCl$_3$ at 20 °C exhibits signals at 89.5, 84.5, 81.7, and −28.4 ppm, shifts which have been attributed to the pyrrole protons. These shifts are similar to those observed for other $S = \frac{5}{2}$ complexes. The appearance of four pyrrole peaks is characteristic for the C_s symmetry of the complex. Protons of the nitrene moiety resonate at 14.3 and 13.7 ppm.

3.5.3. Fe—O—Fe (μ-Oxo) and Fe—OO—Fe (μ-Peroxo) Species

Reactions that form these dimer species were discussed in Sections 2.6.1 and 2.6.2, and the NMR spectra and antiferromagnetic coupling of the dimers were discussed in detail in Section 2.6.4. The reader is referred to these sections.

REFERENCES

Adler, A. D., Longo, F. R., Finarelli, J. D., Goldmacher, J., Assour, J., and Korsakoff, L., 1967, *J. Org. Chem.* **32**:476.

Albert, A., 1971, in *Physical Methods in Heterocyclic Chemistry* (A. R. Katritzky, ed.), Academic Press, New York, **1**:1–108, **3**:1–26.

Anderson, O. P., Kopelov, A. B., and Lavallee, D. K., 1980, *Inorg. Chem.* **19**:2101–2107.

Antipas, A., Buchler, J. W., Gouterman, M., and Smith, P. D., 1978, *J. Am. Chem. Soc.* **100**:3015–3024.

Arasasingham, R. D., Balch, A. L., and Latos-Grazynski, L., 1987, *J. Am. Chem. Soc.* **109**:5846–5847.

Arasasingham, R. D., Balch, A. L., Cornman, C. R., and Latos-Grazynski, L., 1989a, *J. Am. Chem. Soc.* **111**:4357–4363.

Arasasingham, R. D., Cornman, C. R., and Balch, A. L., 1989b, *J. Am. Chem. Soc.* **111**:7800–7805.

Arasasingham, R. D., Balch, A. L., Cornman, C. R., de Ropp, J. S., Eguchi, K., and La Mar, G. N., 1990, *Inorg. Chem.* **29**:1847–1850.

Artaud, I., Gregoire, N., Battioni, J. P., Dupre, D., and Mansuy, D., 1988, *J. Am. Chem. Soc.* **110**:8714–8716.

Artaud, I., Gregoire, N., Leduc, P., and Mansuy, D., 1990, *J. Am. Chem. Soc.* **112**:6899–6905.

Balch, A. L., and Renner, M. W., 1986a, *Inorg. Chem.* **25**:303–307.

Balch, A. L., and Renner, M. W., 1986b, *J. Amer. Chem. Soc.* **108**:2603–2608.

Balch, A. L., Chan, Y.-W., Cheng, R.-J., La Mar, G. N., Latos-Grazynski, L., and Renner, M. W., 1984, *J. Am. Chem. Soc.* **106**:7779–7785.

Balch, A. L., Chan, Y.-W., La Mar, G. N., Latos-Grazynski, L., and Renner, M. W., 1985a, *Inorg. Chem.* **24**:1437.

Balch, A. L., La Mar, G. N., Latos-Grazynski, L., Renner, M. W., and Thanabal, V., 1985b, *J. Am. Chem. Soc.* **107**:3003–3007.

Balch, A. L., Latos-Grazynski, L., and Renner, M. W., 1985c, *J. Am. Chem. Soc.* **107**:2983–2985.

Balch, A. L., La Mar, G. N., Latos-Grazynski, L., and Renner, M. W., 1985d, *Inorg. Chem.* **24**:2432–2436.

Balch, A. L., Cheng, R. J., La Mar, G. N., and Latos-Grazynski, L., 1985e, *Inorg. Chem.* **24**:2651–2656.

Balch, A. L., Hart, R. L., and Latos-Grazynski, L., 1990a, *Inorg. Chem.* **29**:3253–3256.

Balch, A. L., Hart, R. L., Latos-Grazynski, L., and Traylor, T. G., 1990b, *J. Am. Chem. Soc.* **112**:7382–7388.

Balch, A. L., Cornman, C. R., Latos-Grazynski, L., and Olmstead, M. M., 1990c, *J. Am. Chem. Soc.* **112**:7552–7558.

Balch, A. L., Cornman, C. R., Safari, N., and Latos-Grazynski, L., 1990d, *Organometallics* **9**:2420–2421.

Barbush, M., and Dixon, D. W., 1985, *Biochem. Biophys. Res. Commun.* **129**:70–75.

Barkigia, K. M., Chang, C. K., Fajer, J., and Renner, M. W., 1992, *J. Am. Chem. Soc.* **114**:1701–1707.

Battioni, J.-P., Artaud, I., Dupre, D., Leduc, P., and Mansuy, D., 1987, *Inorg. Chem.* **26**:1788–1796.

Behere, D. V., Birdy, R., and Mitra, S., 1982, *Inorg. Chem.* **21**:386–390.

Behere, D. V., and Goff, H. M., 1984, *J. Am. Chem. Soc.* **106**:4945–4950.

Bertini, I., Capozzi, F., Luchinat, C., and Turano, P., 1991, *J. Magn. Reson.* **95**:244–252.

Berzofsky, J. A., Peisach, J., and Blumberg, W. E., 1971, *J. Biol. Chem.* **246**:3367–3377.

Bill, E., Ding, X.-Q., Bominaar, E. L., Trautwein, A., Winkler, H., Handon, D., Weiss, R., Gold, A., Jayaraj, K., Hatfield, W. E., and Kirk, M. L., 1990, *Eur. J. Biochem.* **188**:665–672.

Binstead, R. A., Crossley, M. J., and Hush, N. S., 1991, *Inorg. Chem.* **30**:1259–1264.

Boersma, A. D., and Goff, H. M., 1982, *Inorg. Chem.* **21**:581–586.

Boersma, A. D., and Goff, H. M., 1984, *Inorg. Chem.* **23**:1671–1676.

Bold, T. J., 1978, Ph.D. Thesis, University of California, Davis.

Bondoc, L. L., Chau, M.-H., Price, M. A., and Timkovich, R., 1986, *Biochemistry* **25**:8458–8466.

Bonnett, R., Gale, I. A. D., and Stephenson, G. F., 1967, *J. Chem. Soc. C*, 1168–1172.

Bonnett, R., Czechowski, F., and Latos-Grazynski, L., 1990, *J. Chem. Soc. Chem. Commun.*, 849–851.

Botulinski, A., Buchler, J. W., Lee, Y. J., Scheidt, W. R., and Wicholas, M., 1988, *Inorg. Chem.* **27**:927–933.

Brackett, G. D., Richards, P. L., and Caughey, W. S., 1971, *J. Chem. Phys.* **54**:4383.

Budd, D. L., La Mar, G. N., Langry, K. C., Smith, K. M., and Nayyir-Mazhir, R., 1979, *J. Am. Chem. Soc.* **101**:6091–6096.

Buisson, G., Deronzier, A., Duee, E., Gans, P., Marchon, J.-C., and Regnard, J.-R., 1982, *J. Am. Chem. Soc.* **104**:6793–6795.

Caron, C., Mitschler, A., Rivere, G., Ricard, L., Schappmacher, M., and Weiss, R., 1979, *J. Am. Chem. Soc.* **101**:7401–7402.

Carrington, A., and McLachlan, A. D., 1967, *Introduction to Magnetic Resonance*, Harper and Row, New York, p. 80.

Cavaleiro, J. A. S., Rocha Gonsalves, A. M. d'A., Kenner, G. W., Smith, K. M., Shulman, R. G., Mayer, A., and Yamane, T., 1974a, *J. Chem. Soc., Chem. Commun.*, 392.

Cavaleiro, J. A. S., Rocha Gonsalves, A. M. d'A., Kenner, G. W., and Smith, K. M., 1974b, *J. Chem. Soc., Perkin Trans.* **1**:1771–1781.

Chacko, V. P., and La Mar, G. N., 1982, *J. Am. Chem. Soc.* **104**:7002–7007.

Chatfield, M. J., La Mar, G. N., Balch, A. L., and Lecomte, J. T. J., 1986a, *Biochem. Biophys. Res. Commun.* **135**:309–315.

Chatfield, M. J., La Mar, G. N., Lecomte, J. T. J., Balch, A. L., Smith, K. M., and Langry, K. C., 1986b, *J. Am. Chem. Soc.* **108**:7108–7110.

Chatfield, M. J., La Mar, G. N., and Kauten, R. J., 1987, *Biochemistry* **26**:6939–6950.

Chatfield, M. J., La Mar, G. N., Smith, K. M., Leung, H.-K., and Pandey, R. K., 1988a, *Biochemistry* **27**:1500–1507.

Chatfield, M. J., La Mar, G. N., Parker, W. O., Smith, K. M., Leung, H.-K., and Morris, I. K., 1988b, *J. Am. Chem. Soc.* **110**:6352–6358.

Cheng, R.-J., Latos-Grazynski, L., and Balch, A. L., 1982, *Inorg. Chem.* **21**:2412–2418.

Chestnut, D. B., 1958, *J. Chem. Phys.* **29**:43–47.

Chin, D.-H., Del Gaudio, J., La Mar, G. N., and Balch, A. L., 1977, *J. Am. Chem. Soc.* **99**:5486–5488.

Chin, D.-H., Balch, A. L., and La Mar, G. N., 1980, *J. Am. Chem. Soc.* **102**:1466–1448.

Chin, D. H., Balch, A. L., La Mar, G. N., Latos-Grazynski, L., and Renner, M. W., 1984, *J. Am. Chem. Soc.* **102**:1446.

Cocolios, P., Lagrange, G., and Guilard, R., 1983, *J. Organometallic Chem.* **253**:65–79.

Cohen, I. A., Ostfeld, D., and Lichtenstein, B., 1972, *J. Am. Chem. Soc.* **94**:4522–4525.

Collman, J. P., Brauman, J. I., Doxsee, K. M., Halbert, T. R., Bunnenberg, E., Linder, R. E., La Mar, G. N., Del Gaudio, J., Lang, G., and Spartalian, K., 1980, *J. Am. Chem. Soc.* **102**:4182–4192.

Collman, J. P., Brauman, J. I., Collins, T. J., Iverson, B. L., Lang, G., Pettman, R. B., Sessler, J. L., and Walters, M. A., 1983, *J. Am. Chem. Soc.* **105**:3038–3052.

Crawford, P. W., and Ryan, M. D., 1991, *Inorg. Chim. Acta* **179**:25–33.

Czechowski, F., and Latos-Grazynski, L., 1990, *Naturwissenschaft.* **77**:578–581.

Del Gaudio, J., and La Mar, G. N., 1976, *J. Am. Chem. Soc.* **98**:3014–3015.

Del Gaudio, J., and La Mar, G. N., 1978, *J. Am. Chem. Soc.* **100**:1112–1119.

de Ropp, J. S., and La Mar, G. N., 1991, *J. Am. Chem. Soc.* **113**:4348–4350.

Dixon, D. W., Barbush, M., and Shirazi, A., 1985, *Inorg. Chem.* **24**:1081–1087.

Dolphin, D., Forman, A., Borg, D. C., Fajer, J., and Felton, R. H., 1971, *Proc. Nat. Acad. Sci. USA* **3**:614–618.

Dolphin, D., and Felton, R. H., 1974, *Acct. Chem. Res.* **7**:26–32.

Dolphin, D. H., Sams, J. R., and Tsin, T. B., 1977, *Inorg. Chem.* **16**:711–713.

Dugad, L. B., and Mitra, S., 1984, *Proc. Indian Acad. Sci.* **93**:295–311.

Dugad, L. B., Marathe, V. R., and Mitra, S., 1985, *Proc. Indian Acad. Sci.* **95**:189–205.

Dugad, L. B., Medhi, O. K., and Mitra, S., 1987, *Inorg. Chem.* **26**:1741–1746.

Dugad, L. B., La Mar, G. N., and Unger, S. W., 1990, *J. Am. Chem. Soc.* **112**:1386–1392.

Emerson, S. D., and La Mar, G. N., 1990a, *Biochemistry* **29**:1545–1556.

Emerson, S. D., and La Mar, G. N., 1990b, *Biochemistry* **29**:1556–1566.

Ernst, R. R., Bodenhausen, G., and Wokaun, A., 1987, *Principles of Nuclear Magnetic Resonance in One and Two Dimensions*, Clarendon Press, Oxford, Chapter 8.

Evans, B., Smith, K. M., La Mar, G. N., and Viscio, D. B., 1977, *J. Am. Chem. Soc.* **99**:7070–7073.

Fajer, J., Borg, D. C., Forman, A., Felton, R. H., Vegh, L., and Dolphin, D., 1973, *Ann. New York Acad. Sci.* **206**:349–364.

Fajer, J., and Davis, M. S., 1979, in *The Porphyrins* (D. Dolphin, ed.), Academic Press, New York, **4**:197–256.

Feng, Y., Roder, H., and Englander, S. W., 1990, *Biochemistry* **29**:3494–3504.

Fielding, L., Eaton, G. R., and Eaton, S. S., 1985, *Inorg. Chem.* **24**:2309–2312.

Finzel, B. C., Weber, P. C., Hardman, K. D., and Salemme, F. R., 1985, *J. Mol. Biol.* **186**:627–743.

Fujii, H., and Ichikawa, K., 1992, *Inorg. Chem.* **31**:1110–1112.

Gadsby, P. M. A., and Thomson, A. J., 1990, *J. Am. Chem. Soc.* **112**:5003–5011.

Godziela, G. M., Ridnour, L. A., and Goff, H. M., 1985, *Inorg. Chem.* **24**:1610–1611.

Godziela, G. M., Kramer, S. K., and Goff, H. M., 1986, *Inorg. Chem.* **25**:4286–4288.

Goff, H. M., 1980, *J. Am. Chem. Soc.* **102**:3252–3254.

Goff, H. M., 1983, in *Iron Porphyrins* (A. B. P. Lever and H. B. Gray, eds.), Addison-Wesley, Reading, MA, Part 1, pp. 239–281.

Goff, H. M., 1990, *Biochim. Biophys. Acta* **1037**:356.

Goff, H. M., and La Mar, G. N., 1977, *J. Am. Chem. Soc.* **99**:6599–6606.

Goff, H. M., and Phillippi, M. A., 1983, *J. Am. Chem. Soc.* **105**:7567–7571.

Goff, H. M., and Shimomura, E., 1980, *J. Am. Chem. Soc.* **102**:31–37.

Goff, H. M., La Mar, G. N., and Reed, C. A., 1977, *J. Am. Chem. Soc.* **99**:3641–3646.

Goff, H. M., Shimomura, E. T., Lee, Y. J., and Scheidt, W. R., 1984, *Inorg. Chem.* **23**:315–321.

Gold, A., Jeyaraj, K., Doppelt, P., Weiss, R., Chottard, G., Bill, E., Ding, X., and Trautwein, A. X., 1988, *J. Am. Chem. Soc.* **110**:5756–5761.

Gouterman, M., 1978, in *The Porphyrins* (D. Dolphin, ed.), Academic Press, New York, **3**:1–165.

Gregson, A. K., 1981, *Inorg. Chem.* **20**:81–87.

Griffith, J. S., 1956, *Proc. R. Soc. London, Ser. A* **235**:23–36.

Groves, J. T., Haushalter, R. C., Nakamura, M., Nemo, T. E., and Evans, B. J., 1981, *J. Am. Chem. Soc.* **103**:2884–2886.

Groves, J. T., Quinn, R., McMurry, T. J., Nakamura, M., Lang, G., and Boso, B., 1985, *J. Am. Chem. Soc.* **107**:354–360.

Guilard, R., Perrot, I., Tabard, A., Richard, P., Lecomte, C., Liu, Y.-H., and Kadish, K. M., 1991, *Inorg. Chem.* **30**:27–37.

Gunter, M. J., McLaughlin, G. M., Berry, K. J., Murray, K. S., Irving, M., and Clark, P. E., 1984, *Inorg. Chem.* **23**:283–300.

Gupta, G. P., Lang, G., Reed, C. A., Shelly, K., and Scheidt, W. R., 1987, *J. Chem. Phys.* **86**:5288–5293.

Hambright, P., Turner, A., Cohen, J. S., Lyon, R. C., Katz, A., Neta, P., and Adeyemo, A., 1987, *Inorg. Chim. Acta* **128**:L11–L14.

Hanson, L. K., Chang, C. K., Davis, M. S., and Fajer, J., 1981, *J. Am. Chem. Soc.* **103**:663–670.

Helms, J. H., ter Haar, L. W., Hatfield, W. E., Harris, D. L., Jayaraj, K., Toney, G. E., Gold, A., Mewborn, T. D., and Pemberton, J. R., 1986, *Inorg. Chem.* **25**:2334–2337.

Helpern, J. A., Curtis, J. C., Hearshen, D., Smith, M. B., and Welch, K. M. A., 1987, *Magn. Reson. Med.* **5**:302–305.

Hickman, D. L., and Goff, H. M., 1983, *Inorg. Chem.* **22**:2787–2789.

Hickman, D. L., and Goff, H. M., 1984, *J. Am. Chem. Soc.* **106**:5013–5014.

Hickman, D. L., Shirazi, A., and Goff, H. M., 1985, *Inorg. Chem.* **24**:563–566.

Hickman, D. L., Nanthakumar, A., and Goff, H. M., 1988, *J. Am. Chem. Soc.* **110**:6384–6390.

Hill, H. A. O., and Morallee, K. G., 1972, *J. Am. Chem. Soc.* **94**:731–738.

Hill, H. A. O., Skyte, P. D., Buchler, J. W., Leuken, H., Tonn, M., Gregson, A. K., and Pellizer, G., 1979, *J. Chem. Soc., Chem. Commun.*, 151.

Hobbs, J. D., Larsen, R. W., Meyer, T. E., Hazzard, J. H., Cusanovich, M. A., and Ondrias, M. R., 1990, *Biochemistry* **29**:4166–4174.

Horrocks, W. D., and Greenberg, E. S., 1973, *Biochim. Biophys. Acta* **322**:38–44.

Horrocks, W. D., and Greenberg, E. S., 1974, *Mol. Phys.* **27**:993–999.

Hwang, Y. C., and Dixon, D. W., 1986, *Inorg. Chem.* **25**:3716.

Inniss, D., Soltis, S. M., and Strouse, C. E., 1988, *J. Am. Chem. Soc.* **110**:5644–5650.

Isaac, M. F., Lin, Q., Simonis, U., Suffian, D. J., Wilson, D. L., and Walker, F. A., 1992, *Inorg. Chem.*, submitted.

Ivanca, M. A., Lappin, A. G., and Scheidt, W. R., 1991, *Inorg. Chem.* **30**:711–718.

Jesson, J. P., 1973, in *NMR of Paramagnetic Molecules* (G. N. La Mar, W. D. Horrocks, and R. H. Holm, eds.), Academic Press, New York, pp. 1–53.

Johnson, R. D., Ramaprasad, S., and La Mar, G. N., 1983, *J. Am. Chem. Soc.* **105**:7205–7206.

Jones, R. D., Summerville, D. A., and Basolo, F., 1979, *Chem. Rev.* **79**:139–179.

Kadish, K. M., Rhodes, R. K., Bottomley, L. A., and Goff, H. M., 1981, *Inorg. Chem.* **20**:3195–3200.

Kadish, K. M., Tabard, A., Lee, W., Liu, Y. H., Ratti, C., and Guilard, R., 1991, *Inorg. Chem.* **30**:1542–1549.

Kastner, M. E., Scheidt, W. R., Mashiko, T., and Reed, C. A., 1978, *J. Am. Chem. Soc.* **100**:666–667.

Keating, K. A., de Ropp, J. S., La Mar, G. N., Balch, A. L., Shaiu, F.-Y., and Smith, K. M., 1991, *Inorg. Chem.* **30**:3258–3263.

Keating, K. A., La Mar, G. N., Shaiu, F.-Y., and Smith, K. M., 1992, *J. Am. Chem. Soc.* **114**:in press.

Keller, R., and Wüthrich, K., 1980, *Biochim. Biophys. Acta* **621**:204–217.

Keller, R. M., and Wüthrich, K., 1981, in *Biological Magnetic Resonance* (L. J. Berliner, and J. Reuben, eds.), Plenum Press, New York, 3.

Keller, R., Groudinsky, O., and Wüthrich, K., 1973, *Biochim. Biophys. Acta* **328**:233–238.

Keller, R., Groudinsky, O., and Wüthrich, K., 1976, *Biochim. Biophys. Acta* **427**:497–511.

Keller, R. M., Schejter, A., and Wüthrich, K., 1980, *Biochim. Biophys. Acta* **626**:15–22.

Kellett, P. J., Pawlik, M. J., Taylor, L. F., Thompson, R. G., Levstik, M. A., Anderson, O. P., and Strauss, S. H., 1989, *Inorg. Chem.* **28**:440–447.

Kenner, G. W., and Smith, K. M., 1973, *Ann. New York Acad. Sci.* **206**:138.

Kessel, S. L., and Hendrickson, D. N., 1980, *Inorg. Chem.* **19**:1883–1889.

Kim, Y. O., and Goff, H. M., 1988, *J. Am. Chem. Soc.* **110**:8706–8707.

Kim, Y. O., and Goff, H. M., 1990, *Inorg. Chem.* **29**:3907–3908.

Kimura, K., Peterson, J., Wilson, M., Cookson, D. J., and Williams, R. J. P., 1981, *J. Inorg. Biochem.* **15**:11–25.

Kintner, E. T., and Dawson, J. H., 1991, *Inorg. Chem.* **30**:4892–4897.

Kirner, J. F., Hoard, J. L., and Reed, C. A., 1978, *Abstracts of Papers*, 175th National Meeting of the American Chemical Society, Anaheim, CA, March 13–16, 1978, American Chemical Society, Washington, D.C.

Koerner, R., Aubrecht, K., Tipton, A. R., Norvell, C. J., Mink, L. M., and Walker, F. A., 1992, unpublished.

Komives, E. A., Tew, D., Olmstead, M. M., and Ortiz de Montellano, P. R., 1988, *Inorg. Chem.* **26**:1788–1796.

Kunze, K. L., and Ortiz de Montellano, P. R., 1983, *J. Am. Chem. Soc.* **105**:1380.

Kurland, R. J., and McGarvey, B. R., 1970, *J. Magn. Reson.* **2**:286.

La Mar, G. N., 1979, in *Biological Applications of Magnetic Resonance* (R. G. Shulman, ed.), Academic Press, New York, pp. 305–343.

La Mar, G. N., 1992, personal communication.

La Mar, G. N., and Del Gaudio, J., 1977a, *Bioinorg. Chem.* **2**:207.

La Mar, G. N., and Del Gaudio, J., 1977b, *Adv. Chem. Ser.* **162**:207–226.

La Mar, G. N., and Viscio, D. B., 1974, *J. Am. Chem. Soc.* **96**:7354–7355.

La Mar, G. N., and Walker, F. A., 1972, *J. Am. Chem. Soc.* **94**:8607–8608.

La Mar, G. N., and Walker, F. A., 1973a, *J. Am. Chem. Soc.* **95**:1782–1790.

La Mar, G. N., and Walker, F. A., 1973b, *J. Am. Chem. Soc.* **95**:6950–6956.

La Mar, G. N., and Walker, F. A., 1979, in *The Porphyrins* (D. Dolphin, ed.), Academic Press, New York, **IVB**:57–161.

La Mar, G. N., Frye, J. S., and Satterlee, J. D., 1976, *Biochim. Biophys. Acta* **428**:78–90.

La Mar, G. N., Bold, T. J., and Satterlee, J. D., 1977a, *Biochim. Biophys. Acta* **498**:189–207.

La Mar, G. N., Del Gaudio, J., and Frye, J. S., 1977b, *Biochim. Biophys. Acta* **498**:422–435.

La Mar, G. N., Viscio, D. B., Smith, K. M., Caughey, W. S., and Smith, M. L. 1978a, *J. Am. Chem. Soc.* **100**:8085–8092.

La Mar, G. N., Budd, D. L., Viscio, D. B., Smith, K. M., and Langry, K. C., 1978b, *Proc. Nat. Acad. Sci. USA* **75**:5755–5759.

La Mar, G. N., Minch, M. J., and Frye, J. S., 1981, *J. Am. Chem. Soc.* **103**:5383–5388.

La Mar, G. N., de Ropp, J. S., Latos-Grazynski, L., Balch, A. L., Johnson, R. B., Smith, K. W., Parish, D. W., and Cheng, R. J., 1983, *J. Am. Chem. Soc.* **105**:782–787.

La Mar, G. N., Jackson, J. T., Dugad, L. B., Cusanovich, M. A., and Bartsch, R. G., 1990, *J. Biol. Chem.* **265**:16173–16180.

Lançon, D., Cocolios, P., Guilard, R., and Kadish, K. M., 1984, *Organometallics* **3**:1164–1170.

Latos-Grazynski, L., 1983, *Biochimie* **65**:143–148.

Latos-Grazynski, L., Cheng, R. J., La Mar, G. N., and Balch, A., 1981, *J. Am. Chem. Soc.* **103**:4270–4272.

Latos-Grazynski, L., Cheng, R. J., La Mar, G. N., and Balch, A., 1982, *J. Am. Chem. Soc.* **104**:5922–6000.

Lavallee, D. K., 1987, *The Chemistry and Biochemistry of N-Substituted Porphyrins*, VCH Publishers, New York.

Lee, K.-B., La Mar, G. N., Kehres, L. A., Fujinari, E. M., Smith, K. M., Pochapsky, T. C., and Sligar, S. G., 1990, *Biochemistry* **29**:9623–9631.

Lexa, D., Momenteau, M., and Mispelter, J., 1974, *Biochim. Biophys. Acta* **338**:151–163.

Lexa, D., Mispelter, J., and Saveant, J. M., 1981, *J. Am. Chem. Soc.* **103**:6808–6812.

Lexa, D., Momenteau, M., Saveant, J.-M., and Xu, F., 1985, *Inorg. Chem.* **24**:122.

Li, Z., and Goff, H. M., 1992, *Inorg. Chem.* **31**:1547–1548.

Licoccia, S., Chatfield, M. J., La Mar, G. N., Smith, K. M., Mansfield, K. E., and Anderson, R. R., 1989, *J. Am. Chem. Soc.* **111**:6087–6093.

Lin, Q., Simonis, U., Tipton, A. R., Norvell, C. J., and Walker, F. A., 1992, *Inorg. Chem.* **31**:4216–4217.

Liston, D. J., Kennedy, B. J., Murray, K. S., and West, B. O., 1985, *Inorg. Chem.* **24**:1561–1567.

Loew, G. H., and Herman, Z. S., 1980, *J. Am. Chem. Soc.* **102**:6114–6175.

Longuet-Higgins, H. C., Rector, C. W., and Platt, J. R., 1950, *J. Chem. Phys.* **18**:1174–1181.

Lukat, G., and Goff, H. M., 1990, *Biochim. Biophys. Acta* **1037**:351–359.

Mahy, J. P., Battioni, P., Mansuy, D., Fisher, J., Weiss, R., Mispelter, J., Morgenstern-Badarau, I., and Gans, P., 1984, *J. Am. Chem. Soc.* **106**:1699–1706.

Mahy, J. P., Battioni, P., and Mansuy, D., 1986, *J. Am. Chem. Soc.* **108**:1079–1080.

Mahy, J. P., Battioni, P., Bedi, G., Mansuy, D., Fisher, J., Weiss, R., and Morgenstern-Badarau, I., 1988, *Inorg. Chem.* **27**:353–359.

Malek, A., Latos-Grazynski, L., Bartczak, T. J., and Zadlo, A., 1991, *Inorg. Chem.* **30**:3222–3230.

Maltempo, M. M., 1974, *J. Chem. Phys.* **61**:2540–2547.

Maltempo, M. M., 1975, *Biochim. Biophys. Acta* **379**:95–102.

Maltempo, M. M., and Moss, T. H., 1976, *Quart. Rev. Biophys.* **9**:181–215.

Maltempo, M. M., Moss, T. H., and Cusanovich, M. A., 1974, *Biochim. Biophys. Acta* **342**:290–305.

Mansuy, D., Lange, M., and Chottard, J. C., 1978, *J. Am. Chem. Soc.* **100**:3214–3215.

Marnett, L. J., Weller, P., and Battista, J. R., 1986, in *Cytochrome P-450: Structure, Mechanism, and Biochemistry* (P. R. Ortiz de Montellano, ed.), Plenum Press, New York, pp. 29–76.

Martin, G. E., and Zektzer, A. S., 1988, *Two-Dimensional NMR Methods for Establishing Molecular Connectivity*, VCH Publishers, New York.

Masuda, H., Taga, T., Osaki, K., Sugimoto, H., Yoshida, Z.-I., and Ogoshi, H., 1980, *Inorg. Chem.* **19**:950–955.

Mayer, A., Ogawa, S., Shulman, R. G., Yamane, T., Cavaleiro, J. A. S., Rocha Gonsalves, A. M. d'A., Kenner, G. W., and Smith, K. M., 1974, *J. Mol. Biol.* **86**:749.

Mazumdar, S., 1990, *J. Phys. Chem.* **94**:5947–5953.

Mazumdar, S., 1991, *J. Chem. Soc., Dalton Trans.*, 2091–2096.

Mazumdar, S., and Medhi, O. K., 1990, *J. Chem. Soc., Dalton Trans.*, 2633–2636.

Mazumdar, S., and Mitra, S. J., 1990, *J. Phys. Chem.* **94**:561–566.

Mazumdar, S., Medhi, O. K., and Mitra, S., 1988, *Inorg. Chem.* **27**:2541–2543.

Mazumdar, S., Medhi, O. K., and Mitra, S., 1990, *J. Chem. Soc., Dalton Trans.*, 1057–1061.

Mazumdar, S., Medhi, O. K., and Mitra, S., 1991, *Inorg. Chem.* **30**:700–705.

McConnell, H. M., 1956, *J. Chem. Phys.* **24**:764.

McGarvey, B. R., 1988, *Inorg. Chem.* **27**:4691–4698.

McLachlan, A. D., 1958, *Mol. Phys.* **1**:233.

McLachlan, S. J., La Mar, G. N., and Lee, K. B., 1988, *Biochim. Biophys. Acta* **957**:430.

McMurry, T. J., and Groves, J. T., 1986, in *Cytochrome P-450: Structure, Mechanism, and Biochemistry* (P. R. Ortiz de Montellano, ed.), Plenum Press, New York, pp. 1–28.

Medforth, C. J., Shiau, F.-Y., La Mar, G. N., and Smith, K. M., 1991, *J. Chem. Soc., Chem. Commun.*, 590–592.

Medhi, O. K., Mazumdar, S., and Mitra, S., 1989, *Inorg. Chem.* **28**:3243–3248.

Migita, K., and La Mar, G. N., 1980, *J. Phys. Chem.* **84**:2953–2957.

Minch, M. J., and La Mar, G. N., 1982, *J. Phys. Chem.* **86**:1400–1406.

Mispelter, J., Momenteau, M., and Lhoste, J. M., 1977, *Mol. Phys.* **33**:1715–1728.

Mispelter, J., Momenteau, M., and Lhoste, J. M., 1980, *J. Chem. Phys.* **72**:1003–1012.

Mispelter, J., Momenteau, M., and Lhoste, J. M., 1981, *Biochimie* **63**:911–914.

Mispelter, J., Momenteau, M., Lavalette, D., and Lhoste, J. M., 1983, *J. Am. Chem. Soc.* **105**:5165–5166.

Miyamoto, T. K., Tsuzuki, S., Hasegawa, T., and Sasaki, Y., 1983, *Chem. Lett. (Japan)*, 1587–1588.

Morgan, B., and Dolphin, D., 1987, *Struct. Bonding* **64**:115–203.

Morishima, I., Kitagawa, S., Matsuki, E., and Inubushi, T., 1980, *J. Am. Chem. Soc.* **102**:2429–2437.

Morishima, L., Fujii, H., and Shiro, Y., 1986, *J. Am. Chem. Soc.* **108**:3858–3860.

Moss, T. H., Bearden, A. J., Bartsch, R. G., and Cusanovich, M. A., 1968, *Biochemistry* **7**:1583–1591.

Nakamura, M., 1988, *Chem. Lett. (Japan)*, 453–456.

Nakamura, M., 1989, *Inorg. Chim. Acta* **161**:73–80.

Nakamura, M., and Groves, J. T., 1988, *Tetrahedron* **44**:3225–3230.

Nakamura, M., and Nakamura, N., 1990, *Chem. Lett. (Japan)*, 181–184.

Nakamura, M., and Nakamura, N., 1991, *Chem. Lett. (Japan)*, 1885–1888.

Nanthakumar, A., and Goff, H. M., 1990, *J. Am. Chem. Soc.* **112**:4047–4049.

Nanthakumar, A., and Goff, H. M., 1992, *Inorg. Chem.* **31**:4460–4464.

Neya, S., and Funasaki, N., 1987, *Biochim. Biophys. Acta* **952**:150–157.

Neya, S., Funasaki, N., and Imai, K., 1988, *J. Biol. Chem.* **263**:8810–8815.

Ortiz de Montellano, P. R., 1986, in *Cytochrome P-450: Structure, Mechanism and Biochemistry* (P. R. Ortiz de Montellano, ed.), Plenum Press, New York, pp. 217–272.

Ortiz de Montellano, P. R., and Correia, M. A., 1983, *Ann. Rev. Pharmacol. Toxicol.* **23**:481.

Ortiz de Montellano, P. R., and Reich, N. O., 1986, in *Cytochrome P-450: Structure, Mechanism and Biochemistry* (P. R. Ortiz de Montellano, ed.), Plenum Press, New York, pp. 273–314.

Oster, O., Nereiter, G. W., Clouse, A. O., and Gurd, F. R. N., 1975, *J. Biol. Chem.* **250**:7990–7996.

Oumous, H., Lecomte, C., Protas, J., Cocolios, P., and Guilard, R., 1984, *Polyhedron* **3**:651–659.

Parmely, R. C., and Goff, H. M., 1980, *J. Inorg. Biochem.* **12**:269–280.

Pawlik, M. J., Miller, P. K., Sullivan, E. P., Levstik, M. A., Almond, D. A., and Strauss, S. H., 1988, *J. Am. Chem. Soc.* **110**:3007–3012.

Peisach, J., Blumberg, W. E., and Adler, A., 1973, *Ann. New York Acad. Sci.* **206**:310.

Phillippi, M. A., and Goff, H. M., 1982, *J. Am. Chem. Soc.* **104**:6026–6034.

Phillippi, M. A., Baenziger, N., and Goff, H. M., 1981a, *Inorg. Chem.* **20**:3904–3911.

Phillippi, M. A., Shimomura, E. T., and Goff, H. M., 1981b, *Inorg. Chem.* **20**:1322–1325.

Quinn, R., Nappa, M., and Valentine, J. S., 1982, *J. Am. Chem. Soc.* **104**:2588–2595.

Rafii, K., and Walker, F. A., 1992, unpublished.

Ramaprasad, S., Johnson, R. D., and La Mar, G. N., 1984a, *J. Am. Chem. Soc.* **106**:3632.

Ramaprasad, S., Johnson, R. D., and La Mar, G. N., 1984b, *J. Am. Chem. Soc.* **106**:5330.

Rawlings, J., Stephens, P. J., Nafie, L. A., and Kamen, M. D., 1977, *Biochemistry* **16**:1725–1729.

Reed, C. A., Mashiko, T., Bentley, S. D., Kastner, M. E., Scheidt, W. R., Spartalian, K., and Lang, G., 1979, *J. Am. Chem. Soc.* **101**:2948–2958.

Safo, M. K., Gupta, G. P., Walker, F. A., and Scheidt, W. R., 1991a, *J. Am. Chem. Soc.* **113**:5497–5510.

Safo, M. K., Gupta, G. P., Orosz, R. D., Reed, C. A., and Scheidt, W. R., 1991b, *Inorg. Chim. Acta* **184**:251–258.

Safo, M. K., Gupta, G. P., Watson, C. T., Simonis, U., Walker, F. A., and Scheidt, W. R., 1992, *J. Am. Chem. Soc.* **114**:7066–7075.

Safo, M. K., Walker, F. A., Raitsimring, A., Debrunner, P. G., and Scheidt, W. R., 1992, unpublished.

Sandreczki, T., Ondercin, D., and Kreilick, R. W., 1979, *J. Phys. Chem.* **83**:3388–3393.

Sano, S., Suguira, Y., Maeda, Y., Ogawa, S., and Morishima, I., 1981, *J. Am. Chem. Soc.* **103**:2888–2890.

Satterlee, J. D., 1986, *Ann. Reports NMR Spectrosc.* **17**:79–179.

Satterlee, J. D., 1987, in *Metal Ions in Biological Systems* (H. Sigel, ed.), Marcel Dekker, New York, **21**:121–185.

Satterlee, J. D., and La Mar, G. N., 1976, *J. Am. Chem. Soc.* **98**:2804–2808.

Satterlee, J. D., and Shelnutt, J. A., 1984, *J. Phys. Chem.* **88**:5487–5492.

Satterlee, J. D., La Mar, G. N., and Bold, T. J., 1977, *J. Am. Chem. Soc.* **99**:1088–1093.

Scheidt, W. R., and Gouterman, M., 1983, in *Iron Porphyrins* (A. B. P. Lever, and H. B. Gray, eds.), Addison-Wesley, Reading, MA, Part I, pp. 89–140.

Scheidt, W. R., and Lee, Y. J., 1987, *Struct. Bonding* **64**:1–70.

Scheidt, W. R., and Reed, C. A., 1981, *Chem. Rev.* **81**:543–555.

Scheidt, W. R., and Safo, M. K., 1991, unpublished.

Scheidt, W. R., Geiger, D. K., and Haller, K. J., 1982, *J. Am. Chem. Soc.* **104**:495–499.

Scheidt, W. R., Geiger, D. K., Hayes, R. G., and Lang, G., 1983, *J. Am. Chem. Soc.* **105**:2625–2632.

Scheidt, W. R., Geiger, D. K., Lee, Y. J., Reed, C. A., and Lang, G., 1987, *Inorg. Chem.* **26**:1039–1045.

Scheidt, W. R., Osvath, S. R., Lee, Y. J., Reed, C. A., Shavez, B., and Gupta, G. P., 1989, *Inorg. Chem.* **28**:1591–1595.

Scholz, W. F., Reed, C. A., Lee, Y. J., Scheidt, W. R., and Lang, G., 1982, *J. Am. Chem. Soc.* **104**: 6791–6793.

Shaka, A. J., 1992, personal communication.

Shedbalkar, V. P., Dugad, L. B., Mazumdar, S., and Mitra, S., 1988, *Inorg. Chim. Acta* **148**:17–20.

Shelly, K., Bartczak, T., Scheidt, W. R., and Reed, C. A., 1985, *Inorg. Chem.* **24**:4325–4330.

Shin, K., Kramer, K., and Goff, H., 1987, *Inorg. Chem.* **26**:4103–4106.

Shirazi, A., and Goff, H. M., 1982, *J. Am. Chem. Soc.* **104**:6318–6322.

Shirazi, A., Barbush, M., Ghosh, S., and Dixon, D. W., 1985, *Inorg. Chem.* **24**:2495–2502.

Simonis, U., and Walker, F. A., 1989, *Bioinorg. Chem.* **36**:268.

Simonis, U., and Walker, F. A., 1992, unpublished.

Simonis, U., Brown, B., and Tan, H., 1992a, unpublished.

Simonis, U., Dallas, J. L., and Walker, F. A., 1992b, *Inorg. Chem.*, **31**:5349–5350.

Simonneaux, G., and Sodano, P., 1988a, *J. Organometallic Chem.* **349**:C12–C14.
Simonneaux, G., and Sodano, P., 1988b, *Inorg. Chem.* **27**:3956–3959.
Simonneaux, G., Bondon, A., and Sodano, P., 1987, *Inorg. Chem.* **26**:3636–3638.
Simonneaux, G., Hindre, F., and Le Plouzennec, M., 1989, *Inorg. Chem.* **28**:823–825.
Simonneaux, G., Bondon, A., and Sodano, P., 1990, *Biochim. Biophys. Acta* **1038**:199–203.
Smith, K. M., and Langry, K. C., 1983, *J. Chem. Soc., Perkin Trans.* **1**:439–444.
Smith, K. M., Langry, K. C., and deRopp, J. S., 1979, *J. Chem. Soc. Chem. Comm.* 1001–1003.
Smith, M., and McLendon, G., 1981, *J. Am. Chem. Soc.* **103**:4912–4921.
Snyder, R., and La Mar, G. N., 1977, *J. Am. Chem. Soc.* **99**:7173–7184.
Stolzenberg, A. M., Strauss, S. H., and Holm, R. H., 1981, *J. Am. Chem. Soc.* **103**:4763–4778.
Strauss, S. H., and Pawlik, M. J., 1986, *Inorg. Chem.* **25**:1921–1923.
Strauss, S. H., Silver, M. E., Long, K. M., Thompson, R. G., Hudgens, R. A., Spartalian, K., and Ibers, J. A., 1985, *J. Am. Chem. Soc.* **107**:4207–4215.
Strauss, S. H., Pawlik, M. J., Skowyra, J., Kennedy, J. R., Anderson, O. P., Spartalian, K., and Dye, J. L., 1987, *Inorg. Chem.* **26**:724–730.
Sullivan, E. P., and Strauss, S. H., 1989, *Inorg. Chem.* **28**:3093–3095.
Sullivan, E. P., Grantham, J. D., Thomas, C. S., and Strauss, S. H., 1991, *J. Am. Chem. Soc.* **113**:5264–5270.
Summerville, D. A., Cohen, I. A., Hatano, K., and Scheidt, W. R., 1978, *Inorg. Chem.* **17**:2906–2910.
Suslick, K. S., and Reinert, T. J., 1985, *J. Chem. Ed.* **62**:974–983.
Taylor, C. P. S., 1977, *Biochim. Biophys. Acta* **91**:165–195.
Teraoka, J., and Kitagawa, T., 1980, *J. Phys. Chem.* **84**:1928–1935.
Thanabal, V., de Ropp, J. S., and La Mar, G. N., 1987, *J. Am. Chem. Soc.* **109**:265–272.
Thanabal, V., de Ropp, J. S., and La Mar, G. N., 1988, *J. Am. Chem. Soc.* **110**:3027–3035.
Thomson, A. J., and Gadsby, P. M. A., 1990, *J. Chem. Soc., Dalton Trans.* 1921–1928.
Timkovich, R., 1991, *Inorg. Chem.* **30**:37–42.
Toney, G. E., ter Haar, L. W., Savrin, J. E., Gold, A., Hatfield, W. E., and Sangaiah, R., 1984a, *Inorg. Chem.* **23**:2561–2563.
Toney, G. E., Gold, A., Savrin, J., ter Haar, L. W., Sangaiah, R., and Hatfield, W. E., 1984b, *Inorg. Chem.* **23**:4350–4352.
Traylor, T. G., 1981, *Acc. Chem. Res.* **14**:102–109.
Traylor, T. G., and Berzinis, A. P., 1980, *J. Am. Chem. Soc.* **102**:2844–2846.
Traylor, T. G., Chang, C. K., Geibel, J., Berzinis, A., Mincey, T., and Cannon, J., 1979, *J. Am. Chem. Soc.* **101**:6716–6731.
Trewhella, J., Wright, P. E., and Appleby, C. A., 1980, *Nature (London)* **180**:87.
Unger, S. W., Jue, T., and La Mar, G. N., 1985a, *J. Magn. Reson.* **61**:448–456.
Unger, S. W., LeComte, J. T. J., and La Mar, G. N., 1985b, *J. Magn. Reson.* **64**:521–526.
Viscio, D. B., and La Mar, G. N., 1978a, *J. Am. Chem. Soc.* **100**:8092–8096.
Viscio, D. B., and La Mar, G. N., 1978b, *J. Am. Chem. Soc.* **100**:8096–8100.
Walker, F. A., 1970, *J. Am. Chem. Soc.* **92**:4235–4244.
Walker, F. A., 1980, *J. Am. Chem. Soc.* **102**:3254–3256.
Walker, F. A., 1992, unpublished.
Walker, F. A., and Benson, M., 1982, *J. Phys. Chem.* **86**:3495–3499.
Walker, F. A., and Simonis, U., 1991, *J. Am. Chem. Soc.* **113**:8652–8657.
Walker, F. A., Lo, M. W., and Ree, M. T., 1976, *J. Am. Chem. Soc.* **98**:5552–5560.
Walker, F. A., Balke, V. L., and McDermott, G. A., 1982, *J. Am. Chem. Soc.* **104**:1569–1574.
Walker, F. A., Buehler, J., West, J. T., and Hinds, J. L., 1983, *J. Am. Chem. Soc.* **105**:6923–6929.
Walker, F. A., Reis, D., and Balke, V. L., 1984, *J. Am. Chem. Soc.* **106**:6888–6898.
Walker, F. A., Balke, V. L., and West, J. T., 1985, *J. Am. Chem. Soc.* **107**:1226–1233.

Walker, F. A., Huynh, B. H., Scheidt, W. R., and Osvath, S. R., 1986, *J. Am. Chem. Soc.* **108**:5288–5297.

Walker, F. A., Simonis, U., Zhang, H., Walker, J. M., Ruscitti, T. M., Kipp, C., Amputch, M. A., Castillo, B. V., Cody, S. H., Wilson, D. L., Graul, R. E., Yong, G. J., Tobin, K., West, J. T., and Barichievich, B. A., 1992, *New J. Chem.* **16**:609–620.

Wang, J.-T., Yeh, H. J. C., and Johnson, O. F., 1978, *J. Am. Chem. Soc.* **100**:2400–2405.

Watson, C. T., and Walker, F. A., 1992, unpublished results.

Watson, C. T., Simonis, U., and Walker, F. A., 1992, unpublished.

Weber, P. C., Howard, A., Xuong, N. H., and Salemme, F. R., 1981, *J. Mol. Biol.* **153**:399–424.

Weightman, J. A., Hoyle, N. J., and Williams, R. J. P., 1971, *Biochim. Biophys. Acta* **244**:567–572.

Welborn, C. H., Dolphin, D., and James, B. R., 1981, *J. Am. Chem. Soc.* **103**:2869–2871.

White-Dixon, D., Barbush, M., and Shirazi, A., 1984, *J. Am. Chem. Soc.* **106**:4638–4639.

White-Dixon, D., Barbush, M., and Shirazi, A., 1985, *Inorg. Chem.* **24**:1081–1087.

White-Dixon, D., Woejler, S., Hong, X., and Stolzenberg, A., 1988, *Inorg. Chem.* **27**:3682–3685.

Whitlock, H. W., Hanauer, R., Oester, M. Y., and Bower, B. K., 1969, *J. Am. Chem. Soc.* **91**:7485–7489.

Woon, T. C., Shirazi, A., and Bruice, T. C., 1986, *Inorg. Chem.* **25**:3845–3846.

Wüthrich, K., 1969, *Proc. Nat. Acad. Sci. USA* **63**:1071–1078.

Wüthrich, K., Shulman, R. G., Wyluda, B. J., and Caughey, W. S., 1969, *Proc. Nat. Acad. Sci. USA* **62**:636–643.

Yamamoto, Y., and Fujii, N., 1987, *Chem. Lett. (Japan)* 1703–1706.

Yamamoto, Y., Nanai, N., Chujo, R., and Suzuki, T., 1990, *FEBS Lett.* **264**:113–116.

Yoshimura, T., Suzuki, S., Nakahara, A., Iwasaki, H., Masuko, M., and Marsubara, T., 1985, *Biochim. Biophys. Acta* **831**:267–274.

Yu, B.-S., and Goff, H. M., 1989, *J. Am. Chem. Soc.* **111**:6558–6562.

Yu, C., Unger, S. W., and La Mar, G. N., 1986, *J. Magn. Reson.* **67**:346–350.

Zerner, M., Gouterman, M., and Kobayashi, H., 1966, *Theor. Chim. Acta* **6**:363.

Zhang, H., Simonis, U., and Walker, F. A., 1990, *J. Am. Chem. Soc.* **112**:6124–6126.

Zobrist, M., and La Mar, G. N., 1978, *J. Am. Chem. Soc.* **100**:1944–1946.

5

Proton NMR Studies of Selected Paramagnetic Heme Proteins

J. D. Satterlee, S. Alam, Q. Yi, J. E. Erman, I. Constantinidis, D. J. Russell, and S. J. Moench

1. INTRODUCTION

Among the heme proteins three types have attracted our attention during the past thirteen years. Not all of the heme proteins currently studied in this laboratory are paramagnetic in their native states, but several are. The presence of the heme-containing, redox-active iron ion makes it possible to study paramagnetic forms of all of these proteins. NMR data for three of these proteins will be the topic of this article, which is written from the point of view of relating experiences encountered in this laboratory. One of the first issues faced in reading the NMR literature is realizing that two heme numbering systems are simultaneously in use. The modified Fischer scheme seems to be preferred in the U.S. literature and is also widely employed by porphyrin chemists. The IUB/IUPAC scheme is more widely used outside the U.S. Examples of both are presented in Fig. 1.

J. D. Satterlee, S. Alam, and Q. Yi • Department of Chemistry, Washington State University, Pullman, Washington 99164-4630.   **J. E. Erman** • Department of Chemistry, Northern Illinois University, De Kalb, Illinois 60115.   **I. Constantinidis** • Department of Radiology, Emory University School of Medicine, Atlanta, Georgia 30322.   **D. J. Russell** • Bristol Meyers Squibb, P.O. Box 191, New Brunswick, New Jersey 08903-0191.   **S. J. Moench** • Department of Chemistry, University of Denver, Denver, Colorado 80208.

Biological Magnetic Resonance, Volume 12: NMR of Paramagnetic Molecules, edited by Lawrence J. Berliner and Jacques Reuben. Plenum Press, New York, 1993.

1.1. Cytochrome *c* Peroxidase

Our first work (Satterlee and Erman, 1980) devoted to paramagnetic heme proteins concerned yeast cytochrome *c* peroxidase (C*c*P; E.C. 1.1.11.5) a native ferriheme (Satterlee and Erman, 1980, Bosshard *et al.*, 1991) enzyme found in the mitochondrial intermembrane space of *Saccharomyces cerevisiae*. This enzyme is a soluble globular protein ($M_r \approx 34$ kDa) consisting of a single *b*-type heme (Fig. 1) and single polypeptide. To date its sole implied physiological function is to catalyze the hydrogen peroxide oxidation of ferrocytochrome *c*. In this process the enzyme is oxidized to a paramagnetic intermediate that lies two equivalents (compound-I) above the resting enzyme state by virtue of the initial reaction of the enzyme with hydrogen peroxide. Reducing equivalents are supplied to C*c*P by two ferric cytochrome *c* molecules that form molecular docking complexes with C*c*P during the electron transfer steps (Nicholls and Mochan, 1971;

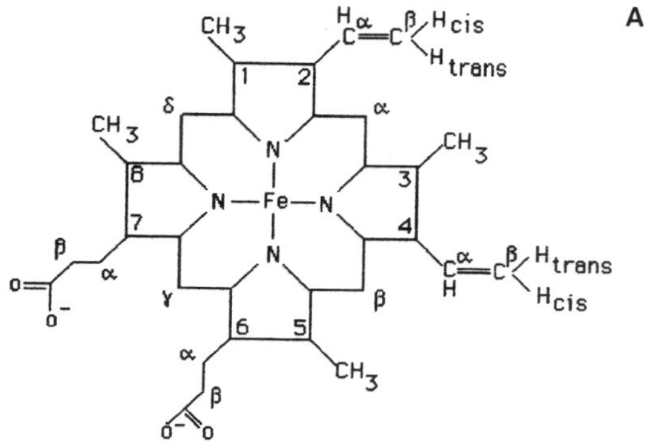

Fe-Protoporphyrin IX--Modified Fischer Numbering System

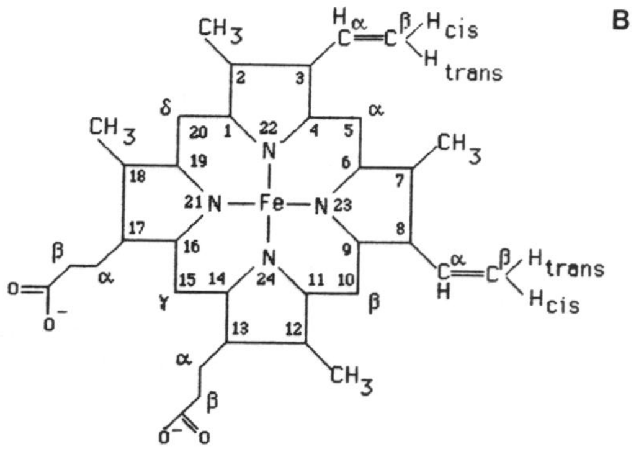

Fe-Protoporphyrin IX--IUB/IUPAC Numbering System

Figure 1. Heme labelling and structures: (A) heme *b*, modified Fischer numbering system; (B) heme *b*, IUPAC/IUB numbering system; (C) heme *b* with proximal histidine coordinated; (D) heme *c* with axially ligated methionine and histidine, typical coordination for *c*-type cytochromes.

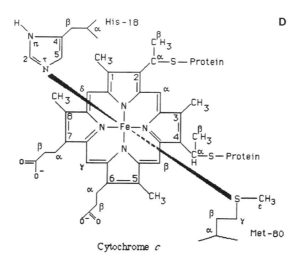

Figure 1. (*Continued*)

Poulos and Finzel, 1984; Poulos and Kraut, 1980). In this article we shall present illustrative examples of NMR information, some of it previously unpublished, about current topics relevant to this enzyme's function: active site proton assignments, characterization of molecular docking complexes, and enzyme aging.

1.2. Cytochromes *c*

Our interest in eukaryotic cytochromes *c* derives only from the fact that two naturally occurring yeast cytochromes *c* (isozyme-1 and isozyme-2) are the physiological redox partners for CcP. In this laboratory three species of cytochrome c have been employed in complex formation studies with CcP: horse, tuna, and yeast. These cytochromes c are all small ($M_r \approx 12.5$ kDa) soluble proteins that possess heme c, which is covalently linked to the polypeptide via thioether linkages (Fig. 1) (Satterlee, 1986, 1987).

Ferrocytochromes c containing heme iron(II) are diamagnetic whereas the oxidized forms of ferricytochromes c, are paramagnetic (Satterlee, 1986, 1987). Although there are extensive sequence homology differences among the four cytochromes c we have used (Tuna, horse, yeast isozyme-2) the overall three-dimensional structures are remarkably similar. In the particular case of yeast iso-1 three exceptional amino acid occurrences compared to horse cytochrome c have potential functional significance: Arg-13 (Lys-13 in horse); Trimethyllysine-72 (Lys in horse); and Cys-102 (Thr-102 in horse).

1.3. *Glycera dibranchiata* Monomer Hemoglobins

The marine annelid *Glycera dibranchiata* possesses both monomeric and polymeric hemoglobins in nucleated erythrocytes (Satterlee, 1991). Three monomer hemoglobins have been isolated, purified, and characterized (Simons and Satterlee, 1991) and all consist of a single *b*-type heme and a single polypeptide. All three proteins exhibit similar sizes ($M_r \approx 16.5$ kDa) and overall structures similar to sperm whale myoglobin, although overall sequence homology comparisons to myoglobins are low ($\sim$24%). Interest in these proteins originates in an exceptional amino acid occurrence. The distal histidine, which occurs at position E-7 in sperm whale myoblobin and is positioned so as to influence ligand binding, is replaced by a distal leucine in the monomer hemoglobins (Fig. 2). This type of amino acid change affects the protein's ligand binding

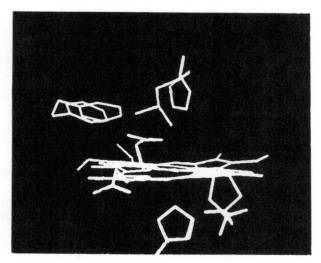

Figure 2. Cystallographic structure comparison of *Glycera dibranchiata* monomer hemoglobin heme active site and sperm whale myoglobin. Detailed view of heme pocket when proximal histidines (lower) are superimposed showing the distal histidine (Mb) and distal leucine (Hb) above the heme and corresponding phenylalanines above and to the left of hemes.

characteristics in comparison to myoblobin. Even though the native ferrous form of these proteins is diamagnetic when ligated, the ferric form of these proteins is paramagnetic in both high-spin and low-spin orbital ground states.

2. BACKGROUND

Perhaps the single greatest impediment to advancing NMR studies on paramagnetic proteins was the problem of making specific resonance assignments. For protein NMR studies, the complexity of spectra obtained from even relatively small, intact paramagnetic proteins provided significant challenges that did not exist in NMR investigations of small paramagnetic metal complexes. Until the advent of synthetic methods useful for selective or semiselective isotope enrichment of b-type hemes (Smith *et al.*, 1983; see also references in Satterlee, 1986, 1987), proposed hyperfine-shifted resonance assignments could be regarded as highly ambiguous.

The reason for this is the fact that nuclei under the influence of a paramagnetic moment experience large observed resonance shifts, termed hyperfine or isotropic shifts, caused by the interactions of nuclei with unpaired electrons. This concept and its theoretical formalism have been described in detail (La Mar *et al.*, 1973; Bertini and Luchinat, 1986; Satterlee, 1990a, b). Principal manifestations of these types of electron–nuclear interactions are that spectral resonance positions may depend upon the magnetic anisotropy, the type of bonding present (contact interaction), or the nuclear position within the three-dimensional molecular structure (pseudocontact or dipolar interaction). For nuclei close to the paramagnetic center these factors dominate the observed resonance shift and overwhelm "chemical" factors that control the NMR spectra of diamagnetic molecules.

These factors lead to the following summary of paramagnetic effects on nuclei exhibiting *strongly hyperfine shifted* resonances in paramagnetic proteins:

1. Simple "chemical shift" correlations are not usually valid.
2. Resonance shifts may exhibit significant temperature dependencies.
3. Nuclear relaxation rates will be enhanced.
4. Resonance linewidth will be significantly larger than in corresponding diamagnetic proteins.

These effects lead to both advantages and disadvantages. Among the advantages is the fact that a paramagnetic heme acts as an intrinsic shift and relaxation reagent that can provide NMR data specific to the molecular active site of a paramagnetic heme protein. Also, the fast nuclear relaxation rates of hyperfine-shifted protons mean that data acquisition for those resonances can be rapid, with typical recycle times for proton NMR experiments in our laboratory ranging between approximately 0.100 and 0.600 s for ferriheme proteins. Disadvantages include the broader resonance lines, which effectively reduce the sensitivity of hyperfine resonances by severely reducing the peak height-to-noise ratio compared to that of narrower resonance lines. In addition, assignment by inspection is no longer possible, and in most cases the hyperfine resonances are so broad that scalar couplings to neighbor nuclei are not obvious (although COSY-type

experiments can be successfully applied; see below). Many of these effects will be illustrated by the spectra accompanying the balance of this chapter.

3.   ASSIGNMENTS

For CcP and the *G. dibranchiata* monomer hemoglobins, proton hyperfine resonance assignments evolved chronologically employing first, deuterium labelled hemes;
second, 1-dimensional nuclear Overhauser effects; and third, 2-dimensional NMR spectroscopy. It is also important to note that additional methods for making proton resonance assignments were developed by others and successfully applied to other
paramagnetic proteins.

3.1.   Deuterium-labeled Hemes

For proteins with *b*-type hemes (Fig. 1) incorporation of selectively isotope-labeled
hemes is usually straightforward, employing standard methods for apoprotein creation
and reconstitution (Teale, 1959). Both CcP and the *G. dibranchiata* monomer hemoglobins were amenable to this procedure, which resulted in assignments of heme proton
resonances from peaks that disappeared in 1-dimensional proton spectra of holoproteins
incorporating specifically deuterium-labeled protohemin (IX) derivatives. Initial proton
resonance assignments were made in this way for high-spin and low-spin forms of CcP
(Satterlee *et al.*, 1983a, b) and high-spin (Constantinidis *et al.*, 1988) and low-spin
(Mintorovitch *et al.*, 1990) forms of the three monomer methemoglobins.

Whereas CcP is well behaved during apoprotein creation and reconstitution, the
Glycera monomer methemoglobins are less so. One of the problems that we have
encountered with the monomer hemoglobins has been the fact that the isolated purified
proteins each show NMR evidence for a major form (~85%) and a minor form (~15%)
that are simultaneously present in solution. Whereas the relative amounts of these
forms is stable from preparation to preparation of the native holoprotein, apoprotein
formation and reconstitution results in alterations in the relative populations of the two
forms, as shown by the NMR spectra in Fig. 3. In this case, proton resonances indicated
by M refer to the minor form of the holoprotein. Clearly, regulating this structural
heterogeneity will be important not only for quantitative NMR studies of the native,
wild-type holoproteins, but for engineered, site-specific mutants as well. Our present
efforts in engineering the *G. dibranchiata* monomer hemoglobins involve an expression
system in *E. coli* strain BL21 (Alam *et al.*, 1992b) and produces primarily the globin.
For further studies the recombinant proteins must be reconstituted with protohemin
(IX). In these cases a reconstitution procedure that generates primarily a single form of
the monomer hemoglobin is highly desirable, and progress toward such a procedure is
being made. However, these examples illustrate the sensitivity of hyperfine shifts to
structural heterogeneity at the molecular level. It is doubtful that this could have been
detected routinely by other bioanalytical methods.

3.2.   1D NOE Experiments

Hyperfine-shifted proton resonance assignments in CcP (Satterlee *et al.*, 1987),
horse and tuna ferricytochromes *c* (Satterlee and Moench, 1987), yeast isozyme-1, yeast

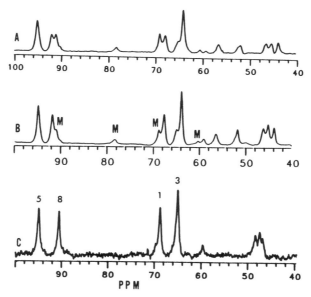

Figure 3. 360-MHz proton NMR spectra of high-frequency hyperfine-shift region of two separate isolations of native *G. dibranchiata* monomer methemyoglobin component II (A, B) and protohemin IX reconstituted component II (C). All spectra were obtained at 22° C, pH = 6.8, 0.1 M phosphate buffer, 0.1 M KCl in 99.9% D₂O. M indicates minor components in most preparations. Heme methyls are labelled in (C).

isozyme-2, and yeast isozyme-1 dimer ferricytochromes *c* (Satterlee *et al.*, 1988; Moench and Satterlee, 1989) and the monomer hemoglobins (Alam and Satterlee, 1992) were expanded using 1-dimensional nuclear Overhauser experiments (Neuhaus and Williamson, 1989). Our application of this method to proton hyperfine resonance assignments was stimulated by the pioneering efforts of Geoffrey Moore's laboratory (Moore and Williams, 1984) and Gerd La Mar's laboratory (Thanabal *et al.*, 1986, 1987a, b, 1988). These two groups realized that despite the relatively short nuclear spin–lattice relaxation times exhibited by hyperfine-shifted proteins, resonance saturation was achievable, thereby enabling the 1D NOE experiments in paramagnetic proteins to be carried out.

Examples of our results are shown for the cyanide-ligated low-spin form of cytochrome *c* peroxidase, C*c*PCN, in Fig. 4. The 1D NOE pulse sequences used in our work are diagrammed in Fig. 5. Initially a sequence was used in which the relaxation delay period was followed by a lower-power decoupler selective pulse or CW irradiation that was adjusted to precisely null the chosen resonance, which, in turn, was directly followed by the nonselective high-power read-pulse and the acquisition time. However, more recently we have used the transmitter for selective irradiation. This method involves rapidly switching the transmitter's frequency of application and where necessary using a Dante sequence for highly selective irradiation. For those with more complete instrumentation than currently available to us irradiation can also be achieved via pulse shaping. Such data are generally presented as difference spectra, as in Fig. 4.

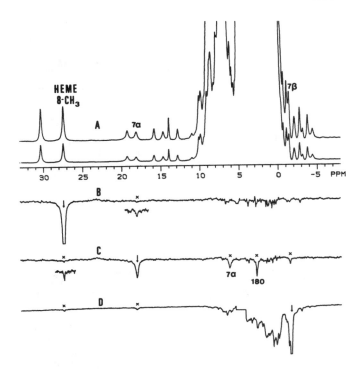

Figure 4. (A) Proton NMR absorption spectrum of CN-ligated cytochrome *c* peroxidase (2.2 mM) at 360 MHz, 23 °C, 0.200 M KNO$_3$, p*D* = 7.0. (B)–(D) NOE difference spectra with irradiation frequency indicated by *arrows* and primary NOEs indicated by *x* for the heme pyrrole IV substituents: 8-CH$_3$, 7-propionate. Insets are *y*-scaled seven times.

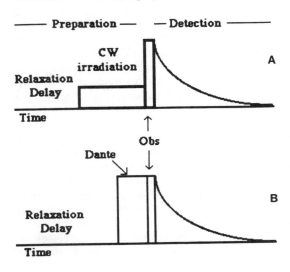

Figure 5. (A) Normal 1-dimensional NOE pulse sequence consisting of decoupler low-power irradiation followed by observe pulse. (B) Modification of normal 1D NOE pulse sequence to perform selective irradiation using the transmitter. Use of DANTE "hard" pulses and on-resonance conditions make possible enhanced selectivity with a total pulse duration under 200 μsec. This sequence is also useful for selective T_1 determination.

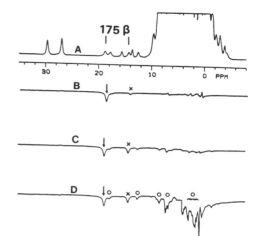

Figure 6. Progress of spin diffusion in CN-ligated cytochrome *c* peroxidase as a function of irradiation time, *D*3. The arrow indicates irradiation frequency, the *x* indicates the primary NOE to a geminal partner proton, open circles indicate regions of secondary NOE at long irradiation time (*D*3). (A) Proton absorption spectrum; (B) *D*3 = 6 ms; (C) *D*3 = 30 ms; (D) *D*3 = 240 ms.

Figures 6 and 7 show the effect of increasing the duration of the selective irradiation period. At shorter irradiation times smaller, primary NOEs are observed, whereas at longer irradiation times not only are primary NOEs enhanced, but spin diffusion (Neuhaus and Williamson, 1989) operates to decrease the NOE specificity, as witnessed by the larger intensities throughout the difference spectrum (Fig. 6). The regime of short irradiation times (<~25 ms for C*c*PCN, Fig. 7) is termed the *presteady state* and the regime of long irradiation times (>~80 ms for C*c*PCN, Fig. 7) is called the *steady state*. In this respect the selective irradiation time in the 1D NOE experiment is analogous to the mixing time in the (2D) NOESY experiment.

It is important to note the general correlation between length of irradiation in 1D NOE experiments (or mixing time in NOESY experiments) and the emergence of spin diffusion, which reduces the NOE specificity. Primary NOEs, useful for detecting nearest-neighbor nuclei, dominate difference spectra in the presteady-state regime. The specific time frame of the presteady-state regime varies with protein size.

One-dimensional NOE experiments are most conveniently analyzed as difference spectra. In the simplest case the difference between two sets of data (free induction decays, or transformed spectra) is taken by subtraction. The two sets constitute irradiation on- and off-resonance. More elaborate subtraction schemes involving data sets of two symmetrically placed (relative to the transmitter) off-resonance irradiation experiments also help minimize subtraction errors, which are due in part to off-resonances effects. Although the choice of which set is subtracted provides two options, it should be noted that NOEs in these medium-sized proteins are negative (Neuhaus and Williamson, 1989). In order to be consistent we prefer to present difference spectra in the form (on-resonance data)–(off-resonance data).

Figures 4 and 8 illustrate some specfic experimental results. Figure 4 shows a series of proton 1D NOE difference spectra of C*c*PCN that were utilized to first assign the hyperfine-shifted protons of heme pyrrole IV (Fig. 1A) substituents (Satterlee *et al.*, 1987). Figure 8 shows how selective single resonance irradiation in 1D NOE experiments can be. The sample in this case is, again, C*c*PCN, but in this case the experiment is

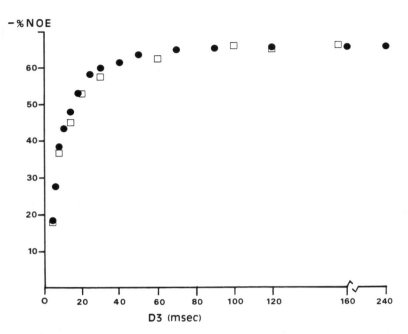

Figure 7. Proton NOE buildup graph of CN-ligated cytochrome *c* peroxidase at 500 MHz (circles, 29 °C, 100 mM KNO$_3$, 12 mM potassium phosphate buffer, p*D* 7.3) and 361 MHz (squares, 23 °C, 200 mM KNO$_3$, p*D* 6.75). The graph is a plot of intensity of the 14.6-ppm resonance as a function of the irradiation time *D*3 of its geminal partner at 19.2 ppm.

carried out with the enzyme dissolved in a buffer solution whose solvent is 90% H$_2$O. In this situation hydrogen–deuterium exchangeable resonances appear in the proton spectrum and two such peaks (indicated by *e*) overlap neighboring nonexchangeable peaks in the high-frequency (Downfield) proton hyperfine-shift region. The exchangeable peaks of interest are irradiated with a high degree of selectivity, as shown by the minimally perturbed intensities of neighboring resonances (compare Fig. 8A–D).

In point of fact, implementation of 1D NOE experiments is easy for low-spin paramagnetic ferriheme proteins wherein proton spin–lattice relaxation times T_1 are comparatively long (greater than approximately 30 ms) and linewidths (measured at half maximum height) are comparatively small (less than approximately 200 Hz at 500 MHz). For high-spin ferriheme proteins that exhibit hyperfine resonances with large linewidths (greater than approximately 200 Hz at 500 MHz) and short nuclear relaxation times (e.g., T_2, $T_1 < 30$ ms), 1D NOE experiments are more challenging (Unger *et al.*, 1985). Examples of 1D NOE difference experiments applied to the high-spin *Glycera dibranchiata* monomer hemoglobin component IV are shown in Fig. 9. The nonselective T_1 values measured for the strongly hyperfine-shifted resonances in this sample were all on the order of 40 ms or less. By comparison with spectra such as those shown in Fig. 4, this spectrum graphically depicts the higher-power irradiation pulse

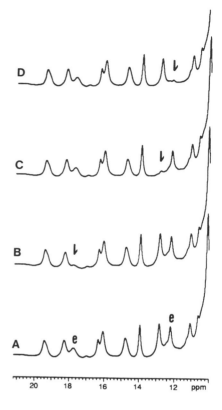

Figure 8. Selective irradiation of overlapping resonances using the transmitter-DANTE sequence (Fig. 5) for a 1D NOE experiment on C*c*PCN in 90% H_2O, 0.17 M KNO_3–0.03 M phosphate buffer, pH = 6.2, 20 °C, 500 MHz. Arrows indicate irradiation position; *e* indicates exchangeable resonances.

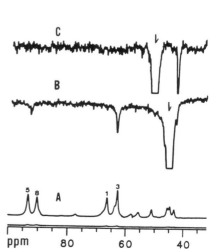

Figure 9. One-dimensional NOE difference spectra of *G. dibranchiata* high-spin monomer methemoglobin component (IV) (pH' = 6.8, 22 °C, 0.1 M phosphate buffer 0.1 M KCl, 99.9% D_2O). (A) 360-MHz proton-absorption spectrum. (B–C) NOE difference spectra with on-resonance irradiation marked by arrow.

required (75 ms duration), the concomitantly larger "spillover" (deviation from irradiation selectivity), and the comparatively small NOEs.

3.3. Results from 2-Dimensional Experiments

Heeding advice offered by Robert Kaptein in 1987 and provided with the experimental examples of Emerson, Yu, and La Mar (Emerson and La Mar, 1990; Yu *et al.*, 1990) our first 2-dimensional proton NMR experiments aimed at making hyperfine resonance assignments were carried out in late 1987. Initially this work was performed on a GE360 (NT console) spectrometer with two proteins: C*c*PCN and a chemically modified yeast iso-1 ferricytochrome *c* (Cys-102 S-methylated). These initial experiments were ultimately expanded and work was accomplished on commercially available instruments from Bruker, JEOL, and Varian (Busse *et al.*, 1990; Satterlee and Erman, 1991; Satterlee *et al.*, 1991). Illustrative of this, Fig. 10 presents a 500-MHz phase-sensitive NOESY contour map of C*c*PCN, while Fig. 11 compares results from proton homonuclear phase-sensitive COSY and magnitude COSY experiments on C*c*PCN.

The combination of these results confirmed many previous proton assignments (Satterlee *et al.*, 1987), added even more assignments to the literature, and provided insights for implementation of 2-dimensional methods to medium-sized (by NMR standards) paramagnetic proteins. Furthermore, these are now being used to make proton assignments in the *G. dibranchiata* monomer cyanide-ligated methemoglobins. As with results from others (Emerson and La Mar, 1990; Yu *et al.*, 1990; de Ropp *et al.*, 1991; Yamamoto *et al.*, 1991) our data on C*c*PCN, C*c*PCN mutants, and the *Glycera*

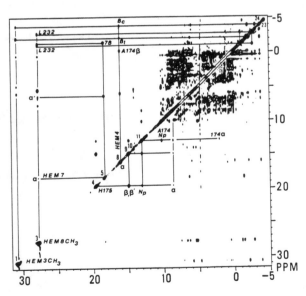

Figure 10. 500-MHz phase-sensitive NOESY contour spectrum of C*c*PCN (pH' = 6.6, 23 °C, 0.17 M KNO$_3$–0.03 M phosphate buffer in 99.9% D$_2$O; mixing time = 50 ms).

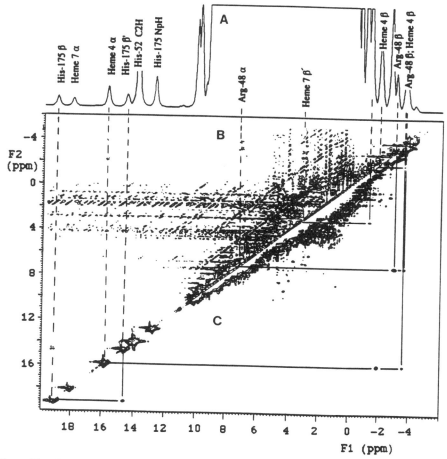

Figure 11. 500-MHz proton 2-dimensional bond-correlated spectra for C*c*PCN. (A) Reference projection, (B) phase-sensitive COSY, symmetrized proton contour plot, (C) MCOSY, symmetrized contour plot. Region shown is 20 ppm to −6 ppm with spectra taken at 27 °C, pH′ = 7.0.

dibranchiata monomer methemoglobins indicate the following generalizations for hyperfine-shifted proton resonances:

1. Homonuclear NOESY and magnitude COSY provide the highest levels of cross-peak detectability. In processing the phase-sensitive NOESY experiment, options such as displaying only positive contours (or only negative contours) can be valuable for enhancing cross-peak detection. In some cases, *p*-type NOESY experiments with absolute value processing can also be valuable in assigning hyperfine proton resonances.

2. Phase sensitive homonuclear coherence transfer experiments produce reduced-intensity cross peaks compared to the MCOSY results (Fig. 11), probably from the cancellation of overlapping antiphase components in broad peaks. The detection of scalar-coupled spin systems even for broad hyperfine-shifted resonances displaying no obvious spin–spin splittings (because the linewidth $\Delta v_{1/2} \gg J$) is readily achievable using MCOSY.

3. Multiquantum filtered homonuclear proton COSY experiments are, in our experience, not optimal for detecting broad resonances and their associated cross peaks. They are, however, useful for tracing extended connectivities including the more typically diamagnetic proton resonances within the paramagnetic protein.

4. Relayed Coherence Transfer (RCT) experiments are, in our experience, ineffective at mapping out spin systems involving hyperfine-shifted resonances. This apparently results from the rapid decay of nonequilibrium magnetization during the relay period as a result of short nuclear relaxation times.

5. TOCSY experiments provide a challenge for two reasons. First, the short nuclear spin–spin relaxation times T_2 in our proteins apparently compromises cross-peak intensity during the spin-locking interval. Second, in order to provide a suitable spin-locking field over the entire proton spectrum of a paramagnetic protein (*circa* 20–50 kHz), high power levels are required, which can produce significant sample heating. Even with pulsed spin-lock sequences designed specifically to cover wide spectral widths while minimizing thermal effects, we have achieved only mixed results so far. Our experiences with ROESY are similar to those with TOCSY, leading us to consider alternative strategies for implementing these experiemnts.

4. RELATIVE ATTRIBUTES OF SELECTED 1D AND 2D METHODS WITH C*c*P: CYTOCHROME *c* COMPLEXES AS EXAMPLES

In the case where one requires scalar coupling data, the most viable experiment is magnitude COSY. There is no comparably efficient 1D experiment from which to derive such information in most paramagnetic proteins for which $\Delta v_{1/2} \gg J$, thereby masking scalar coupling in 1D absorption spectra. When nuclear dipole–dipole couplings are required, NOESY and 1D NOE experiments both are viable. One potential drawback in using 2-dimensional experiments applied to larger paramagnetic proteins concerns digital resolution in one or both dimensions. Whereas data matrices of $4K \times 4K$ complex points have been acquired, commonly available storage media and time constraints realistically limit one to sizes of $2K \times 2K$ complex data points or less. As discussed below, use of such large data sets is really necessary only for spectra in which resolution of overlapping hyperfine resonances is important or in the case where connectivities to narrow resonances are desired. Note that these data marices are typically larger than would normally be encountered for corresponding experiments in diamagnetic molecules because (in part) small spectral widths are required in diamagnetic molecules. For example, square data matrices involving 256–512 complex points are frequently appropriate in 2D spectroscopy of diamagnetic molecules.

To put this in perspective, a processed spectrum containing 2K real points distributed over a 20-kHz spectral width in one dimension, typically needed for a paramagnetic protein, corresponds to a digital resolution of ≈ 10 Hz/point. Given the varying range of spectral widths encountered for paramagnetic proteins, digital resolution can be significantly worse (as much as 50 Hz/point or more) and still yield usable spectra. Whereas this situation may not negatively affect interpretation of data from well-separated hyperfine-shifted resonances, as shown in Figs. 10 and 11, the −2 ppm to 10 ppm region of a contour map can be quite indistinct under these conditions (Fig. 10). When information within this region is desired, one can carry out additional experiments with reduced spectral widths, but this is also potentially complicated by hyperfine-shifted resonances "folded" into the new spectral window. Digital resolution considerations can even be important when overlapping or closely spaced hyperfine-shifted resonances are of interest. An example of the lack of resolution along the diagonal is shown in Fig. 12

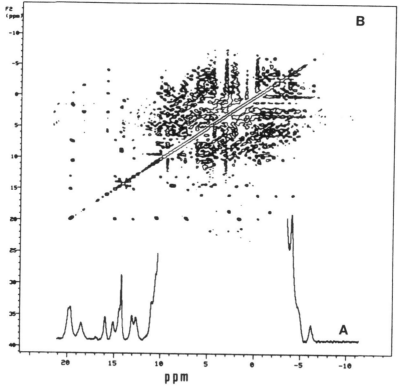

Figure 12. 500-MHz spectra of the *in situ* 1 : 1 noncovalent complex of CcP and tuna ferricytochrome *c* (pH′ = 6.4, 20 °C, 0.01 M KNO₃, 99.9% D₂O). (A) Partial reference spectrum. (B) Corresponding NOESY contour plot.

for closely grouped proton hyperfine resonances in the C*c*PCN : ferricytochrome *c* (tuna) complex.

A second factor that renders 2D experiments in paramagnetic proteins more challenging compared to those for diamagnetic molecules concerns "detectabiliy." This is related to sensitivity or signal-to-noise ratio issues for resonances of identical integrated intensities, but of greatly differing linewidths. An example that illustrates this problem is given by the high-frequency region of the 1-dimensional proton absorption spectrum of the 1 : 1 C*c*P : ferricytochrome *c* (horse) complex shown in Fig. 13. In this complex C*c*P has a high-spin ferriheme and displays four very broad heme methyl resonances at 54–85 ppm whose integrated intensities equal three protons each. However, the resolved heme 8-methyl and 3-methyl resonances of the horse ferricytochrome *c* in this spectrum, each also corresponding to a three-proton integrated intensity, have much narrower linewidths and are significantly taller. They are much narrower due to the low-spin electronic ground state of ferricytochrome *c*, and this results in a much greater peak height-to-noise level ratio than for the C*c*P peaks. As a result, the ferricytochrome *c* proton resonances are much easier to detect in 2D spectra. Their peak heights are further removed from the noise baseline than are those for the C*c*P peaks.

Whereas 2D experiments have the advantage of providing comprehensive data for a molecule in a single experiment, our experiences indicate that they are not a universal panacea. Aside from the resolution and detectability factors already mentioned, efficient implementation of 2D experiments generally requires fairly high protein concentrations. For small, highly soluble proteins like ferricytochromes *c* for which concentrations of 6 mM or greater can be achieved this is not a serious problem. However, for larger, less soluble proteins and for those exhibiting broader hyperfine resonances, longer acquisitions (12–24 hrs) may be required to gather data on hyperfine-shifted proton resonances even under rapid pulsing conditions (i.e., 5–10 S^{-1}).

There are also specific instances when 1D NOE experiments may have advantages over NOESY experiments for paramagnetic proteins.

Selectivity. When data is required for a limited set of resonances or a single resonance. In shorter experiment periods 1D NOE experiments may be advantageous for

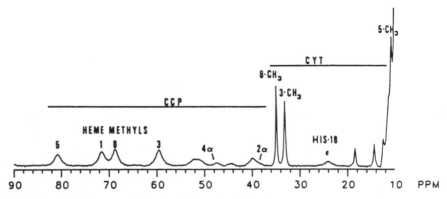

Figure 13. High-frequency hyperfine shift region of the 1 : 1 *in situ* noncovalent complex of C*c*P and horse ferricytochrome *c* (22 °C, pH′ = 6.4, 0.01 M KNO$_3$).

detecting typically weak (1–5%) primary NOEs involving broad peaks and may also be useful in factoring out NOEs from overlapping resonances.

Efficiency. When a limited set of connectivities is desired. Under such circumstances 1D NOE experiments can usually be carried out much more rapidly than a full 2D data set collection.

Resolution. Combined with the two previous factors, when NOE connectivities from hyperfine shifted protons involve overlapping peaks in the crowded 10 to −2 ppm region. Here, the greater digital resolution possible in the 1D NOE experiment could be beneficial. However, a major disadvantage of 1D NOE measurements comes from the loss of selectivity encountered in the 10 to −2 ppm proton spectral region. Two-dimensional experiments are the techniques of preference for tracing connectivities in this spectral region.

5. ILLUSTRATIVE APPLICATIONS

5.1. Elucidating the Distal Hydrogen Bonding Network Involving the Catalytic Triad of CcP

As mentioned earlier and shown in Fig. 14, the downfield hyperfine-shift region in the proton NMR spectrum of CcPCN depends on the isotopic enrichment level of the solvent water. When a buffer solution using 90% deionized distilled tap water is used, additional isotope-exchangeable resonances are observed between 11 and 29 ppm compared to spectra taken in 99.9% D_2O solvent (peaks 2, 6, 7, 12). Considering the heme electronic structure and the physical structure of the CcP heme pocket (Poulos and

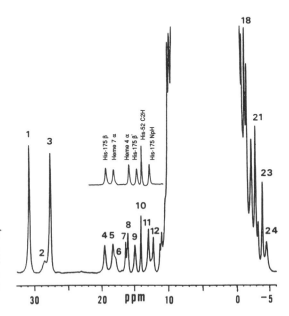

Figure 14. Proton NMR spectra of CcPCN (same conditions as in Fig. 8) in 99.9% D_2O (upper inset) and 90% H_2O–10% D_2O (full spectrum). Deuterium-exchangeable resonances are 2, 6, 7, 12.

Kraut, 1980; Finzel *et al.*, 1984) reasonable proton assignments were proposed for H/D exchangeable protons of Trp-51, His-52, and Arg-48, amino acids implicated in catalysis (Satterlee and Erman, 1991).

One-dimensional NOE experiments, essentially completed before 1987 (Fig. 15), and NOESY experiments, essentially completed by late 1989, were carried out on fully active wild-type enzyme, which had been cyanide-ligated and maintained in 85–90% H_2O buffer solutions for NMR studies (Satterlee and Erman, 1991; Satterlee *et al.*, 1991). The picture that evolved from that work was one of a distal hydrogen bonding network involving Trp-51, His-52, and Arg-48. Those assignments in CcPCN supported the view that similar hydrogen bonding interactions were part of the enzyme's hydrogen peroxide decomposition mechanism, as proposed originally by Kraut, Poulos, and coworkers (Finzel *et al.*, 1984; Bosshard *et al.*, 1991). Those original proton assignments, made between 1987 and 1989, have largely been confirmed by more recent work (Banci *et al.*, 1991). This more recent work, however, proposed alternate assignments (as Arg-48 protons) for peak 6 and peak 12 in Fig. 14. Further careful work in this laboratory casts doubt on these alternate assignments. Careful 1D NOE experiments, with irradiation selectivity as shown in Fig. 8 and subsequent 2D NOESY experiments have so far not revealed the connectivity of peaks 6 and 12 that was the basis for the alternative Arg-48 assignments. Furthermore, using site-specific CcP mutants (Alam *et al.*, 1992a) designed to probe the distal H-bond network has resulted in data that reinforce the original assignments.

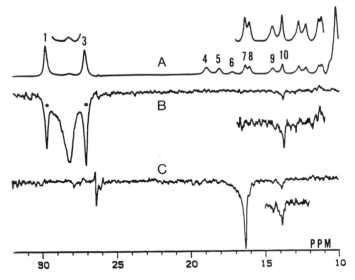

Figure 15. One-dimensional NOE difference experiments at 360 MHz on CcPCN in 90% H_2O–10% D_2O, 0.2 M KNO$_3$, pD = 6.8, 25 °C: (A) absorption spectrum, no irradiation, peaks numbered as in Fig. 14. Insets are vertical expansions; (B) peak 2 irradiated, showing connectivity to peak 10; decoupler spillover to peaks 1 and 3 noted by circles; (C) peak 7 irradiated, showing connectivity to peak 10. In D_2O no connectivity between peak 10 and other hyperfine-shifted resonances is found. Feature at ~27 ppm is carrier position.

5.2. Is There Such a Phenomenon as CcP Aging?

An issue arose beginning in 1987 that concerned the integrity and similarity of CcP preparations from different laboratories. Minor, region-specific, primary-sequence differences have been noted in CcP preparations (Bosshard *et al.*, 1991), which seemingly depend upon the local strain of Red Star™ yeast used in the preparation. Preparative methods also differ between labs. We have examined several of these different preparations and recombinant wild-type CcP (CcPMI) and found essentially identical proton hyperfine-shift patterns.

However, one important question concerns the proposal that CcP could be "aged" by improper isolation and storage (Smulevich *et al.*, 1986; Yonetani and Anni, 1987; Bosshard *et al.*,1991). In this context "aging"refers to a type of native CcP demonstrating a mixture of 5- and 6-coordination. Proton NMR spectra are quite sensitive to ferriheme ligation and spin-state changes, making this an obvious method by which to study this phenomenon.

Although it has been demonstrated that structural changes in CcP can be indicated from changes in optical spectra, the contention that only "aged" CcP demonstrates pH-dependent optical and NMR spectra is an incomplete conclusion (Vitello *et al.*, 1990). An extensive study revealed that CcP optical spectra and kinetics results depended upon the identity of the salt present in solution, the method of isolation, and the method of storage (Vitello *et al.*, 1990). Figure 16 shows that in potassium phosphate solution (0.1 M), native CcP exhibits resonance broadening and slight resonance shifts between pH 7.5 and pH 4.90. By comparison, an identically prepared and purified CcP solution in 0.1 M potassium nitrate solution exhibits a much more pH-dependent NMR spectrum

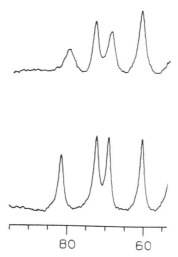

Figure 16.   High-frequency proton hyperfine shift region of native wild-type CcP showing the four heme methyl resonances. Conditions: 0.10 M potassium phosphate, 22 °C; top, pH′ = 4.90; bottom, pH′ = 7.25.

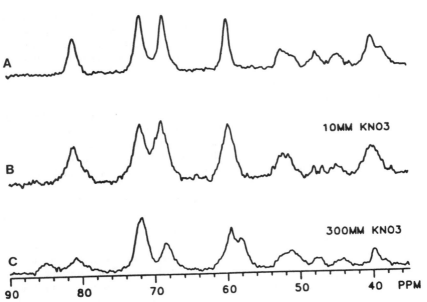

Figure 17. High-frequency proton hyperfine shift region of native wild-type CcP (in D_2O, 360 MHz, 22 °C): (A) in 0.300 M KNO_3, pH' = 7.0; (B) in 0.010 M KNO_3, pH' = 5.90; (C) in 0.300 M KNO_3, pH' = 5.90.

(Satterlee and Erman, 1980), with two active site forms simultaneously present in solution below a pH of approximately 5.8. Also, the pH-dependent proton NMR spectrum of CcP in KNO_3 solution is reversible with ionic strength (Fig. 17). These results cast doubt on the use of pH-dependent spectral properties of CcP as a true indicator of "aged" enzyme. Instead, what these results add to is the conclusion from previous work (Vitello *et al.*, 1990) that the active site of CcP is sensitive to the type and concentration of salt in the enzyme solution as well as the pH of the solution.

5.3. NMR Exchange Dynamics in CcP: Ferricytochrome *c* Complexes

Proton NMR spectroscopy is particularly sensitive to complex formation between CcP and ferricytochrome *c*. This type of complex is transiently formed during enzymatic electron transfer steps (Bosshard *et al.*, 1991), but such complex formation is not easily detected by other spectroscopies (Erman and Vitello, 1980).

Recently, we have found large species-specific differences in complex-induced proton NMR shifts for horse, tuna, and yeast ferricytochromes *c* binding to CcP (Moench *et al.*, 1992). The results distinguish between complexes formed with physiological (yeast isozyme-1 and isozyme-2) or nonphysiological (tuna, horse) ferricytochromes *c*, with the physiological docking partners experiencing lager and more extensive complex-induced NMR shifts. One result of this work is that the yeast isozyme-1 ferricytochrome *c* heme 3-methyl resonance can be detected simultaneously in free and bound forms, corresponding to a slow exchange regime in NMR dynamics (Fig. 18a). This has stimulated additional work both to define the rate law for this exchange and to quantitate the exchange dynamics. It is important to emphasize that results so far show that this system clearly does not satisfy first-order kinetic criteria (Fig. 18). Work in progress has allowed a preliminary estimate of the preexchange lifetime for CcP-bound yeast iso-1 ferricytochrome *c* of 4.0 msec at 5 °C. This estimate is currently being refined by inversion transfer experiments. However, even now it is clear that the ability to resolve free and CcP-bound ferricytochrome *c* forms in the 500-MHz proton spectrum occurs only at low total protein concentrations and low temperature. For example, the spectrum

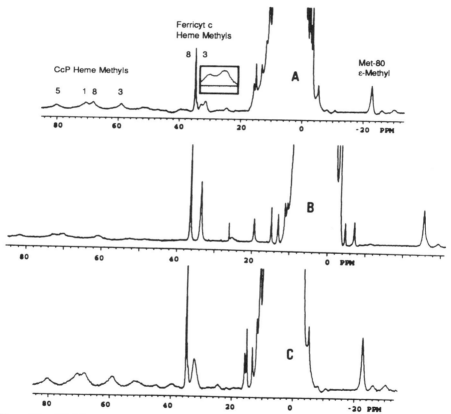

Figure 18. 500-MHz proton NMR spectra of noncovalent complexes of CcP (A) with yeast iso-1 ferricytochrome *c*, where [CcP] = 0.10 mM, [Cyt *c*] = 0.26 mM, and insert is expansion of the heme 3-methyl resonances; (B) with horse ferricytochrome *c*, where [CcP] = 0.12 mM, [Cyt *c*] = 0.26 mM; (C) with yeast iso-1 ferricytochrome *c*, where [CcP] = 1.7 mM, [Cyt *c*] = 4.1 mM. All samples were run in 10-mM KNO_3–D_2O solutions, 27 °C, pH' = 6.55. (A) and (C) are plotted on identical horizontal scales; (C) has a different horizontal scale; the ferricytochrome *c* heme methyl resonances are aligned.

shown in Fig. 18A resulted from a sample in which the C*c*P concentration was 100 μM. At higher total protein concentrations or higher temperatures a fast-exchange NMR regime occurs (Fig. 18C).

5.4. Proton Relaxation Measurements

It has been clear to us since approximately 1984 that nonselective proton T_1 measurements on C*c*PCN hyperfine resonances demonstarted nonlinear behavior in semilog analyses (Satterlee 1986). This data has been reproduced consistently at different fields and also occurs for a small protein, the *Glycera dibranchiata* monomer hemoglobin component IV. What this means is that the concept of a T_1 for the hyperfine-shifted protons is not valid, although we and others have estimated T_1 values (called T_1^{app}) from the initial, linear portions of semilog plots of nonselective proton inversion–recovery data (Satterlee *et al.*, 1991). These data reveal to us the presence of significant amounts of spin diffusion in both the larger C*c*PCN and smaller Hb(IV)CN, which manifests itself in the semilog relaxation plots as early as 40 ms.

These results and the fact that optimal conditions for NOESY experiments are dependent upon the factor ρ (the intrinsic, i.e., selective, nuclear T_1 (Neuhaus and Williamson, 1989)), has stimulated us to begin comparative selective and nonselective relaxation measurements. The type of data collected are shown in Fig. 19, which is a series of stacked plots of *Glycera* monomer component IV HbCN, showing a selected high-frequency region of hyperfine shifts and the selective inversion of one resonance. Results from similar data on all resonances in this region reveal that the selective T_1^{app} values (estimated from initial linear portions of semilog plots) are uniformly much shorter than the nonselective values. So far we have found selective constants to range

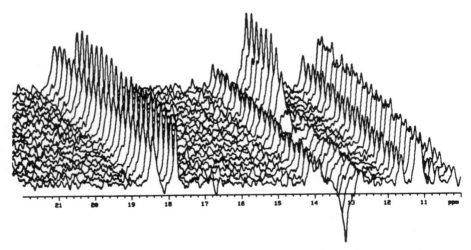

Figure 19. Stack plot of proton selective inversion–recovery experiments for *Glycera* Hb component IV used in measuring selective T_1. Conditions: 500 MHz, 0.10 M KCl, 0.10 M potassium phosphate buffer in 99.9% D_2O, 20 °C, pH' = 6.8.

between 32% and 83% of the nonselective constants. This work is being expanded to CcPCN and several of its mutants that display anomalous linewidths for purposes of attempting to understand proton relaxation in paramagnetic heme proteins and optimizing NOESY acquisitions.

ACKNOWLEDGMENTS. The unpublished experimental work that appears in this article was supported by the following grants: NSF (DMB8716459 to J.E.E.; DMB8716511 and MCB9018982 to J.D.S.), NIH (HL01758, GM 47645, and RR0631401 to J.D.S.) and Battelle Pacific Northwest Laboratory.

REFERENCES

Alam, S., and Satterlee, J. D., 1992, unpublished.

Alam, S., Satterlee, J. D., Mauro, J. M., and Poulos, T. L., 1992a, unpublished.

Alam, S., Simons, P., and Satterlee, J. D., 1992b, unpublished.

Banci, L., Bertini, I., Turano, P., Ferrer, J. C., and Mauk, A. G., 1991, *Inorg. Chem.* **30**:4510–4516.

Bertini, I., and Luchinat, C., 1986, *NMR of Paramagnetic Molecules in Biological Systems*, Benjamin Cummings, Menlo Park, CA.

Bosshard, H. R., Anni, H., and Yonetani, T., 1991, in *Peroxidases in Chemistry and Biology* (J. Everse, K. E. Everse, and M. B. Grisham, eds.), vol. II, pp. 52–84, CRC Press, Boca Raton, FL.

Busse, S. C., Moench, S. J., and Satterlee, J. D., 1990, *Biophys. J.* **58**:45–51.

Constantinidis, I., Satterlee, J. D., Pandey, R., Lewis, H., and Smith, K. M., 1988, *Biochemistry* **27**:3069–3076.

de Ropp, J. S., Yu, L. P., and La Mar, G. N., 1991, *J. Biomolec. NMR* **1**:175–190

Emerson, S. D., and La Mar, G. N., 1990, *Biochemistry* **29**:1545–1556.

Erman, J. E., and Vitello, L. B., 1980, *J. Biol. Chem.* **259**:6224–6228.

Finzel, B. C., Poulos, T. L., and Kraut, J., 1984, *J. Biol. Chem.* **259**:13027–13036.

La Mar, G. N., Horrocks, W. DeW., and Holm, R. H., eds., 1973, *NMR of Paramagnetic Molecules*, Academic Press, New York.

Mintorovitch, J., Satterlee, J. D., Pandey, R., Lewis, H., and Smith, K. M., 1990, *Inorg. Chim. Acta* **170**:157–159.

Moench, S. J., Chroni, S., Lou, B. S., Enman, J. E., and Satterlee, J. D., 1992, *Biochemistry* **31**:3661–3670.

Moench, S. J., and Satterlee, J D., 1989, *J. Biol. Chem.* **264**:9923–9931.

Moore, G. R., and Williams, G., 1984, *Biochim. Biophys. Acta* **788**:147–150.

Neuhaus, D., and Williamson, M., 1989, *The Nuclear Overhauser Effect in Structural and Conformational Analysis*, VCH, New York.

Nicholls, P., and Mochan, E., 1971, *Biochem. J.* **121**:55–67.

Poulos, T. L., and Finzel, B. C., 1984, *Pept. Protein Rev.* **4**:115–171.

Poulos, T. L., and Kraut, J., 1980, *J. Biol. Chem.* **255**:8199–8205.

Satterlee, J. D., 1986, *Ann. Reports on NMR Spectroscopy* **17**:79–178.

Sattelee, J. D., 1987, in *Metal Ions in Biological Systems* (H. Sigel, ed.), Vol. 21, pp. 51–109, Marcel Dekker, New York.

Satterlee, J. D., 1990a, *Concepts in Magn. Reson.* **2**:69–79.

Satterlee, J. D., 1990b, *Concepts in Magn. Reson.* **2**:119–129.

Satterlee, J. D., 1991, in *Structure and Function of Invertebrate Oxygen Carriers* (S. Vinogradov and O. Kapp, eds.), pp. 171–177, Springer-Verlag, New York.

Satterlee, J. D., and Erman, J. E., 1980, *Arch. Biochem. Biophys.* **202**:608–616.

Satterlee, J. D., and Erman, J. E., 1991, *Biochemistry* **30**:4390–4405.

Satterlee, J. D., and Moench, S. J., 1987, *Biophys. J.* **52**:101–107.

Satterlee, J. D., Erman, J. E., La Mar, G. N., Smith, K. M., and Langry, K. C., 1983a, *Biochem. Biophys. Acta* **743**:246–255.

Satterlee, J. D., Erman, J. E., La Mar, G. N., Smith, K. M., and Langry, K. C., 1983b, *J. Am. Chem. Soc.* **105**:2099–2105.

Satterlee, J. D., Erman, J. E., and de Ropp, J. S., 1987, *J. Biol. Chem.* **262**:11578–11583.

Satterlee, J. D., Moench, S. J., and Avizonis, D., 1988, *Biochim. Biophys. Acta* **952**:317–324.

Satterlee, J. D., Russell, D. J., and Erman, J. E., 1991, *Biochemistry* **30**:9072–9077.

Simons, P., and Satterlee, J. D., 1991, in *Structure and Function of Invertebrate Oxygen Carriers* (S. Vinogradov and O. Kapp, eds.), pp. 227–232, Springer-Verlag, New York.

Smith, K. W., Fuginari, E. M., Langry, K. C., Parish, D. W., and Tabba, H. D., 1983, *J. Am. Chem. Soc.* **105**:6638–6646.

Smulevich, G., Evangelista-Kirkup, R., English, A., and Spiro, T. G., 1986, *Biochemistry* **25**:4426–4430.

Teale, F. W. J., 1959, *Biochim. Biophys. Acta* **35**:543–546.

Thanabal, V., de Ropp, J. S., and La Mar, G. N., 1986, *J. Am. Chem. Soc.* **106**:4438–4444.

Thanabal, V., de Ropp, J. S., and La Mar, G. N., 1987a, *J. Am. Chem. Soc.* **109**:265–272.

Thanabal, V., de Ropp, J. S., and La Mar, G. N., 1987b, *J. Am. Chem. Soc.* **109**:7516–7525.

Thanabal, V., de Ropp, J. S., and La Mar, G. N., 1988, *J. Am. Chem. Soc.* **110**:3027–3035.

Unger, S. W., Le Comte, J. T. J., and La Mar, G. N., 1985, *J. Magn. Reson.* **64**:521–526.

Vitello, L. B., Huang, M., and Erman, J. E., 1990, *Biochemistry* **29**:4283–4288.

Yamamoto, Y., Chûjo, R., and Suzuki, T., 1991, *Eur. J. Biochem.* **198**:285–291.

Yonetani, T., and Anni, H., 1987, *J. Biol. Chem.* **262**:9547–9553.

Yu, L. P., La Mar, G. N., and Rajarathnam, K., 1990, *J. Am. Chem. Soc.* **112**:9527–9534.

6

Heteronuclear Magnetic Resonance
Applications to Biological and Related Paramagnetic Molecules

Joël Mispelter, Michel Momenteau, and Jean-Marc Lhoste

> *Les corps faiblement magnétiques ont un coefficient d'aimantation indépendant de l'intensité du champ*; mais ces corps se comportent tout autrement au point de vue des changements produits par la température. La loi de variation du coefficient d'aimantation a une allure hyperbolique et le plus souvent *le coefficient d'aimantation spécifique varie simplement en raison inverse de la température absolue.*
>
> —PIERRE CURIE (1895)

1. INTRODUCTION

1.1. General Overview

Nuclear magnetic resonance spectroscopy of paramagnetic molecules has proven to be a powerful tool in investigating the geometric and electronic structures of biological molecules and, more generally, of chemical compounds incorporating paramagnetic metal ions. The magnetic perturbations due to the unpaired electrons are such that nuclei located in the near vicinity of the paramagnetic center exhibit unusual chemical shifts and relaxation rates which distinguish them from the diamagnetic background.

Joël Mispelter, Michel Momenteau, and Jean-Marc Lhoste • Institut Curie, Section de Biologie, Centre Universitaire, 91405 Orsay Cedex, Paris, France.

Biological Magnetic Resonance, Volume 12: NMR of Paramagnetic Molecules, edited by Lawrence J. Berliner and Jacques Reuben. Plenum Press, New York, 1993.

The proton is by far the most popular nucleus for NMR investigations of structural and electronic properties of paramagnetic metalloproteins (Bertini *et al.*, 1989), probably because of its high detection sensitivity and high natural abundance. On the contrary, the inherently lower sensitivity of heteronuclei is unfavorable for NMR, and the low natural abundance of many nuclei of biological interest (^{13}C, ^{15}N, ^{17}O) considerably increases the sensitivity problem. These drawbacks are partly overcome, by the use of labeled proteins, but at the price of extensive (and expensive) biochemical syntheses. Thus, despite its potential interest, heteronuclear magnetic resonance dealing with large paramagnetic proteins has been to date the subject of only a few studies, using ^{13}C-labeled biomolecules (for example Khalifha *et al.*, 1984; Sankar *et al.*, 1987) or at the natural abundance level (Yamamoto *et al.*, 1988).

On the other hand, heteronuclear paramagnetic resonance has been extensively applied for a long time (McGarvey and Kurland, 1973) to NMR of small ligands. Many investigations have been concerned with the measurement of exchange rates between complex and solvent and the determination of coordination numbers for paramagnetic ions (Dwek, 1973). In a similar approach, interactions of ions with amino acids and peptides have been studied by detailed analysis of relaxation data (Led, 1985; Henry *et al.*, 1986; Gotsis and Fiat, 1987a, b). In practically all cases it is noteworthy that the combined use of both proton and heteronuclear NMR data is the basis of the analysis (Kushnir and Navon, 1984; Haddy and Sharp, 1989; Haddy *et al.*, 1989).

Other applications used small ligands which bind directly to the active sites of enzymes. In this respect, most work has used ^{13}C, ^{15}N, ^{19}F, and ^{17}O NMR. The ^{13}C NMR resonance of bounded ^{13}C-labeled CN$^-$ ion to low-spin ($S = \frac{1}{2}$) iron(III) synthetic models can actually be observed (Goff, 1977), but is hardly detectable in hemoproteins. On the contrary, ^{15}N NMR of ^{15}N-labeled CN$^-$ has been successfully observed for a number of hemoproteins (Morishima and Inubushi, 1978; Behere *et al.*, 1985, 1986; Lukat and Goff, 1990), since it corresponds to an atom not directly bounded to the paramagnetic metal. It should be noted, however, that the first observation of bound imidazole ^{15}N resonance in low-spin iron(III) porphyrins is very recent (Yamamoto *et al.*, 1989b). The cyanide ion has proved to be a sensitive probe for *trans* as well as *cis* ligand effects on the electronic structure of the prosthetic group in hemoproteins (Morishima and Inubushi, 1978). It has also been used to probe the heme environment in the ferric states of various hemoproteins, including horseradish peroxidase, lacto-peroxidase, chloroperoxidase (Behere *et al.*, 1985), P-450-CAM (Lukat and Goff, 1990) and ferricytochromes *c* (Behere *et al.*, 1986). In the last case, the hemoproteins showed remarkable conservation of the heme environment among organisms of diverse origins.

Fluorinated substrates have been used as a sensitive probe for axial ligation of the camphor-metabolizing cytochrome P-450-CAM. In that case, binding of the potent CO inhibitor molecule to the sixth coordinating position of the iron(II) has been followed using ^{19}F NMR of fluorosubstituted camphor molecules and observing the resulting spin-state transition from paramagnetic to diamagnetic. In this way, it was demonstrated that CO binds simultaneously with the camphor substrate molecule to the protein (Crull *et al.*, 1989).

In a similar way, binding of O$_2$ (a paramagnetic molecule) to deoxymyoglobin, hemoglobin, and model molecules (which are also paramagnetic complexes) has been studied using ^{17}O NMR in order to elucidate the nature of iron–dioxygen binding. The observation of two well-resolved ^{17}O resonances in the latter diamagnetic complexes in solution was interpreted (Gerothanasis and Momenteau, 1987; Gerothanasis et al., 1989) in favor of the end-on angular structure proposed by Pauling (1964) for oxyhemoglobin. In addition, the absence of any paramagnetic effect (temperature dependence of the shifts) rules out a possible paramagnetism for the oxyhemoglobins (Cerdonio et al., 1977). More recently, Odfield et al. (1991) reported similar conclusions obtained by solid-state ^{17}O NMR recorded, for the first time, for oxymyoglobin and oxyhemoglobin.

Deuterium can directly replace the proton in molecules without extensive electronic perturbation. This substitution resulted in much narrower linewidth (Reuben and Fiat, 1969; Johnson and Everett, 1972; Everett and Johnson, 1972), allowing the recording of spectra for complexes that incorporate metal ions with long electron spin-relaxation time (for example Godziela and Goff, 1986). However, in comparing ^{2}H NMR spectra with ^{1}H NMR spectra one should keep in mind that isotopic effects are much larger for paramagnetic than for diamagnetic complexes. Hebendanz et al. (1989) reported primary isotope shift values up to 2.33 ppm for perdeuterated metallocene and a secondary isotopic shift of 0.71 ppm (observed in partially deuterated molecules). ^{2}H-labeled hemins have been used as a probe for five or six coordination at the heme iron binding site in ferric high-spin hemoproteins (Morishima et al., 1985). Assignment of hyperfine-shifted resonance has also been performed using reconstituted hemoproteins with deuterium-labeled hemins (La Mar et al., 1980; Satterlee et al., 1983; La Mar et al., 1988). This method will probably be replaced by 2D-correlation NMR methods, which are now currently applied to paramagnetic molecules (Yu et al., 1986; Yamamoto, 1987; Yamamoto et al., 1988; Jenkis and Lauffer, 1988; Yamamoto et al., 1990b; Satterlee and Erman, 1991; Satterlee et al., 1991; Bertini et al., 1991; de Ropp and La Mar, 1991; Skjeldal et al., 1991; Peyton, 1991; La Mar and de Ropp, Chapter 1 of this volume) since it has been recognized that coherent (Dodrell et al., 1980) as well as incoherent (Johnson et al., 1983; Barbush and Dixon, 1985) magnetization transfer is still observable among hyperfine-shifted resonances. Finally, interesting applications of ^{2}H NMR take advantages of the quadrupole moment of the deuterium nucleus for measuring the anisotropic components of the magnetic susceptibility of diamagnetic (Lohman and McLean, 1981, and references therein) and paramagnetic molecules (Domaille, 1980) in solution (see also Section 2.2.4).

1.2. Specificity of Heteronuclear Magnetic Resonance in Paramagnetic Molecules

Except for deuterium, the specificity of heteronuclear magnetic resonance compared to that of the proton in paramagnetic molecules is threefold.

The first specificity concerns the immediate electronic vicinity of the nucleus. The main difference between proton and heteronucleus is the possibility of unpaired spin occupying non-s orbitals centered at the nucleus. This leads to complications in the interpretation of paramagnetic NMR properties, but is also a source of information about the detailed electronic structure of the complex.

The second specificity is related to the position of the heteronucleus inside the chemical structure of the ligand. This provides unique local probes of the ligand–core electronic structure.

The third specificity relates to the lower gyromagnetic ratio of heteronuclei. This was already considered in some details by McGarvey and Kurland (1973) and need no further discussion. It is sufficient to say that one expects roughly a narrower line for heteronuclei than for the proton, in principle in the ratio of $(\gamma_N/\gamma_H)^2$. In fact, in most cases, the line narrowing is (fortunately!) not what was expected, due to the effect of local unpaired spin. This also provides useful information about the electronic structure of the complex.

In this chapter, we shall focus our attention on the utility of heteronuclei (specifically, ^{13}C) as probes of the detailed electronic structure of stable complexes. For this, one must keep in mind that proton NMR data complement heteronuclear NMR instead of being a separate method of investigation.

2. ^{13}C NMR OF PARAMAGNETIC MOLECULES

This section describes, first on a general level, the perturbations of NMR characteristics (shift and relaxation) by electron–nucleus magnetic coupling. Then, these effects will be illustrated in the last subsection using hemes as examples. In view of the complementary nature of paramagnetic heteronuclear and proton NMR, both will be discussed together.

2.1. Backgrounds

All the paramagnetic effects on NMR are due to the magnetic interactions between the unpaired electrons and the nuclear spin magnetic moments. These effects are significant at varying distances from the paramagnetic center, depending on the extension of unpaired spin into the chemical structure of the complex. The unpaired electron spin may be entirely delocalized on a radical or concentrated on an ion center, but it shows itself by both "direct" contact and through-space dipolar interactions with the ligand nuclei. The former interaction results from spin delocalization and/or from spin polarization mechanisms leading to nonzero unpaired electron spin at the nucleus position. Due to the much larger magnetic moment of the electron spin as compared to that of the nuclear spin, the effects may be very large and obscure the usual nuclear spin–spin interactions involved in diamagnetic NMR.

Both contact and through-space interactions may deeply perturb chemical shifts and nuclear relaxation properties, but by different mechanisms. Thus, combining these data allows one to factorize the paramagnetic perturbations on the nuclear spin, providing information on the electron spin delocalization mechanisms and ultimately on the electronic interactions between the paramagnetic center and its surrounding ligands. In addition, the through-space dipolar coupling provides structural information.

In many respects, paramagnetic NMR is a complementary method to other magnetic resonance spectroscopies such as electron paramagnetic resonance (EPR) and electron nuclear double resonance (ENDOR). Their spectral characteristics are governed by the same electron–nucleus hyperfine interaction described by the Hamiltonian (Kurland and McGarvey, 1970)

$$\mathcal{H}_N = g_N\beta_N\mathbf{I}(\mathbf{A}_F + \mathbf{A}_D + \mathbf{A}_L) \tag{1}$$

where $\mathbf{I}$ is the nuclear spin operator and $\mathbf{A}_F$, $\mathbf{A}_D$, and $\mathbf{A}_L$ are the operators representing, respectively, the Fermi contact and the dipolar spin–spin and nuclear spin–orbital interactions. These operators involve the spin $\mathbf{s}_i$ and orbital moment $\mathbf{l}_i$ of each unpaired electron:

$$\mathbf{A}_F = (8\pi/3)g_e\beta \sum_i \delta(r_i)\mathbf{s}_i$$

$$\mathbf{A}_D = g_e\beta \sum_i [3(\mathbf{s}_i \cdot \mathbf{r}_i)\mathbf{r}_i - r_i^2\mathbf{s}_i]r_i^{-5}$$

$$\mathbf{A}_L = g_e\beta \sum_i r_i^{-3}\mathbf{l}_i$$

(2)

where $\mathbf{r}_i$ is the electron–nucleus vector of length r_i.

The hyperfine interaction shows itself differently depending on the magnetic moment involved in the resonant spectroscopy. To consider only the case of nuclear spin $I = \frac{1}{2}$, electron spins "see" the two possible orientations of I as two distinct populations of molecules. This results in the well-known hyperfine splitting of the EPR and ENDOR lines. In the latter case, the spectrum results from a perturbation of the EPR lines when the nuclear spin transitions are irradiated but the result is similar. A splitting is observed which gives the hyperfine coupling constants directly.

The effect of the hyperfine interaction on NMR is not as straightforward. Instead of a splitting of lines, one observes a temperature-dependent shift of the resonances and a very variable broadening of the lines. The shift results from the fact that nuclear spins "see" only the thermal average of the total electron spin instead of separate populations of molecules bearing in one case an up and in the other a down orientation of the electron spin with respect to the static magnetic field. Indeed, the electron spin orientation changes rapidly; in other words, the electron relaxation time is generally much shorter than the inverse of the hyperfine interaction energy. Typically, electron spin relaxation time ranges from 10^{-13} to 10^{-6} s, the latter value being beyond the limits of NMR observation due to extremely large broadening. In contrast, the former value corresponds to complexes exhibiting very sharp NMR lines but is unfavorable for EPR spectroscopy.

The indirect expression of the hyperfine interactions in NMR spectroscopy as a temperature-dependent shift requires a lot of computational effort in order to get the relevant information from the data. This inevitable computational step is based on more or less accurate theories, and interpretation of the data may sometimes be tedious. In fact, the advantages of the high resolution provided by liquid-state NMR, the possibility of direct assignment of the resonance lines to specific nuclei in the chemical structure, and the complementarity with EPR and ENDOR spectroscopies with respect to visibility justify the efforts to get the hyperfine terms from the NMR data.

The description of paramagnetic effects on the NMR properties has been the subject of many treatises and is therefore well documented. The purpose of this section is mainly to give a condensed and simple presentation based on well-known physical concepts. The argument will not be always as rigorous as it should be, but the validity of the results is assured by theories developed by others. However, particular attention will be paid to the approximations made. Finally, the discussion will be limited to $I = \frac{1}{2}$ nuclear spins (more specifically, ^{13}C and ^{1}H). The description of the perturbations on nuclear relaxation will be correct only for $S = \frac{1}{2}$ spin and isotropic electron g values. Nevertheless, the conclusions may be applied to higher spins, but with some caution (Bertini *et al.*, 1985, 1990; Banci *et al.*, 1986; Vasavada and Nageswara Rao, 1989).

2.2. Paramagnetic Shifts

2.2.1. Introduction

The paramagnetic shift, often called the "isotropic shift" in the literature, is the difference between the observed shifts for the paramagnetic complex and for a suitable diamagnetic molecule of the same chemical structure:

$$\delta_{\text{para}} = \delta_{\text{obs}} - \delta_{\text{dia}}$$

In the following, δ will be positive if the shift is to low field. Note, however, that the reverse convention is sometimes used.

A general equation has been derived by Kurland and McGarvey (1970) for the paramagnetic shift for complexes in solution:

$$\frac{\Delta B}{B_0} = -3kT \sum_{\Gamma n} \exp(-E_\Gamma/kT)^{-1}$$

$$\times \sum_{\alpha = x,y,z} \left[\sum_{\Gamma n, \Gamma m} \exp(-E_\Gamma/kT) \langle \Gamma n | \mu_\alpha | \Gamma m \rangle \langle \Gamma m | A_{N\alpha} | \Gamma n \rangle \right.$$

$$- kT \sum_{\Gamma n, \Gamma m'} \langle \Gamma n | \mu_\alpha | \Gamma m' \rangle \langle \Gamma m' | A_{N\alpha} | \Gamma n \rangle$$

$$\left. \times \frac{\exp(-E_\Gamma/kT) - \exp(-E_{\Gamma'}/kT)}{(E_\Gamma - E_{\Gamma'})} \right] \qquad (3)$$

where $|\Gamma n\rangle$ are the eigenfunctions of the zero-field Hamiltonian (including the crystal field, electron repulsion interactions and the spin–orbit coupling) associated with the energy E_Γ. The index n refers to the Kramers degeneracy of the levels and μ_α is proportional to the Zeeman operator:

$$\mu_\alpha = -\beta(L_\alpha + g_e S_\alpha) \qquad (\alpha = x, y, z) \qquad (4)$$

$A_{N\alpha}$ is the component of the hyperfine interaction operator along the α axis.

This equation is based on the following assumptions: the electron spin relaxation rate is much greater than the nuclear–electron hyperfine interaction and the nuclear Zeeman interaction is much larger than the hyperfine interaction so that the nuclear spin is quantized along the static magnetic field B_0.

Equation (3) is obtained by writing the energy of nuclear–electron hyperfine interaction as

$$E = -g_N\beta_N\mathbf{A}_N \cdot \mathbf{I} = -\boldsymbol{\mu}_N \cdot \Delta\mathbf{B} = -g_N\beta_N I_\alpha \Delta B_\alpha \tag{5}$$

where $\alpha(=x, y,$ or $z)$ is the orientation of B_0 with respect to the molecular axes. The additional magnetic field created at the nucleus position by the electron spin ΔB_α is obtained from an average over all thermally populated states:

$$\Delta B_\alpha = \langle \mathbf{A}_N \rangle \cdot \frac{\mathbf{B}_0}{B_0} \tag{6}$$

The next step is to evaluate $\langle \mathbf{A}_N \rangle$ for each orientation of the complex with respect to B_0. Kurland and McGarvey (1970) used the density matrix formalism (Slichter, 1963) to obtain

$$\langle \mathbf{A}_N \rangle = \frac{\text{tr}(\rho \mathbf{A}_N)}{\text{tr}(\rho)} \tag{7}$$

and expanded the density operator to first order in the electronic Zeeman energy.

Golding and Stubbs (1977) averaged directly the expectation value of the hyperfine interaction over a Boltzmann distribution, and ΔB_α is obtained from

$$\Delta B_\alpha = \frac{1}{g_N\beta_N B_0}\left\{ \frac{\partial^2}{\partial B_\alpha \partial I_\alpha} \sum_n \langle \Psi_n | \mathcal{H}_N | \Psi_n \rangle \exp(-E_n/kT) \bigg/ \sum_n \exp(-E_n/kT) \right\} \tag{8}$$

In that equation, Ψ_n and E_n are the eigenfunctions and eigenvalues of the Hamiltonian *including* the electron Zeeman interaction. $\mathcal{H}_N$ is the hyperfine interaction operator, Eq. (1), and the derivatives are evaluated at $B_\alpha = 0$.

For complexes in solution, one must average ΔB_α over all orientations (Vega and Fiat, 1972) since it depends generally on the orientation of the static magnetic field with respect to the molecular axes:

$$\Delta B = \frac{1}{3} \sum_{\alpha = x,y,z} \Delta B_\alpha \tag{9}$$

In the following, the relevant interactions contributing to the paramagnetic shift are separated as contact and dipolar contributions corresponding, respectively, to $\mathbf{A}_F$ and $\mathbf{A}_D + \mathbf{A}_L$ of the hyperfine coupling Hamiltonian (Eq. 1). The latter originates from both ligand-centered and metal-centered dipolar coupling. This separation is permitted whenever the complex consists of a central paramagnetic ion surrounded by smoothly covalent ligands. In fact, the ligand-centered contribution is often incorporated into the so-called contact shift. In the following, the contact shift refers to the Fermi interaction.

2.2.2. The Fermi Contact Interaction

2.2.2a. The Contact Shift. The Fermi contact interaction $\mathbf{A}_F$ results from unpaired electron spin in s atomic orbitals and is proportional to the electron density at the nucleus. This interaction is very efficient. For example, the coupling energy between an unpaired electron spin in a $1s$ hydrogen orbital and its nucleus corresponds to $A/h = 1420$ MHz (Carrington and McLachlan, 1967). It can be directly observed in the EPR spectra of monoatomic hydrogen radical as a splitting of 507 G.

A pictorial description (Swift, 1973) of the effects of the contact interaction gives a clear idea of the origin of the temperature dependence of the paramagnetic shifts and of the line broadening. For the simplest case of a complex with $S = \frac{1}{2}$, these effects can be analyzed on the basis of a nucleus which undergoes fast exchange between two sites corresponding to molecules with, respectively, $S_z = \frac{1}{2}$ and $S_z = -\frac{1}{2}$. Due to fast exchange only one NMR line can be observed, instead of a doublet. But the two sites are not equally populated. They differ by a small amount, on the order of $\Delta E/kT$, where ΔE is the energy difference between the two electron spin levels and kT is the thermal fluctuation energy. Then the collapsed NMR line is shifted with respect to the midpoint of the doublet (Fig. 1) by

$$\Delta\nu = \frac{\Delta E}{4kT}\frac{A}{h} \tag{10}$$

At X band, the EPR resonant frequency of the free electron is on the order of 0.3 cm^{-1}. The frequency shift of an NMR line observed at 300 K ($kT = 208$ cm^{-1}) in the same magnetic field will be $3.6 \; 10^{-4} \times A/h$, that is, about 360 Hz for a typical interaction of 1 MHz. This corresponds to a shift (to low field if A is positive) of 23.5 ppm for a proton, far above the potential resolution of NMR. Thus, paramagnetic NMR appears to be a very sensitive method for detection of delocalized unpaired electron spin at any position of a ligand molecule. But the simplicity of Eq. (10), which shows an ideal situation, is obscured by the presence of other mechanisms that contribute to the observed shift and, furthermore, by the contribution of thermally accessible states for which the hyperfine coupling constant is not necessarily the same.

To proceed further into the analysis, several stages of approximation can be invoked. The first one is applicable to complexes with no residual angular momentum,

$$\langle \Gamma n | L_a | \Gamma m' \rangle = 0 \qquad \forall \Gamma n, \Gamma m' \quad \text{and} \quad \alpha = x, y, z$$

In that case, one deduces easily from Eq. (3) that the contact shift is directly proportional to the average magnetic susceptibility χ:

$$\Delta B/B_0 = \frac{A_F}{g_e|\beta|g_N\beta_N}\frac{\chi}{N} \tag{11}$$

This holds even in case of anisotropic susceptibility but assumes that all the A_F associated with each electronic configurations contributing to the magnetic properties of the complex are the same.

In the case of a well-defined spin state S, it follows that

$$\frac{\Delta B}{B_0} = \frac{A_F|\gamma_e/\gamma_N|S(S+1)}{3kT} \tag{12}$$

the well-known generalization of Eq. (10). In principle, it could be used only for complexes the magnetic properties of which follow a simple T^{-1} (Curie law) behavior with a zero intercept at infinite temperature ($T^{-1} \to 0$).

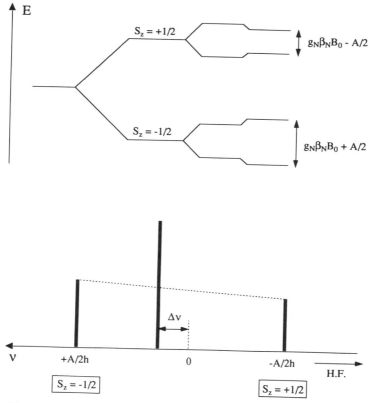

Figure 1. Energy levels of an electron–nuclear spin system interacting with a static magnetic field B_0 (upper). The lower diagram shows the nuclear spin transitions. The transition at high frequency (to the left in conventional NMR) is slightly more intense than the other one, due to the higher population of the corresponding ($S_z = -\frac{1}{2}$) electron spin level. Fast exchange between the two sites ($S_z = -\frac{1}{2}$ and $S_z = +\frac{1}{2}$) results in only one NMR line, shifted to the left.

In cases where the temperature dependence of the paramagnetic shift exhibits a non-Curie behavior and/or noticeable nonzero intercept, Eq. (3), or similar forms, must be used. The intermediate-spin ($S = 1$) ferrous porphyrins are one example of such a situation and will be discussed later.

2.2.2b. Hyperfine Coupling Constant in Terms of Electronic Molecular Properties (La Mar, 1973). When considering complexes which involve an aromatic ligand, the hyperfine coupling constant A can be separated into two distinct contributions:

$$A = A^\pi + A^\sigma \qquad (13)$$

where A^π describes the contribution from π-spin densities which reside in $2p_z$ orbitals. They contribute indirectly to the Fermi contact term by $\pi-\sigma$ spin polarization (Carrington and McLachlan, 1967). On the other hand, A^σ corresponds to the mixing of σ

molecular orbitals with symmetry-adapted metal orbitals containing unpaired electron in such a way that unpaired spin can be found in s type orbitals of the ligand.

The problem of separating the contributions A^π and A^σ is left for a subsequent section. We shall discuss here the interpretation of A^π in terms of spin delocalization. The π–σ polarization mechanisms inside a C—H bond have been fully described during the earlier observations of hyperfine coupling in the EPR spectra of organic radicals (McConnell and Chesnut, 1958; Karplus and Fraenkel, 1961; Strom *et al.*, 1972). For a proton, A^π is related to the spin density residing in the $2p$ orbital of the attached carbon by the well-known relation (McConnell, 1958)

$$A^\pi = Q^H_{CH}\rho^\pi_C \tag{14}$$

where Q^H_{CH} is an empirical parameter, which depends slightly on the molecular structure. For an interpretation of the paramagnetic shift, it is sufficient in virtually all cases to use $Q^H_{CH} = -63$ MHz. Note that ρ^π_C is the normalized spin density ($\sum_\rho = 1$, independent on the expectation value of S^2).

A similar expression holds for heteroatoms, but both contributions of the spin polarization of in-plane sp^2 bonds and the contributions from the neighbor atoms (ρ^π_{Xi}) must be accounted for. Thus (Karplus and Fraenkel, 1961)

$$A^\pi = \left(S_C + \sum_i Q_{CXi}\right)\rho^\pi_C + \sum_i Q_{XiC}\rho^\pi_{Xi} \tag{15}$$

where the first term reflects spin polarization of the $2s$ orbital and hybridized σ bonds by the unpaired spin residing in the $2p_z$ orbital of the central atom. The second term involves the effect of spin densities on attached atoms. For practical uses, Eq. (15) is written as:

$$A^\pi = Q_1\rho^\pi_C + \sum_i Q_i\rho^\pi_{Ci} \tag{16}$$

2.2.3. The Dipolar Interactions

Calculation of the paramagnetic shift resulting from the remaining terms $A_D + A_L$ of the hyperfine interaction can be performed in a way similar to that for obtaining the general Eq. (3). Due to the r^{-3} dependence, the dipolar contribution can be separated into contributions from the unpaired spin residing on the metal center and from unpaired spin delocalized onto the ligand atomic orbitals. Each contribution will be discussed separately.

2.2.3a. The So-called "Pseudocontact" Shift from the Metal-centered Dipolar Coupling. The energy of interaction between the magnetic moment $\boldsymbol{\mu}$ of the electron spin and the magnetic moment of the nuclear spin (which is assumed to be aligned with B_0) is, within the point dipole approximation, given by

$$E_{dip} = \mu_N\{\langle\boldsymbol{\mu}\cdot\mathbf{k}\rangle - 3\langle(\boldsymbol{\mu}\cdot\boldsymbol{\sigma})(\boldsymbol{\sigma}\cdot\mathbf{k})\rangle\}/r^3 \tag{17}$$

where $\langle \ \rangle$ indicates the average of the operator taken over all thermally accessible states, and $\mathbf{k}$ and $\boldsymbol{\sigma}$ are unitary vectors along, respectively, $\mathbf{B}_0$ and the vector $\mathbf{r}$ linking the electron and the nucleus. Assuming $\mathbf{B}_0$ is along one axis of the molecular system (x, y, z) that diagonalizes the magnetic susceptibility tensor, the macroscopic magnetic moment $\mathbf{M}$ is aligned parallel to the static field. Hence

$$\langle \boldsymbol{\mu} \cdot \mathbf{k} \rangle = \langle \mu_a \rangle = \frac{M\alpha}{N} = \frac{\chi_{a\alpha} B_0}{N} \tag{18}$$

where $\chi_{a\alpha}$ is a principal component of the paramagnetic susceptibility tensor ($\alpha = x, y, z$) and N is Avogadro's number.

For each canonical orientation of B_0 (Fig. 2), one obtains from Eq. (17)

$$B_0//z: E_{\mathrm{dip}}/B_0 = \mu_N(\chi_{zz}/N)\{1 - 3\cos^2\theta\}/r^3$$

$$B_0//y: E_{\mathrm{dip}}/B_0 = \mu_N(\chi_{yy}/N)\{1 - 3\cos^2\xi\}/r^3 \tag{19}$$

$$B_0//x: E_{\mathrm{dip}}/B_0 = \mu_N(\chi_{xx}/N)\{1 - 3\cos^2\zeta\}/r^3$$

where θ, ξ, and ζ define the orientation of $\mathbf{r}$ in the (x, y, z) molecular frame. The corresponding paramagnetic shift is obtained by averaging over all orientations the magnetic field resulting from this interaction ($E_{\mathrm{dip}} = -\boldsymbol{\mu}_N \cdot \Delta\mathbf{B}$):

$$\frac{\Delta B}{B_0} = -\left(\frac{1}{3N}\right) \sum_{\alpha = x,y,z} \frac{\chi_{a\alpha}(1 - 3\cos^2\Omega_\alpha)}{r^3} \qquad (\Omega_\alpha = \theta, \xi, \zeta) \tag{20}$$

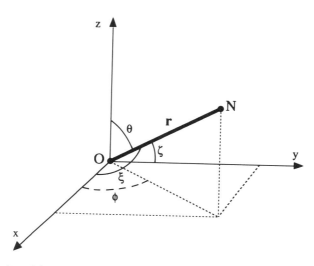

Figure 2. Location of the nucleus (N) in the molecular frame in which the magnetic susceptibility tensor is diagonalized.

from which one deduces the well-known equation (Kurland and McGarvey, 1970) for the pseudocontact shift:

$$\frac{\Delta B}{B_0} = -\frac{1}{3Nr^3}\{[\chi_{zz} - \tfrac{1}{2}(\chi_{xx} + \chi_{yy})](1 - 3\cos^2\theta) + \tfrac{3}{2}\sin^2\theta\cos 2\phi(\chi_{yy} - \chi_{xx})\} \quad (21)$$

where θ and ϕ are defined in Fig. 2.

In cases where the susceptibility tensor is axially symmetric, Eq. (21) simplifies to

$$\frac{\Delta B}{B_0} = -\left(\frac{1}{3N}\right)(\chi_\parallel - \chi_\perp)\left[\frac{(1 - 3\cos^2\theta)}{r^3}\right] \quad (22)$$

At this point, it should be noted that the point dipole approximation breaks down for "short" distances (0.2 to 0.3 nm) from the metal (Golding *et al.*, 1976a, 1982). The limit depends on several characteristics of the complex. For nuclei close to the ion, the paramagnetic shift may exhibit very complicated variations with the nucleus position (Golding and Stubbs, 1979). Another consequence is a residual pseudocontact contribution even in cases of a high symmetry such as octahedral (Golding and Stubbs, 1980), which comes from high-order terms (r^{-5} and r^{-7}) in the development of the dipolar interaction (Golding *et al.*, 1976b). These terms decrease rapidly, however, as one moves away from the paramagnetic ion.

2.2.3b. The Ligand-centered Dipolar Contribution to the Pseudocontact Shift. This contribution results from the dipolar coupling of the nuclear spin magnetic moment with the unpaired electron spin delocalized into the ligand orbitals. This effect is already accounted for in general Eq. (3) provided the $|\Gamma n\rangle$ include a linear combination of ligand atomic orbitals. In this approximation, the matrix elements of $A_D + A_L$ involve one- and two-center integrals. The one-center integrals corresponds to the dipolar coupling of spin residing generally in the $2p$ orbitals (the dipolar contribution from s-type orbitals is zero) attached to the nucleus of interest. The two-center integrals correspond to dipolar coupling arising from the metal (which have been taken into account in the last paragraph) and from the unpaired spin residing in neighboring atomic orbitals attached to a different nucleus.

Let us estimate the one-center contribution of an unpaired electron spin density ρ^π residing in a $2p_z$ carbon orbital of a planar ligand. The relevant dipolar interaction is described by a traceless tensor of axial symmetry $(2\Pi, -\Pi, -\Pi)$. Its principal components (Fig. 3) are about 214, -107, and -107 MHz (Morton and Preston, 1978) for unit spin density. Due to magnetic anisotropy the corresponding interaction is not averaged to zero in solution.

The magnetic field resulting from the electron–nuclear coupling can be obtained by analogy with the last calculation from

$$E_{\text{dip}} = \mathbf{I} \cdot \mathbf{T} \cdot \langle \mathbf{S} \rangle = -\boldsymbol{\mu}_N \cdot \Delta\mathbf{B} = -g_N\beta_N\mathbf{I} \cdot \Delta\mathbf{B} \quad (23)$$

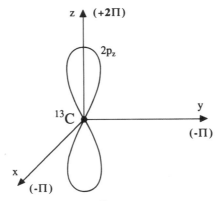

Figure 3. The dipolar hyperfine coupling of a ^{13}C nucleus with an unpaired spin residing in a $2p_z$ atomic orbital.

where $\langle S \rangle$ is the thermal average of the electron spin operator and $\mathbf{T}$ is the dipolar coupling tensor. Then

$$\Delta B_\alpha = -(T_{\alpha\alpha}/g_N\beta_N)\langle S_\alpha \rangle \qquad \alpha = x, y, z$$
$$T_{xx} = T_{yy} = -\Pi\rho^\pi, \; T_{zz} = 2\Pi\rho^\pi \tag{24}$$

and, by averaging over all orientations:

$$\frac{\Delta B}{B_0} = -\frac{[2\Pi\langle S_z \rangle - \Pi(\langle S_x \rangle + \langle S_y \rangle)]\rho^\pi}{3g_N\beta_N B_0} \tag{25}$$

In case of a complex where all thermally accessible states belong to the same multiplicity and assuming an isotropic g tensor, one obtains in case of axial symmetry

$$\frac{\Delta B}{B_0} = -\frac{2\Pi}{g_N\beta_N g_e|\beta|}\frac{(\chi_\parallel - \chi_\perp)}{3N}\rho^\pi \tag{26}$$

For a carbon atom located in the plane perpendicular to the symmetry axis ($\theta = \pi/2$) at 0.5 nm from the metal, the ligand-centered contribution has the opposite sign to that of the metal-centered contribution, and the ratio of the two contribution is:

$$\delta_{\text{local}}/\delta_{\text{metal}} = -\frac{2\Pi}{g_N\beta_N g_e|\beta|}\frac{\rho^\pi}{r^3} \tag{27}$$

which, for 1% unpaired electron spin, amounts to -14. The absolute value of the ratio decreases to about 3 for $r = 0.3$ nm, which corresponds also to the limit where the point dipole approximation may break down (Golding and Stubbs, 1979). But, even at this distance, the ligand-centered contribution dominates the dipolar contribution to the paramagnetic shift for a carbon bearing in its $2p_z$ orbital a typical unpaired electron spin of 1%.

Protons are located much farther from the metal ion than from the unpaired spin contained in the $2p_z$ orbital of the attached carbon atom. The corresponding dipolar contribution is, however, often neglected, without justification.

From Eq. (24) one obtains

$$\frac{\Delta B_a}{B_0} = \left(\frac{1}{3B_0}\right) \sum_{a=x,y,z} \frac{T_{aa}}{g_N\beta_N\langle S_a \rangle} \tag{28}$$

where T_{aa} are the components of the dipolar hyperfine tensor in the C–H bond and expressed along the principal magnetic axes. McGarvey (1988) points out the importance of this term for iron porphyrin complexes. Using the point dipole approximation, he gives these components, for a unit spin density, as

$$\frac{T_{xx}}{g_N\beta_N} = \frac{g_e\beta}{r^3} (3\cos^2\alpha - 1)$$

$$\frac{T_{yy}}{g_N\beta_N} = \frac{g_e\beta}{r^3} (3\sin^2\alpha - 1) \tag{29}$$

$$\frac{T_{zz}}{g_N\beta_N} = -\frac{g_e\beta}{r^3}$$

where it is assumed that the C—H bond lies in the x, y plane and makes an angle α with the x axis. It would be better to use empirical values obtained from radicals (McConnell *et al.*, 1960) and triplet states of aromatic molecules (Hutchison, 1967; Clarke and Hutchison, 1971). Then, for a unit spin density in the carbon $2p_z$ orbital, the traceless dipolar tensor has approximately the following principal components:

$$+35 \text{ MHz along the C—H bond}$$

$$-35 \text{ MHz along } y \tag{30}$$

$$0 \text{ MHz perpendicular to the plane}$$

These components are close to the values derived from integration of A_D over an atomic carbon $2p_z$ orbital (respectively +43, −38, −5 MHz, McConnell and Strathdee, 1959; Barfield, 1970, 1971). Furthermore the empirical values take into account a possible dipolar contribution of unpaired spin contained in sp^2 orbitals and resulting from π–σ spin polarization in the C—H bond. According to Eqs. (28) and (30), in case of an axially symmetric complex the local dipolar contribution to the paramagnetic shift for the in-plane protons is, fortuitously, averaged to zero if in addition the symmetry axis is parallel to the zero component of the dipolar coupling. However, in other situations such as the rhombic case or when considering protons in axial ligands, *the local contribution may be of the same order of magnitude as that of the metal-centered dipolar shift* (McGarvey, 1984).

When considering a complex with isotropic susceptibility, the dipolar interaction is averaged to zero as would be expected. In fact, this contribution may also be important in cases of octahedral symmetry (Kurland and McGarvey, 1970). The nonzero contribution observed in that case arises from the nuclear-spin–electron-orbital interaction A_L which has not been considered here but was taken implicitly into account for the metal-centered dipolar coupling term. This interaction contributes to the dipolar term when the orbital angular momentum of p electrons within the ligand is not quenched. Regarding the hemes, this interaction does not contribute to the ligand-centered shift since one is concerned only with $2p_z$ atomic orbitals.

2.2.4. Factorization of the Dipolar and Contact Contributions

When we are concerned with metallic complexes incorporating a well-localized paramagnetic center, the three contributions described above factor naturally into two terms providing information of different kinds. The first, the metal-centered dipolar term, relates to structural information, while the hyperfine interaction corresponding to the ligand-centered and the Fermi contributions relates to the electronic spin distribution over the ligand orbitals. Factorization of the last two contributions, and subsequently into the π- and σ-spin density contributions, is more difficult and sometimes requires additional information related to the hyperfine coupling. The relaxation rates of the nuclear magnetization are useful data. A theoretical model of the complex obtained from crystal field or molecular orbital theories may also be needed. Some of the existing methods of factorization is described in such models.

Our first task is to evaluate the metal-centered dipolar contribution. It involves geometric factors $F(r, \theta)$ and $G(r, \theta, \phi)$, depending on the location of the nucleus of interest with respect to the magnetic axes of the molecules as well as the components of the susceptibility tensor

$$\frac{\Delta B}{B_0} = -\frac{1}{3N}\{[\chi_{zz} - \tfrac{1}{2}(\chi_{xx} + \chi_{yy})]F(r, \theta) + (\chi_{xx} - \chi_{yy})G(r, \theta, \phi)\} \tag{31}$$

In all cases, the geometry of the complex must be known. It remains to determine the three principal components of the susceptibility tensor. A valuable approach (Horrocks and Hall, 1971) is to use magnetic susceptibility data when it is available. However, one must keep in mind that these data are obtained in the solid state, while NMR experiments are carried out in solution. For example, this approach has given rise to much controversy concerning the assignment of the ground state for the intermediate spin $(S = 1)$ four-coordinated ferrous porphyrins (see below). Another approach has been used frequently but should be considered with extreme caution. It uses the spectroscopic $g_{\alpha\alpha}$ factors obtained from EPR spectroscopy and related to the susceptibility tensor components by

$$\chi_{\alpha\alpha} = \frac{\beta^2 S(S + 1)}{3kT} g_{\alpha\alpha}^2 \qquad \alpha = x, y, z \tag{32}$$

In addition to the fact that the EPR spectra are recorded in the solid state (often in frozen solutions) and at very low temperature (down to 4 K or lower), Eq. (32) holds for a complex with only one thermally populated multiplet of spin S and no second-order Zeeman effects.

In cases of axial symmetry the problem is considerably simplified since only two parameters are involved, a geometric factor and the magnetic susceptibility anisotropy. The most obvious method consists in finding a nucleus insulated from spin delocalization but rigidly attached to the complex for a reliable evaluation of the geometric factor $F(r, \theta)$. If this condition is fulfilled, the magnetic susceptibility can be easily calculated and the dipolar shift for all other nuclei of the ligand can be evaluated. Conversely, one can calculate the geometric factor of a nucleus when chemical arguments have ascertained the absence of any delocalized spin densities at the corresponding position. This is probably one of the most powerful applications of paramagnetic NMR for structural investigations of metalloproteins.

Several methods exist for determining whether the paramagnetic shift measured for a proton is solely of dipolar origin. The first is to substitute a methyl group (Goff, 1983) or an heteroatom such as a fluorine nucleus (Mispelter *et al.*, 1977, 1978) for the proton of interest. The fluorine substitution method is particularly sensitive since the hyperfine coupling within the C—F bond is of opposite sign and about double that of the proton (Anderson *et al.*, 1960; Kaplan *et al.*, 1965; Icli and Kreilick, 1971). Furthermore it is proportional to the π-spin density residing on the attached carbon atom (Sinclair and Kivelson, 1968; Mispelter *et al.*, 1971a, b). Another advantage of this method is that the geometric factor for the fluorine nucleus in the substituted molecule is defined with the same accuracy as that of proton in the unsubstituted molecule. In contrast, methyl groups exhibit fast rotation leading to a less accurate estimation of $F(r, \theta)$. These methods of substitution are, however, of limited use since the substituted ligands are not always available.

Another method, particularly suitable for the tetraphenylporphyrins (Fig. 4), has been designed and extensively used by the La Mar group (La Mar and Walker, 1973; Goff, 1983). These complexes incorporate phenyl substituents at the meso position which have an orthogonal orientation with respect to the porphyrin plane. As a consequence, delocalization of unpaired π-spin density into the phenyl groups may be limited, except for radical species (Hickman *et al.*, 1988; Nanthakumar and Goff, 1991) that exhibit large unpaired spin densities at the meso position. Hence, the paramagnetic shifts for the phenyl protons and ^{13}C nuclei should be directly proportional to the geometric factor (Fig. 5).

Another interesting method has been recently proposed by Yamamoto *et al.* (1989a) for axially symmetric hemins and for hemoproteins with rhombic symmetry (Yamamoto *et al.*, 1990a). The authors use simultaneously the NMR data for the protons and ^{13}C of the four pyrrolic methyl groups of natural hemes (Fig. 6). The method relies on the fact that the local hyperfine contribution to the shift (Fermi and ligand-centered dipolar) for both the methyl carbon and attached protons is proportional to the π-spin density residing in the $2p_z$ orbital of the pyrrolic carbon atom to which the methyl group is bonded. Thus, a plot of the ^{13}C shifts versus the proton shifts would be a straight line with a zero intercept (Fig. 7) unless the metal-centered dipolar shift does contribute. Any deviation is therefore assigned to this contribution. The metal-centered dipolar shift for the protons and carbon of the heme methyl groups is systematically calculated for all possible orientations of a magnetic tensor and subtracted from the experimental data to obtain a proton–carbon "contact" shift correlation. The procedure is repeated

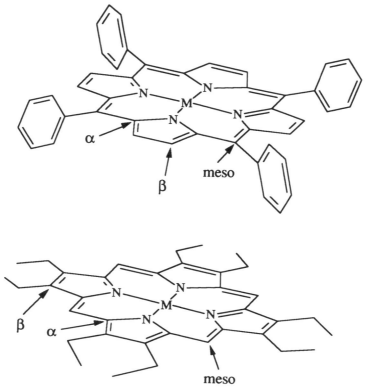

Figure 4. Usual porphyrin macrocycles: tetraphenylporphyrin (top) and octaethylporphyrin (bottom).

until the deviation from a straight line passing through the origin is minimized. By this method, Yamamoto *et al.* (1990a), determined the magnetic axes of the prosthetic group of the cyanide complex of ferric sperm whale myoglobin. The magnetic susceptibility of this low-spin ($S = \frac{1}{2}$) complex is not axially symmetric and the apparent success of their method seems encouraging. However, they report a magnetic tensor orientation differing from previously reported results (Emerson and La Mar, 1990, and references therein) but gave no explanation for this discrepancy. The results by Emerson and La Mar (1990) are based on 41 known dipolar shifts and on the magnetic anisotropy already determined by Horrocks and Greenberg (1973), and should therefore be considered with confidence. Nevertheless, the method of Yamamoto *et al.* (1990a) should also be considered but requires a more extensive evaluation for establishing its reliability.

Finally we will mention a method taking advantage of the deuterium quadrupole. This is an extension of a method initially developed for diamagnetic molecules (Lohman and McLean, 1981, and references therein) allowing a direct determination of the anisotropic susceptibility components *in solution*. This method (Domaille, 1980) is based on the partial alignment of magnetically anisotropic molecules along the static magnetic

δ para, ppm

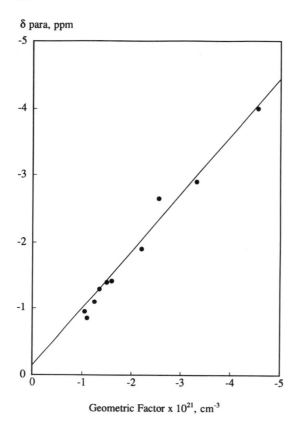

Geometric Factor x 10²¹, cm⁻³

Figure 5. Correlation of phenyl proton and carbon paramagnetic shifts with their respective geometric factor for low-spin ($S = \frac{1}{2}$) iron(III) tetraphenylporphyrin complexes (Goff, 1981).

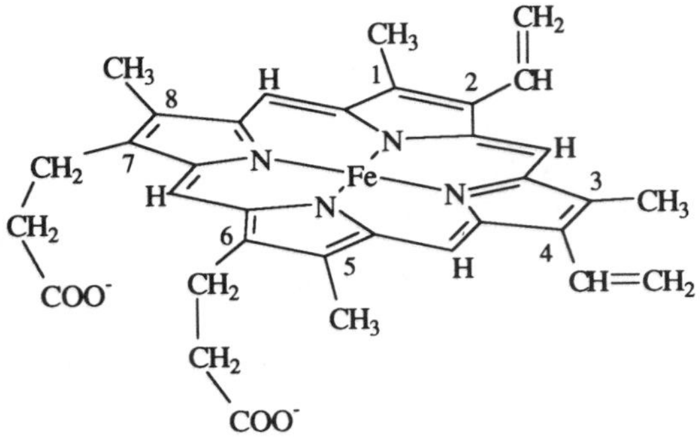

Figure 6. Chemical structure of natural heme (protoporphyrin IX).

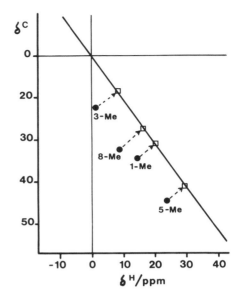

Figure 7. Plots of paramagnetic shift for the heme methyl carbon versus that for the attached proton, for all the heme methyl groups (see Fig. 6 for numbering) in cyanide metmyoglobin. The straight line corresponds to the correlation obtained after subtracting the metal-centered dipolar shift contribution (Yamamoto *et al.*, 1990a).

field. The degree of orientation is given by the quadrupolar splitting of the deuterium NMR lines, from which one obtains $\Delta\chi_\parallel = \chi_{zz} - \frac{1}{2}(\chi_{xx} + \chi_{yy})$ and $\Delta\chi_\perp = \chi_{xx} - \chi_{yy}$. It has been implemented with success for a number of paramagnetic complexes incorporating deuterium-labeled ligands (Domaille, 1980; Strauss *et al.*, 1987, and references therein).

2.3. Nuclear Magnetic Relaxation by the Unpaired Electron Spins

2.3.1. Introduction

Relaxation of the macroscopic nuclear magnetic moment is described by two characteristic times T_1 and T_2. They measure, respectively, the speed with which the longitudinal component reaches its equilibrium value, corresponding to the Boltzmann population of the spin levels, and the speed with which the transverse components decrease to zero due to the loss of coherence in the precession of individual spins.

For diamagnetic molecules in solution, these characteristic times are on the order of 1 s. Relaxation results from the fluctuating magnetic fields created mainly by the thermal agitation of the microscopic magnetic moments of nuclear spin. These fluctuations are caused by Brownian motion in liquids and lattice vibrations in solids, as well as intramolecular reorientations, chemical exchange processes, and so on.

For paramagnetic complexes, the strong magnetic moment of unpaired electrons creates much larger magnetic fields than nuclear spins do. As a consequence, nuclear magnetic relaxation is more efficient in the presence of unpaired electron magnetic moments. The characteristic times T_1 and T_2 decrease to less than 1 ms for complexes in solution. In addition, there exists a new motion causing fluctuations in the magnetic fields resulting from the hyperfine interaction: the electron spin relaxation itself, which corresponds to random changes of the orientation of the individual microscopic magnetic moments with respect to the static magnetic field.

For transition metal complexes, the electron spin relaxation time depends on the metal, its redox state, and its environment and spans a large range of values from 0.1 ps to 10 ns. For radical species it can be as long as 1 μs. As a consequence, the NMR spectra of paramagnetic molecules exhibit a large sampling of resolution.

A comprehensive analysis of the relaxation properties of nuclear spins requires knowledge of the mechanisms of interactions creating magnetic fields (the only efficient way to achieve nuclear magnetic relaxation) and of the fluctuations modulating these magnetic fields. As for the shifts, the relaxation rate for a nucleus embedded in a paramagnetic molecule is made up of two parts:

$$\frac{1}{T_{i\text{obs}}} = \frac{1}{T_{i\text{dia}}} + \frac{1}{T_{i\text{para}}} \qquad i = 1, 2$$

where $1/T_{i\text{dia}}$ are the relaxation rates in a diamagnetic reference compound. The fluctuating magnetic fields relevant to the paramagnetic contribution results from the hyperfine interaction already introduced for the description of paramagnetic shifts. Obviously, the motions that cause magnetic fields to fluctuate depend on the mechanism of interaction. To a first approximation, when considering isotropic complexes with no zero-field splitting, the Fermi contact interaction A_F is modulated by chemical exchange and by electron spin relaxation but not by rotation, in contrast to the dipolar coupling.

As discussed in the analysis of shifts, the local dipolar interaction (ligand-centered) contributes to the relaxation process of a nucleus close to large unpaired spin delocalized into the ligand. Thus, this section will follow a format similar to that for paramagnetic shifts. First, the Fermi contact and the dipolar interaction with the metal electron spin will be considered; this relates to the Bloembergen–Solomon equations. Second, the ligand-centered contribution will be discussed in some detail.

It should be mentioned at this point that the approximations used here are drastic and probably fit in only a few cases, although they are of common use. In fact, in many cases they are sufficient for an estimate, at least when it is possible to assign to the complex a well-defined spin state. But the reader is urged to gain more details from Banci in Chapter 2 of this volume. Here we shall focus our attention on the specificity of ^{13}C nuclear relaxation as due to local spin densities, but assuming an isotropic complex.

2.3.2. The Bloembergen–Solomon Equations

For the particular case in which the dipolar interaction with the electron spin is modulated by Brownian motion and for isotropic paramagnetic complexes (no zero-field splitting and isotropic g factor), the longitudinal and transverse relaxation rates are given by the well-known equations (Solomon, 1955; Bloembergen, 1957a, b;

Solomon and Bloembergen, 1956), which in their complete form are

$$
\frac{1}{T_{1\text{para}}} = \frac{S(S+1)\gamma_I^2 g^2 \beta^2}{15 r^6} \left[\frac{2\tau_{c2}}{1+(\omega_s-\omega_I)^2 \tau_{c2}^2} + \frac{6\tau_{c1}}{1+\omega_I^2 \tau_{c1}^2} + \frac{12\tau_{c2}}{1+(\omega_s+\omega_I)^2 \tau_{c2}^2} \right]
$$

$$
+ \frac{S(S+1)}{3}\left(\frac{A}{\hbar}\right)^2 \left[\frac{2\tau_{e2}}{1+(\omega_s-\omega_I)^2 \tau_{e2}^2} \right] \tag{33}
$$

$$
\frac{1}{T_{2\text{para}}} = \frac{S(S+1)\gamma_I^2 g^2 \beta^2}{15 r^6} \times \left[4\tau_{c1} + \frac{\tau_{c2}}{1+(\omega_s-\omega_I)^2 \tau_{c2}^2} + \frac{3\tau_{c1}}{1+\omega_I^2 \tau_{c1}^2} \right.
$$

$$
\left. + \frac{6\tau_{c2}}{1+\omega_s^2 \tau_{c2}^2} + \frac{6\tau_{c2}}{1+(\omega_s+\omega_I)^2 \tau_{c2}^2} \right] + \frac{S(S+1)}{3}\left(\frac{A}{\hbar}\right)^2
$$

$$
\times \left[\tau_{e1} + \frac{\tau_{e2}}{1+(\omega_s-\omega_I)^2 \tau_{e2}^2} \right] \tag{34}
$$

The first term in both equations relates to the fluctuation of the dipolar interaction with the metal electron spin. The correlation time for this modulation is related to the correlation times for the various contributing motions by

$$
\frac{1}{\tau_{c1,2}} = \frac{1}{T_{1,2e}} + \frac{1}{\tau_R} + \frac{1}{\tau_M} + \frac{1}{\tau_{ex}} + \frac{1}{\tau_{int}} \tag{35}
$$

where $T_{1,2e}$ are the longitudinal (T_{1e}) and transverse (T_{2e}) electronic relaxation times, τ_R is the molecular rotational time, τ_M and τ_{ex} are, respectively, the chemical and electron exchange time, and τ_{int} relates to internal rearrangement of the molecular structure. The terms in brackets in Eqs. (33) and (34) are the spectral density function proportional to the power of the efficient components in the magnetic noise created by the motional fluctuations.

The efficient frequencies for relaxation are those involving a change of nuclear spin orientation (efficient for T_1 and T_2) and the static components at $\omega = 0$ (for T_2). In addition to the usual ω_I components inducing transitions among the nuclear spin states, the two transitions at $\omega_S + \omega_I$ and $\omega_S - \omega_I$ (Fig. 8) are efficient for both T_1 and T_2 since they involve changes of nuclear spin orientation. In contrast, transitions at ω_S are inefficient for T_1 but provide an efficient transverse relaxation pathway due to the broadening of the nuclear energy levels resulting from the finite lifetime of the corresponding states.

The second term of the Bloembergen–Solomon equations relates to relaxation processes due to modulation of the Fermi contact interaction. Due to its isotropic nature, this interaction is not modulated by rotational motions; the corresponding correlation time is given by the same expression as for the dipolar interaction correlation time, with τ_R dropping out

$$
\frac{1}{\tau_{e1,2}} = \frac{1}{T_{1,2e}} + \frac{1}{\tau_M} + \frac{1}{\tau_{ex}} + \frac{1}{\tau_{int}} \tag{36}
$$

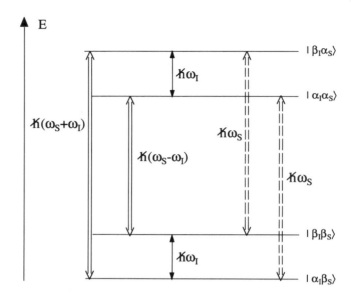

Figure 8. Energy levels for an electron (S)–nucleus (I) coupled two-spin system. Transitions indicated by full lines are important for both T_1 and T_2 relaxation processes. Transitions represented by broken lines do not involve a nuclear spin change. They are efficient only for T_2.

In fact, this is an approximation since the Fermi contact interaction, although isotropic, may give rise to anisotropic magnetic fields (McGarvey and Nagy, 1987). In other words, the magnetic field ΔB_α [Eq. (6)] resulting from this interaction may depend on the orientation of the complex with respect to the static field B_0 in the same way as the components of $\langle S \rangle$. As a result, it could be modulated by rotational motion.

In practice, one usually assumes that $T_{1e} = T_{2e}$, so that

$$\tau_{c1} = \tau_{c2} = \tau_c \quad \text{and} \quad \tau_{e1} = \tau_{e2} = \tau_e$$

Then, recalling that the electron spin resonance frequency $\omega_S/2\pi$ is much larger than the nuclear resonance frequency $\omega_I/2\pi$, the usable Bloembergen–Solomon equations reduce to:

$$\frac{1}{T_{1\text{para}}} = \frac{S(S+1)\gamma_I^2 g^2 \beta^2}{15 r^6}\left[\frac{6\tau_c}{1 + \omega_I^2 \tau_c^2} + \frac{14\tau_c}{1 + \omega_S^2 \tau_c^2}\right] + \frac{S(S+1)}{3}\left(\frac{A}{\hbar}\right)^2\left[\frac{2\tau_e}{1 + \omega_S^2 \tau_e^2}\right] \quad (37)$$

$$\frac{1}{T_{2\text{para}}} = \frac{S(S+1)\gamma_I^2 g^2 \beta^2}{15 r^6}\left[4\tau_c + \frac{3\tau_c}{1 + \omega_I^2 \tau_c^2} + \frac{13\tau_c}{1 + \omega_S^2 \tau_c^2}\right]$$

$$+ \frac{S(S+1)}{3}\left(\frac{A}{\hbar}\right)^2\left[\tau_e + \frac{\tau_e}{1 + \omega_S^2 \tau_e^2}\right] \quad (38)$$

For a well-defined metallic complex without chemical exchange or internal rearrangement, only two motions compete with each other, the molecular rotation and the electron spin relaxation. For most macromolecules, NMR spectra exhibit well-resolved hyperfine-shifted resonances, and the correlation time for nuclear relaxation is dominated by the electronic relaxation time τ_S (i.e., $\tau_R \gg \tau_S$). However, for very large τ_R/τ_S ratios, rotation becomes dominant, at least for T_2, through the so-called Curie-spin relaxation process (Guéron, 1975; Vega and Fiat, 1976) and is relevant for metalloproteins. Since this mechanism will be described in other chapters of this volume, we restrict our discussion to whether or not it could be neglected.

The ratio of the Curie-spin contribution to the dipolar hyperfine contribution, Eq. (38), is (Guéron, 1975)

$$x = \frac{12}{7} \frac{(g\beta)^2 S(S+1)}{(3kT)^2} \frac{\tau_R}{\tau_S} B_0^2 \tag{39}$$

or $x = 3.8 \times 10^{-6} S(S+1)(\tau_R/\tau_S)B_0^2$, where B_0 is in Tesla. For $S = \frac{1}{2}$ and a static field of 14.1 T (corresponding to a proton resonant frequency of 600 MHz) Curie-spin relaxation becomes dominant ($x > 1$) if $\tau_R/\tau_S > 1.8 \times 10^3$. This condition is fulfilled with macromolecules of at least 30 to 40 kDa. As a matter of fact, this relaxation behavior has been well demonstrated for the proton NMR spectra of myoglobin and hemoglobin (Johnson et al., 1977). For the iron porphyrin complexes considered in the next section, the corresponding contribution is very small and can be neglected.

The Bloembergen–Solomon equations (33)–(34) suggest (Reuben and Fiat, 1969) that heteronuclear paramagnetic NMR spectra could be better resolved than proton NMR spectra due to the γ_N^2 proportionality of both the dipolar and contact contributions. In fact, quadrupolar relaxation for $I > \frac{1}{2}$ spins, an increased hyperfine coupling constant A for heteroatoms (McGarvey and Kurland, 1973), and a dipolar coupling with local unpaired spin densities often result in unexpectedly broadened NMR lines. This complication is, however, a source of information, particularly with respect to the unpaired spin layout in the ligand.

2.3.3. Nuclear Magnetic Relaxation by Delocalized Unpaired Spin

The Bloembergen–Solomon equations account for the effect of spin delocalization through the contact contribution $A/\hbar$. However, it has been recognized for a long time that the dipolar interaction with unpaired spins delocalized in ligand atomic orbitals may be an important, if not the dominant, mechanism of nuclear relaxation. Qualitatively, this could be understood from the r^{-6} dependence of the dipolar contribution to the relaxation rates. For example, when a nucleus is located at 0.5 nm from the metal ion, unpaired spins localized at 0.1 nm are 15,625 (5^6) times as efficient as the metal unpaired spins for the dipolar relaxation process.

In an approach similar to that for the quantification of the hyperfine coupling constant, estimation of the contribution of the ligand unpaired spins to the nuclear relaxation requires a density matrix formalism using molecular orbital calculations (Gottlieb et al., 1977; Kowalewski et al., 1981). Gottlieb et al. (1977) obtained for the longitudinal relaxation rate an expression which may be written as

$$\frac{1}{T_1} = \frac{1}{15} \gamma_N^2 g_e^2 \beta^2 S(S + 1) \Delta^2 \Sigma^d J_1(\omega, \tau).$$ (40)

Basically, Eq. (40) is similar to the Bloembergen–Solomon equations with Δ^2 replacing $1/r^6$. $\Sigma^d J_1(\omega, \tau)$ is the spectral density function for the random modulation of the dipolar interaction, identical to the term within brackets in Eqs. (33) and (37). Equation (40) was derived under the assumption of general spin S, but Gottlieb *et al.* (1977) point out the possibility that it may not have complete generality for spin $S > \frac{1}{2}$. In fact, the limits of applicability of their theory are those applying to the Bloembergen–Solomon theory.

In the Gottlieb *et al.* (1977) theory, Δ^2 involves the sum of squared components of the hyperfine interaction, written as

$$\Delta^2 = (4\pi/5) \sum_{v = -2,2} \left| \sum_i \rho_i \int |F_2^v(\mathbf{r}')| |u_i(\mathbf{r})|^2 \, d\mathbf{r} \right|^2$$ (41)

where ρ_i is the diagonal element of the spin density matrix associated with the atomic orbital $u_i(\mathbf{r})$ centered at $\mathbf{r}$. The nondiagonal elements of the density matrix, corresponding to the "overlap spin population," are not considered in this calculation. This is a usual approximation made in analyzing hyperfine interaction in radicals. $F_2^v(\mathbf{r}')$, $v = -2$ to $+2$, are spatial components of the electron–nuclear dipolar Hamiltonian centered at the nucleus position $(\mathbf{r}')$. These components are integrated over the unpaired spin containing atomic orbitals $u_i(\mathbf{r})$. The merit of this approach is to avoid the point dipole approximation. All analytical expressions for the required integrals are given by Gottlieb *et al.* (1977), making use of Eq. (41) a simple job. Nevertheless, we present here a more intuitive but fully equivalent approach closer to the experimenters' concepts. In any case, the reader can verify that the results are consistent with a complete integration of Eq. (41). It should be noted in addition that one assumes the point dipole approximation holds for the metal-centered contribution.

An equivalent expression for Δ^2 could be obtained in terms of the total hyperfine coupling tensor components, $D_{\alpha\beta}$, as (Van Broekhoven *et al.*, 1971; Hendricks and DeBoer, 1975; Ronfard-Haret and Chachaty, 1978; Quaegebeur *et al.*, 1979)

$$\Delta^2 = (1/6) \sum_{\alpha,\beta} D_{\alpha\beta}^2 \qquad \alpha, \beta = x, y, z$$ (42)

Let us consider the example of a carbon nucleus interacting with unpaired electrons residing on the metal (assumed to be a point dipole) and unpaired π-spin densities located in its $2p_z$ atomic orbital (Fig. 9). Both tensors are diagonal in the same frame, and the total hyperfine tensor has the following principal components:

$$D_{zz} = 2\Pi - D$$

$$D_{yy} = -\Pi + 2D$$ (43)

$$D_{xx} = -\Pi - D$$

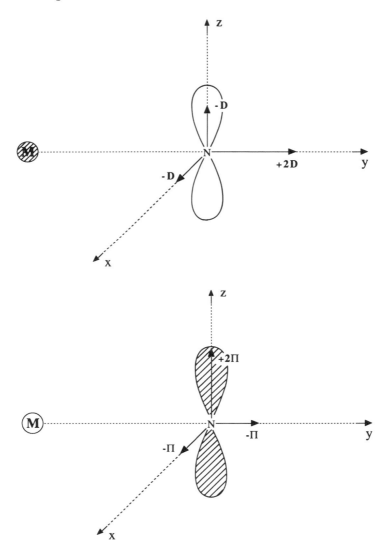

Figure 9. Relative orientation of the components of the metal-centered (upper) and of the ligand-centered (lower) dipolar coupling tensors.

Then

$$\Delta^2 = (1/6)[(2\Pi - D)^2 + (-\Pi + 2D)^2 + (\Pi + D)^2] = \Pi^2 + D^2 - D\Pi \qquad (44)$$

The terms Π^2 and D^2 refer to the separate effects from the dipolar coupling with, respectively, the local $2p$ spin density and the metal spins. The third term $D\Pi$ has been sometimes wrongly neglected. It results from electron correlation between the metal and the ligand. In other words, the dipolar contribution to the relaxation rates from each center does not simply add, rather, the total components of the dipolar coupling must be calculated before they are squared and averaged over all orientations.

The following values can be assigned to D and Π:

$$D = \rho_M r_M^{-3} \tag{45}$$

$$\Pi = 2/5\langle r^{-3}\rangle_{2p}\rho_C^\pi, \tag{46}$$

where ρ_C^π is the normalized spin density residing in the $2p_z$ orbital (usually on the order of 1% or less) and $\langle r^{-3}\rangle_{2p}$ is averaged over Hartree–Fock–Slater atomic orbitals (Morton and Preston, 1978). For a carbon atom, $\langle r^{-3}\rangle_{2p}$ is 1.34×10^{25} cm^{-3}, ρ_M is the spin density residing on the metal ion (close to 1 for most transition metal complexes), and r_M is the metal–nucleus distance. Hence

$$\Delta^2 = [\rho_M^2 r_M^{-6} + 4/25\langle r^{-3}\rangle_{2p}^2(\rho_C^\pi)^2 - 2/5\langle r^{-3}\rangle_{2p}r_M^{-3}\rho_C^\pi\rho_M] \tag{47}$$

Figure 10 shows the dependence of Δ^2 as a function of r_M for realistic values for ρ_C^π, assuming $\rho_M = 1$. For distances larger than 0.3 nm, the local hyperfine coupling dominates the dipolar relaxation provided ρ_C^π is larger than about 0.1%. For smaller ρ_C^π values, the Bloembergen–Solomon equations practically applies for all distances relevant for chemical species of interest. This puts a limit to the detectability of unpaired π-spin densities from relaxation data of heteronuclei. It should be also mentioned that the point dipole approximation breaks down for a distance r_M less than 0.3 to 0.4 nm as already pointed out in the calculation of the metal-centered dipolar shift. The limit also depends on the nature of the complex and on the desired accuracy. It can be readily shown that for a spherical ion the limit is close to the ion center, whereas Gottlieb *et al.* (1977) showed that for an $S = \frac{1}{2}$ iron(III) complex the error is 15% at a distance of 0.4 nm from the metal. An accurate estimation of the metal-centered dipolar contribution requires the integration of the dipolar operator over "true" atomic orbitals. In practice, when concerned with heteronuclei, the relaxation rates are mainly dominated by the contribution from local interactions (i.e., the second term and, to a lesser extent, the third term of Eq. (47)); the error due to the point dipole approximation in estimating the first term is therefore attenuated.

The dipolar contribution to the proton relaxation can be estimated in a similar way. Let T_{xx}, T_{yy}, and T_{zz} be the principal components of the dipolar hyperfine tensor in a C—H bond assumed along the y axis and due to unpaired spin residing in the attached carbon $2p_z$ orbital. Since the x or y principal axis of this tensor does not necessarily coincide with the metal–proton axis, one must first express the tensor in a new rotated frame using

$$\mathbf{T}' = \mathbf{R} \cdot \mathbf{T} \cdot \mathbf{R}^{-1}, \tag{48}$$

where $\mathbf{R}$ is the rotation matrix

$$\mathbf{R} = \begin{pmatrix} \cos\varphi & -\sin\varphi & 0 \\ \sin\varphi & \cos\varphi & 0 \\ 0 & 0 & 1 \end{pmatrix} \tag{49}$$

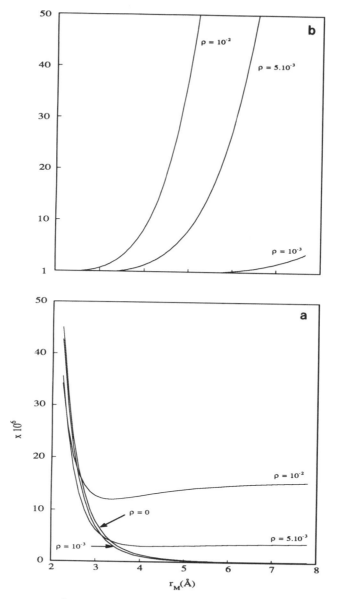

Figure 10. (a) Plot of Δ^2, Eq. (47), corresponding to the dipolar contribution to the carbon nucleus magnetic relaxation versus the metal–nucleus distance (r_M) in the presence of $2p_z$ spin density ρ; $\rho = 0$ corresponds to the Bloembergen–Solomon approximation. (b) Ratio of the total dipolar carbon relaxation rate to the Bloembergen–Solomon dipolar contribution. For $\rho = 10^{-3}$, the ligand-centered contribution is hardly detectable unless r_M is larger than 0.6 nm. For $\rho = 10^{-2}$, the ligand-centered contribution dominates the nuclear relaxation in most cases.

and φ is the angle between the C—H bond and the metal–proton line (Fig. 11). Using the fact that the tensor **T** is traceless ($T_{zz} = -T_{xx} - T_{yy}$), one obtains from Eq. (42)

$$\Delta^2 = \rho_M^2 r_M^{-6} + (\rho_C^\pi)^2 (T_{xx}^2 + T_{yy}^2 + T_{xx}T_{yy})/3 + \rho_C^\pi \rho_M r_M^{-3} (T_{xx} \sin^2 \varphi + T_{yy} \cos^2 \varphi) \quad (50)$$

It should be mentioned that in Eq. (50) the components of **T** are expressed in cm^{-3}. The next step is to estimate these components for a unit spin density. They can be obtained, in a way similar to that for estimation of the ligand-centered shift, from integration of the dipolar coupling operator over suitable atomic orbitals (McConnell and Strathdee, 1959; Barfield, 1970, 1971) or from empirical values deduced from experimental data. Gottlieb *et al.* (1977) give the necessary analytical expressions, considering Slater type orbitals. Using their notation, the components of **T** are

$$T_{xx} = -\frac{(A_{2p} + B_{2p})}{R_{C—H}^3}$$

$$T_{yy} = \frac{2A_{2p}}{R_{C—H}^3} \quad (51)$$

$$T_{zz} = \frac{-A_{2p} + B_{2p}}{R_{C—H}^3}$$

where $R_{C—H}$ is the carbon-proton bond length.

If one prefers using the empirical values of Eq. (30), then

$$T_{xx} = -4.43 \times 10^{23} \text{ cm}^{-3}$$

$$T_{yy} = +4.43 \times 10^{23} \text{ cm}^{-3} \quad (52)$$

$$T_{zz} = 0$$

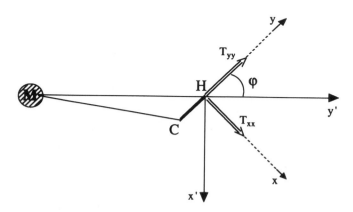

Figure 11. Orientation of the in-plane components of the hyperfine coupling within the C—H bond with respect to the dipolar coupling originating from the metal. The z components of both tensors (not shown on the figure) coincide.

Using these empirical values, Δ^2 is plotted in Fig. 12 as a function of r_M and for different values of ρ_C^π. As for ^{13}C, the Bloembergen–Solomon equations break down for $\rho_C^\pi > 10^{-3}$, and the difference becomes significant for metal–proton distances larger than 0.4 to 0.5 nm (Fig. 12). This range corresponds just to the iron–proton distances of peripheral protons in hemes.

This mechanism of relaxation provides a useful probe for delocalized spin densities on the ligand which adds up to the quantification of paramagnetic shifts. Combination of both kinds of data allows one to separate π and σ delocalization mechanisms for transition metal complexes, as will be illustrated now for iron porphyrins.

2.4. An Example: ^{13}C NMR of the Paramagnetic Iron(II) and Iron(III) Porphyrins

2.4.1. Introduction

In this section, we shall use selected results regarding iron porphyrins in order to illustrate striking features of ^{13}C NMR of paramagnetic molecules. The discussion will be limited to the two most frequent redox states of hemes in hemoproteins, namely the d^6 iron(II) and d^5 iron(III). For more details, the reader may refer to one of the reviews that have appeared in the literature during the last decade (for example Goff, 1983; Bertini and Luchinat, 1986; Walker and Simonis, Chapter 4 of this volume).

After the pioneering work of Wüthrich and Baumann (1973) dealing with low-spin ($S = \frac{1}{2}$) complexes of iron(III) porphyrins, ^{13}C NMR of paramagnetic iron porphyrins was developed mainly during the early 1980s in order to understand the relationships among electronic structure and iron redox and coordination states. This complements the more extensively used proton NMR. More recently, paramagnetic ^{13}C NMR has been used for an investigation of the structural properties of the prosthetic group and of neighboring residues in hemoproteins. In this section, we examine whether the electronic structure of these paramagnetic complexes can be elucidated by combined use of proton and ^{13}C NMR data. The minimum data required and the limitations of this approach will also be addressed.

2.4.2. ^{13}C NMR as a Probe for the Spin State

Figure 13 shows typical ^{13}C NMR spectra recorded for the five paramagnetic states of iron(III) and iron(II) porphyrins. Although these spectra may vary slightly with the chemical structure of the porphyrin ligand, the main features remain identical.

The high-spin states [Fe(III), $S = \frac{5}{2}$, and Fe(II), $S = 2$] can be distinguished from the other paramagnetic states on the basis of the observed chemical shifts, considering only the two porphyrin α and β core carbons (Fig. 4). For both high-spin states, the paramagnetic shifts are indeed very large, about 1000 ppm to low field. This is in contrast with those observed for the low-spin [Fe(III), $S = \frac{1}{2}$] and intermediate-spin [Fe(II), $S = 1$] complexes. In those cases, the paramagnetic shifts for the α and β carbons are one order of magnitude smaller and to low or high field.

The so-called quantum mixed spin state is intermediate between the other two groups. The α and β carbon shifts reflect the degree of mixing between two electronic configurations ($^6A_1$, $S = \frac{5}{2}$, and $^4A_2$, $S = \frac{3}{2}$) for which the occupancy of the iron

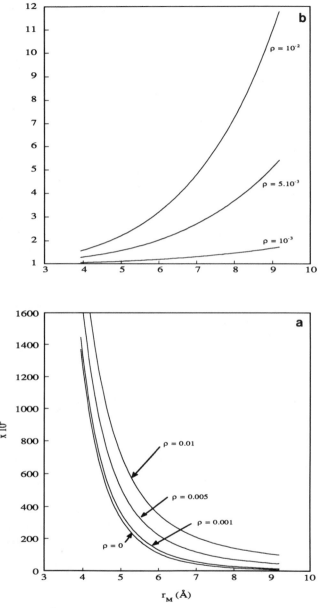

Figure 12. (a) Plot of Δ^2, Eq. (50), corresponding to the dipolar contribution to the proton-nucleus magnetic relaxation versus the metal-nucleus distance r_M in the presence of $2p_z$ spin density ρ; $\rho = 0$ corresponds to the Bloembergen–Solomon approximation. (b) Ratio of the total dipolar proton relaxation rate to the Bloembergen–Solomon dipolar contribution.

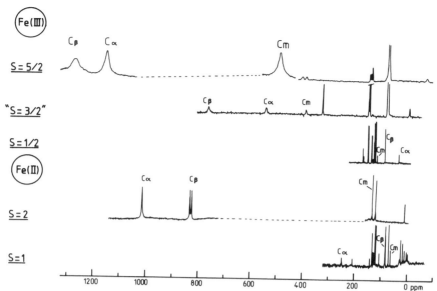

Figure 13. Typical ^{13}C NMR spectra for the five paramagnetic states of iron(III) and iron(II) porphyrin complexes. Only the resonances corresponding to the α, β, and *meso* carbons are indicated. All spectra were recorded at 25 MHz and 307 K, except the "intermediate spin" ($S = \frac{3}{2}$) spectrum (299 K). From Boersma and Goff (1982).

$d_{x^2} - d_{y^2}$ orbital differs by the presence or absence of an unpaired electron spin. It should be noted that the assignment to a quantum-mechanical mixture of states (Maltempo, 1974) rather than to a spin state equilibrium could not be based on the observation of paramagnetic shifts at only one temperature. In fact, the ground state for these complexes has been established by a number of physical methods (Reed et al., 1979; Spartalian et al., 1979) including paramagnetic NMR of the proton (Dugad et al., 1985) and of ^{13}C (Boersma and Goff, 1982; Toney et al., 1984).

In contrast to the α and β carbons, the third porphyrin core carbon (meso-C, Fig. 4) exhibits in most cases a small shift and, for the high-spin Fe(III), only a moderate shift to low field.

A qualitative interpretation of these spectral features can be derived from a simple analysis of delocalization pathways for the unpaired electron spin depending on the metal d orbital occupancy (Fig. 14) for the five paramagnetic states (see also Walker and Simonis, Chapter 4 of this volume).

For the two high-spin configurations, the $d_{x^2} - d_{y^2}$ orbital is occupied by an unpaired electron spin. This orbital, of σ symmetry, mixes with molecular porphyrin orbitals of the same symmetry. Because of this, small unpaired electron spin densities exist in s orbitals of the ligand atoms and lead to a large Fermi contact interaction. One can estimate that only 0.022% of unpaired spin (normalized to unity) residing in a $2s$ carbon orbital explains a 1000 ppm shift to low field for an $S = \frac{5}{2}$ complex at 300 K. Furthermore, it results from symmetry arguments that the σ molecular orbital which

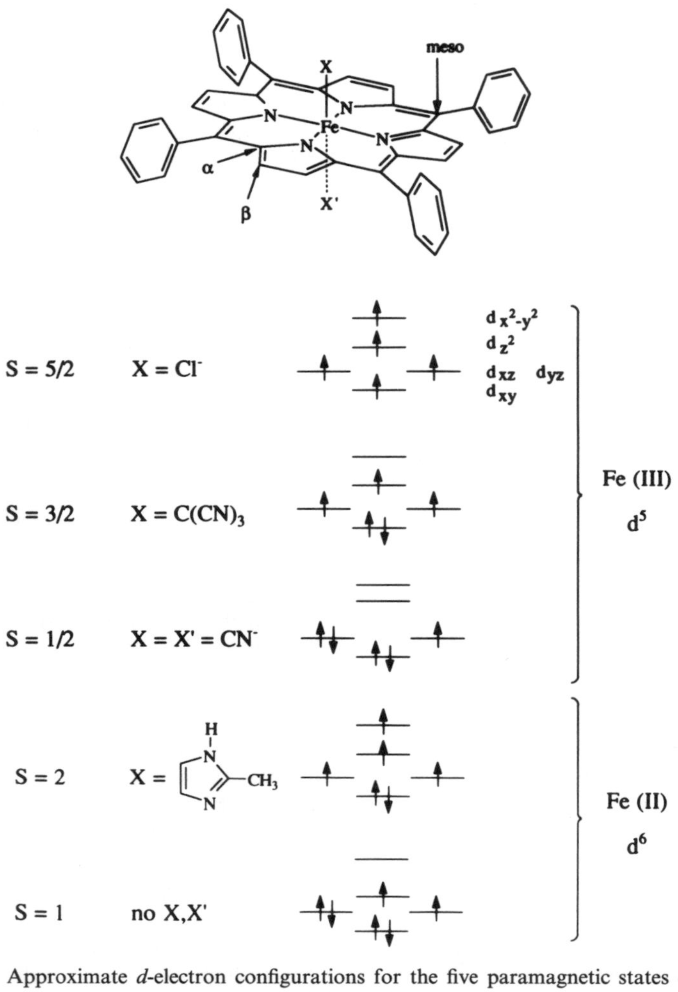

Figure 14. Approximate d-electron configurations for the five paramagnetic states of iron(III) and iron(II) porphyrins.

mixes with the iron $d_{x^2} - d_{y^2}$ orbital possesses nodes at the meso carbon position. This explains the small shift of the corresponding nucleus for all cases. In fact, the meso-carbon contact shift results solely from a $\pi-\sigma$ polarization mechanism. Such a mechanism is also at the root of the α and β carbon contact shifts for the low-spin and intermediate-spin states. The sole π-spin delocalization resulting from the mixing of d_{xz} and d_{yz} with π-molecular orbitals is much less efficient in inducing unpaired spin into the carbon $2s$ orbital. Indeed, a 100-ppm shift for an $S = 1$ spin state at 300 K requires 0.36% of unpaired spin in the corresponding carbon $2p_z$ orbital.

^{13}C NMR paramagnetic shifts allow an immediate assignment of the electronic configuration of the complex in an easier way than does proton NMR. Indeed, the proton paramagnetic shifts are much more sensitive to other mechanisms than the simple direct delocalization of unpaired spin. Metal-centered dipolar coupling as well as $\pi-\sigma$ polarization mechanisms partially hide direct σ delocalization mechanisms. Although it has been recognized that the direction of proton shifts as compared with those observed for substituted methyl groups allows an assignment of the delocalization pathways (Goff, 1983), the observation of the ^{13}C NMR shifts affords a more direct assignment. Nevertheless, these arguments cannot be generalized for a quantitative analysis. In particular, the carbon NMR shifts of the "π-complexes" do not result exclusively from the Fermi contact interaction, and the contributions from other mechanisms must be considered just as well for a quantitative interpretation of the observed shifts.

2.4.3. The High-Spin States

2.4.3a. Ferric High-Spin State $(S = \frac{5}{2})$. Let us consider the case of high-spin Fe(III) tetraphenylporphyrin (Fig. 4), for which all the relevant data are available (Mispelter *et al.*, 1981). The temperature dependence of the paramagnetic shifts for the three carbon resonances of interest, corresponding to the α, β, and meso carbons of the porphyrin core, follows a well defined Curie law with zero intercept at infinite temperature (Fig. 15). This is relatively exceptional for hemes, but contrasts with the curvature observed in the temperature dependence of the shifts for protons in the same complex (Walker and La Mar, 1973; La Mar *et al.*, 1973; Behere *et al.*, 1982) which results from the pseudocontact contribution.

The magnetic susceptibility for high-spin hemes may be expressed as (Kurland and McGarvey, 1970)

$$\chi_\perp - \chi_\parallel = \frac{112\beta^2 DN}{3(kT)^2} \tag{53}$$

$$\chi = \frac{35\beta^2 N}{3kT} \tag{54}$$

where D is the zero-field splitting (ZFS) parameter (Fig. 16) and the high-temperature approximation $(D \ll kT)$ is assumed. It is noteworthy that the ZFS for these complexes result from spin–orbit coupling of the $^6A_1$ ground state with a close-lying $^4A_2$ quartet state (Han *et al.*, 1972; Birdy *et al.*, 1983). However, the $^6A_1$ ground state assignment implies no residual electronic orbital moment to first order, hence an isotropic g factor. Furthermore, excited states are not thermally accessible. Hence, from Eqs. (11) and (54), one obtains the contact contribution as

$$\delta_{\text{cont}} = \frac{35\beta}{6\gamma_N kT} \frac{A}{\hbar} \tag{55}$$

Obviously, the contact contribution to the shift follows a Curie law. This is in contrast with the metal-centered dipolar contribution, which exhibits a $1/T^2$ temperature dependence:

$$\delta_{\text{dip}}^M = -\frac{112\beta^2 D}{9(kT)^2} \frac{3\cos^2\theta - 1}{r^3} \tag{56}$$

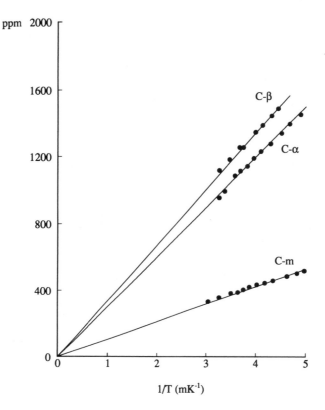

Figure 15. Temperature dependence of the carbon paramagnetic shifts for high-spin ($S = \frac{5}{2}$) iron(III) tetraphenylporphyrin (Mispelter *et al.*, 1981).

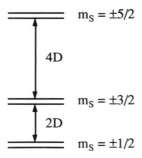

Figure 16. Zero-field energy levels for the $S = \frac{5}{2}(^6A_1)$ ground state coupled by spin–orbit interaction with a $^4A_2$ excited state.

The ligand-centered contribution can be estimated directly from Eq. (26) as

$$\delta_{\text{dip}}^L = -\frac{56\beta D}{9\gamma_N(kT)^2} \frac{2\Pi}{\hbar} \rho_C^\pi \tag{57}$$

showing the same $1/T^2$ dependence as for the metal-centered contribution.

The linearity and zero intercept observed for the temperature dependence of the carbon paramagnetic shifts suggest that the contact contribution largely dominates (Behere and Mitra, 1982). As a matter of fact, the metal- and ligand-centered dipolar contribution was estimated as not larger than, respectively, 4% and 16% of the total shift (Mispelter *et al.*, 1981). Furthermore, the opposite sign for these contributions is in favor of the cancellation of their effects. For completeness, however, all contributions have been considered in the analysis of the NMR data.

Due to direct σ spin delocalization from the iron $d_{x^2} - d_{y^2}$ iron orbital, analysis of the shifts in terms of π spin density is not directly possible. But when considering relaxation rates (linewidths in that case), further experimental data are available. The linewidth at half height are assumed to be related to the transverse relaxation rate by

$$\lambda = \frac{1}{\pi T_2} \tag{58}$$

and $1/T_2$ is given by an equation similar to Eq. (40):

$$\frac{1}{T_2} = \frac{1}{15} \gamma_N^2 g_e^2 \beta^2 [S(S+1)]\Delta^2 \Sigma^d J_2(\omega, \tau) + \frac{S(S+1)}{3} \frac{A}{\hbar^2} \Sigma^c J_2(\omega, \tau) \tag{59}$$

Where the appropriate spectral density functions for, respectively, the dipolar and contact interactions are

$$\Sigma^d J_2(\omega, \tau) = [7 + 13/(1 + \omega_S^2\tau_S^2)]\tau_S \tag{60}$$

$$\Sigma^c J_2(\omega, \tau) = [1 + 1/(1 + \omega_S^2\tau_S^2)]\tau_S \tag{61}$$

Δ^2 is given by Eqs. (47) and (50) for the carbon nuclei and protons, respectively. To complete the relationships between the NMR data and the π-electron spin distribution, one must consider the McConnell and Karplus–Fraenkel equations [Eqs. (14) and (15)]. The contact shifts for the three carbons and the pyrrolic protons of the porphyrin ring are written as

$$A_{C\beta}^\pi/h = Q_1\rho_{C\beta}^\pi + Q_2(\rho_{C\beta}^\pi + \rho_{C\alpha}^\pi) \tag{62}$$

$$A_{Cm}^\pi/h = Q_1'\rho_{Cm}^\pi + 2Q_2\rho_{C\alpha}^\pi \tag{63}$$

$$A_{C\alpha}^\pi/h = Q_1'\rho_{C\alpha}^\pi + Q_2(\rho_{C\beta}^\pi + \rho_{Cm}^\pi + \rho_N^\pi) \tag{64}$$

$$A_H^\pi/h = Q_{CH}^H\rho_{C\beta}^\pi \tag{65}$$

where $Q_1 = 99.6$ MHz, $Q_1' = 85.5$ MHz, $Q_2 = -39$ MHz, and $Q_{CH}^H = -63$ MHz.

Six experimental data are available: the linewidths for the three carbons and the β-proton and the paramagnetic shifts for the β and meso-carbon. These are related to seven unknowns: the unpaired spin densities $\rho_{C\alpha}^{\pi}$, $\rho_{C\beta}^{\pi}$, ρ_{Cm}^{π}, and ρ_{Fe}, the σ contribution to the Fermi contact shifts (A_{β}^{σ} and A_{m}^{σ}), and the electron spin relaxation time τ_S. It should be noted that the paramagnetic shift for the α carbon has not been included, since this would add two unknowns (ρ_N^{π} and A_{α}^{σ}). Due to lack of one parameter, the additional assumption that A_m^{σ} is zero has been made. This assumption is justified from the symmetry properties of the σ molecular orbital, which mixes with the iron $d_{x^2} - d_{y^2}$ iron atomic orbital of B_{1g} symmetry in D_{4h}. The MO belonging to this symmetry must have indeed a node at the meso position. Although the equations relating the paramagnetic shifts and linewidths to the unknowns are not linear, it was possible to obtain a unique set of solutions for the ρs, A^{σ}s, and τ_S. The complete factorization of the contributions to the paramagnetic shifts and linewidths resulting from this analysis is reported in Table 1. The following facts are noteworthy. The contribution to the linewidth originating from the dipolar coupling with the metal is relatively small, except for the proton, although it is not the dominant contribution even in that case. The hyperfine interaction with π-unpaired spin densities is the dominant mechanisms of the relaxation process for the meso and β carbons as well as for the β proton. The ligand-centered contribution to the relaxation rate for α carbon results, in fact, from the counterbalancing effect of the cross $[-(2/5)\langle r^{-3}\rangle_{2p} r_M^{-3} \rho_C^{\pi} \rho_M]$ and one-center $[(4/25)\langle r^{-3}\rangle_{2p}^2 (\rho_C^{\pi})^2]$ terms.

The success of this approach in deriving the π-electron spin distribution and in separating all the contributions to NMR properties is possible because of the large contribution of π-spin densities to the relaxation rates. However, estimates of these effects (Figs. 10 and 12) already suggest that for unpaired spin densities on the order

TABLE 1

Spin Density Values (Normalized to Unity) and Contribution to the Paramagnetic Shift (ppm, Positive to Low-Field) and Linewidth (Hz) of the ^{13}C and β Proton Resonances in High-Spin ($S = 5/2$) Iron(III) Porphyrin (chloro-iron(III) tetraphenylporphyrin)[a]

	Resonance			
Parameter	$\alpha(^{13}C)$[b]	$\beta(^{13}C)$	$m(^{13}C)$	$\beta(^1H)$
ρ_C^{π}	0.5×10^{-2}	0.92×10^{-2}	0.88×10^{-2}	—
δ_{dip}^M	24	9	17	5
δ_{dip}^L	−38	−70	−67	0
$\delta_{cont}\begin{cases}\pi\\\sigma\end{cases}$	< -350 / > 1400	430 / 781	424 / 0	−174 / 233
δ_{cont} (total)	1024	1211	424	59
λ_M^c	136	18	66	84
$\lambda_M + \lambda_{LM}$[d]	16	467	333	151
λ_{cont}[e]	298	415	51	16
λ_{total}	450	900	450	251

[a] Mispelter *et al.*, 1981. Other parameters are $\tau_s = 10.5 \times 10^{-2}$ s, $\rho_{Fe} = 3.5/5$, $D = -7600$ ppm/unit spin, and $T = 307$ K.
[b] See Fig. 4 for numbering.
[c] Metal-centered dipolar contribution to the linewidth.
[d] Total ligand-centered dipolar contribution to the linewidth.
[e] Contact contribution to the linewidth.

of 0.1% or smaller, the relaxation rates for carbon nuclei and protons closely follow the Bloembergen–Solomon equations. The high-spin ferrous porphyrin complexes exemplify this situation.

2.4.3b. Ferrous High-Spin State ($S = 2$). The ^{13}C NMR spectra of high-spin iron(II) porphyrin derivatives has been published by Shirazi *et al.* (1983). But to our knowledge no detailed analysis of the paramagnetic carbon shifts and no relaxation data are available from the literature. The availability of ^{13}C-enriched tetraphenylporphyrin derivatives has allowed us to measure the longitudinal relaxation rates ($1/T_1$) for the core carbons and for the β pyrrolic protons (Mispelter, 1981). Figure 17 reports these data together with the ^{13}C spectrum obtained from iron(II) ^{13}C-enriched derivatives axially coordinated by one molecule of 2-methylimidazole. The spectrum, in full agreement with the assignment by Shirazi *et al.* (1983), clearly shows that the dominant contribution to the paramagnetic shifts for the α and β carbons is from A^σ. In contrast, the meso-carbon paramagnetic shift is very small, consistent with the assumption that $A^\sigma_{Cm} = 0$ and that negligible unpaired π spin resides at this position (Goff and La Mar, 1977). Furthermore, the very good resolution of the spectrum suggests that the electron spin relaxation is fast but gives no quantitative information regarding the π-spin distribution in the porphyrin ligand. In fact, the longitudinal relaxation time for the three carbons and the β-pyrrolic protons closely follows predictions of the Bloembergen–Solomon equations, providing clear evidence that all π-spin densities are smaller by one order of magnitude than those detected in the parent iron(III) complex.

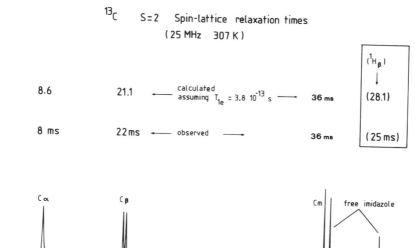

Figure 17. Typical ^{13}C NMR spectrum for high-spin ($S = 2$) iron(II) porphyrin and spin–lattice relaxation data, at 25 MHz and 307 K. On the right, the relaxation data for the β-pyrrolic proton are indicated in parentheses. Note the disruption in the chemical shift scale. The T_1 values were calculated using the Bloembergen–Solomon approximation, assuming an electron spin T_{1e} which fits the C-*meso* relaxation time.

At this stage, it is interesting to note that in this case the proton data are more efficient in detecting contributions from π-unpaired spins. The meso proton shift of the high-spin iron(II) octaethylporphyrin (Fig. 4) was estimated as between 8 and 13 ppm at 307 K. The two values correspond to the assumption that $A^{\sigma}_{Cm} = 0$ and $A^{\sigma}_{Cm} \neq 0$, respectively (Mispelter *et al.*, 1978). The corresponding π-spin density values (ρ^{π}_{Cm}) are 10^{-3} and 6×10^{-4}. They fall below the range of detectable unpaired spin by ^{13}C nuclear relaxation, in agreement with the present data. It should also be noted that these values are not in contradiction with a small meso-carbon shift ($\delta_{Cm} = 5.7$ ppm at 299 K, Shirazi *et al.*, 1983). If only π-spin densities contribute to the meso-carbon shift, then δ_{Cm} can be small and either to low or high field since A^{π} is the sum of the opposite contributions from ρ^{π}_{Cm} and ρ^{π}_{Ca} [Eq. (63)]. Regarding the β pyrrolic position, ^{13}C NMR study of labeled vinyl groups in deoxymyoglobin (Sankar *et al.*, 1987) has confirmed that π bonding should be small for these complexes.

These examples show the power and the limitations of heteronuclear relaxation rates to investigate the π-electron spin distribution of paramagnetic molecules. When combined with the paramagnetic shifts and proton NMR data, a quantitative analysis can be made. For "π-complexes" where direct σ delocalization does not take place, one could take the relaxation data out of consideration. This approach to deriving the electronic structure will be illustrated by the following cases of low-spin ferric and intermediate-spin ferrous porphyrins.

2.4.4. The "π-Complexes"

2.4.4a. Ferric Low-Spin State ($S = \frac{1}{2}$). ^{13}C NMR spectra of low-spin iron(III) porphyrins have been analyzed in terms of ligand π-electron spin distribution by Wüthrich and Baumann (1973, 1974) and by Goff (1981, 1983). Both groups used mainly the same approach. But in their earlier work Wüthrich and Baumann encountered difficulties in an attempt to estimate the metal-centered dipolar contribution to the paramagnetic shifts. This step, however, is essential for an analysis of the contact and ligand-centered dipolar contribution in terms of ligand π-spin densities. Goff (1981) estimated the metal-centered dipolar contribution using the correlation of the phenyl proton and carbon paramagnetic shifts with their geometrical factors in tetraphenylporphyrin. The good correlation (Fig. 5) suggests that a small unpaired π-spin density resides at the meso position, ensuring negligible contact contribution for the phenyl proton and carbon paramagnetic shifts. This was confirmed subsequently by the analysis of carbon shifts. Subtracting the metal-centered dipolar contribution from the paramagnetic shift, one obtains contributions depending solely on the unpaired spin distribution:

$$\delta_{\text{para}} - \delta^M_{\text{dip}} = \delta^L_{\text{dip}} + \delta_{\text{cont}} \tag{66}$$

The ligand-centered dipolar shift (δ^L_{dip}) is assumed to be related to the π-spin density at the observed carbon [Eq. (25)] by a constant D such that

$$\delta^L_{\text{dip}} = D\rho^{\pi}_C \tag{67}$$

This assumption is justified if one neglects the effects of neighboring and overlapping spin densities. However, it must be realized that the empirical D value cannot be transferred to other complexes exhibiting different magnetic properties since it depends on the components of $\langle S \rangle$. As a matter of fact, the ligand-centered dipolar shift for in-plane carbons was found to be of the same sign as the metal-centered contribution for cyanide and 1-methyl-imidazole *bis*-liganded low-spin iron(III) porphyrins, in contrast with the prediction from Eq. (27) and with the estimate for the intermediate-spin iron(II) complex (see below). The reason for this contradiction is not clear.

The contact contribution was assumed to follow the Curie law, Eq. (12):

$$\delta_{\text{cont}} = A_N^\pi |\gamma_e/\gamma_N|[S(S+1)/3kT] \tag{68}$$

where N stands for protons (H) or carbons (C). The hyperfine coupling constant A is given by Eqs. (62)–(65). This assumption is rather surprising, due to the non-Curie behavior of the paramagnetic shifts. But, since both the proton and the carbon Fermi contact shifts are expected to follow a similar temperature function $K(T)$, the assumption of a Curie-law, $K(T) = S(S+1)/3kT$, must influence the absolute π-spin density values but not their ratio. In contrast, the D value must suffer from similar uncertainties as the absolute π-spin density values. This can be understood from the following arguments.

Let r_α, r_m, and r_N be the ratios of, respectively, the π-spin densities at the α carbon, meso carbon, and nitrogen to the π-spin density at the β-carbon ($r_\alpha = \rho_{C\alpha}^\pi/\rho_{C\beta}^\pi$, $r_m = \rho_{Cm}^\pi/\rho_{C\beta}^\pi$, $r_N = \rho_N^\pi/\rho_{C\beta}^\pi$). The following equations can be obtained when considering the three core carbons and β-pyrrolic proton paramagnetic shifts:

$$\delta_{\text{cont},H\beta}(T) = Q_{CH}^H \rho_{C\beta}^\pi |\gamma_e/\gamma_H| K(T) \tag{69}$$

$$\delta_{\text{dip},C\beta}^L + \delta_{\text{cont},C\beta} = D(T)\rho_{C\beta}^\pi + [q_1 + q_2(1 + r_m)]|\gamma_H/\gamma_C|\delta_{\text{cont},H\beta}(T) \tag{70}$$

$$\delta_{\text{dip},Cm}^L + \delta_{\text{cont},Cm} = D(T)\rho_{C\beta}^\pi r_m + (q_1' r_m + 2q_2 r_\alpha)|\gamma_H/\gamma_C|\delta_{\text{cont},H\beta}(T) \tag{71}$$

$$\delta_{\text{dip},C\alpha}^L + \delta_{\text{cont},C\alpha} = D(T)\rho_{C\beta}^\pi r_\alpha + [q_1' r_\alpha + q_2(1 + r_m + r_N)]$$
$$\times |\gamma_H/\gamma_C|\delta_{\text{cont},H\beta}(T) \tag{72}$$

where $K(T)$ is assumed to be the temperature-functional dependence of the Fermi contact shift and q_i are the carbon Q_i values divided by Q_{CH}^H. Resolution of Eqs. (69)–(72) requires additional data. In the present case this was the paramagnetic shift for the meso proton of a parent complex, namely, the porphin (Wüthrich and Baumann, 1973) or the octaethylporphyrin (Goff, 1981) low-spin iron(III) derivatives. Thus

$$\delta_{\text{cont},Hm}(T) = Q_{CH}^H \rho_{Cm}^\pi |\gamma_e/\gamma_H| K(T) \tag{73}$$

From Eqs. (69) and (73) one deduces r_m. Then, the three remaining unknowns $D(T)\rho_{C\beta}^\pi$, r_α, and r_N may be obtained from Eqs. (70)–(72). Finally, the $\rho_{C\beta}^\pi$ value, hence the D value, is obtained from Eq. (69) assuming a Curie law.

The method described in the literature follows this scheme. The authors derived the spin densities $\rho_{C\beta}^\pi$ and ρ_{Cm}^π from, respectively, the β- and meso-proton contact shifts, using Eqs. (68) and (73) and replacing $K(T)$ by $[S(S+1)/3kT]$ (Curie law). Then the

TABLE 2
π-Spin Density Values and Empirical D Parameter Used in Interpreting the [13]C
NMR Data for Low-Spin ($S = 1/2$) Iron(III) Porphyrins[a]

$\rho^{\pi}_{C\beta}$	$\rho^{\pi}_{C\alpha}$	ρ^{π}_{Cm}	ρ^{π}_{CN}	D^b
1.2×10^{-2}	0.55×10^{-2}	-0.15×10^{-2}	10^{-2}	-6550

[a] Goff, 1981.
[b] Ligand-centered dipolar shift (in ppm) for a unit spin density and a temperature of 300 K, Eq. (67).

D value and $\rho^{\pi}_{C\alpha}$ were obtained from two equations describing the β- and meso-carbon shift. Finally, ρ^{π}_{N} was obtained from the α-carbon shift. These values are reported in Table 2. It should be mentioned that the ρs are obtained from the shifts at a temperature of 299 K (Goff, 1981). Using a set of shifts measured at another temperature should give a different set of values due to the assumption of a Curie law.

Attempts have been made to derive the hyperfine coupling constant independently from relaxation measurements (Von Goldammer and Zorn, 1976; Von Goldammer *et al.*, 1976). The proton relaxation data were interpreted in terms of an unexpected temperature-dependent hyperfine coupling constant A. This allowed the authors to explain an anomalous behavior of the paramagnetic shifts not fully accounted for by a second-order Zeeman effect. Curiously, no such temperature dependence of A could be found from ^{15}N NMR data (Von Goldammer, 1979). It should be noted, however, that the ligand-centered dipolar contribution has not been taken into account in the analysis of relaxation data, which is unreasonable for complexes exhibiting large π-spin densities. An anomalous behavior of the paramagnetic shifts can arise from aggregation. More recent measurements (Mazumdar *et al.*, 1990) performed on hemes dispersed in a micellar solution ensuring that the paramagnetic shifts were free from aggregation effects still showed similar deviations from a Curie behavior. In that case, these deviations have been fully explained by considering second-order Zeeman effects (Mazumdar *et al.*, 1990; Horrocks and Greenberg, 1973, 1974).

Although the ρ values should be considered only as estimates (Goff, 1981), these data can be interpreted with confidence in terms of a ligand-to-metal π-charge transfer. Such a charge transfer involves the highest $3e(\pi)$ filled orbital of the porphyrin ligand. From simple Hückel theory (Longuet-Higgins *et al.*, 1950) this orbital exhibits large spin population on the nitrogen atom (9×10^{-2}) and at the β-pyrrol position (6.25×10^{-2}). A smaller spin density is expected at the α position (1.7×10^{-2}) and a zero spin density at the meso position.

Newly characterized low-spin ferric isocyanide tetraphenylporphyrin complexes (Simonneaux *et al.*, 1989) exhibit interesting magnetic and electronic properties amenable to ^{13}C examination. Specifically, insofar as the proton data were correctly interpreted, the magnetic anisotropy is under the level of detectability and the β-proton shift is surprisingly small (1.28 ppm to low field at 298 K). The phenyl substituent proton shifts are much larger and of alternate sign: -7.06, $+6.12$, and -4.42 ppm for the *ortho*, *meta*, and *para* protons, respectively. This was interpreted in terms of large spin density at the meso position, indicating a π-charge transfer reversal with respect to *bis*-imidazole low-spin complexes for which small (and negative) spin densities were detected at the same position. Indeed, a metal-to-ligand π-charge transfer involves the lowest $4e(\pi^{*})$

vacant porphyrin orbital. Simple Hückel MO theory predicts the largest spin population at the meso position (0.1). A definite conclusion regarding the reversal of π-charge transfer would come from ^{13}C NMR, providing further new data for D values which are expected to be very small for these complexes due to their apparent magnetic isotropy.

    2.4.4b. Ferrous Intermediate-Spin State ($S = 1$). This iron(II) spin state has been the subject of continuous interest during the past two decades, both experimentally and theoretically. Magnetic susceptibility measurements (Kobayashi and Yanagawa, 1972; Brault and Rougee, 1973; Collman *et al.*, 1975; Boyd *et al.*, 1979; Strauss *et al.*, 1987), ^{1}H and ^{2}H NMR (Goff *et al.*, 1977; Mispelter *et al.*, 1977, 1980; Strauss *et al.*, 1985, 1987; Strauss and Pawlik, 1986; Medhi *et al.*, 1989), Mössbauer spectroscopy (Collman *et al.*, 1975; Lang *et al.*, 1978), and electron density maps determined by X-ray diffraction (Coppens, 1989; Li *et al.*, 1990) agree for an intermediate spin state ($S = 1$) assignment but sometimes disagree regarding the assignment of a specific electron configuration to the ground state for these square-planar iron(II) complexes. Most theoretical calculations (Kashiwagi and Obara, 1981; Obara and Kashiwagi, 1982; Sontum *et al.*, 1983; Rohmer, 1985; Edwards *et al.*, 1986) showed similar difficulties in predicting the ground state, due to the presence of many close-lying states requiring high accuracy to assign the ground state definitively. It should be noted that most of these theoretical calculations did not consider the effect of spin–orbit coupling, which is the relevant mechanism for explaining the magnetic properties of these complexes.

    In fact, there are at least two candidates for the ground state of these complexes, corresponding to the following d-electron configurations: $d_{xy}^2 d_{z^2}^2 (d_{xz}, d_{yz})^2 [^3A_{2g}]$ and $d_{xy}^2 d_{z^2}^1 (d_{xz}, d_{yz})^3 [^3E_gA]$. This result was obtained independently by three different methods, Mössbauer spectroscopy (Lang *et al.*, 1978), magnetic susceptibility (Boyd *et al.*, 1979), and proton NMR (Mispelter *et al.*, 1980), providing some confidence with respect to its validity. A third electron configuration, $d_{xy}^1 d_{z^2}^2 (d_{xz}, d_{yz})^4 [^3B_{2g}]$, has also been considered to explain the magnetic susceptibility data measured over a wide range of temperatures (Boyd *et al.*, 1979). However, consideration of this configuration was found unnecessary for interpreting the NMR data (Mispelter *et al.*, 1980; McGarvey, 1988). In addition, McGarvey (1988) included the $d_{xy}^2 d_{z^2}^2 (d_{xz}, d_{yz})^3 [^3E_gB]$ state which mixes with ^3E_gA by configurational interaction (Obara and Kashiwaga, 1982). We found this complication unnecessary for an interpretation of our results. On the contrary, extensive CI calculations are necessary for accurate results obtained from *ab initio* self-consistent field calculations (Rohmer, 1985).

    The two main configurations, ^3E_g and $^3A_{2g}$, are strongly coupled by spin–orbit interaction resulting in large residual orbital momentum components. Thus, the magnetic properties of the complex must depend critically on the relative energy of these two states. This energy is very sensitive to the strength of the ligand field along the z axis as well as to the chemical nature of the macrocycle. For a ^3E_g ground state, one expects a larger $\chi_\parallel$ than $\chi_\perp$ due to unquenched L_z, while for a $^3A_{2g}$ ground state, the magnetic susceptibility should be isotropic and equal to the spin-only value. In contrast, for the actual case of iron(II) porphyrins, mixing by spin–orbit coupling between both configurations induces residual L components corresponding to an unusually large average magnetic susceptibility and to a large anisotropy, either positive ($\chi_\perp > \chi_\parallel$) or negative ($\chi_\parallel > \chi_\perp$), depending on whether $^3A_{2g}$ or ^3E_g lies lowest. Most of the iron(II) porphyrins exhibit a large anisotropy with $\chi_\perp > \chi_\parallel$, but due to the critical dependence on the relative energy spacing Δ between $^3A_{2g}$ and ^3E_g, one can observe a cancelling followed by an inversion of the magnetic anisotropy for the same complex. This explains the strange temperature dependence already observed (Mispelter *et al.*, 1980) for some "basket-handle" porphyrins (Momenteau *et al.*, 1979; Momenteau, 1986).

Since the pseudocontact paramagnetic shift is proportional to the magnetic aniso-tropy, NMR is a sensitive probe for the $^3A_{2g}$–^3E_g energy spacing Δ. As a result, the $^3A_{2g}$ state has already been found to be lowest, with the ^3E_g state lying 350 to 1150 cm^{-1} higher in energy, depending on the porphyrin ligand. The corresponding variation in magnetic anisotropy was as large as 40% at a temperature of 300 K. For complexes incorporating aliphatic "handles," a reversal of energy ordering was observed at low temperatures, with ^3E_g lying 1000 cm^{-1} below $^3A_{2g}$, resulting in a magnetic anisotropy close to zero. An in-plane rhombic distortion also induces large perturbations of the magnetic properties (Strauss *et al.*, 1985, 1987; Strauss and Pawlik, 1986) resulting in linear T^{-1} dependence of the shifts but with very large nonzero intercepts at infinite temperature attributed to a large "temperature-independent paramagnetism" (Strauss *et al.*, 1985; McGarvey, 1988).

We shall focus our attention on the ^{13}C NMR of D_{4h} porphyrin complexes, illustrat-ing the analysis of NMR data for complexes exhibiting strong deviations from a Curie behavior. Figure 18 shows the Curie plot for the three porphyrin core carbon resonances and Fig. 19 illustrates a typical spectrum (Mispelter, 1980). Noteworthy is the hyperfine shift for the α carbon exhibiting an apparent anti-Curie behavior. In fact, this strange behavior results from the balancing effects of large contributions with nonzero intercept at infinite temperature. It should be mentioned at this point that these ^{13}C NMR data were obtained for the "backet-handle" porphyrin shown in Fig. 20. The handles ensure the necessary long-term stability for data averaging and preclude any intermolecular association. A good test for these essential properties is provided by the excellent agree-ment between calculated and observed metal-centered dipolar shifts (Fig. 21). This ensures a constant well-defined iron–ligand field throughout the investigated tempera-ture range. In the present case, the proton at the *para* position of the meso-phenyl groups (Mispelter *et al.*, 1977) and the carbon nuclei of the "handles" (Fig. 20) were the probes for the measurement of the metal-centered dipolar shift.

Analysis of the proton paramagnetic shifts using a theoretical model based on the ligand field theory outlined above and on the Kurland–McGarvey equation (3) has been done for axially symmetric complexes (Mispelter *et al.*, 1980). Recently, the theory was extended by McGarvey (1988) to the more general case of rhombic symmetry. The theory can be applied just as well to an analysis of the carbon shifts, using the appropri-ate equations for the components $A_{N\alpha}$ of the hyperfine coupling operator in Eq. (3):

$$A_{N\alpha} = [A_F(Q, \rho) + T_{\alpha\alpha}\rho_C^\pi]S_\alpha \qquad \alpha = x, y, z \qquad (74)$$

In Eq. (74), $A_F(Q, \rho)$ is the contact coupling constant given by Eqs. (62)–(65). The corresponding term contributes to the Fermi contact shift. $T_{\alpha\alpha}$ ($\alpha = x, y, z$) are the anisotropic components of the hyperfine coupling constant, given by Eqs. (24) and (30) for the carbon nuclei and proton, respectively. This term can be omitted for the proton in cases of axial symmetry (Section 2.2.3b). It contributes to the ligand-centered carbon shift, Eq. (67).

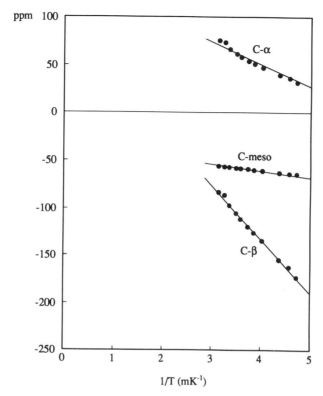

Figure 18. Temperature dependence of the paramagnetic carbon shifts for four-coordinated intermediate-spin ($S = 1$) iron(II) porphyrin. The straight lines are calculated from the phenomenological analysis of the shifts (see text). The carbon numbering corresponds to that given in Fig. 20.

Since McGarvey (1988) gave all the required analytical expressions for doing the calculations, we mention only some specific points which deserve attention. Although his theory applies in case of axial symmetry, it is preferable to use a specifically adapted basis set in order to simplify the spin–orbit matrix and to avoid troubles in calculating the z component of the paramagnetic shift. Furthermore, McGarvey (1988) included several refinements we do not include in the present calculation (no orbital reduction factor and no configurational interaction mixing) since they do not improve the results significantly. Then only one adjustable parameter is needed, the energy separation Δ between the $^3A_{2g}$ and ^3E_g states. Finally, in order to avoid any confusion, the convention of McGarvey (1988) referencing the energies to the $^3A_{2g}$ level is used.

Δ is obtained from the best fit for the metal-centered dipolar shift. The corresponding value for the complex of interest (Fig. 20) is $350\ cm^{-1}$, in agreement with earlier results obtained from proton NMR data (Mispelter et al., 1980). Then, using the π-spin densities as adjustable parameters, the best fit (Table 3) to the paramagnetic shifts for the protons and carbon nuclei is obtained as shown in Figs. 22 and 23, respectively.

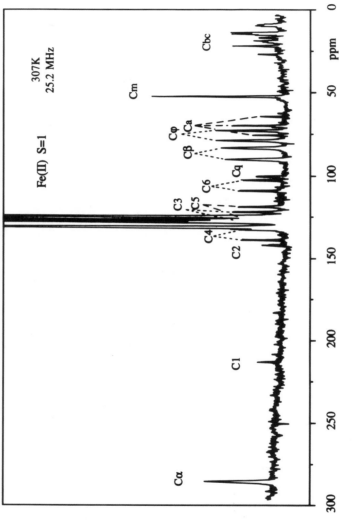

Figure 19. Typical ^{13}C NMR spectrum for four-coordinated intermediate-spin ($S = 1$) iron(II) porphyrin (see Fig. 20 for numbering) at 307 K, 25.2 MHz.

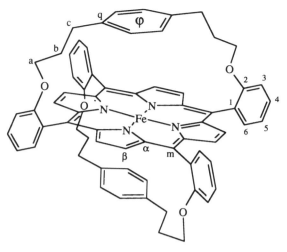

Figure 20. Chemical structure of the basket-handle porphyrin used to record the temperature dependence of the ^{13}C shifts.

As already mentioned (Mispelter *et al.*, 1980; McGarvey, 1988) the theory does not predict accurately the contact contribution to the proton shifts (Fig. 22). This disagreement is more apparent for the a carbon paramagnetic shift (Fig. 23). All the refinements to the theory known to date have been unable to improve the results. McGarvey (1988) suggests that the Qs may be temperature dependent due to some dynamical processes. In his opinion, there is no other possible explanation. A definite conclusion therefore requires additional experimental data. For example, relaxation data, if correctly interpreted, could provide additional arguments in favor of the McGarvey proposal. Other possibilities, such as a dynamic Jahn–Teller effect, may also be explored.

The paramagnetic shifts may be in turn analyzed using a phenomenological approach closely related to that described above for low-spin complexes. As already noted, the first step is to subtract from the hyperfine shifts the metal-centered dipolar contribution. For the porphyrin core carbons, this contribution amounts to more than 100 ppm. The second step is to assume that the temperature-functional dependence $K(T)$ of the contact shift is the same for protons and carbon nuclei [Eqs. (69)–(72)] and that the ρs are temperature-independent. This assumption allows one to search for a set of variables $(D(T)\rho_{C\beta}^{\pi}, r_a, r_m,$ and $r_N)$ which best fit the experimental data within the whole temperature range investigated when incorporated in Eqs. (70)–(72). A least-squares search procedure could be programmed, but a correlation between the carbons and proton contact shifts (Fig. 24) provides a direct estimate for r_a and r_m. Indeed, if T_0 is the temperature at which $\delta_{\text{cont},H\beta}$ goes through zero, the carbon shift at this temperature (note that T_0 has no physical meaning) is, from Eqs. (70)–(72),

$$\delta_{\text{dip},C}^{L} + \delta_{\text{cont},C} = D(T_0)\rho_C^{\pi} \tag{75}$$

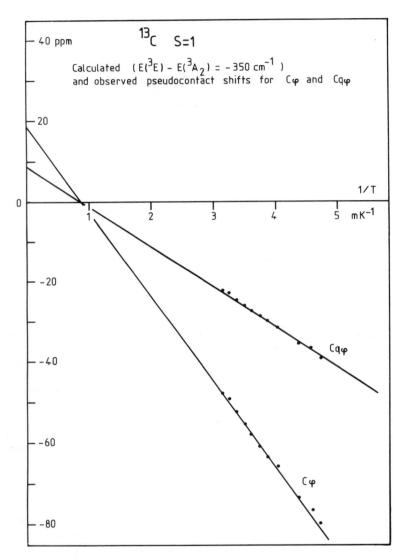

Figure 21. Temperature dependence of paramagnetic shifts for selected "handle" carbon resonances (see Fig. 20 for numbering). The straight lines correspond to the best fit using the ligand field theory described in the text, with $\Delta = 350\ \mathrm{cm}^{-1}$. Note that the calculation was limited to the experimental temperature range and extrapolated linearly outside that range.

TABLE 3
Best-Fit Parameters for Calculating the NMR Shifts for Square–
Planar Intermediate-Spin ($S = 1$) Iron(II) Porphyrins

	Ligand field theory ($\Delta = +350$ cm^{-1})	Phenomenological approach
$\rho_{C\beta}^{\pi}$ [a]	0.74×10^{-2}	0.74×10^{-2}
$\rho_{C\alpha}^{\pi}$	1.72×10^{-2}	1.86×10^{-2}
ρ_{CN}^{π}	2.35×10^{-2}	2.82×10^{-2}
ρ_{Cm}^{π} [b]	-0.47×10^{-2}	-0.47×10^{-2}
D^{c}	$-12,000$	-9600

[a] The spin densities are normalized to unity, not to $2S$.
[b] The *meso* carbon shift was found not very sensitive to ρ_{Cm}^{π}. The indicated value is that obtained from r_m (see the text). An additional hyperfine coupling constant of about 1 MHz was essential to fit the *meso* carbon shift. The origin of this additional contribution is unknown.
[c] D is the ligand-centered dipolar contribution to the carbon shift [Eq. (67)]. It is given in ppm for a unit spin density and a temperature of 300 K.

These values are obtained from the intercept to $\delta_{\text{cont},H\beta} = 0$ in the carbon–proton shift correlation (Fig. 24) from which one deduces r_α and r_m, provided $D(T_0)$ is not zero. It is noteworthy that the known large and negative value for ρ_{Cm}^{π} is confirmed, as well as the large π-spin density at the α carbon (Table 3) already deduced from the theoretical analysis. The final fits are given in Fig. 18 as straight lines, showing the good agreement with experimental data, throughout the temperature range.

This procedure is usable due to the non-Curie behavior of the hyperfine shifts and is based entirely on the fact that $D(T_0) \neq 0$. The procedure should not work if the paramagnetic shifts extrapolate linearly to zero at infinite temperature. For those cases exhibiting a perfect Curie law this procedure is, however, needless.

Figures 25–27 show the factorization of the carbon paramagnetic shifts into their various contributions. As expected, the sign of the ligand-centered dipolar shift is opposite to that of the metal-centered dipolar contribution, except for the meso-carbon, since its $2p_z$ orbital contains a negative π-spin density. The large spin density at the α carbon is in agreement with a visible broadening of the corresponding resonance (Fig. 19). However, no relaxation data are available for an in-depth investigation. The absolute values of the π-spin density reported in Table 3 are derived from the $\rho_{C\beta}^{\pi}$ value obtained from the theoretical calculations. Both methods give very similar results confirming the previous statement that the spin density ratios are relatively independent on the theory used. But the absolute values depend critically on the accuracy of the theory interpreting the β-pyrrolic proton data.

2.5. Concluding Remarks

The paramagnetic contribution to both NMR shift and relaxation is governed by the hyperfine coupling between the electron spin and the nuclear magnetic moment. This hyperfine interaction is made up of isotropic and anisotropic contributions, which cannot be directly distinguished due to the fact that NMR is usually recorded in the liquid state, leaving only an average over all orientations. The Fermi contact interaction is isotropic. However, the anisotropy of the thermally averaged components of the total

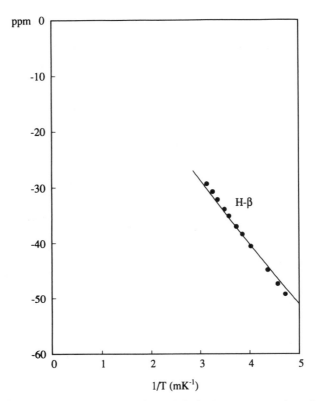

Figure 22. Calculated (full line) and experimental (points) temperature dependence for the β-pyrrolic proton resonance in four-coordinated intermediate-spin ($S = 1$) iron(II) porphyrin. The calculation was done using the ligand field theory described in the text, with $\rho^{\pi}_{C\beta} = 0.74 \, 10^{-2}$, corresponding to the best fit to the experimental data.

spin $\langle S \rangle$ implies that the corresponding magnetic perturbing field is anisotropic. This has no real effect on the observed shifts recorded in the liquid state (except a deviation from the Curie law) but this may be a broadening factor. The purely dipolar hyperfine interaction also contributes to the shift through the anisotropy of $\langle S \rangle$. The dipolar coupling involves unpaired electron spin residing on two distinct centers, the metal and the ligand orbitals. Both these contributions also contribute to the nuclear magnetic relaxation rates. Noteworthy is the difference in contribution for the proton and for heteronuclei. Thus, combining heteronuclear and ^{1}H data opens the way to the factorization of the various contributions, allowing an interpretation in terms of the electron spin layout on the ligand, providing information on the electronic structure of the complex. It should be noted that the term "ligand" assumes a complex including a well-localized paramagnetic center coordinated by ligand(s). However, the argumentations presented here apply as well to spin-delocalized species such as organic radicals.

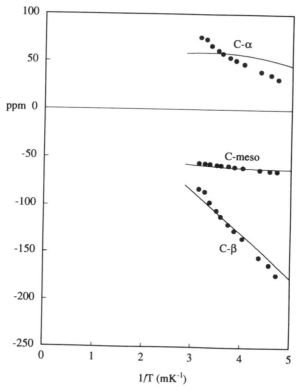

Figure 23. Calculated (full lines) and experimental (points) temperature dependence for the α, β, and *meso* carbon resonances in four-coordinated intermediate-spin ($S = 1$) iron(II) porphyrin. The parameters used in the calculation (Table 3) correspond to the best fit to the experimental data.

Analyses of the NMR parameters are feasible in all cases providing a sufficient number of data have been acquired. The minimum is recording a temperature dependence of the paramagnetic shifts, as for all studies dealing with paramagnetic species. As noted above, proton data are essential, giving in addition an absolute reference for the calculation of spin densities residing in the ligand atomic orbitals. Relaxation rates may be useful and complement the paramagnetic shift data especially when direct σ spin delocalization into the ligand takes place.

If the magnetic properties of the complex do not comply with a Curie law, a correlation between the proton contact shifts and the "local" hyperfine contribution to the carbon paramagnetic shifts allows a direct determination of the π-spin density ratios. More generally, the anisotropic contribution to the hyperfine coupling of the carbon nucleus magnetic moment with the electron spin can be obtained. This is legitimate as long as the ligand-centered anisotropic contribution to the proton shift can be neglected. As shown above, this implies that the relative orientation of the local hyperfine tensor and the magnetic susceptibility principal axes fulfill a number of conditions. Thus,

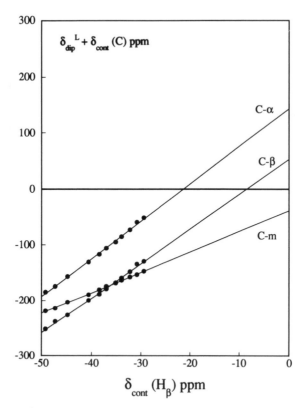

Figure 24. Correlation of the carbon "contact" shift (including the ligand-centered dipolar contribution) with the β-pyrrolic proton contact shift in four-coordinated intermediate-spin ($S = 1$) iron(II) porphyrin. The solid lines are linear least-square fits to the data. Note that the intercepts at $\delta_{cont}(H_\beta) = 0$ give the ratio $r_m(=\rho^\pi_{Cm}/\rho^\pi_{C\beta})$ and $r_a(=\rho^\pi_{Ca}/\rho^\pi_{C\beta})$ as, respectively, -0.64 and 2.5.

caution must be exercised before one draws any conclusion from such a correlation. It must be mentioned that this method applies just as well to complexes where direct σ spin delocalization contributes to the paramagnetic shifts.

If, not by chance, the magnetic susceptibility of the complex follows a Curie law, the correlation between the proton contact and carbon hyperfine shifts must be a straight line extrapolating to the origin. Any deviation indicates an incorrectly estimated metal-centered dipolar contribution or, likely, a nonnegligible ligand-centered dipolar shift for the proton. More generally, such a deviation would result from the anisotropic contribution to the hyperfine coupling involving the proton. At this point, one may remark that the factorization method of Yamamoto *et al.* (1989a, 1990a) relates to a similar approach, except that they consider the total paramagnetic shift, including the dipolar contribution, and that the varying parameter is the π-spin density rather than the temperature. The observed deviations from a straight line for the proton–carbon correlation can be entirely assigned to the metal-centered dipolar shift since they consider the

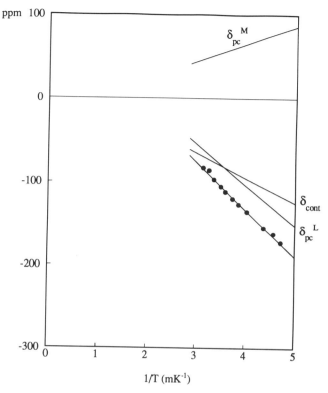

Figure 25. Factorization of the β-carbon (see Fig. 20 for numbering) paramagnetic shift for $S = 1$ iron(II) porphyrin into the Fermi contact (δ_{cont}), metal-centered (δ_{pc}^M), and ligand-centered (δ_{pc}^L) dipolar contributions.

carbon and attached protons of substituted methyl groups. In that case, the ^{13}C local hyperfine interaction depends solely on the π-spin density ρ_C^π, residing in the $2p_z$ atomic orbital of the carbon atom to which the methyl group is attached. When ρ_C^π goes to zero, both the proton contact and the carbon "local" hyperfine shift must cancel out.

Finally, for complexes exhibiting no magnetic anisotropy, a direct interpretation of the observed paramagnetic shifts in terms of Fermi contact interaction is feasible. However, separating the π and σ contributions calls for relaxation data, which in those cases could be analyzed without any difficulties. In other cases, relaxation data should be analyzed keeping in mind the effects coming from the magnetic anisotropy and zero-field splitting (for $S > \frac{1}{2}$) (Bergen *et al.*, 1979; Bertini *et al.*, 1984, 1985, 1990; Benetis *et al.*, 1984; Banci *et al.*, 1986; Vasavada and Nageswara Rao, 1989; Fukui *et al.*, 1990). In any case, they usefully complement the shift data, allowing a direct confirmation of the presence or absence of π-spin densities at the observed carbon positions.

A combined use of both proton and carbon NMR data thus allows in practically all cases an interpretation of the hyperfine shift data in terms of the electronic structure

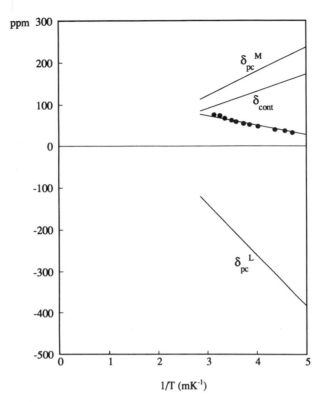

Figure 26. Factorization of the α-carbon (see Fig. 20 for numbering) paramagnetic shift for $S = 1$ iron(II) porphyrin into the Fermi contact (δ_{cont}), metal-centered (δ_{pc}^M), and ligand-centered (δ_{pc}^L) dipolar contributions.

of the complexes. The only requirement was a careful adaptation of the analysis to the magnetic properties of the complex.

Due to the significant improvement in the technical characteristics of NMR equipment during the past few years and to the rapidly growing number of 2D correlation experiments involving paramagnetic molecules (Yu *et al.*, 1986; Yamamoto, 1987; Yamamoto *et al.*, 1988, 1990b; Jenkis and Lauffer, 1988; Satterlee and Erman, 1991; Satterlee *et al.*, 1991; Bertini *et al.*, 1991; de Ropp and La Mar, 1991; Skjeldal *et al.*, 1991; Peyton, 1991; La Mar and de Ropp, Chapter 1 of this volume) one could expect a significant increase in the number of data pertaining to heteronuclear NMR of paramagnetic biological molecules. Furthermore, the development of modern multidimensional techniques (Wüthrich, 1986) attempting to resolve the three-dimensional structure of proteins up to at least 25 kDa requires availability of ^{13}C and ^{15}N labeled proteins (Kay *et al.*, 1990). For some of these, one could be confronted with the interpretation of paramagnetic heteronuclear NMR data. We hope this chapter will provide the necessary elements and make the reader willing to analyze these data in depth.

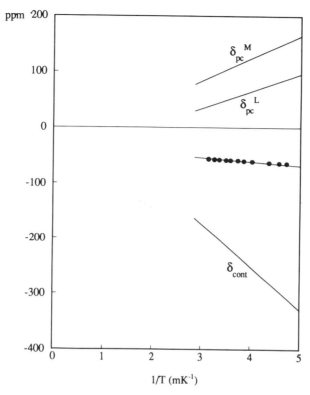

Figure 27. Factorization of the *meso*-carbon (see Fig. 20 for numbering) paramagnetic shift for $S = 1$ iron(II) porphyrin into the Fermi contact (δ_{cont}), metal-centered (δ_{pc}^M), and ligand-centered (δ_{pc}^L) dipolar contributions.

REFERENCES

Anderson, P. H., Frank, P. J., and Gutowski, H. S., 1960, *J. Chem. Phys.* **32**:196.

Banci, L., Bertini, I., Briganti, F., and Luchinat, C., 1986, *J. Magn. Reson.* **66**:58.

Barbush, M., and Dixon, D. W., 1985, *Biochem. Biophys. Res. Commun.* **129**:70.

Barfield, M., 1970, *J. Chem. Phys.* **53**:3836.

Barfield, M., 1971, *J. Chem. Phys.* **55**:4682.

Behere, D. V., and Mitra, S., 1982, *Proc. Indian Acad. Sci.* **91**:145.

Behere, D. V., Birdy, R., and Mitra, S., 1982, *Inorg. Chem.* **21**:386.

Behere, D. V., Gonzales-Vergara, E., and Goff, H. M., 1985, *Biochim. Biophys. Acta* **832**:319.

Behere, D. V., Ales, D. C., and Goff, H. M., 1986, *Biochim. Biophys. Acta* **871**:285.

Benetis, N., Kowalewski, J., Nordenskiöld, L., Wennerström, H., and Westlund, P. O., 1984, *J. Magn. Reson.* **58**:261.

Bergen, H. A., Golding, R. M., and Stubbs, L. C., 1979, *Mol. Phys.* **37**:1371.

Bertini, I., and Luchinat, C., 1986, *NMR of Paramagnetic Molecules in Biological Systems*, Physical Bioinorganic Chemistry, Vol. 3, Benjamin-Cummings, Menlo Park, CA.

Bertini, I., Luchinat, C., Mancini, M., and Spina, G., 1984, *J. Magn. Reson.* **59**:213.
Bertini, I., Luchinat, C., and Kowalewski, J., 1985, *J. Magn. Reson.* **62**:235.
Bertini, I., Banci, L., and Luchinat, C., 1989, *Methods Enzymol.* **177**:246.
Bertini, I., Luchinat, C., and Vasavada, K. V., 1990, *J. Magn. Reson.* **89**:243.
Bertini, I., Briganti, F., Luchinat, C., Messori, L., Monnanni, R., Scozzafava, A., and Vallini, G., 1991, *FEBS Lett.* **289**:253.
Birdy, R., Behere, D. V., and Mitra, S., 1983, *J. Chem. Phys.* **78**:1453.
Bloembergen, N., 1957a, *J. Chem. Phys.* **27**:572.
Bloembergen, N., 1957b, *J. Chem. Phys.* **27**:595.
Boersma, A. D., and Goff, H. M., 1982, *Inorg. Chem.* **21**:581.
Boyd, P. D. W., Buckingham, D. A., McMeeking, R. F., and Mitra, S., 1979, *Inorg. Chem.* **18**:3585.
Brault, D., and Rougee, M., 1973, *Nature (London)* **241**:19.
Carrington, A., and McLachlan, A. D., 1967, *Introduction to Magnetic Resonance*, Harper and Row, New York.
Cerdonio, M., Congiu-Castellano, A., Mogno, F., Pispisa, B., Romani, G. L., and Vitale, S., 1977, *Proc. Natl. Acad. Sci. USA* **74**:398.
Clarke, R. H., and Hutchison, C. A., 1971, *J. Chem. Phys.* **54**:2962.
Collman, J. P., Hoard, J. L., Kim, N., Lang, G., and Reed, C. A., 1975, *J. Am. Chem. Soc.* **97**:2676.
Coppens, P., 1989, *J. Phys. Chem.* **93**:7979.
Crull, G. B., Nardo, J. V., and Dawson, J. H., 1989, *FEBS Lett.* **254**:39.
Curie, P., 1895, *Annales de Chimie et de Physique*, 7e Série **5**:289.
de Ropp, J. S., and La Mar, G. N., 1991, *J. Am. Chem. Soc.* **113**:4348.
Dodrell, D. M., Bergen, H., Thomas, D., Pegg, D. T., and Bendall, M. B., 1980, *J. Magn. Reson.* **40**:591.
Domaille, P. J., 1980, *J. Am. Chem. Soc.* **102**:5392.
Dugad, L. B., Marathe, V. R., and Mitra, S., 1985, *Proc. Indian Acad. Sci.* **95**:189.
Dwek, R. A., 1973, *NMR in Biochemistry*, Clarendon Press, Oxford.
Edwards, W. D., Weiner, B., and Zerner, M. C., 1986, *J. Am. Chem. Soc.* **108**:2196.
Emerson, S. D., and La Mar, G. N., 1990, *Biochemistry* **29**:1556.
Everett, G. W., and Johnson, A., 1972, *J. Am. Chem. Soc.* **94**:6397.
Fukui, H., Miura, K., and Matsuda, H., 1990, *J. Magn. Reson.* **88**:311.
Gerothanassis, I. P., and Momenteau, M., 1987, *J. Am. Chem. Soc.* **109**:6944.
Gerothanassis, I. P., Momenteau, M., and Loock, B., 1989, *J. Am. Chem. Soc.* **111**:7006.
Godziela, G. M., and Goff, H. M., 1986, *J. Am. Chem. Soc.* **108**:2237.
Goff, H. M., 1981, *J. Am. Chem. Soc.* **103**:3714.
Goff, H. M., 1983, in *Iron Porphyrins, Part I* (A. B. P. Lever and H. B. Gray, eds.), Chap. 4, Addison-Wesley, Reading, Massachusetts.
Goff, H. M., and La Mar, G. N., 1977, *J. Am. Chem. Soc.* **99**:6599.
Goff, H. M., La Mar, G. N., and Reed, C. A., 1977, *J. Am. Chem. Soc.* **99**:3641.
Golding, R. M., and Stubbs, L. C., 1977, *Proc. Roy. Soc. Lond.* **A354**:223.
Golding, R. M., and Stubbs, L. C., 1979, *J. Magn. Reson.* **33**:627.
Golding, R. M., and Stubbs, L. C., 1980, *J. Magn. Reson.* **40**:115.
Golding, R. M., Pascual, R. O., and Stubbs, L. C., 1976a, *Mol. Phys.* **31**:1933.
Golding, R. M., Pascual, R. O., and Vrbancich, J., 1976b, *Mol. Phys.* **31**:731.
Golding, R. M., Pascual, R. O., and McGarvey, B. R., 1982, *J. Magn. Reson.* **46**:30.
Gotsis, E. D., and Fiat, D., 1987a, *Polyhedron* **6**:2053.
Gotsis, E. D., and Fiat, D., 1987b, *Magn. Reson. Chem.* **25**:407.
Gottlieb, H. P. W., Barfield, M., and Dodrell, D. M., 1977, *J. Chem. Phys.* **67**:3785.
Guéron, M., 1975, *J. Magn. Reson.* **19**:58.
Haddy, A. E., and Sharp, R. R., 1989, *Biochemistry* **28**:3656.

Haddy, A. E., Frasch, W. D., and Sharp, R. R., 1989, *Biochemistry* **28**:3664.

Han, P. S., Das, T. P., and Rettig, M. F., 1972, *J. Chem. Phys.* **56**:3862.

Hebendanz, N., Köhler, F. H., and Scherbaum, F., 1989, *Magn. Reson. Chem.* **27**:798.

Hendriks, B. M. P., and De Boer, E., 1975, *Mol. Phys.* **29**:129.

Henry, B., Boubel, J. C., and Delpuech, J. J., 1986, *Inorg. Chem.* **25**:623.

Hickman, D. L., Nanthakumar, A., and Goff, H. M., 1988, *J. Am. Chem. Soc.* **110**:6384.

Horrocks, W. D., and Greenberg, E. S., 1973, *Biochim. Biophys. Acta* **322**:38.

Horrocks, W. D., and Greenberg, E. S., 1974, *Mol. Phys.* **27**:993.

Horrocks, W. D., and Hall, D. DeW., 1971, *Inorg. Chem.* **10**:2368.

Hutchison, C. A., 1967, in *The Triplet State* (A. B. Zahlan, ed.), p. 63, Cambridge University Press, Cambridge.

Icli, S., and Kreilick, R. W., 1971, *J. Phys. Chem.* **75**:3462.

Jenkins, B. G., and Lauffer, R. B., 1988, *J. Magn. Reson.* **80**:328.

Johnson, E., and Everett, G. W., 1972, *J. Am. Chem. Soc.* **94**:1419.

Johnson, M. E., Fung, L. W. M., and Ho, C., 1977, *J. Am. Chem. Soc.* **99**:1245.

Johnson, R. D., Ramaprasad, S., and LaMar, G. N., 1983, *J. Am. Chem. Soc.* **105**:7205.

Kaplan, M., Bolton, J. R., and Fraenkel, G. K., 1965, *J. Chem. Phys.* **42**:955.

Karplus, M., and Fraenkel, G. K., 1961, *J. Chem. Phys.* **35**:1312.

Kashiwagi, H., and Obara, S., 1981, *Internat. J. Quantum Chem.* **20**:843.

Kay, L. E., Clore, G. M., Bax, A., and Gronenborn, A. M., 1990, *Science* **249**:411.

Khalifah, R. G., Rogers, J. I., Harmon, P., Morely, P. J., and Carroll, S. B., 1984, *Biochemistry* **23**:3129.

Kobayashi, H., and Yanagawa, Y., 1972, *Bull. Chem. Soc. Japan* **45**:450.

Kowalewski, J., Laaksonen, A., Nordenskiöld, L., and Blomberg, M., 1981, *J. Chem. Phys.* **74**:2927.

Kurland, R. J., and McGarvey, B. R., 1970, *J. Magn. Reson.* **2**:286.

Kushnir, T., and Navon, G., 1984, *J. Magn. Reson.* **56**:373.

La Mar, G. N., 1973, in *NMR of Paramagnetic Molecules* (G. N. LaMar, W. De W. Horrocks, and R. H. Holm, eds.), Chap. 3, Academic Press, New York.

La Mar, G. N., and Walker, F. A., 1973, *J. Am. Chem. Soc.* **95**:1728.

La Mar, G. N., Eaton, G. R., Holm, R. H., and Walker, F. A., 1973, *J. Am. Chem. Soc.* **95**:63.

La Mar, G. N., Budd, D. L., Smith, K. M., and Langry, K. C., 1980, *J. Am. Chem. Soc.* **102**:1822.

La Mar, G. N., Jue, T., Nagai, K., Smith, K. M., Yamamoto, Y., Kauten, R. J., Thanabal, V., Langry, K. C., Pandey, R. K., and Leung, H. K., 1988, *Biochim. Biophys. Acta* **952**:131.

Lang, S., Spartalian, K., Reed, C. A., and Collman, J. P., 1978, *J. Chem. Phys.* **69**:5424.

Led, J. J., 1985, *J. Am. Chem. Soc.* **107**:6755.

Li, N., Su, Z., Coppens, P., and Landrum, J., 1990, *J. Am. Chem. Soc.* **112**:7294.

Lohman, J. A. B., and McLean, C., 1981, *J. Magn. Reson.* **42**:5.

Longuet-Higgins, H. C., Rector, C. W., and Platt, J. R., 1950, *J. Chem. Phys.* **18**:1174.

Lukat, G. S., and Goff, H. M., 1990, *Biochim. Biophys. Acta* **1037**:351.

Maltempo, M. M., 1974, *J. Chem. Phys.* **61**:2540.

Mazumdar, S., Medhi, O. K., and Mitra, S., 1990, *J. Chem. Soc., Dalton Trans.* **1990**:1057.

McConnell, H. M., 1958, *J. Chem. Phys.* **28**:1188.

McConnell, H. M., and Chesnut, D. B., 1958, *J. Chem. Phys.* **28**:107.

McConnell, H. M., and Strathdee, J., 1959, *Mol. Phys.* **2**:129.

McConnell, H. M., Heller, C., Cole, T., and Fessenden, R. W., 1960, *J. Am. Chem. Soc.* **82**:766.

McGarvey, B. R., 1984, *Can. J. Chem.* **62**:1349.

McGarvey, B. R., 1988, *Inorg. Chem.* **27**:4691.

McGarvey, B. R., and Kurland, R. J., 1973, in *NMR of Paramagnetic Molecules* (G. N. LaMar, W. De W. Horrocks, and R. H. Holm, eds.), Chap. 14, Academic Press, New York.

McGarvey, B. R., and Nagy, S., 1987, *Inorg.Chem.* **26**:4198.

Mehdi, O. K., Mazumdar, S., and Mitra, S., 1989, *Inorg. Chem.* **28**:3243.

Mispelter, J., 1980, unpublished.

Mispelter, J., 1981, unpublished.

Mispelter, J., Grivet, J. Ph., and Lhoste, J. M., 1971a, *Mol. Phys.* **21**:999.

Mispelter, J., Grivet, J. Ph., and Lhoste, J. M., 1971b, *Mol. Phys.* **21**:1015.

Mispelter, J., Momenteau, M., and Lhoste, J. M., 1977, *Mol. Phys.* **33**:1715.

Mispelter, J., Momenteau, M., and Lhoste, J. M., 1978, *Chem. Phys. Lett.* **57**:405.

Mispelter, J., Momenteau, M., and Lhoste, J. M., 1980, *J. Chem. Phys.* **72**:1003.

Mispelter, J., Momenteau, M., and Lhoste, J. M., 1981, *J. Chem. Soc., Dalton Trans.* **1981**:1729.

Momenteau, M., 1986, *Pure Appl. Chem.* **58**:1493.

Momenteau, M., Loock, B., Mispelter, J., and Bisagni, E., 1979, *Nouv. J. Chim.* **3**:77.

Morishima, I., and Inubushi, T., 1978, *J. Am. Chem. Soc.* **100**:3568.

Morishima, I., Shiro, Y., and Wakino, T., 1985, *J. Am. Chem. Soc.* **107**:1063.

Morton, J. R., and Preston, K. F., 1978, *J. Magn. Reson.* **30**:577.

Nanthakumar, A., and Goff, H. M., 1991, *Inorg. Chem.* **30**:4460.

Obara, S., and Kashiwagi, H., 1982, *J. Chem. Phys.* **77**:3155.

Oldfield, E., Lee, H. C., Coretsopoulos, C., Adebodun, F., Park, K. D., Yang, P., Chung, J., and Phillips, B., 1991, *J. Am. Chem. Soc.* **113**:8680.

Pauling, L., 1964, *Nature (London)* **203**:182.

Peyton, D. H., 1991, *Biochem. Biophys. Res. Commun.* **175**:515.

Quaegebeur, J. P., Chachaty, C., and Yasukawa, T., 1979, *Mol. Phys.* **37**:409.

Reed, C. A., Mashiko, T., Bentley, S. P., Kastner, M. E., Scheidt, W. R., Spartalian, K., and Lang, G., 1979, *J. Am. Chem. Soc.* **101**:2948.

Reuben, J., and Fiat, D., 1969, *J. Am. Chem. Soc.* **91**:1242.

Rohmer, M. M., 1985, *Chem. Phys. Lett.* **116**:44.

Ronfard-Haret, J. C., and Chachaty, C., 1978, *J. Phys. Chem.* **82**:1541.

Sankar, S. S., La Mar, G. N., Smith, K. M., and Fujinari, E. M., 1987, *Biochim. Biophys. Acta* **912**:220.

Satterlee, J. D., and Erman, J. E., 1991, *Biochemistry* **30**:4398.

Satterlee, J. D., Erman, J. E., La Mar, G. N., Smith, K. M., and Langry, K. C., 1983, *J. Am. Chem. Soc.* **105**:2099.

Satterlee, J. D., Russell, D. J., and Erman, J. E., 1991, *Biochemistry* **30**:9072.

Shirazi, A., Leum, E., and Goff, H. M., 1983, *Inorg. Chem.* **22**:360.

Simonneaux, G., Hindre, F., and Le Plouzennec, M., 1989, *Inorg. Chem.* **28**:823.

Sinclair, J., and Kivelson, D., 1968, *J. Am. Chem. Soc.* **90**:5074.

Skjeldal, L., Westler, W. M., Oh, B. H., Krezel, A. M., Holden, H. M., Jacobson, B. L., Rayment, I., and Markley, J. L., 1991, *Biochemistry* **30**:7363.

Slichter, C. P., 1963, *Principles of Magnetic Resonance*, Harper and Row, New York, p. 127.

Solomon, I., 1955, *Phys. Rev.* **99**:559.

Solomon, I., and Bloembergen, N., 1956, *J. Chem. Phys.* **25**:261.

Sontum, S. F., Case, D. A., and Karplus, M., 1983, *J. Chem. Phys.* **79**:2881.

Spartalian, K., Lang, G., and Reed, C. A., 1979, *J. Chem. Phys.* **71**:1832.

Strauss, S. H., and Pawlik, M. J., 1986, *Inorg. Chem.* **25**:1921.

Strauss, S. H., Silver, M. E., Long, K. M., Thompson, R. G., Hudgens, R. A., Spartalian, K., and Ibers, J. A., 1985, *J. Am. Chem. Soc.* **107**:4207.

Strauss, S. H., Long, K. M., Magerstädt, M., and Gansow, O. A., 1987, *Inorg. Chem.* **26**:1185.

Strom, E. T., Underwood, G. R., and Jurkowitz, D., 1972, *Mol. Phys.* **24**:901.

Swift, 1973, in *NMR of Paramagnetic Molecules* (G. N. LaMar, W. De W. Horrocks, and R. H. Holm, eds.), Chap. 2, Academic Press, New York.

Toney, G. E., Gold, A., Savrin, J., TerHaar, L. W., Sangaiah, R., and Hatfield, W. E., 1984, *Inorg. Chem.* **23**: 4350.

Van Broekhoven, Hendriks, B. M. P., and DeBoer, E., 1971, *J. Chem. Phys.* **54**:1988.
Vasavada, K. V., and Nageswara Rao, B. D., 1989, *J. Magn. Reson.* **81**:275.
Vega, A. J., and Fiat, D., 1972, *Pure and Appl. Chem.* **32**:307.
Vega, A. J., and Fiat, D., 1976, *Mol. Phys.* **31**:347.
Von Goldammer, E., 1979, *Z. Naturforsch.* **34C**:1106.
Von Goldammer, E., and Zorn, H., 1976, *Mol. Phys.* **32**:1423.
Von Goldammer, E., Zorn, H., and Daniels, A., 1976, *J. Magn. Reson.* **23**:199.
Walker, F. A., and La Mar, G. N., 1973, *Ann. Acad. Sci.* **206**:328.
Wüthrich, K., 1986, *NMR of Proteins and Nucleic Acides*, Wiley, New York.
Wüthrich, K., and Baumann, R., 1973, *Helv. Chim. Acta* **56**:585.
Wüthrich, K., and Baumann, R., 1974, *Helv. Chim. Acta* **57**:336.
Yamamoto, Y., 1987, *FEBS Lett.* **222**:115.
Yamamoto, Y., Nanai, N., Inoue, Y., and Chûjô, R., 1988, *Biochem. Biophys. Res. Commun.* **151**:262.
Yamamoto, Y., Nanai, N., Inoue, Y., and Chûjô, R., 1989a, *Bull. Chem. Soc. Japan.* **62**:1771.
Yamamoto, Y., Nanai, N., Inoue, Y., and Chûjô, R., 1989b, *J. Chem. Soc., Chem. Commun.* **1989**:1419.
Yamamoto, Y., Nanai, N., and Chûjô, R., 1990a, *J. Chem. Soc., Chem. Commun.* **1990**:1556.
Yamamoto, Y., Nanai, N., Chûjô, R., and Suzuki, T., 1990b, *FEBS Lett.* **264**:113.
Yu, C., Unger, S. W., and La Mar, G. N., 1986, *J. Magn. Reson.* **67**:346.

7

NMR of Polymetallic Systems in Proteins

Claudio Luchinat and Stefano Ciurli

1. OCCURRENCE AND ROLE OF POLYMETALLIC SYSTEMS IN BIOLOGICAL MOLECULES

1.1. Introduction

The importance of metal ions in the chemistry of biological processes is nowadays widely recognized, and the field of bioinorganic chemistry is developing rapidly. The discovery of an increasing number of metalloproteins that have active sites consisting of more than one metal center has sparked interest in this special field, because the magnetic, spectroscopic, and chemical properties of these metal clusters are markedly different from those observed for single-metal sites.

One fascinating aspect of these systems directly connected with NMR is the intricacy of their electronic structures. Magnetic coupling provides new energy levels for the systems, which result in peculiar alterations of the electron–nucleus coupling. Dramatic changes in the electron relaxation times may also occur, either because a slowly relaxing metal ion can be relaxed by a fast-relaxing coupled ion or because (in systems with more than two metal ions) completely new relaxation pathways may become operative. All these properties, which are characteristic of exchange-coupled systems, may be both elucidated and exploited by NMR to reach a better understanding of the relationship between structure and biological function. This type of analysis takes great advantage of the magnetic properties of these clusters, which determine an extreme

Claudio Luchinat and Stefano Ciurli • Institute of Agricultural Chemistry, University of Bologna, 40127 Bologna, Italy.

Biological Magnetic Resonance, Volume 12: NMR of Paramagnetic Molecules, edited by Lawrence J. Berliner and Jacques Reuben. Plenum Press, New York, 1993.

sensitivity of the isotropic shift to their electronic structure, which is in turn very sensitive to small structural variations of the protein backbone around these systems.

In this introductory section the biological systems containing two or more metal ions are reviewed. Many of these systems are paramagnetic, while some are diamagnetic. In all cases, metal substitution allows the introduction of paramagnetic probes different from the natural ones, thereby multiplying the possibilities of investigation. Only a few of these systems have so far been thoroughly investigated by NMR. Nevertheless, we thought that a brief presentation of the large variety of systems available might stimulate more researchers to enter the field. After this introductory section, a discussion is given of the theory underlying the NMR of magnetically coupled systems. Finally, selected examples are given. In these last sections we hope to show the reader the potentiality of modern NMR spectroscopy, especially considering the powerful 2D NMR techniques recently applied to large paramagnetic molecules, and the uniqueness of the NMR approach in the elucidation of the electronic properties of these systems.

1.2. Iron Proteins

An important class of iron proteins contain iron–sulfur clusters. These proteins are found in bacteria, fungi, algae, plants, and animals (Thomson, 1985). Their main role is biological electron transfer (Thauer and Schoenheit, 1982) but, as we will see, in an increasing number of cases they have also been shown to perform catalysis of redox reactions such as nitrogen, hydrogen, sulfite, and nitrite reductions, and even a direct involvement in substrate binding and hydrolysis catalysis has been demonstrated. The structurally known coordination units associated with the clusters contained in these proteins are shown in Fig. 1.

1.2.1. 2Fe–2S Cores

The two-Fe site of spinach ferredoxin (Fd), Fig. 1A, was established spectroscopically in 1966 (Britzinger et al., 1966), whereas the X-ray structure from the blue-green algae *Spirulina platensis* was reported in 1980 (Fukuyama et al., 1980), followed, shortly after, by the structure of the same cluster from *Aphanothece sacrum* (Tsukihara et al., 1981). The [2Fe–2S] core is bound to the protein backbone by four cysteine residues, providing a distorted tetrahedral geometry for the Fe ions.

These proteins were originally named "plant ferredoxins," but they have also been found recently in bacteria and animals. They are very important in photosynthesis, where they carry out electron transport in the reduction path from photosystem I to $NADP^+$. They are also known to transfer electrons between cytochrome b and cytochrome c_1 in mitochondrial respiration (Sykes, 1991). The oxidized protein bears two high-spin Fe^{3+} ions, whereas the reduced protein ($E'_0 = -200$–450 mV) contains one Fe^{3+} and one Fe^{2+} ion (Palmer et al., 1971).

Another type of protein containing the [2Fe–2S] core is the so-called Rieske-type ferredoxins (Rieske et al., 1964), membrane-bound redox centers found in mitochondria, whose main characteristics are atypical spectroscopic properties and a much higher redox potential ($E'_0 = +150$–$+350$ mV) (Fee et al., 1984), as a consequence of the substitution of two cysteine residues with two hystidines bound to (presumably) the same

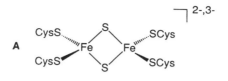

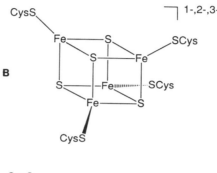

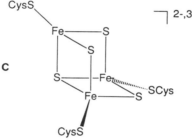

Figure 1. Schemes of the structurally known coordination units in iron–sulfur clusters in proteins. (A) 2Fe–2S clusters; (B) 4Fe–4S clusters; (C) 3Fe–4S clusters.

iron ion (Cline *et al.*, 1985), a hypothesis supported by recent ENDOR studies (Gurbiel *et al.*, 1989).

1.2.2. 4Fe–4S Cores

Proteins containing the $[Fe_4S_4]$ cluster, shown in Fig. 1B, are among the most widespread metalloprotein electron-carriers found in nature (Thomson, 1985). They occur at three different core oxidation levels, $[Fe_4S_4]^{3+}$, $[Fe_4S_4]^{2+}$, and $[Fe_4S_4]^{1+}$. Formally, the $[Fe_4S_4]^{3+}$ cluster contains three Fe^{3+} and one Fe^{2+} ion, the $[Fe_4S_4]^{2+}$ cluster contains two Fe^{3+} and two Fe^{2+} ions, and the $[Fe_4S_4]^{1+}$ cluster contains one Fe^{3+} and three Fe^{2+} ions. *In vivo*, only one redox couple is active, and the $1+/2+$ couple found in ferredoxins (Fd) is the most common. The ferredoxins have a range of redox potential ($E_0' = -250$---650 mV) (Yoch and Carithers, 1979; Armstrong et al., 1988) much lower than that observed for a small class of [Fe_4S_4] cluster–containing proteins, the so-called high-potential iron–sulfur proteins (HiPIP), in which the $2+/3+$ couple is employed ($E_0' = +50$--$+450$ mV) (Meyer et al., 1983). This large variation in redox potential must derive from either a difference in the structure of the cluster core or a different protein environment around the cluster. Efforts to determine the X-ray structure of a number of high- and low-potential ferredoxins have been directed to the solution of this problem. The solid-state structures of the four-Fe site of HiPIP from *Chromatium vinosum* (Carter et al., 1974; Carter, 1977) and *Ectothiorhodospira halophila I* (Breiter et al., 1991) have been determined and compared with the four-Fe site of ferredoxins from *Peptococcus aerogenes* (Adman et al., 1973, 1976), *Azotobacter vinelandii* (Stout et al., 1988; Stout, 1989), and *Bacillus thermoproteolyticus* (Fukuyama et al., 1988, 1989). The almost identical structural features of the cluster cores in all these cases lead to the conclusion that the cluster environment causes the potentials to be so diverse. The fact that the cluster is more buried in the case of the high-potential ferredoxins (Orme-Johnson et. al., 1983), that the number of hydrogen bonds between the protein backbone and the sulfide ions of the cluster is higher in the low-potential ferredoxins (Sheridan et al., 1981), and that the environment of HiPIP is more hydrophobic than that of ferredoxins (Backes et al., 1991) have all been considered to be responsible for the potential shift from low- to high-potential 4Fe–4S proteins. It is very likely that a combination of all these factors is actually what makes the difference.

The amino acid sequences of more than 30 [Fe_4S_4] ferredoxins reveal that in almost all of them four cysteinyl residues are bound to the four Fe atoms (Bruschi and Guerlesquin, 1988) but, in five instances, evidence exists for unconventional ligation at one iron subsite. In ferredoxins from *Sulfolobus acidocaldarius* (Minami et al., 1983), *Thermoplasma acidophilum* (Wakabayashi et al., 1983), *Desulfovibrio africanus* (Bovier-Lapierre et al., 1987), *Desulfovibrio vulgaris* (Okawara et al., 1988), and *Pyrococcus furiosus* (Aono et al., 1989; Conover et al., 1990a), a Cys–Asp replacement occurs. In the case of *P. furiosus* this replacement allows the elimination of the iron ion bound to the carboxylate ligand to give a three-Fe core (Conover et al., 1990b).

1.2.3.   3Fe–4S Cores

Trinuclear clusters have been detected in more than twenty proteins as well as a number of enzymes (Beinert and Thomson, 1983). The structure of the voided-cubane three-Fe site shown in Fig. 1C has been established in *A. vinelandii* Fd I (Stout et al., 1988; Stout, 1989), *Desulfovibrio gigas* Fd II (Kissinger et al., 1988, 1989), and pig heart aconitase (Robbins and Stout, 1989). The main conclusion drawn from these structural studies is that the macromolecular backbone supports in all cases a voided-cubane [Fe_3S_4] fragment in which the metric features of the core are not significantly different from analogous parameters found in other protein-bound [Fe_4S_4] clusters (Kissinger et al., 1988).

In most cases, the [Fe_4S_4] cluster–containing protein from which the [Fe_3S_4] core is obtained upon aerobic isolation of the proteins or by oxidation with ferricyanide is the agent of electron transfer, but the same core conversion process has also been carried out in enzymes such as aconitase, beef heart succinate–ubiquinone oxidoreductase,

Escherichia coli nitrate reductase, *E. coli* fumarate reductase, and succinate dehydrogenase (Holm *et al.*, 1990).

In the case of aconitase, an enzyme that converts citrate to isocitrate via the intermediate *cis*-aconitate in the Krebs cycle, the three-Fe protein represents the inactive form of the enzyme, which can be rendered active by reconstitution of the four-Fe form (Emptage, 1988). The crystal structures of the active and inactive enzymes have shown the peculiarity of this enzyme, that is, the absence of a fourth cysteinate ligand to the iron ion that can be reversibly inserted. Instead, a molecule of water or a hydroxyl ion has been detected (Robbins and Stout, 1989).

1.2.4. 6Fe–6S Cores

A fifth type of cluster core, the $[Fe_6S_6]$ prismane-like cluster, has been shown to be stable in solution when bound to biologically relevant ligands (Kanatzidis *et al.*, 1985), and it has been proposed as the active site of a hydrogenase and in a protein, whose role is still unknown, both isolated from *Desulfovibrio vulgaris* (Hagen *et al.*, 1989). In the latter case this conclusion is supported by EPR spectra of the protein, which are essentially analogous to those observed for the $[6Fe–6S]^{3+}$ core of synthetic prismane-like clusters (Kanatzidis *et al.*, 1985). Similar EPR spectra have been also found in *D. vulgaris* dissimilatory sulfite reductase (Pierik and Hagen, 1991).

1.2.5. Sulfite and Nitrite Reductases

E. coli sulfite reductase and spinach nitrite reductase are the most thoroughly studied enzymes of this type (Siegel, 1978). They catalyze the six-electron reduction of sulfite to sulfide, and of nitrite to ammonia.

In the case of *E. coli* sulfite reductase, the active site consists of one molecule of an isobacteriochlorin-type heme termed siroheme (Murphy *et al.*, 1973; Murphy and Siegel, 1973; Scott *et al.*, 1978), one $[Fe_4S_4]$ cluster (Siegel *et al.*, 1982), and one single binding site for substrate (Rueger and Siegel, 1976). These three prosthetic groups constitute an electron transport chain, in which electrons travel from the cluster to the heme system and finally to the substrate.

A 3-Å resolution crystal structure of this site has been reported, and a model is shown in Fig. 2 (McRee *et al.*, 1986). In this structure the $[Fe_4S_4]$ cluster is very close to the siroheme group, the distance between the siroheme Fe and the nearest cluster Fe atom being 4.4 Å. The structure reveals the presence of a bridging ligand between the siroheme and the cluster, and a model involving a cysteine sulfur that binds both the siroheme Fe and a cluster Fe has been proposed. The siroheme would thus have an accessible binding site on the opposite side with respect to the cluster. Spectroscopic evidence has been obtained that suggests that the active site of spinach nitrite reductase has the same spatial distribution of the same prosthetic groups (Lancaster *et al.*, 1979; Hirasawa-Soga and Tamura, 1981).

The oxidized form of the enzyme contains a Fe^{3+} siroheme and a $[Fe_4S_4]^{2+}$ cluster core. In the semireduced enzyme, the cluster is in the same oxidation state as before, whereas the siroheme iron is in the 2+ oxidation state. The catalytically active form of the enzyme is the reduced state, which contains a $[Fe_4S_4]^{1+}$ cluster and a siroheme

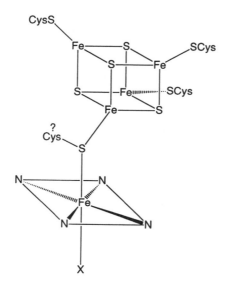

Figure 2. Arrangement of the prosthetic groups in the active site of *E. coli* sulfite reductase.

moiety identical to the semireduced form. In all these states, the cluster and the siroheme are weakly exchange-coupled (Christner *et al.*, 1983a, b,1984).

These features support the hypothesis that the substrate is bound and transformed at the level of the siroheme iron, whereby the cluster would function as storage and direct source of electrons needed for substrate reduction.

1.2.6. Oxo-bridged Dinuclear Sites

Oxo-bridged dinuclear iron proteins are characterized by the presence of the Fe—O—Fe structural unit (Vincent *et al.*, 1990). This type of metal center is found in a large variety of metallobiomolecules, carrying out such diverse functions as oxygen activation and insertion in C—H bonds, oxygen transport, electron transfer, and phosphate ester hydrolysis. Even if the rationale for the wide distribution of this cluster could be found in the high stability of this structural unit in aerobic aqueous systems containing iron in the 3+ oxidation state, the capability of enzymes to utilize it in very diverse reactions is amazing. The structures of the proposed or established clusters present in hemerythrin (Stenkamp *et al.*, 1984), ribonucleotide reductase (Norlund *et al.*, 1990), and purple acid phosphatase (Vincent *et al.*, 1990), are schematically given in Fig. 3a, b, and c, respectively.

Hemerythrin (Hr) is the best-characterized protein containing the Fe₂O cluster. It is found in some classes of marine invertebrates, where it performs an oxygen-carrier function (Wilkins and Harrington, 1983; Wilkins and Wilkins, 1987).

Ribonucleotide reductase catalyzes the transformation of ribonucleotide phosphates to deoxyribonucleotides, thus representing a fundamental step in the process of DNA synthesis (Sjoberg and Graslund, 1983). Three types of ribonucleotide reductases

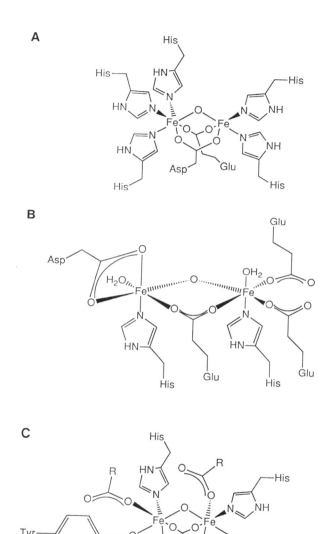

Figure 3. Structures of the proposed or established clusters present in iron-oxo proteins. (A) Hemerythrin; (B) ribonucleotide reductase; (C) purple acid phosphatase.

(RR) have been found: one type contains a cobalt macrocycle (vitamin B_{12}), whereas the other two require a manganese or an iron cofactor (Willing *et al.*, 1988). The iron-containing enzyme has been found in animals, bacteria, and virus-infected mammalian cells.

Purple acid phosphatases catalyze the hydrolysis of phosphate esters under acidic conditions and are characterized by an intense pink or violet coloration (Antanaitis and Aisen, 1983; Doi *et al.*, 1988). They have been isolated from mammals, plants, and microbes. Mammalian purple acid phosphatases readily hydrolyze aryl phosphates, di- and trinucleotides, phosphoproteins, pyrophosphate, and a few other activated phosphates, whereas plant enzymes preferentially hydrolyze nucleotides, including mononucleotides, rather than aryl phosphates. In mammals, purple acid phosphatases are believed to play a role in the degradation of phosphorylated erythrocyte proteins (Vincent *et al.*, 1990).

Methane monooxygenase catalyzes the conversion of methane to methanol in some methanotrophic bacteria (Dalton, 1980). Two types of methane monooxygenases exist, a membrane-bound enzyme containing copper (Tonge *et al.*, 1977) and a soluble enzyme containing an Fe—O—Fe dinuclear cluster (Vincent *et al.*, 1990). [*Note added in proof*: Recent studies on methane monooxygenase are consistent with the presence of an alkoxo, hydroxo, or monodentate carboxylate bridge, instead of an oxo bridge (De Witt *et al.*, 1991).]

Finally, *Desulfovibrio vulgaris* rubrerythrin is a protein of unknown function containing two rubredoxin-like FeS_4 centers and one dinuclear site similar to that in hemerythrin (Que and True, 1990). The main properties of all iron-oxo proteins have recently been reviewed (Que and True, 1990; Vincent *et al.*, 1990) and will not be further discussed here.

1.2.7. Ferritin

Ferritin is a protein found in plants, animals, and some bacteria; its main function is the storage of iron (Theil, 1988; Brock, 1985; Aisen and Listowsky, 1980).

In mammals, ferritin is constituted by a spherical protein coat of 24 subunits of about 20 kDa, inside which up to 4500 ferric iron ions can be arranged in a structure characterized by oxo- and phosphate-bridged iron clusters. In the early stages of iron-core formation, almost all the iron atoms are interacting with the protein coat and oxo- and carboxylate-bridged iron dimers and trimers have been detected by EPR (Chasteen *et al.*, 1985), EXAFS (Yang *et al.*, 1987), and Mössbauer (Yang *et al.*, 1987) studies. When the inner core is well formed, its order depends upon the type of organism and the amount of phosphate (Theil, 1988).

Even if the iron-storage role of ferritin is widely recognized, the mechanism by which iron enters the core or is released from the core *in vivo* is poorly understood.

1.3. Copper Proteins

Binuclear copper systems, consisting of pairs of Cu(II) ions lying in close proximity, are utilized by biological systems for either oxygen transport or substrate hydroxylation, that is, for functions that are very analogous to those displayed by dinuclear iron centers, thus showing the close link between the biological activity of Fe and Cu.

1.3.1. Hemocyanin (Hc)

This protein, containing a coupled binuclear-copper active site, is found in arthropods and molluscs (Van Holde and Miller, 1982). In arthropods, Hc has a dioxygen carrier role, whereas in molluscs, in addition to being a respiratory protein, it exhibits catalase activity, that is, it catalyzes the disproportionation of hydrogen peroxide to water and dioxygen (Gaykema *et al.*, 1984; Linzen *et al.*, 1985; Gaykema *et al.*, 1985; Ellerton *et al.*, 1983).

Hemocyanin exists in the oxy, deoxy, and *met* forms. Resonance Raman spectroscopy has shown that in oxyHc the molecule of bound oxygen has actually been reduced to peroxide, thus putting the copper ions into the 2+ oxidation state (Freedman *et al.*, 1976; Eichman *et al.*, 1978). By displacement of peroxide from oxyHc, the *met* form, containing two Cu^{2+} ions, is obtained. In the deoxyHc, two Cu^{1+} are present. The crystal structure of the deoxyHc from *Palinurus interruptus* has been determined (Linzen *et al.*, 1985); its oxygen binding site consists of a dinuclear copper center with each of the copper atoms bound to the protein backbone through three histidine imidazole groups (Fig. 4A). The Cu–Cu distance is 3.7 Å, with no evidence for any bridging ligand. To know exactly the mode of dioxygen binding to this dimer, it would be necessary to have the crystal structure of the oxyHc form, which is unfortunately not available. However, the structure of a synthetic analog is known (Fig. 4B), in which the dioxygen molecule is held between the two copper ions in a side-on fashion (Kitajima *et al.*, 1989). This model presents strikingly similar spectral properties to oxyHc, and it is believed to basically reproduce the binding mode in the biological system.

Much less is known about the structure of molluscan hemocyanins, which are enormous molecules, each containing ten or twenty 400-kDa subunits. Each subunit is composed of seven or eight covalently linked functional units bearing one dinuclear copper site each (Van Holde and Miller, 1982).

1.3.2. Tyrosinase

This enzyme is widely distributed in nature, being found in prokaryotic and eukaryotic sources such as bacteria, fungii, plants, insects, and vertebrates (Lerch, 1981; Solomon, 1981; Robb, 1984). It is involved in the biosynthesis of melanins and other polyphenolic compounds, where it catalyzes the *ortho*-hydroxylation of monophenols and the oxydation of *ortho*-diphenols to *ortho*-quinones (Jolley *et al.*, 1972, 1974; Makino and Mason, 1973; Makino *et al.*, 1974).

The enzyme isolated from the mushroom *Agaricus bisporus* contains two pairs of antiferromagnetically coupled Cu(II) ions in its resting form (Makino *et al.*, 1974; Bouchilloux *et al.*, 1963), which can be reduced by two electrons ($E'_0 = +360$ mV) (Makino *et al.*, 1974). Resting tyrosinases from *A. bisporus*, *Neurospora*, and *Streptomyces glaucescens* react with hydrogen peroxide to give an oxygenated form of the enzyme whose spectral properties are very similar to those displayed by oxyHc (Wilcox *et al.*, 1985), thus suggesting the presence of the same active site in all these tyrosinases of such diverse phylogenetic origin (Jolley *et al.*, 1973).

The active site exists in the *met* [Cu(II)—Cu(II)], deoxy [Cu(I)—Cu(I)], and oxy [Cu(II)—O_2^{2-}—Cu(II)] forms, and is the last form which is supposed to hydroxylate the substrate (Makino and Mason, 1973).

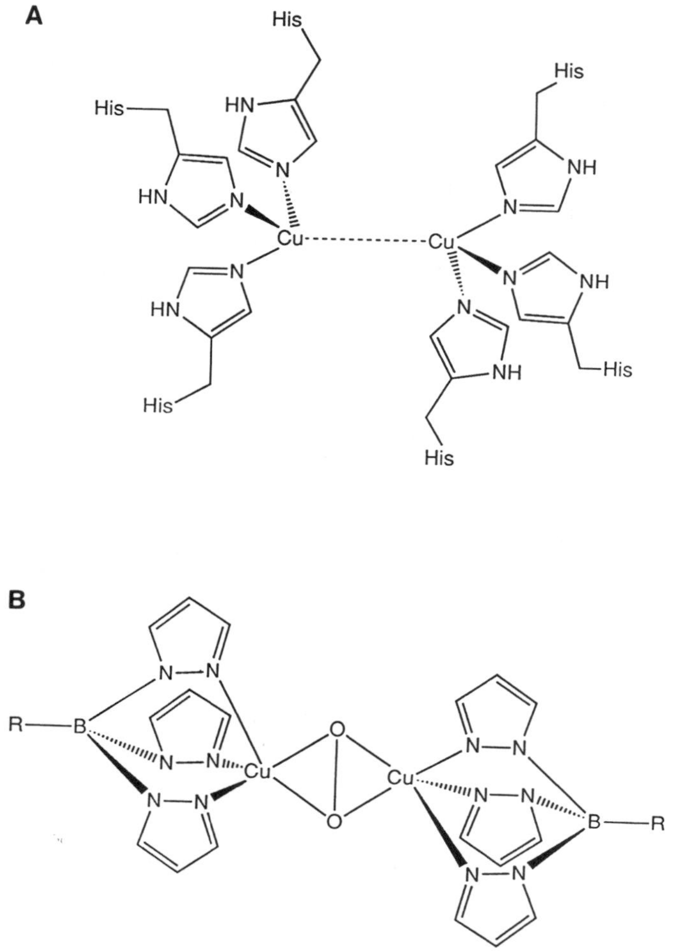

Figure 4. (A) Schematic structure of the active site of hemocyanin from *Palinurus interruptus* (Linzen *et al.*, 1985); (B) Schematic structure of a synthetic model of oxyhemocyanin (Kitajima *et al.*, 1989).

1.3.3. Laccase

Laccases have been isolated from a variety of Asian lac trees and from fungii (Urbach, 1981). They catalyze the oxidation of various substrates, even though they are classified by their ability to oxidize *para*-diphenols to *para*-quinones.

They contain both mononuclear and polynuclear copper centers. Dinuclear centers have been proposed to occur as strongly antiferromagnetically coupled pairs of Cu(II) ions, with $J \simeq 500$ cm^{-1}, probably a consequence of multiple ligand bridging between the two copper ions (Fee *et al.*, 1969).

1.3.4. Ceruloplasmin

Ceruloplasmin (Frieden, 1981) is a multifunctional protein found in the plasma of vertebrates (Holmberg and Laurell, 1948), where it performs a copper transport role (Hsieh and Frieden, 1975) in addition to its involvement in the mobilization of iron into the serum (Frieden, 1979), in the destruction of superoxide radicals (Goldstein *et al.*, 1979), and in the regulation of the serum concentration of biogenic amines (Barrass *et al.*, 1974).

The protein contains six or seven copper atoms and it has been proposed that some of them are present as diamagnetic, antiferromagnetically coupled dinuclear copper(II) centers (Baici *et al.*, 1979; Andréasson and Vänngård, 1970).

1.3.5. Ascorbate Oxidase

Ascorbate oxidase is a plant enzyme which catalyzes the oxidation of *L*-ascorbate to dehydroascorbate, a process which is carried out together with the reduction of dioxygen to water (Frieden, 1981). The X-ray structure has been recently solved (Messerschmidt *et al.*, 1989). The enzyme contains a trinuclear copper center which is the site of oxygen binding and reduction. This center is composed of a copper ion lying 11 Å away from a dinuclear copper site.

1.3.6. Cytochrome *c* Oxidase

Cytochrome *c* oxidase acts as the terminal oxidant in the mitochondrial respiratory chain, and it catalyzes the four-electron reduction of dioxygen to water and the one-electron oxidation of cytochrome *c* from the Fe(II) to the Fe(III) form.

Each functional unit of the enzyme contains two types of heme prosthetic groups (heme *a* and heme a_3) and two or three copper ions (Cu_a and Cu_b) (Wikström *et al.*, 1981). The enzyme consists of up to thirteen subunits (Kadenbach *et al.*, 1983) with heme *a*, heme a_3, and Cu_b residing in subunit I (Winter *et al.*, 1980) and Cu_a in subunit II (Martin *et al.*, 1988). In addition, one zinc ion, one magnesium ion, and additional copper (Cu_x) are intrinsic components of the enzyme (Pan *et al.*, 1991). Heme a_3 and Cu_b are strongly coupled magnetically, producing a unique and still not completely understood system (Thomson *et al.*, 1977).

1.3.7. Superoxide Dismutase

Superoxide dismutase is a metalloenzyme containing copper and zinc, found in almost all eukaryotic cells and in some bacteria, where it catalyzes the disproportionation of superoxide to dioxygen and hydrogen peroxide, thus functioning as a protective agent against the effects of superoxide (Oberley, 1982).

Most studies have been carried out on the bovine enzyme, which has been shown, by X-ray diffraction, to consist of two subunits, each containing an active site in which a Cu^{2+} and a Zn^{2+} ions are found in close proximity, bridged by the imidazolate ring of a histidyl residue (Fig. 5) (Tainer *et al.*, 1983). The physical and catalytic properties of the enzyme isolated from other sources appear to be quite similar. The site of reactivity for superoxide with the oxidized form of the enzyme is the Cu(II) ion (Valentine and Pantoliano, 1981).

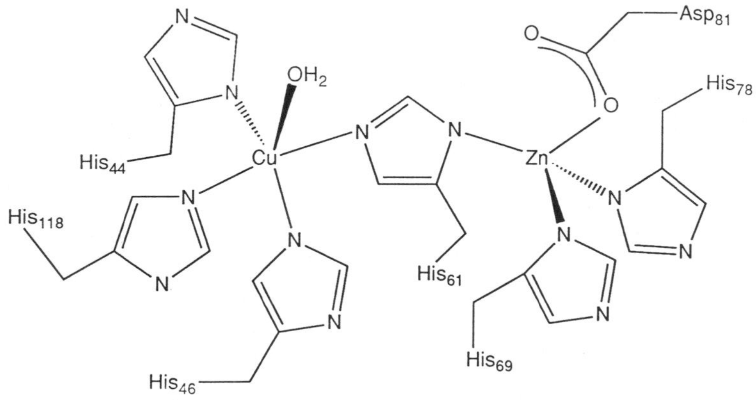

Figure 5. Schematic structure of the active site of Cu_2Zn_2 superoxide dismutase. The labeling is that of the bovine isoenzyme (Tainer *et al.*, 1983).

1.3.8. Nitrous Oxide Reductase

Nitrous oxide reductases are copper-dependent enzymes that catalyze the conversion of N_2O to N_2, a key step of the microbial nitrogen cycle (Kroneck and Zumft, 1991). The enzyme from *Pseudomonas stutzeri* has been most extensively studied (Kroneck *et al.*, 1990) and it has been shown to be composed of two identical subunits containing four copper ions each (Coyle *et al.*, 1985). The EPR spectrum of the reduced, catalytically active form of the enzyme is consistent with a mixed-valence delocalized copper dimer $(Cu^{+1.5}$—$Cu^{+1.5})$ (Kroneck *et al.*, 1988) with the copper ions having cysteine coordination.

1.4. Proteins Containing Other Metals

1.4.1. Photosystem II

Photosystem II is an enzyme that catalyzes the oxidation of water and the reduction of plastoquinone. In addition, the enzyme creates a pH gradient across the thylakoid membrane (Brudvig, 1987, 1988).

Four manganese atoms are functionally associated with photosystem II (Andréasson and Vänngård, 1988), a number consistent with the periodicity of four in the yield of O_2 in a series of flashes, giving rise to five states (S_0 through S_4) characterized by increasing light-induced charge-separation levels (Joliot and Kok, 1975). The S_4 state releases one molecule of dioxygen, forming the S_0 state again. X-ray absorption studies of the S_1 and S_2 states indicate that the Mn ions are probably hexacoordinated and bound to light atoms such as oxygen or nitrogen (Guiles *et al.*, 1987). The EXAFS data are very similar to those of a mixed-valence Mn(III)—Mn(IV) μ-oxo dimer and are consistent with a structure in which two μ-oxo dimers are very close to each other (Guiles *et al.*, 1987). XANES (Goodin *et al.*, 1984) and electronic absorption studies

(Lavergne, 1987) are also consistent with progressive oxidation of the four Mn ions going from S_0 to S_3, which is the state that coordinates water. Both ferro- and antiferromagnetic coupling must be present in the Mn cluster to account for the EPR features of the S_2 state (De Paula et al., 1986), and this information, together with the conclusions drawn from EXAFS (Guiles et al., 1987), has stimulated investigators to propose the mechanism illustrated in Fig. 6 (Brudvig and Crabtree, 1986). According to this hypothesis, photosystem II would consist of an oxo- or hydroxide-bridged Mn_4O_4 cubane-like structure in the S_0, S_1, and S_2 states, with progressive oxidation of the (initially four) Mn^{3+} ions to two Mn^{3+} and two Mn^{4+} ions. In the S_3 state the cubane core would incorporate two O^{2-} or OH^- ions from two water molecules to afford a Mn_4O_6 adamantane–like structure, the further oxidation of which would give rise to the S_4 state that spontaneously loses dioxygen, regenerating the S_0 state.

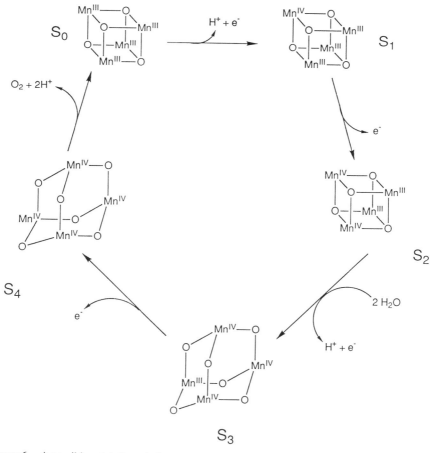

Figure 6. A possible catalytic cycle for water oxidation in photosystem II (Brudvig and Crabtree, 1986).

1.4.2. Catalase

Catalases are enzymes that catalyze the conversion of hydrogen peroxide to water and dioxygen. Usually they contain heme centers (Kono and Fridovich, 1982) but also manganese-containing catalases have been isolated from the aerobic lactic acid bacteria *Thermoleophilum album* (Allgood and Perry, 1986) *Lactobacillus plantarum* (Beyer and Fridovich, 1985), and *Thermus thermophilus* (Baranyn *et al.*, 1986). EPR spectroscopy (Beryer and Fridovich, 1985; Fronko *et al.*, 1988) and crystallographic properties of these enzymes (Que and True, 1990) show the presence of two close-lying $(3.6 \pm 0.3 \text{ Å})$ manganese ions. The enzyme utilizes a $Mn(II)/Mn(II)-Mn(III)/Mn(III)$ catalytic redox couple.

1.4.3. Manganese Ribonucleotide Reductase

A type of ribonucleotide reductase requiring neither cobalt nor iron, but manganese, has been discovered in the bacteria *Brevibacterium ammoniagenes* and *Micrococcus luteus* (Schimpff-Weiland and Follmann, 1981). The similarity of their electronic spectra to those of manganese catalases and synthetic Mn_2O complexes (Willing *et al.*, 1988) has suggested the presence of a dinuclear manganese active site (Sheats *et al.*, 1987).

1.4.4. Metallothioneins

These small proteins play the role of metal transport in many organisms (Vasak and Kägi, 1983). They bind principally Zn, Cd, and Cu, but are also to transport other metals (Vasak, 1986). Mammalian proteins possess twenty conserved cysteines, all of them involved in the coordination of the metal ions. The crystal and solution structure of rat liver metallothionein shows that seven cadmium ions cluster in two different sites, with four ions in one aggregate and three ions in the other (Furey *et al.*, 1986; Frey *et al.*, 1985; Braun *et al.*, 1986; Wagner *et al.*, 1987). In both the three- and the four-metal sites, the cysteine residues act both as terminal and bridging ligands (Fig. 7).

1.4.5. Nitrogenases

Nitrogenase is an enzyme found both in free-living and in symbiotic prokaryotic microorganisms, and *in vivo* it catalyzes the six-electron reduction of dinitrogen to ammonia and the two-electron reduction of hydrogen ions to hydrogen. The current state of understanding of nitrogenase has been summarized in detail (Stiefel *et al.*, 1988). The enzyme consists of two components, the Fe protein, which contains one [4Fe–4S] cluster, and the MoFe protein, which contains two identical, dissociable clusters, named M-centers, of approximate composition $MoFe_{6-7}S_{8-10}$ and considered to be the catalytic sites, and another type of Fe–S clusters, named P-clusters, whose structure is believed to involve [4Fe–4S] cores. A recent crystallographic study of the MoFe protein from *C. pasterianum* has revealed the spatial distribution of the metal clusters with respect to each other (Bolin *et al.*, 1991). The M centers are found to be separated by roughly 70 Å, whereas the distance between either M-center and the P-clusters is 19 Å. These data rule out proposed models for the chemistry of dinitrogen reduction claiming the involvement of both Mo atoms in the binding and reduction of a single substrate molecule.

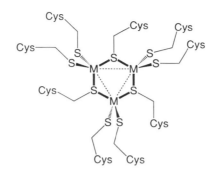

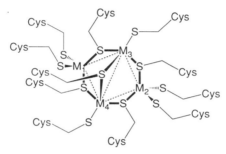

Figure 7. Schematic structures of the three- and four-metal sites in M_7-metallothioneins (Wagner *et al.*, 1987).

An alternative type of nitrogenase from *Azotobacter chroococcum* and *A. vinelandii* has been shown to contain vanadium, and a cofactor of approximate composition VFe_6S_5 has been extracted and partially characterized (Robson *et al.*, 1986; Dilworth *et al.*, 1987). Recently, a third type of nitrogenase containing only Fe and S has been reported (Pan *et al.*, 1989).

1.4.6. Nickel Proteins

Nickel has been considered to be biologically relevant only in the last few years with the recognition of the biological role of this element in four types of enzymes: hydrogenases (Cammack *et al.*, 1988), methyl coenzyme M reductase, ureases, and carbon monoxide dehydrogenase (Hausinger, 1987; Walsh and Orme-Johnson, 1987).

Ureases are nickel-containing enzymes that catalyze the hydrolysis of urea in plants and in bacteria. Microbial ureases are important in human health and in soil nitrogen management (Mobley and Hausinger, 1989). Variable temperature MCD studies have revealed similar electronic and magnetic properties for the nickel centers in plants and bacteria ureases, indicating the presence of Ni(II) octahedrally coordinated to oxygen ligands (Kiick *et al.*, 1990). These studies support the proposal of a binuclear Ni(II) active site for ureases (Clark and Wilcox, 1989) in which the two nickel ions are in equilibrium between ferromagnetic and antiferromagnetic coupling via the protonation/deprotonation process of a bridging ligand.

Nickel is also present in carbon monoxide dehydrogenase, an enzyme that catalyzes the interconversion of CO and CO_2 in several acetogenic and methanogenic bacteria (Hausinger, 1987). The observation of broadening of the EPR spectrum of this enzyme enriched with ^{61}Ni, ^{13}C, and ^{57}Fe indicated that the CO carbon and the enzyme-bound nickel and iron ions are all spin-coupled in a complex which is responsible for the EPR spectrum (Stephens et al., 1989). The hypothesis that this coupled system is a $NiFe_3S_4$ cubane-type cluster has been tested by comparing the X-ray absorption spectra of a synthetic cluster containing this core (Ciurli et al., 1990, 1992) and carbon monoxide dehydrogenase from *Rhodospirillum rubrum* (Tan et al., 1992). The results of these experiments point to a nickel cluster different from that found in the synthetic model but support the hypothesis of a nickel center lying near, and bridged to, an iron–sulfur cluster.

2. PARAMAGNETISM IN MAGNETICALLY COUPLED SYSTEMS

2.1. Exchange Interactions

Magnetic coupling between two paramagnetic centers (e.g., two paramagnetic metal ions) occurs because of bonding interaction, i.e., orbital overlap, and/or because of through-space interaction. In spin Hamiltonian formalism used in this volume (Stevens, 1976), the magnetic interaction is described by a term of the type

$$\mathscr{H} = \mathbf{S}_1 \cdot \mathbf{J}_{12} \cdot \mathbf{S}_2 \tag{1}$$

where $\mathbf{S}_1$ and $\mathbf{S}_2$ are the spin angular momentum operators for the individual metal ions and $\mathbf{J}_{12}$ is a second-rank tensor containing all the information about the coupling between S_1 and S_2. As a result of the magnetic coupling, the system can be described by a new set of spin states labeled according to a new total spin quantum number S. The latter is defined according to the addition rules for angular momentum:

$$|S_1 - S_2| \le S \le S_1 + S_2 \tag{2}$$

The energy ordering and spacing of the various S states depends on the type of interaction. The magnetic properties of the system ultimately depend on the S value of the ground state and on how accessible the excited states are at the temperature of interest. For typical room-temperature NMR experiments, the population of the excited states cannot be usually neglected.

Like any second-rank tensor, $\mathbf{J}_{12}$ in Eq. (1) can be decomposed into the sum of an antisymmetric ($\mathbf{A}_{12}$) and a symmetric ($\mathbf{S}_{12}$) tensor; the first is traceless, the second can be made traceless by further decomposing it into a traceless tensor $\mathbf{D}_{12}$ and a scalar J_{12} equal to one-third of its trace. This decomposition leads to rewriting Eq. (1) as

$$\mathscr{H} = J_{12}\mathbf{S}_1 \cdot \mathbf{S}_2 + \mathbf{d}_{12} \cdot \mathbf{S}_1 \times \mathbf{S}_2 + \mathbf{S}_1 \cdot \mathbf{D}_{12} \cdot \mathbf{S}_2 \tag{3}$$

where the first term describes the isotropic part of the interaction, the second term describes the antisymmetric part of the interaction (written as the dot product of a polar vector $\mathbf{d}_{12}$ with the vector product of $\mathbf{S}_1$ and $\mathbf{S}_2$), and the third term describes the anisotropic part of the interaction (Bencini and Gatteschi, 1990).

Both through-bond and through-space interactions can be decomposed in the same way, giving a total of six conceptually different contributions to the magnetic coupling. However, through bond—or exchange—interactions are largely isotropic, so that the first term in Eq. (3) is usually dominant. On the other hand, through-space interactions are usually well described, except at very short distances, by dipole–dipole interactions which, neglecting second-order effects, are represented by symmetric traceless tensors; as a consequence, for through-space interactions, the third term is usually dominant. When both through-bond and through-space interactions are important, it still often happens that the isotropic part of the interaction is the largest (Bencini and Gatteschi, 1990). Under such conditions the magnetic coupling Hamiltonian reduces to

$$\mathcal{H} = J_{12}\mathbf{S}_1 \cdot \mathbf{S}_2 \tag{4}$$

It should be pointed out that a satisfactory characterization of an exchange-coupled system from, for instance, the EPR point of view always requires consideration of at least the anisotropic part of the interaction (Stevens, 1985). However, from the point of view of nuclei magnetically coupled to one or both exchange-coupled metal ions, the requirements are less stringent and Hamiltonian (4) is always a good approximation (Banci *et al.*, 1991d). On the other hand, at variance with low-temperature EPR (Bencini and Gatteschi, 1990), careful considerations of the excited states is required.

The eigenvalues of Eq. (4) are

$$E(S) = \tfrac{1}{2}J_{12}[S(S+1) - S_1(S_1+1) - S_2(S_2+1)] \tag{5}$$

each S state being $(2S+1)$-fold degenerate. Note that the terms containing the individual spin quantum numbers provide only a constant energy offset. The energy separations between two adjacent states S and $S-1$ are

$$\Delta E = \tfrac{1}{2}J_{12}[S(S+1) - (S-1)(S-1+1)] = SJ_{12} \tag{6}$$

The order of the levels will thus be either of increasing or of decreasing S, depending on the sign of J_{12}. A positive J_{12} means antiferromagnetic coupling; a negative J_{12}, ferromagnetic coupling. Equations (5) and (6) hold for any value of S_1 and S_2 and for both like and unlike spins. Figure 8 shows a few representative situations. It is useful to recall that for antiferromagnetic coupling between like spins the ground state is diamagnetic, but the compound is not diamagnetic unless J_{12} is so large that the contribution from population of the paramagnetic excited states can be neglected. For unlike spins, even the ground state is paramagnetic, independently of the sign and magnitude of J_{12}.

A pictorial representation of the origin of the ferromagnetic and antiferromagnetic contributions can be obtained by taking as an example $S_1 = S_2 = \tfrac{1}{2}$: when exchange coupling is present, the system can be in either a singlet ($S = 0$) or a triplet ($S = 1$) state. By viewing the exchange interaction as a bonding interaction, in the MO language we will have a bonding and an antibonding orbital, resulting from the overlap of two atomic orbitals with one electron each, separated by an energy proportional to the

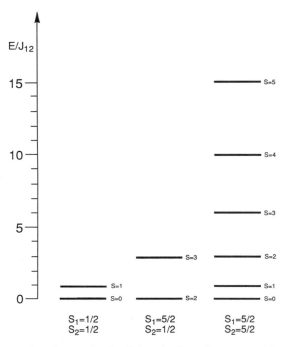

Figure 8. Some examples of energy-level splittings in dimetallic systems. Note the much larger spreading of the energy ladder for large S values.

strength of the bond (Fig. 9). If the bond is strong, the two electrons are paired in the bonding orbital (Fig. 9A); if the bond is so weak that the energy separation is smaller than the electron pairing energy, the electrons are distributed one per orbital, according to Hund's rule (Fig. 9B). So $S = 0$ will be the ground state in the first case, while $S = 1$ will be the ground state in the second case. In many cases of practical interest, bonding will occur not by direct overlap of metal orbitals but rather through a bridging ligand. This type of bonding interaction is usually termed superexchange interaction (Bencini and Gatteschi, 1990; Ginsberg, 1971).

2.2. Double Exchange

This contribution to the description of magnetic interactions between paramagnetic centers is less well known, but its importance has recently been recognized (Papaefthymiou *et al.*, 1987), especially in relation to some of the biological systems discussed here. Double exchange is present only in mixed-valence systems, i.e., in systems constituted by two unlike spins residing on two metal ions of the same kind and in the same—or similar—chemical environment (Anderson and Hasegawa, 1955). The simplest example is a homodimetallic system with the two metal ions in identical chemical environments and with different oxidation states, both ions being paramagnetic; $M^{n+}-M^{(n+1)+}$.

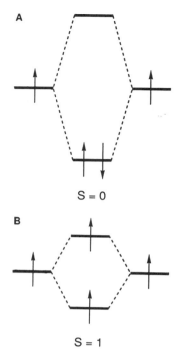

Figure 9. Exchange coupling viewed as a strong (A, antiferromagnetic) or weak (B, ferromagnetic) bonding interaction.

Examples include Ni^+-Ni^{2+}, $Fe^{2+}-Fe^{3+}$, $Cr^{2+}-Cr^{3+}$, etc. By taking Ni^+-Ni^{2+} as an example, we can visualize the $S = \frac{3}{2}$ electronic state of the system (Fig. 10A) as an $S_1 = \frac{1}{2}$ state coupled to an $S_2 = 1$ state. However, the same situation is obtained by reversing the oxidation states, since the two centers are identical. The possibility then exists that the extra electron may actually be resonating between the two centers. Resonance contributes to the stabilization of the system. The term double exchange originates from the originally proposed pathway for electron delocalization through a bridging ligand, involving transfer of the extra electron to the ligand and simultaneous transfer of a ligand electron to the other metal (Zener, 1951). Figure 10B shows the $S = \frac{1}{2}$ state of the system, i.e., with the $M_{S_1} = -\frac{1}{2}$ state. In this case, transfer of the extra electron from metal 2 to metal 1 is forbidden by the "wrong" spin state of the latter. As a consequence, no resonance stabilization is possible. In other words, double exchange tends to stabilize states with largest S. If double exchange is very strong, it may result in an overall ferromagnetic coupling, with the state with largest S lying lowest in energy.

The double exchange term has been recently formally derived in the spin Hamiltonian formalism (Blondin and Girerd, 1990). If we define $|^1S_1{}^1S_2S\rangle = |1\rangle$ and $|^2S_1{}^2S_2S\rangle = |2\rangle$ as the two equivalent eigenstates of the Heisenberg Hamiltonian (4) obtained when the extra electron is on site 1 or 2, respectively, the complete Hamiltonian

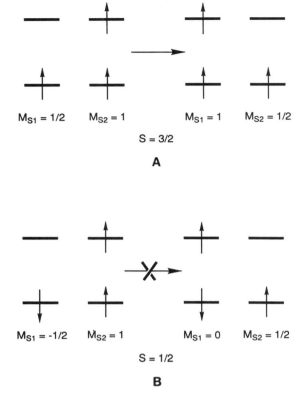

Figure 10. Double exchange is permitted in a high-spin (A) but not in a low-spin (B) configuration of the spin pair.

becomes

$$\mathcal{H} = [J_{12}{}^1\mathbf{S}_1 \cdot {}^1\mathbf{S}_2]\mathbf{O}_1' + [J_{12}{}^2\mathbf{S}_1 \cdot {}^2\mathbf{S}_2]\mathbf{O}_2 + \tfrac{1}{2}B_{12}\mathbf{V}_{12}\mathbf{T}_{12} \tag{7}$$

where $^j\mathbf{S}_i$ represents the spin angular momentum operator S_i when the extra electron is on site j, $\mathbf{O}_j$ is the occupation operator, giving 1 as eigenvalue when applied to $|j\rangle$ and 0 when applied to $|i\rangle$, B_{12} is the double exchange parameter, $\mathbf{T}_{12}$ is the transfer operator that converts $|1\rangle$ into $|2\rangle$ and *vice versa*, and $\mathbf{V}_{12}$ is an operator producing as eigenvalue $(2S + 1)$. The eigenvalues of (7), omitting the energy shift terms, are

$$E(S) = \tfrac{1}{2}[J_{12}S(S + 1) \pm B_{12}(2S + 1)] \tag{8}$$

Figure 11 shows the variation in the energies of the various S states for an $S_1 = 2$, $S_2 = \tfrac{5}{2}$ system in which antiferromagnetic coupling is present (positive J_{12}, cf. Fig. 8) as a function of the B_{12}/J_{12} ratio. It appears that, for $B_{12}/J_{12} > 4.5$, the

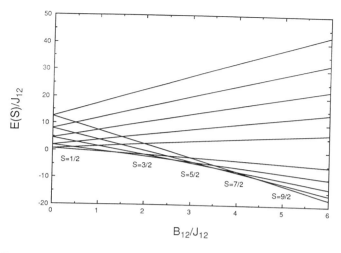

Figure 11. Energy levels in a $S_1 = 2$, $S_2 = \frac{5}{2}$ system in which antiferromagnetic coupling and double exchange are present as a function of B_{12}/J_{12}. The changes in ground state are indicated (Münck *et al.*, 1988).

$S = S_1 + S_2 = \frac{9}{2}$ state becomes the ground state (Münck *et al.*, 1988). It is also worth noting that the spacing of the levels is no longer a simple function of J_{12}.

3. NMR PARAMETERS IN EXCHANGE-COUPLED DIMETALLIC SYSTEMS

3.1. Hyperfine Shift

As already discussed in detail in other chapters of this volume, the hyperfine shift experienced by a nucleus coupled to an isolated spin S_1 is proportional to the expectation value of S_{1z} (McConnell and Chestnut, 1958; Bertini and Luchinat, 1986):

$$\frac{\Delta \nu}{\nu_0} = -\frac{A_1}{\hbar \gamma_N B_0} \langle S_{1z} \rangle \tag{9}$$

where A_1 is the isotropic hyperfine coupling constant between the nucleus and spin S_1 and the other symbols have the usual meaning. For a thermally isolated S_1 multiplet and electron Zeeman splitting very small with respect to kT, $\langle S_{1z} \rangle$ is given by

$$\langle S_{1z} \rangle = -S_1(S_1 + 1) \frac{g_e \mu_B B_0}{3kT} \tag{10}$$

and the hyperfine shift becomes

$$\frac{\Delta \nu}{\nu_0} = \frac{g_e \mu_B}{3\hbar \gamma_N kT} A_1 S_1 (S_1 + 1) \tag{11}$$

When excited states are accessible, the expectation value should be evaluated over all the S_i states, each weighted according to its multiplicity $(2S_i + 1)$ and Boltzmann population. The overall shift can thus be written (Dunham *et al.*, 1971) as

$$\frac{\Delta \nu}{\nu_0} = -\frac{1}{\hbar \gamma B_0} \sum_i A_i \langle S_{iz} \rangle$$

$$= \frac{g_e \mu_B}{3\hbar \gamma_N kT} \sum_i A_i S_i (S_i + 1) \frac{(2S_i + 1) \exp(-E_i/kT)}{\sum_i (2S_i + 1) \exp(-E_i/kT)} \tag{12}$$

Equation (12) is of general validity. It also holds in the case of an exchange-coupled system, in which case the S_i levels are those originated by the coupling between individual S_1 and S_2 ions. As it is, however, it can be of little use, because it requires that one know the value of the isotropic hyperfine coupling constant A_i between the nucleus and each S_i multiplet. On the other hand, what are usually known, or can be guessed from appropriate model systems, are the hyperfine coupling constants A_1 and A_2 between the nucleus and the isolated metal ions. We will show here how the A_i values can be related to the A_1 and A_2 values in noncoupled systems.

For simplicity, and without loss of generality, we can assume that the nucleus of interest interacts only with metal 1 of the pair. If the nucleus interacts with both ions, the overall shift will be given by the sum of the term containing A_1 and of an analogous term containing A_2.

The A_i values in Eq. (12) now represent the hyperfine coupling constants experienced by the nucleus with the ith state of the pair. We have assumed that, in our system, the nucleus actually interacts only with metal 1 with hyperfine coupling constant A_1. Therefore, we are interested in the fractional contribution to this coupling arising from the fractional contribution of metal 1 to each A_i value. It can be shown that each A_i is related to A_1 by the equation (Scaringe *et al.*, 1978)

$$A_i = A_1 C_{i1} \tag{13}$$

where the C_{in} coefficients ($n = 1, 2$ according to which metal ion in the pair interacts with the nucleus) are given by

$$C_{in} = \frac{S_i(S_i + 1) \pm [S_1(S_1 + 1) - S_2(S_2 + 1)]}{2S_i(S_i + 1)} \tag{14}$$

Therefore, Eq. (12) becomes

$$\frac{\Delta \nu}{\nu_0} = \frac{g_e \mu_B}{3\hbar \gamma_N kT} A_1 \sum_i C_{i1} S_i (S_i + 1) \frac{(2S_i + 1) \exp(-E_i/kT)}{\sum_i (2S_i + 1) \exp(-E_i/kT)} \tag{15}$$

where we have achieved the important goal of having only one hyperfine coupling parameter, A_1. Equation (15) is of fundamental importance to the understanding of the many peculiar properties of the NMR of coupled systems (Banci *et al.*, 1990c).

The C_{in} coefficients calculated from Eq. (14) for the most common dimetallic pairs are reported in Table 1. It can be noted that, for the smallest S, $C_{i1} < 0$ (for $S_1 < S_2$). Because of this fact, the contribution to the hyperfine shift of protons sensing metal 1, arising from the state with smallest S, is reversed in sign with respect to an uncoupled situation. If the coupling is antiferromagnetic and strong, i.e., if the state with lowest S lies lowest and the excited states are far apart in energy, then the actual hyperfine shift can have a reversed sign. For weak antiferromagnetic coupling, the overall hyperfine shift may not be reversed at room temperature, because several excited states with positive C_{i1} values are appreciably populated. However, the negative contribution of the ground state increases with decreasing temperature, causing an overall *decrease* in hyperfine shift with decreasing temperature. This behavior is opposite to that predicted by the Curie law (McConnell and Chestnut, 1958) and qualitatively observed in uncoupled systems. This key feature of the NMR of coupled systems can be exploited, as will be shown in Sections 5 and 7–9, to obtain detailed information on the electronic structure of the system. In the high-temperature limit, i.e., when all S states are equally populated, it can easily be shown that Eq. (15) reduces to Eq. (11) for the shift in uncoupled systems (Banci *et al.*, 1991d).

3.2. Relaxation Rates

The detailed equations for nuclear relaxation in paramagnetic systems can be found in Chapter 2 of this volume. For the coupling of the nucleus with a spin S_1 they have the general form

$$T_{1,2}^{-1} = KA_1^2 S_1(S_1 + 1)f_{1,2}(\omega, \tau_c) \tag{16}$$

where the subscript 1, 2 indicates the longitudinal or transverse relaxation rate; K is a constant; A_1 is the hyperfine coupling constant, which depends on the type of relaxation contribution (dipolar or contact); and f is a sum of spectral density terms depending on the nuclear and electron Larmor frequencies and on the correlation time τ_c. For dipolar relaxation τ_c is dominated by the shortest among the rotational correlation time τ_r, the electronic relaxation time τ_s, and the chemical exchange time τ_M; for contact relaxation, by the shortest of τ_s and τ_M (Banci *et al.*, 1991d).

By following a line of reasoning analogous to that used for the hyperfine shift, it can be shown that the relaxation equations need to be modified by summing up contributions from all the S states of the pair, weighted according to their population, and with the inclusion of the *square* of the coefficients reported in Table 1 (Banci *et al.*, 1990c, 1991a, d):

$$T_{1,2}^{-1} = KA_1^2 \sum_i C_{i1}^2 S_i(S_i + 1) \frac{(2S_i + 1) \exp(-E_i/kT)}{\sum_i (2S_i + 1) \exp(-E_i/kT)} f_{1,2}(\omega, \tau_{ci}) \tag{17}$$

TABLE 1
C_{i12} Coefficients Relating the Hyperfine Coupling of an S_1 or S_2 Spin System in a Monomer to That in a Coupled System, for Each S_i Spin Level

| | S_2 | | | | | | | | | | | | | | | | | | | | |
| | $\frac{1}{2}$ | | | 1 | | | $\frac{3}{2}$ | | | 2 | | | $\frac{5}{2}$ | | | 3 | | | $\frac{7}{2}$ | | |
S_1	S	C_1	C_2	S	C_1	C_2	S	C_1	C_2	S	C_1	C_2	S	C_1	C_2	S	C_1	C_2	S	C_1	C_2
$\frac{1}{2}$	1	$\frac{1}{2}$	$\frac{1}{2}$	$\frac{3}{2}$	$\frac{1}{3}$	$\frac{2}{3}$	2	$\frac{1}{4}$	$\frac{3}{4}$	$\frac{5}{2}$	$\frac{1}{5}$	$\frac{4}{5}$	3	$\frac{1}{6}$	$\frac{5}{6}$	$\frac{7}{2}$	$\frac{1}{7}$	$\frac{6}{7}$	4	$\frac{1}{8}$	$\frac{7}{8}$
	0	—	—	$\frac{1}{2}$	$-\frac{1}{3}$	$\frac{4}{3}$	1	$-\frac{1}{4}$	$\frac{5}{4}$	$\frac{3}{2}$	$-\frac{1}{5}$	$\frac{6}{5}$	2	$-\frac{1}{6}$	$\frac{7}{6}$	$\frac{5}{2}$	$-\frac{1}{7}$	$\frac{8}{7}$	3	$-\frac{1}{8}$	$\frac{9}{8}$
1				2	$\frac{1}{2}$	$\frac{1}{2}$	$\frac{5}{2}$	$\frac{2}{5}$	$\frac{3}{5}$	3	$\frac{1}{3}$	$\frac{2}{3}$	$\frac{7}{2}$	$\frac{2}{7}$	$\frac{5}{7}$	4	$\frac{1}{4}$	$\frac{3}{4}$	$\frac{9}{2}$	$\frac{2}{9}$	$\frac{7}{9}$
				1	$\frac{1}{2}$	$\frac{1}{2}$	$\frac{3}{2}$	$\frac{4}{15}$	$\frac{11}{15}$	2	$\frac{1}{6}$	$\frac{5}{6}$	$\frac{5}{2}$	$\frac{4}{35}$	$\frac{31}{35}$	3	$\frac{1}{12}$	$\frac{11}{12}$	$\frac{7}{2}$	$\frac{4}{63}$	$\frac{59}{63}$
				0	—	—	$\frac{1}{2}$	$-\frac{2}{3}$	$\frac{5}{3}$	1	$-\frac{1}{2}$	$\frac{3}{2}$	$\frac{3}{2}$	$-\frac{2}{5}$	$\frac{7}{5}$	2	$-\frac{1}{3}$	$\frac{4}{3}$	$\frac{5}{2}$	$-\frac{2}{7}$	$\frac{9}{7}$
$\frac{3}{2}$							3	$\frac{1}{2}$	$\frac{1}{2}$	$\frac{7}{2}$	$\frac{3}{7}$	$\frac{4}{7}$	4	$\frac{3}{8}$	$\frac{5}{8}$	$\frac{9}{2}$	$\frac{1}{3}$	$\frac{2}{3}$	5	$\frac{3}{10}$	$\frac{7}{10}$
							2	$\frac{1}{2}$	$\frac{1}{2}$	$\frac{5}{2}$	$\frac{13}{35}$	$\frac{22}{35}$	3	$\frac{7}{24}$	$\frac{17}{24}$	$\frac{7}{2}$	$\frac{5}{21}$	$\frac{16}{21}$	4	$\frac{1}{5}$	$\frac{4}{5}$
							1	$\frac{1}{2}$	$\frac{1}{2}$	$\frac{3}{2}$	$\frac{1}{5}$	$\frac{4}{5}$	2	$\frac{1}{12}$	$\frac{11}{12}$	$\frac{5}{2}$	$\frac{1}{35}$	$\frac{34}{35}$	3	0	1
							0	—	—	$\frac{1}{2}$	-1	2	1	$-\frac{3}{4}$	$\frac{7}{4}$	$\frac{3}{2}$	$-\frac{3}{5}$	$\frac{8}{5}$	2	$-\frac{1}{2}$	$\frac{3}{2}$
2										4	$\frac{1}{2}$	$\frac{1}{2}$	$\frac{9}{2}$	$\frac{4}{9}$	$\frac{5}{9}$	5	$\frac{2}{5}$	$\frac{3}{5}$	$\frac{11}{2}$	$\frac{4}{11}$	$\frac{7}{11}$
										3	$\frac{1}{2}$	$\frac{1}{2}$	$\frac{7}{2}$	$\frac{26}{63}$	$\frac{37}{63}$	4	$\frac{7}{20}$	$\frac{13}{20}$	$\frac{9}{2}$	$\frac{10}{33}$	$\frac{23}{33}$
										2	$\frac{1}{2}$	$\frac{1}{2}$	$\frac{5}{2}$	$\frac{12}{35}$	$\frac{23}{35}$	3	$\frac{1}{4}$	$\frac{3}{4}$	$\frac{7}{2}$	$\frac{4}{21}$	$\frac{17}{21}$
										1	$\frac{1}{2}$	$\frac{1}{2}$	$\frac{3}{2}$	$\frac{2}{15}$	$\frac{13}{15}$	2	0	1	$\frac{5}{2}$	$-\frac{2}{35}$	$\frac{37}{35}$
										0	—	—	$\frac{1}{2}$	$-\frac{4}{3}$	$\frac{7}{3}$	1	-1	2	$\frac{3}{2}$	$-\frac{4}{5}$	$\frac{9}{5}$
$\frac{5}{2}$													5	$\frac{1}{2}$	$\frac{1}{2}$	$\frac{11}{2}$	$\frac{5}{11}$	$\frac{6}{11}$	6	$\frac{5}{12}$	$\frac{7}{12}$
													4	$\frac{1}{2}$	$\frac{1}{2}$	$\frac{9}{2}$	$\frac{43}{99}$	$\frac{56}{99}$	5	$\frac{23}{60}$	$\frac{37}{60}$
													3	$\frac{1}{2}$	$\frac{1}{2}$	$\frac{7}{2}$	$\frac{25}{63}$	$\frac{38}{63}$	4	$\frac{13}{40}$	$\frac{27}{40}$
													2	$\frac{1}{2}$	$\frac{1}{2}$	$\frac{5}{2}$	$\frac{11}{35}$	$\frac{24}{35}$	3	$\frac{5}{24}$	$\frac{19}{24}$
													1	$\frac{1}{2}$	$\frac{1}{2}$	$\frac{3}{2}$	$\frac{1}{15}$	$\frac{14}{15}$	2	$-\frac{1}{12}$	$\frac{13}{12}$
													0	—	—	$\frac{1}{2}$	$-\frac{5}{3}$	$\frac{8}{3}$	1	$-\frac{5}{4}$	$\frac{9}{4}$
3																6	$\frac{1}{2}$	$\frac{1}{2}$	$\frac{13}{2}$	$\frac{6}{13}$	$\frac{7}{13}$
																5	$\frac{1}{2}$	$\frac{1}{2}$	$\frac{11}{2}$	$\frac{64}{143}$	$\frac{79}{143}$
																4	$\frac{1}{2}$	$\frac{1}{2}$	$\frac{9}{2}$	$\frac{14}{33}$	$\frac{19}{33}$
																3	$\frac{1}{2}$	$\frac{1}{2}$	$\frac{7}{2}$	$\frac{8}{21}$	$\frac{13}{21}$
																2	$\frac{1}{2}$	$\frac{1}{2}$	$\frac{5}{2}$	$\frac{2}{7}$	$\frac{5}{7}$
																1	$\frac{1}{2}$	$\frac{1}{2}$	$\frac{3}{2}$	0	1
																0	—	—	$\frac{1}{2}$	-2	3
$\frac{7}{2}$																			7	$\frac{1}{2}$	$\frac{1}{2}$
																			6	$\frac{1}{2}$	$\frac{1}{2}$
																			5	$\frac{1}{2}$	$\frac{1}{2}$
																			4	$\frac{1}{2}$	$\frac{1}{2}$
																			3	$\frac{1}{2}$	$\frac{1}{2}$
																			2	$\frac{1}{2}$	$\frac{1}{2}$
																			1	$\frac{1}{2}$	$\frac{1}{2}$
																			0	—	—

TABLE 2

X_{12} Coefficients Relating the Nuclear Relaxation Enhancements Caused by Isolated Spins S_1 or S_2 with Those Caused by the Same Spins in an Exchange Coupled System

S_1		$\frac{1}{2}$	1	$\frac{3}{2}$	2	$\frac{5}{2}$	3	$\frac{7}{2}$
$\frac{1}{2}$	X_1	0.5	0.407407	0.375	0.36	0.351852	0.346939	0.343750
	X_2	0.5	0.777778	0.875	0.92	0.944444	0.959184	0.968750
1	X_1		0.5	0.422222	0.388889	0.371429	0.361111	0.354497
	X_2		0.5	0.691852	0.796296	0.856327	0.893519	0.918031
$\frac{3}{2}$	X_1			0.5	0.433143	0.400463	0.381769	0.37
	X_2			0.5	0.645714	0.743056	0.806803	0.85
2	X_1				0.5	0.441411	0.410	0.390823
	X_2				0.5	0.616968	0.705	0.767932
$\frac{5}{2}$	X_1					0.5	0.447860	0.417917
	X_2					0.5	0.597398	0.676622
3	X_1						0.5	0.453023
	X_2						0.5	0.583256
$\frac{7}{2}$	X_1							0.5
	X_2							0.5

where all the symbols have the same meaning as in Eq. (12) and (16). It should be noted immediately that, at variance with Eq. (15), Eq. (17) does *not* reduce to the Eq. (16) for uncoupled systems in the high-temperature limit. Therefore, the presence of magnetic coupling, no matter how small, alters the nuclear relaxation rates.[1] As can be predicted from Table 1, the effect can be sizable. In the high temperature limit, Eq. (17) can be simplified (Owens *et al.*, 1986) to

$$T_{1,2}^{-1} = KA_1^2 S_1(S_1 + 1)X_1 f_{1,2}(\omega, \tau_c) \qquad (18)$$

Table 2 reports the $X_{1,2}$ values calculated for the most common cases. All $X_{1,2}$ values are smaller than 1, indicating that a reduction of $T_{1,2}^{-1}$ is always obtained with respect to the uncoupled system. A decrease of paramagnetic effect on nuclear transverse relaxation rate is always desirable in paramagnetic systems, so it is already clear that coupled systems offer an advantage over isolated metal ions. However, we see from Table 2 that the $X_{1,2}$ values are never smaller than 0.3, so that paramagnetic line broadening will never decrease by more than a factor of three.

In Chapter 2 of this volume it is shown that different paramagnetic metal ions can have dramatically different nuclear relaxation effects, depending on their different electronic relaxation times τ_s. In macromolecules, τ_s usually dominates the correlation time, so nuclear relaxation effects differing by three to four orders of magnitude are possible (Banci *et al.*, 1991d). Ions such as high-spin cobalt(II) or iron(II), low-spin iron(III), and the lanthanides except gadolinium(III) (the so-called shift reagents) have very short τ_s and produce very little nuclear relaxation enhancements, whereas ions like copper(II), manganese(II), and gadolinium(III) (the so-called relaxation reagents) have

---

[1] The actual lower limit for magnetic coupling to be effective from the point of view of nuclear relaxation is $J \geq g_e \mu_B B_0$. This is a very small threshold, because in most cases of interest magnetic coupling is much larger.

long τ_s and produce dramatic nuclear relaxation enhancements, preventing the observation of NMR signals from nuclei close to the metal binding site in the protein.

In principle, when two metal ions are magnetically coupled their electronic relaxation times may be altered by the coupling. The behavior in several situations may be theoretically predicted. If the two metal ions are identical, to first order (i.e., by neglecting the anisotropic and antisymmetric parts of the interaction) there is no effect on τ_s (Bencini and Gatteschi, 1990; Banci et al., 1991d). If the two metal ions are different, then the one with shorter τ_s (let us say metal 2) will provide a further relaxation pathway for slow-relaxing metal 1, thereby shortening its electronic relaxation time (Bertini et al., 1985b, d, 1987; Banci et al., 1990c, 1991a). The equations for electronic relaxation in coupled systems are also reported in Chapter 2 of this volume. Qualitatively, the relaxation enhancement is proportional to the electronic relaxation time of the fast-relaxing metal ion and to the square of J_{12}. If the coupling is strong, a limit will be approached at which the electronic relaxation time of metal 1 will become as short as that of metal 2 (Banci et al., 1991d). Magnetic coupling can thus turn a relaxation reagent, considered unsuitable for high-resolution NMR into a shift reagent.

To summarize, magnetic coupling between two paramagnetic metal ions is expected to alter both the nuclear hyperfine shifts and relaxation rates. Relaxation rates are altered even for relatively small couplings, whereas hyperfine shifts are altered if J_{12} is not negligible with respect to kT. Striking examples of the two situations are given in Sections 4 and 5.

4. EXPLOITATION OF WEAK EXCHANGE COUPLING IN DIMETALLIC SYSTEMS: Cu$_2$M$_2$ SUPEROXIDE DISMUTASE (M=Co, Ni)

As already outlined in Section 1.3.7, superoxide dismutase (SOD) contains copper(II) as the catalytic center. The effect of copper(II) on nuclear relaxation is dramatic. From the τ_s value of 2×10^{-9} s of copper(II) in SOD, estimated from NMRD measurements (Bertini et al., 1985; see also Chapter 2 of this volume), a linewidth of $\simeq 10$ kHz for a proton at 5 Å from the metal can be predicted. These values obviously prevent the observation of ^{1}H NMR signals from the histidine ligand protons and therefore prevent one from obtaining the first-hand information on the active site that is available from metalloproteins containing more favorable metal ions. However, the zinc(II) ion in SOD can be replaced by copper(II), cobalt(II), or nickel(II) among other divalent cations, with virtually full retention of catalytic activity, since the active site's copper(II) is minimally perturbed by the substitution (Valentine and Pantoliano, 1981). Here we have, therefore, the possibility of exploiting magnetic coupling to decrease the nuclear relaxing properties of copper(II). We discard copper(II) as a partner because it is expected (and experimentally verified) (Bertini et al., 1988a; see also Chapter 2 of this volume) not to alter τ_s of the catalytic copper center and thus to cause a decrease of only a factor of two in the relaxation rate of nuclei interacting with the catalytic center (Owens et al., 1986), cf. Eq. (16) and Table 2.

High-spin cobalt(II) and nickel(II) are both good candidates, having τ_s values around 10^{-11} s (Banci et al., 1991d; Bertini and Luchinat, 1986). Indeed, the NMR spectra of Cu$_2^I$M$_2$SOD (i.e., with the catalytic copper ion in the reduced, diamagnetic,

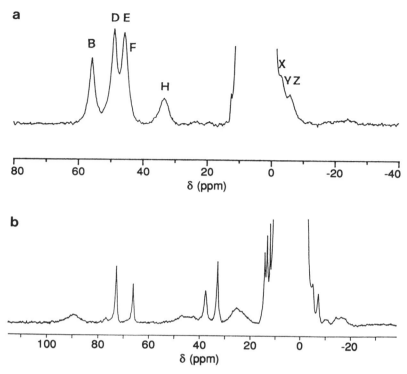

Figure 12. ^{1}NMR spectra at 200 MHz and 300 K of (a) Cu(I)$_2$Co(II)$_2$SOD (Bertini *et al.*, 1985e; 1991c) and (b) Cu(I)$_2$Ni(II)$_2$SOD (Ming *et al.*, 1988a).

form) give reasonably well-resolved signals arising from the ligands of the zinc site (Bertini *et al.*, 1985e, 1991c; Ming *et al.*, 1988a) (Fig. 12). Exchange coupling between copper(II) and paramagnetic metals at the zinc site occurs through a superexchange mechanism *via* the imidazolate bridge of His 61. J_{12} values of 33 cm^{-1} (Morgenstern-Badarau *et al.*, 1986) and 52 cm^{-1} (Fee and Briggs, 1975) have been measured for copper(II)–cobalt(II) and copper(II)–copper(II), respectively. The energy of the first excited state for the copper(II)–nickel(II) derivative is expected to be in the same range. The energy levels of the coupled systems are shown in Fig. 13. We are clearly in the situation $\hbar\tau_s^{-1} \ll J_{12} \ll kT$. Therefore, we expect very small effects on the hyperfine shift values and their temperature dependence, but still dramatic effects on nuclear relaxation rates.

These considerations prompted Bertini *et al.* to look for hyperfine-shifted ^{1}H NMR signals in the spectrum of Cu$_2$Co$_2$SOD (Bertini *et al.*, 1985c). The spectrum is shown in Fig. 14. As many as nineteen reasonably sharp signals are detected outside the diamagnetic region, with a temperature dependence shown in Fig. 15. It is apparent, by referring to the active-site structure of Fig. 5 (Tainer *et al.*, 1983), that these signals cannot arise from the ligands of the zinc site alone. Many of them must arise from the copper(II) site as well, immediately confirming the theoretical expectations.

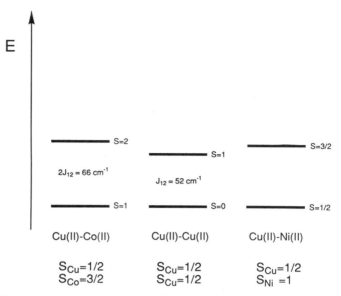

Figure 13. Energy levels arising from magnetic coupling in dimetallic derivatives of SOD. The Cu(II)–Ni(II) case is drawn on an arbitrary scale.

Since 1985, several studies have been performed on this system (Banci *et al.*, 1987a, b, 1988a, b, 1989a, b, c, 1990a, b, d; Bertini *et al.*, 1988b, c, 1989a; Ming *et al.*, 1988b), lately including 2D NMR experiments (Banci *et al.*, 1992a). These studies have yielded a full assignment of the hyperfine-shifted signals to individual ligand protons of both copper and cobalt sites (Table 3). It is interesting to note that the signals from the copper(II) site are, on the average, even sharper than those from the cobalt site. For instance, the protons in *ortho*-like position with respect to the coordinating nitrogen, which are the closest to the paramagnetic center, have linewidths around 300 Hz in the copper site (at 400 MHz) and are broad beyond detection in the cobalt site (a lower limit for the linewidth can be estimated around 5000 Hz). Smaller relaxation rates for protons coordinated to copper are at first glance surprising, but can be accounted for

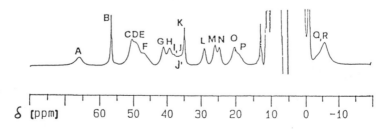

Figure 14. ^{1}H NMR spectrum at 300 MHz and 300 K of Cu(II)$_2$Co(II)$_2$SOD (Bertini *et al.*, 1985c).

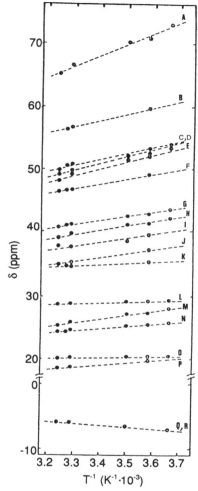

Figure 15. Chemical shifts versus $1/T$ for hyperfine-shifted proton signals in $Cu(II)_2Co(II)_2SOD$ (Bertini *et al.*, 1985c).

by considering that, in the limit of equal τ_s values for the two metal ions, the relaxation enhancement caused by copper(II) over that caused by cobalt(II) should be given by the ratio between $X_1S_1(S_1 + 1)$ and $X_2S_2(S_2 + 1)$, i.e., 0.086 [Eq. (16)]. This observation thus constitutes a striking experimental verification of the theory of nuclear relaxation in coupled systems (Banci *et al.*, 1991d). It should be added that, with increasing magnetic field, the difference in linewidths becomes even more remarkable (Banci *et al.*, 1987b). This is due to the onset of Curie relaxation (Gueron, 1975; Vega and Fiat, 1976) (cf. Chapter 2 of this volume), which depends on the square of $S(S + 1)$ for each metal ion and therefore is much more dramatic for cobalt(II) than for copper(II).

TABLE 3
Assignment of the Hyperfine-shifted Signals of
Cu(II)$_2$Co(II)$_2$SOD to Specific Protein Residues

Signal	Proton
A	Hδ2 (His 61)
B	Hδ1 (His 118)
C	Hε2 (His 44)
D	Hδ2 (His 69)
E	Hδ2 (His 78)
F, J	Hε2 (His 78), Hε2 (His 69)
G	Hδ2 (His 44)
H	Hε1 (His 118)
I, J'	Hβ1 (Asp 81), Hβ2 (Asp 81)
K	Hδ1 (His 46)
L	Hδ2 (His 46)
M	Hε1 (His 44)
N	Hδ2 (His 118)
O	Hε1 (His 46)
P	Hβ1 (His 44)
Q	Hβ1 (His 69)
R	Hβ2 (His 44)

As expected, the temperature dependence of the hyperfine shifts does not show interesting features (Fig. 15) (Bertini *et al.*, 1985c). All signals qualitatively follow a Curie-type dependence, although the intercepts deviate considerably from zero. On the other hand, nonlinear $1/T$ behaviors are almost a rule even in monometallic systems because of zero-field splitting for ions with $S > \frac{1}{2}$ and, in more general terms, because the electron magnetic moments do not arise only from pure electron spins but also contain orbital contributions (Bertini and Luchinat, 1986).

The ^{1}H NMR spectrum of Cu$_2$Ni$_2$SOD is shown in Fig. 16 (Ming *et al.*, 1988a). Overall, the signals are even sharper than those of Cu$_2$Co$_2$COD, probably because of the shorter τ_s for nickel(II) than for cobalt(II) in a pseudotetrahedral environment

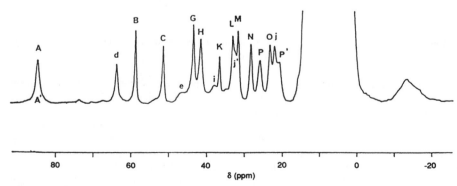

Figure 16. ^{1}H NMR spectrum at 300 MHz and 303 K of Cu(II)$_2$Ni(II)$_2$SOD (Ming *et al.*, 1988a).

(Banci *et al.*, 1991d; Bertini and Luchinat, 1986) (cf. Chapter 2 of this volume). A full assignment of the signals using 2D techniques is also being completed for this derivative. A feature of nickel(II) in this system that is interesting from the theoretical point of view is that in $Cu_2^INi_2SOD$ the nuclear relaxation rates increase dramatically with magnetic field (Ming *et al.*, 1988a). Since both T_1 and T_2 are affected, the origin cannot be Curie relaxation, which affects only T_2. Instead, it can be explained by a field dependence of τ_s for nickel, which is theoretically understood and experimentally verified in nickel-containing proteins (Banci *et al.*, 1991d). In Cu_2Ni_2SOD this effect is quenched (Bertini *et al.*, 1992d). Apparently, in the coupled system new electron relaxation mechanisms are operative which contribute to the overall shortening of τ_s and are field-independent. A reasonable guess is that the new mechanisms depend on the presence of the low-lying excited states made available by magnetic coupling (cf. Fig. 13) and can thus be referred to as a "solid state" type of mechanisms (Banci *et al.*, 1991d) (see Chapter 2 of this volume).

As far as the biochemistry is concerned, the assignment of the active-site signals allowed NMR to be used to monitor changes occurring in the active site upon binding of monovalent anions, as models for the superoxide anion (Banci *et al.*, 1987a, 1988b, 1989b, c, 1990a; Ming *et al.*, 1988b), changes in pH (Banci *et al.*, 1991e, 1992b) and, even more interestingly, upon substitution of specific residues by site-directed mutagenesis (Banci *et al.*, 1988a, 1990b, 1991e, 1992b; Bertini *et al.*, 1989a). It has been learned, for instance, that binding of strong ligands like cyanide and azide causes one of the copper ligands, His 46, to be virtually detached from coordination to the metal (Banci *et al.*, 1990a) (Fig. 17); as a consequence, the overall geometry shifts toward a square pyramid with three histidine and the exogenous ligand in the basal plane and His 46 loosely interacting in the fifth coordination position. The evidence comes from the substantial decrease in hyperfine shifts experienced by all the ring protons belonging to His 46 (Bertini *et al.*, 1985c). Titration of SOD with anions can be easily followed by NMR, and the data analysis yield very accurate anion binding constants (Bertini *et al.*, 1988b). Finally, the NMR spectra are substantially pH-independent up to pH > 10 (Bertini *et al.*, 1985c), indicating that the metal environment is maintained all over the pH range of activity of the enzyme. The NMR spectra of the mutants show that it is possible for the copper coordination geometry to vary within a relatively broad range, and even to lose the coordinated water (Bertini *et al.*, 1989a) giving a nearly square–planar environment, without affecting the enzymatic activity. On the other hand, it is learned that site-directed modifications that dramatically affect the activity may not

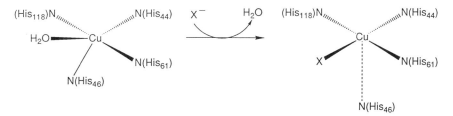

Figure 17. Binding of strong anionic ligands to copper(II) in SOD causes water removal and weakening of the Cu–His 46 bond (Banci *et al.*, 1990a).

cause appreciable structural changes but, as indicated by molecular dynamics calculations (Banci *et al.*, 1992c) simply affect the distribution of electric charge in the vicinity of the active site and the mobility of the active-site cavity.

To summarize, replacement of the zinc ion by paramagnetic metal ions in SOD provides derivatives in which the native catalytic metal is turned into a shift reagent by a drastic reduction of its τ_s. The system has all the advantages of a shift reagent without the disadvantages encountered when a shift reagent is *substituted* for the catalytically active metal. Moreover it has the further advantage that it maintains $S = \frac{1}{2}$ (shift reagents often have large S) and that the nuclear relaxation enhancement is intrinsically reduced by an additional factor of three because of magnetic coupling (Table 2).

5. EXPLOITATION OF STRONG EXCHANGE COUPLING IN DIMETALLIC SYSTEMS: Fe_2S_2 FERREDOXINS

5.1. The ^1H NMR Spectra of Fe_2S_2 Ferredoxins

Ferredoxins have attracted the interest of NMR researchers since the early seventies (Poe *et al.*, 1970, 1971; Phillips *et al.*, 1970a; Dunham *et al.*, 1971), i.e., even before FT instruments became commonly used. The interest was dictated by the fact that, with the low- or medium-field instruments available at that time, diamagnetic proteins gave too little resolution in the ^1H NMR spectra to be of much practical use. Ferredoxins, together with heme proteins, which are extensively covered in other chapters of this volume, provided several resolved, hyperfine-shifted signals, whose analysis constituted the basis for the first guesses on the structure–function relationship of these systems. When the *bis*-(μ-sulfide) bridged structure depicted in Fig. 1A was established for Fe_2S_2 ferredoxins (Britzinger *et al.*, 1966; Cammack *et al.*, 1971), it was apparent that magnetic coupling had to play an important role in defining their electronic, and hence their NMR, properties. It was soon predicted that in the oxidized iron(III)–iron(III) form, the $S = 0$ ground state originating from antiferromagnetic coupling would have caused a marked antiCurie temperature dependence of the hyperfine shifts (Glickson *et al.*, 1971). The ^1H NMR spectra reported in Fig. 18, although taken almost twenty years later (Banci *et al.*, 1990c), show the same feature observed originally, i.e., a broad composite downfield signal accounting for eight protons and with a marked antiCurie temperature dependence. The obvious assignment was that the eight signals arise from the β-CH_2 protons of the four cysteines coordinated pairwise to the two metal ions (Glickson *et al.*, 1971). The two metal ions are not equivalent by symmetry, and the protons in each β-CH_2 pair are also not equivalent, but apparently the differences are accidentally small and do not allow resolution of the individual signals. Today we know that the signals are broad because the system consists of two high-spin iron(III) ions. High-spin iron(III) is in a ^6A state and therefore is expected to have rather long electronic relaxation times (see Chapter 2 of this volume). On the other hand, as discussed in Section 3, magnetic coupling in a homodimer does not allow at first approximation for additional electron relaxation mechanisms. As a matter of fact, in oxidized rubredoxin, which contains one iron(III) ion coordinated by four cysteine sulfurs in a similar pseudotetrahedral environment, ^1H NMR signals are too broad to be detected (Phillips *et al.*, 1970b). Apparently, besides the factor-of-two reduction expected upon coupling

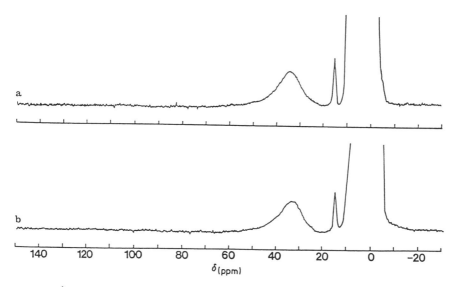

Figure 18. ^{1}H NMR spectra at 200 MHz and 303 K of oxidized Fe_2S_2 ferredoxins from (a) spinach, (b) red algae *Porphyra umbilicalis* (Banci *et al.*, 1990c).

(Table 2) the electronic relaxation times in the dimer are intrinsically slightly shorter.

Upon reduction, yielding an iron(II)–iron(III) system, several sharp signals appear (Fig. 19a). In particular, six resonances were originally observed in the region 15–45 ppm downfield (Dunham *et al.*, 1971), four of them experiencing antiCurie temperature dependence. Among various hypotheses, it was proposed that the four antiCurie resonances could arise from the β-CH_2 protons of the two cysteines coordinated to iron(II) and that the four β-CH_2 resonances belonging to cysteines coordinated to iron(III) were

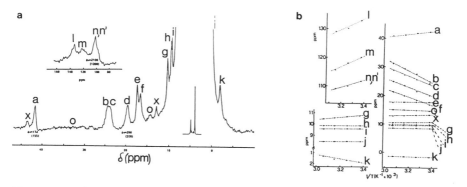

Figure 19. (a) ^{1}H NMR specrum at 300 MHz and 297 K of reduced Fe_2S_2 ferredoxin from spinach and (b) chemical shifts versus $1/T$ for the hyperfine-shifted signals (Bertini *et al.*, 1984).

too far downfield to be detectable (Dunham *et al.*, 1971). This proposal originated from theoretical considerations analogous to those developed in Section 3: in the presence of antiferromagnetic coupling, the coefficient with which iron(II) contributes to the $S = \frac{1}{2}$ ground state is negative, whereas that of iron(III) is positive. At room temperature, the contribution from excited states counterbalances that of the ground state for iron(II), giving small shifts, compared to those of iron(III), and antiCurie temperature dependence.

^{1}H NMR spectra of various Fe_2S_2 ferredoxins, all similarly inconsistent with theoretical predictions, continued to appear over the years (Glickson *et al.*, 1971; Salmeen and Palmer, 1972; Anderson *et al.*, 1975; Chan and Markley, 1983; Takahashi *et al.*, 1981; Nagayama *et al.*, 1983) until Bertini *et al.* (1984) undertook a careful search over a wide spectral region and were able to find four additional broad signals, beyond 100 ppm downfield, with a Curie-type temperature dependence (Fig. 19). This finding constituted the first proof of the validity of the theoretical model (Dunham *et al.*, 1971) and, at the same time, provided a powerful tool for the specific assignment of the oxidation state of the individual iron ions in these and more complex systems (see Sections 7–9).

It is worth analyzing the spectrum in Fig. 19 in more detail in the light of the theoretical considerations developed in Sections 2 and 3. To do so, we report in Fig. 20 the theoretical inverse temperature dependence for protons sensing the iron(III) ions and protons sensing the iron(II) ions according to Eq. (15) (Banci *et al.*, 1990c). We have used the coefficients of Table 1 and assumed that $A_1 = A_2 = A$, i.e., that the proton–electron hyperfine coupling in the absence of exchange coupling between the two metal ions is the same for both iron(II) and iron(III). We have taken $A/\hbar = 1$ MHz, which is a reasonable value for a cysteine ligand. The temperature dependence is reported against J_{12}/kT to underline that what is important is the ratio between the two quantities rather than the absolute value of either [cf. Eq. (15)]. It can be easily appreciated that the

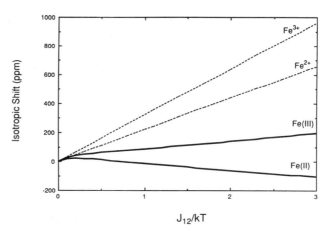

Figure 20. Theoretical dependence of chemical shifts on J_{12}/kT for protons sensing iron(III) and iron (II) in an antiferromagnetically coupled dimer. The straight dashed lines show the behavior in the absence of magnetic coupling (Curie behavior).

calculated curves for both iron(III) and iron(II) sites are in very good agreement with the experimental observations (Fig. 19) in terms of absolute values of the hyperfine shifts and temperature dependences. Apparently, a situation very similar to the experimental one at room temperature is obtained for $J_{12}/kT \simeq 1$, i.e., a large downfield shift and Curie-type temperature dependence for the iron(III) site and a small downfield shift and antiCurie-type temperature dependence for the iron(II) site. For smaller J_{12}/kT values, the antiCurie behavior of the signals of the iron(II) site attenuates, and eventually the curve merges with that of iron(III) and with that of uncoupled systems (dashed lines). For larger values, the hyperfine shifts for iron(II) become negative and the temperature dependence becomes of the Curie type. However, this is not true Curie behavior: on the contrary, the behavior is so strongly antiCurie that the signs of the shifts are reversed. We term this behavior "pseudoCurie." It is not experimentally observed in plant and bacterial Fe_2S_2 systems (Banci et al., 1990c; Skjeldal et al., 1991a), whereas it has been recently reported that mammalian ferredoxins may show such behavior (Skjeldal et al., 1991b). A pseudoCurie behavior is more commonly encountered in Fe_4S_4 systems (see later).

5.2. Identification of the Oxidation State of Individual Iron Ions Within the Protein Frame

The classification of the ^1H NMR signals of Fe_2S_2 ferredoxins in two sets, belonging to cysteines coordinated in one case to iron(III) and in the other to iron(II), opened the way to a study aimed at identifying which of the two iron ions is in the reduced state within the protein frame (Dugad et al., 1990). The study was performed by using the known X-ray structure of one of these proteins and 1D NOE techniques (see Chapters 1 and 2 of this volume), which allowed investigators to find dipolar connectivities, either intercysteine or between β-CH_2 cysteine protons and other protons unequivocally assignable to specific residues in the neighborhood of the cluster (Dugad et al., 1990). This is schematically illustrated in Fig. 21. This study permitted the identification of the iron ion of reduced in Fe_2S_2 ferredoxins as the one closer to the surface of the protein. It remains to be seen if this property is connected with the function of the protein as an electron-transport system and with the electron-transfer pathway within the protein.

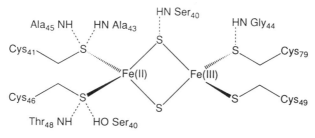

Figure 21. Sequence-specific assignment of cysteine protons in Fe_2S_2 ferredoxin from *Spirulina platensis*. As a consequence, the oxidation state of the individual iron ions is established (Dugad et al., 1990).

It is worth stressing that no other spectroscopic technique but NMR is presently able to provide such a "valence-specific" assignment. X-ray analysis of proteins does not yet have the resolution necessary to detect the small increases in metal–donor distances expected upon reduction of iron(III) to iron(II). Mössbauer spectroscopy is certainly able to distinguish between iron(II) and iron(III), but not to relate the observation to individual iron ions in the protein.

5.3. Exchange and Double Exchange

A comment is in order on the possible presence and relevance of double exchange in reduced Fe_2S_2 ferredoxins. Magnetic susceptibility measurements have provided J_{12} values of about 400 cm^{-1} for oxidized ferredoxins and 200 cm^{-1} for reduced ferredoxins (Palmer et al., 1971). These values are obtained by measuring the temperature dependence of the magnetic susceptibility down to very low temperatures and then fitting the data to an energy-level scheme of the type of Eq. (5).[2] This is certainly correct for the oxidized iron(III)–iron(III) case, but in the reduced iron(II)–iron(III) case the possibility that double exchange is present should also be taken into account (Zener, 1951). In this case, a B_{12}-containing term should be added and the energies would be given by Eq. (8). Both magnetic susceptibility studies (Palmer et al., 1971) and EPR spectroscopy at low temperatures (Gibson et al., 1966) demonstrate that the ground state is $S = \frac{1}{2}$. Therefore, by referring to Fig. 11, one can see clearly that double exchange, if present, is not so strong as to overcome effect of antiferromagnetic exchange coupling. To maintain an $S = \frac{1}{2}$ ground state, the B_{12}/J_{12} ratio must be smaller than unity. The actual value of B_{12} is very difficult to estimate through magnetic measurements (Blondin and Girerd, 1990). Equation (8) has been derived in the simple case of a symmetric dimer, i.e., for a case in which the two states with the extra electron on one or the other center have the same energy. This is not the case in real systems like proteins, obviously lacking any symmetry. It is easily conceivable that the two metals in Fe_2S_2 ferredoxins may have reduction potentials differing by tens or even hundreds of millivolts. A 100-mV difference in reduction potential translates into an energy difference ΔE of $\simeq 800$ cm^{-1}. It has been shown that, in the limit $\Delta E \gg B_{12}(2S + 1)$, double exchange lowers the energies by $-B_{12}^2(S + \frac{1}{2})^2/\Delta E$. By dropping the S-independent terms, which produce only an offset in all energy levels, this contribution becomes $-(B_{12}^2/\Delta E)S(S + 1)$, i.e., indistinguishable from a ferromagnetic contribution to J_{12}. Therefore, the J_{12} values obtained from a fitting of magnetic susceptibility data neglecting double exchange should actually be regarded as effective values J_{eff} incorporating additional contributions from double exchange. This is the experimental situation in Fe_2S_2 systems. There are a few dimetallic systems other than Fe–S clusters in which the double exchange contribution is dominant (Ding et al., 1990; Que and True, 1990).

What more do we learn from NMR of reduced Fe_2S_2 ferredoxins with respect to the information available from EPR, magnetic susceptibility studies, and Mössbauer

---

[2] Actually, a quantitative treatment requires the inclusion of other terms besides electron Zeeman and exchange coupling in the Hamiltonian. For instance, in the case of iron(II), zero-field splitting cannot be neglected. We have omitted these terms in the present discussion to focus on the interplay between exchange and double exchange.

spectroscopy? All these techniques point to

1. an $S = \frac{1}{2}$ ground state,
2. localized valences, and
3. small contribution of double exchange to the energies of the system.

However, all this information is obtained at low temperature, where the $S = \frac{1}{2}$ ground state is essentially the only populated state. Nothing is known about ΔE and about the dynamics of the electron exchange between the two states at higher temperature. The observation of a striking difference in the temperature dependence of the ^{1}H NMR signals from the two metal sites around room temperature demonstrates that

1. the valences are still localized;
2. if the electron exchange rate between the two sites is slow on the NMR time scale (which is unlikely), then one of the two situations must be predominant because no other species is apparent in the NMR spectrum (an upper limit of 5% for the undetected amount of the other species can be estimated, translating to a $\Delta E \geq 600$ cm^{-1});
3. if the electron exchange is fast, then again the contribution from the other species to the average spectrum must be small, or else the divergence in the temperature dependence would be quenched.

6. EXCHANGE-COUPLED POLYMETALLIC SYSTEMS

For a system of n paramagnetic centers characterized by spins $S_1, S_2, \ldots, S_n$, the general exchange coupling scheme is given by the spin Hamiltonian:

$$\mathcal{H} = \sum_{i \neq j} J_{ij} \mathbf{S}_i \cdot \mathbf{S}_j \tag{19}$$

We have seen in Section 2 that an analytical solution is easily obtained for $n = 2$. For $n = 3$, analytical solutions can be obtained only for $J_{12} \neq J_{13} = J_{23}$ (or for all J values equal). For $n = 4$ analytical solutions can be obtained for $J_{12} \neq J_{34} \neq J_{13} = J_{14} = J_{23} = J_{24}$ or for $J_{12} \neq J_{13} = J_{23} \neq J_{14} = J_{24} = J_{34}$ (or for more of them being equal, up to all six J values equal). In these cases, the total spin S can be subdivided into subspins, which still give good quantum numbers, namely S_{12} and S for $n = 3$ and either S_{12}, S_{34}, and S or S_{12}, S_{123}, and S for $n = 4$.

For $n = 3$, Hamiltonian (19) can thus be rewritten as

$$\mathcal{H} = J(\mathbf{S}_1 \cdot \mathbf{S}_2 + \mathbf{S}_2 \cdot \mathbf{S}_3 + \mathbf{S}_1 \cdot \mathbf{S}_3) + \Delta J_{12} \mathbf{S}_1 \cdot \mathbf{S}_2 \tag{20}$$

and the energies are given by

$$E(S_{12}, S) = \frac{1}{2}[JS(S + 1) + \Delta J_{12} S_{12}(S_{12} + 1)] \tag{21}$$

where $J = J_{13} = J_{23}$ and $\Delta J_{12} = J_{12} - J$.

Analogously, for $n = 4$, the first of the two situations which give analytical solutions is

$$\mathscr{H} = J(\mathbf{S}_1 \cdot \mathbf{S}_2 + \mathbf{S}_1 \cdot \mathbf{S}_3 + \mathbf{S}_1 \cdot \mathbf{S}_4 + \mathbf{S}_2 \cdot \mathbf{S}_3 + \mathbf{S}_2 \cdot \mathbf{S}_4 + \mathbf{S}_3 \cdot \mathbf{S}_4)$$
$$+ \Delta J_{12}\mathbf{S}_1 \cdot \mathbf{S}_2 + \Delta J_{34}\mathbf{S}_3 \cdot \mathbf{S}_4 \tag{22}$$

with relative energies

$$E(S_{12}, S_{34}, S) = \tfrac{1}{2}[JS(S + 1) + \Delta J_{12}S_{12}(S_{12} + 1) + \Delta J_{34}S_{34}(S_{34} + 1)] \tag{23}$$

where $J = J_{13} = J_{14} = J_{23} = J_{24}$ and $\Delta J_{12} = J_{12} - J$ and $\Delta J_{34} = J_{34} - J$. The second is

$$\mathscr{H} = J(\mathbf{S}_1 \cdot \mathbf{S}_2 + \mathbf{S}_1 \cdot \mathbf{S}_3 + \mathbf{S}_1 \cdot \mathbf{S}_4 + \mathbf{S}_2 \cdot \mathbf{S}_3 + \mathbf{S}_2 \cdot \mathbf{S}_4 + \mathbf{S}_3 \cdot \mathbf{S}_4) + \Delta J_{12}\mathbf{S}_1 \cdot \mathbf{S}_2$$
$$+ \Delta J_{123}(\mathbf{S}_1 \cdot \mathbf{S}_2 + \mathbf{S}_1 \cdot \mathbf{S}_3 + \mathbf{S}_2 \cdot \mathbf{S}_3) \tag{24}$$

with relative energies

$$E(S_{12}, S_{123}, S) = \tfrac{1}{2}[JS(S + 1) + \Delta J_{123}[S_{123}(S_{123} + 1) - S_{12}(S_{12} + 1)]$$
$$+ \Delta J_{12}S_{12}(S_{12} + 1)] \tag{25}$$

where $J = J_{14} = J_{24} = J_{34}$ and $\Delta J_{123} = J_{13}(=J_{23}) - J$ and $\Delta J_{12} = J_{12} - J_{13}(=J_{23})$. In all other cases numerical solutions can be obtained with the aid of computer programs.

Of course, if the paramagnetic centers in the polymetallic systems are constituted by metal ions of the same kind but differing in oxidation state, double exchange may be present. In principle, as many double exchange B_{ij} parameters as J_{ij} parameters should be taken into account. However, before introducing this complication in the treatment, we would like to illustrate a phenomenon peculiar to polymetallic systems— it is absent in dimetallic systems—that conceptually interferes with the behavior of double exchange by mimicking many of its experimental consequences.

6.1. Spin Frustration

Consider a trimetallic system constituted by one high-spin iron(II) ($S_1 = 2$) and two high-spin iron(III) ($S_2 = S_3 = \tfrac{5}{2}$) ions (Fig. 22). This case corresponds to reduced Fe_3S_4 ferredoxin, whose cluster geometry is illustrated in Fig. 1C. If we initally assume that $J_{12} = J_{13} = J_{23} = J$, the energies of the system are given by $\tfrac{1}{2}JS(S + 1)$, with S ranging from 0 to 7. For antiferromagnetic coupling (positive J) the $S = 0$ state is lowest. However, if one of the J values is smaller (for instance, $\Delta J_{12} < 0$), Eq. (21) shows that the ground state tends to be the one with largest S_{12} and smallest S, i.e., $S_{12} = \tfrac{9}{2}$ and $S = 2$. Figure 23 shows the variation in the energy of the various S states as a function of $\Delta J_{12}/J$. The $S = 2$ state becomes the ground state for $\Delta J_{12}/J < -0.44$. In other words, the system behaves as if there were ferromagnetic coupling between spins S_1 and S_2. Of course, all J's are antiferromagnetic, but the two larger couplings force the third pair (the S_1–S_2 pair) to align the spins with maximum multiplicity. This

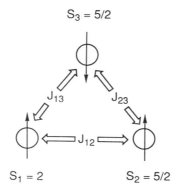

Figure 22. Spin frustration in a trimetallic system.

phenomenon is called "spin frustration." Pictorially, one can figure out that, in a triangle, if S_1 is antiparallel to S_3 and S_3 is antiparallel to S_2, S_2 cannot be antiparallel to S_1 (Fig. 22).

6.2. Double Exchange in Fe_3S_4 Systems

Experimentally, reduced Fe_3S_4 clusters in proteins display an $S = 2$ ground state (Thomson *et al.*, 1981). Mössbauer data (Papaefthymiou *et al.*, 1987) indicate that one of the two ferric ions has a true +3 oxidation state while the other, together with the ferrous ion, constitutes a mixed-valence pair with oxidation state +2.5. A mixed-valence pair has complete delocalization of the electron over the two centers. It therefore seemed reasonable to introduce a double exchange term in the Hamiltonian for the system,

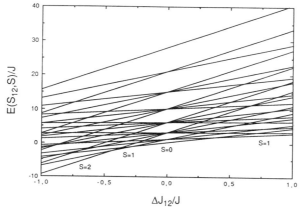

Figure 23. Energy levels in an $S_1 = 2$, $S_2 = \frac{5}{2}$, $S_3 = \frac{5}{2}$ system antiferromagnetically coupled as a function of $\Delta J_{12}/J$. The changes in ground state are indicated.

which results in the following energy levels (Münck *et al.*, 1988):

$$E(S_{12}, S) = \tfrac{1}{2}[JS(S + 1) + B_{12}(2S_{12} + 1)] \tag{26}$$

The proper $S = 2$ ground state has already been obtained for $B_{12}/J > 2$ (Fig. 24). We recall here that double exchange becomes dominant in dimers only for $B_{12}/J > 4.5$. The different behavior of trimers is the result of spin frustration that, even with all J values equal, tends to destabilize the antiferromagnetic coupling on the pair chosen to display double exchange. At this point it should be clear that there is no obvious way to distinguish between double exchange and J-inequivalence as the driving force for the partial delocalization observed in Fe_3S_4 systems [cf. Eqs. (3) and (8)]. It is true that the presence of double exchange is demonstrated by the electron delocalization over one spin pair shown by Mössbauer spectroscopy; on the other hand, the magnitude of B_{12} is totally unkown, and may not be sufficient to cause ferromagnetic coupling in the S_1–S_2 pair if it were not helped by some J-inequivalence. As already pointed out, proteins lack any symmetry, and there are no reasons for the three J values to be equal. Further-more, if double exchange were dominant, it is not clear why it should not be able to delocalize the electron over all three centers. Full delocalization has been shown to be possible from a theoretical point of view by including double exchange terms all over the triangle (only numerical solutions are possible) and, in this approach, partial delocal-ization can be achieved only by introducing a sizable asymmetry into the system. Again, asymmetry will certainly introduce J-inequivalence. The existing NMR data (Nagayama *et al.*, 1983; Sweeney, 1981; Moura *et al.*, 1977) are in a qualitative agreement with the above picture. Compared with Fe_4S_4 clusters (see below), the $S = 2$ ground state causes the lines to be rather broad, making a detailed analysis more difficult. [*Note added in proof*: The NMR spectra of proteins containing only one Fe_3S_4 core have been recorded since 1991, and the signals corresponding to the β-CH_2 protons of cluster-bound cysteines have been assigned on the basis of the theory discussed above (*Biochemistry*

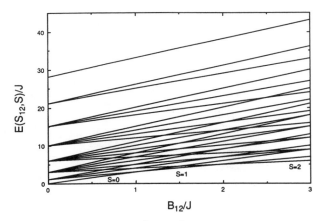

Figure 24.   Energy levels in an $S_1 = 2$, $S_2 = \tfrac{5}{2}$, $S_3 = \tfrac{5}{2}$ system antiferromagnetically coupled in the presence of double exchange within the S_1–S_2 pair, as a function of B_{12}/J. The changes in ground state are indicated.

31:952, 1992; Moura *et al.*, personal communication). Thus, it is easy to predict that the Fe_3S_4 cluster containing proteins will be elegantly understood as has been done with Fe_4S_4 proteins.]

Larger clusters can be discussed following the same guidelines. In general, partial electron delocalization is observed, which requires the introduction of double exchange over at least one pair. Again, there is large covariance between B_{ij} and $-\Delta J_{ij}$ for that pair, so that it is difficult to estimate the magnitude of either by magnetic measurements alone.

We should end this section by mentioning that double exchange should give rise to electronic transitions, called intervalence bands, whose energies and intensities should allow a quantitative estimate of B (Blondin and Girerd, 1990). Given the importance that this issue has acquired over the past few years, it is conceivable that measurements of this kind will become more and more available, allowing a better quantitative characterization of the magnetic properties of these systems.

7. NMR OF Fe_4S_4 HIGH-POTENTIAL IRON–SULFUR PROTEINS

As already pointed out in Section 1, the peculiarity of HiPIP that distinguishes them from all other 4Fe–4S proteins is the much higher redox potential at which they perform electron transfer processes in bacteria. This feature must arise from some small but distinctive difference in the protein environments around the cluster. We will show here how NMR is a uniquely useful method for providing both a detailed mapping of the amino acid residues surrounding the cluster and an understanding of the electronic distribution in the cluster core and of the magnetic coupling among the iron ions.

In the ^1H NMR spectrum of the oxidized protein from *C. vinosum*, shown in Fig. 25B, seven hyperfine-shifted signals appear downfield of the diamagnetic region and two upfield (Cowan and Sola, 1990a; Bertini *et al.*, 1991b, 1992a). The β-CH_2 protons of the cysteine residues bound to the four iron ions of the cluster account for eight out of nine signals, whereas the ninth must arise from another proton in the vicinity of the paramagnetic cluster. The geminal protons do not exhibit identical isotropic shifts because of the different protein environment. ^1H NOE experiments on these signals have allowed the assignment of the pairs a'–c', b'–d', h'–i', and f'–g' as geminal couples of the cysteine ligands (Cowan and Sola, 1990a) and of signal e' as the α-CH proton of the cysteine residue to which signals b' and d' also belong (Bertini *et al.*, 1991b).

Analogous NOE experiments conducted on the reduced form of the HiPIP, the NMR spectrum of which is presented in Fig. 25A, have allowed the pairwise assignment of the cysteine geminal β-CH_2 protons as the pairs a–y, b–z, c–w, and d–e (Bertini *et al.*, 1991b).

Monodimensional saturation transfer and bidimensional EXSY NMR experiments have allowed a full correlation between each signal in the oxidized and reduced protein, as shown in Fig. 25 (Bertini *et al.*, 1991b). Thus, signals a, b, c, d, e, v, w, y, and z of the reduced HiPIP correspond to signals i', a', b', f', e', d', h', and c' of the oxidized form, respectively. The temperature dependence of the signals in the reduced protein shows that all the paramagnetic signals have antiCurie behavior (Bertini *et al.*, 1991b). This can be explained by recalling that the cluster has an $S = 0$ ground state arising from antiferromagnetic exchange coupling among the iron ions of the cluster. The

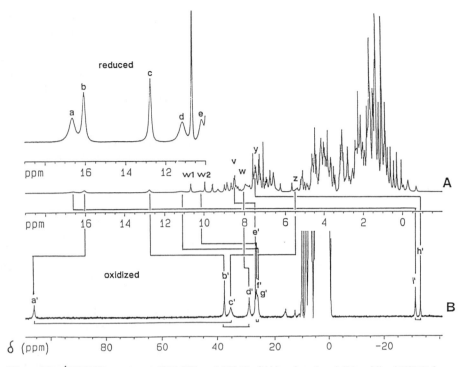

Figure 25. ^{1}H NMR spectra at 600 MHz and 300 K of (A) reduced and (B) oxidized HiPIP from
C. vinosum. The EXSY and NOE connectivities are indicated (Bertini *et al.*, 1992a).

increase in temperature determines an increase in the population of more highly excited
paramagnetic levels. The temperature dependence of the eight signals of the β-CH$_2$
protons of the cysteine ligands in the oxidized protein (Fig. 26) shows that four of the
downfield-shifted signals decrease their shift with increasing temperature, in a Curie-
type behavior; two of the downfield shifted signals display an antiCurie behavior; and
the two upfield signals, again, apparently follow a Curie behavior (Bertini *et al.*, 1991b).
We will show here how this temperature dependence can be explained on the basis of
the theoretical treatment previously illustrated, and how, together with information
coming from Mössbauer spectroscopy, it can permit the assignment of the oxidation
states of the individual iron ions in the cluster.

^{57}Fe Mössbauer spectroscopy, the best method for identifying the oxidation state
of iron ions, has revealed that the cluster contains a pair of Fe^{3+} and a pair of Fe$^{2.5+}$
ions (Moss *et al.*, 1968; Dickson *et al.*, 1974; Middleton *et al.*, 1980; Papaefthymiou *et
al.*, 1986). EPR spectroscopy has established that the electron spin ground state is
$S = \frac{1}{2}$ (Antanaitis and Moss, 1975; Peisach *et al.*, 1977). This ground state must arise
from the magnetic coupling of the individual subspins of each iron ion. Hyperfine
splitting, caused by coupling of the nuclear spin with the unpaired electron(s), can be
observed in Mössbauer spectroscopy if the spectra are recorded at very low temperature

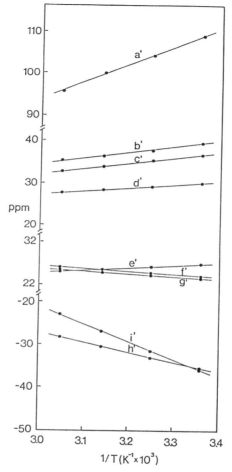

Figure 26. Chemical shifts versus $1/T$ dependence of the hyperfine-shifted signals of oxidized HiPIP from *C. vinosum* (Bertini *et al.*, 1991b).

in the presence of an additional, externally generated, magnetic field B_0 strong enough to split the electronic spin levels in such a way as to freeze the electrons in their ground spin states and to put all the clusters into the fundamental total spin state. The sign of the internal hyperfine magnetic field B_{int}, which is larger than B_0, can be obtained by observing how the hyperfine splitting varies in external fields of various magnitudes: if the splitting increases with increasing B_0, the resultant flux density is given by $|B_0 + B_{int}|$ and B_{int} is positive; on the other hand, if the splitting decreases with increasing B_0, then the flux density is given by $|B_0 - B_{int}|$ and B_{int} is negative. For spin-uncoupled ^{57}Fe, B_{int} is always negative (Gibb, 1976).

In spin-coupled systems such as the Fe_4S_4 clusters, the electronic intracluster spin–spin coupling is much greater than the nuclear Zeeman coupling, so that information on the way electron subspins couple among themselves within the cluster core can be gained from the knowledge of the sign of the hyperfine coupling constants of the various iron subsites. In fact, the sign of the internal hyperfine magnetic field B_{int} depends on whether the electronic subspin is parallel (as in monomeric species) or antiparallel to the nuclear spin of the iron subsites, thus being negative in the former and positive in the latter case. In electron spin-coupled systems, the larger subspin aligns itself with the external field, generating a negative hyperfine coupling constant A, whereas the smaller subspin aligns parallel or antiparallel to the external field depending on whether the electron spin-coupling is ferro- or antiferromagnetic, thus showing a negative or positive hyperfine constant A, respectively (Gibb, 1976). We recall here, as discussed in Section 3.1, that for antiferromagnetic coupling, for the smallest S, C_{i1} is always negative, and the corresponding hyperfine coupling constant behaves in the same way.

The ferric and mixed-valence pair observed in the Mössbauer spectra constitute two subsets characterized by a positive and a negative hyperfine constant, respectively (Middleton *et al.*, 1980). This indicates that the $S = \frac{1}{2}$ overall electronic spin ground state is produced by antiferromagnetic coupling between two subspins S_{12} and S_{34} and that the subspin of the mixed-valence pair is larger than the subspin of the ferric pair (Middleton *et al.*, 1980). As for the possible values for S_{12} and S_{34}, the pairs $|\frac{9}{2}, 4\rangle$ and $|\frac{7}{2}, 3\rangle$ would be consistent with the values of the observed hyperfine constants, which are both relatively large with respect to monomeric iron complexes, for which $S = 0-\frac{5}{2}$ (Noodleman, 1988; Mauesca *et al.*, 1991; Banci *et al.*, 1991c). The two Fe atoms within each couple are not resolvable by this technique and have been considered equivalent. On the other hand, NMR spectroscopy, taking advantage of the paramagnetic ground state and of the overall antiferromagnetism of oxidized cores of HiPIP, can ultimately provide the means to distinguish, locate, and characterize all the Fe subsites individually.

In reduced HiPIP the cluster is in the $[Fe_4S_4]^{2+}$ oxidation level, formally constituted by two Fe^{3+} and two Fe^{2+} ions. The ground spin state is diamagnetic ($S = 0$), and Mössbauer spectroscopy of *C. vinosum* HiPIP suggests that all the iron ions are actually in the $+2.5$ oxidation state, with data coming from magnetic Mössbauer spectra consistent with the presence of two pairs of antiferromagnetically coupled pairs of iron ions (Moss *et al.*, 1968; Dickson *et al.*, 1974; Middleton *et al.*, 1980; Papaefthymiou *et al.*, 1986).

On the basis of the information regarding the oxidation states and the sign of the hyperfine constant for each iron ion, which, as we have seen, is derived from Mössbauer spectroscopy, it is possible to identify the signals belonging to pairs of methylene protons of cysteines bound to the mixed-valence pair or to the ferric pair. Thus, the downfield signals showing Curie behavior must correspond to cysteines bound to the mixed-valence pair—which, as seen before, has been shown by magnetic Mössbauer spectroscopy to display a higher spin ground state than the ferric pair, and thus must have a negative hyperfine constant. The β-CH_2 protons attached to the cysteines bound to the ferric pair must display a positive hyperfine constant because of the smaller spin ground state, and the remaining four signals must thus move to higher resonant fields by decreasing the temperature, consistently with the negative C_{ij} coefficients of the lower-lying spin levels. In the case of the two signals with positive isotropic shifts, the temperature dependence is thus of an antiCurie type because of the decrease in the isotropic shift with decreasing temperature, whereas for the two signals with negative isotropic shifts the behavior has been termed "pseudoCurie" because it tends to increase the (upfield) isotropic shift with decreasing temperature (Bertini *et al.*, 1991b). The behavior of the ferric pair is analogous to that observed, and discussed in Section 5, for the ferrous ion Fe_2S_2 ferredoxins.

On the basis of the concepts so far developed, we can move on to illustrate the strategy that has been followed to reach the final step in the investigation of HiPIP coming from diverse microorganisms, that is, the determination of which cysteine residue is bound to a specific iron ion whose electronic structure has been elucidated. This final accomplishment can be achieved only by NMR, also utilizing complementary data coming from the amino acid sequence, the high-resolution X-ray structure of the protein, and Mössbauer spectroscopy, indicating the iron ions' oxidation states and the relative sign of the hyperfine constants.

The steps that have been followed are as follows:

1. *Assignment of the hyperfine shifted signals to the geminal β-CH_2 proton pairs belonging to the four cysteines bound to the cluster iron ions.* This information can be gained by 1D NOE difference spectra obtained by selective saturation of these signals and/or by 2D NOESY and COSY spectra in the oxidized and reduced states of the protein. Obtaining a full correlation between each hyperfine-shifted signal in the oxidized and reduced protein through 1D and 2D NMR saturation transfer experiments may also be useful.

2. *Determination of the temperature dependence of the hyperfine-shifted signals in the oxidized protein.* Together with the theoretical treatment outline above, this step allows us to correlate the oxidation state of each iron ion with the temperature dependence of the signals of the protons belonging to the residues bound to each iron ion.

3. *Detection of* 1D NOEs *from the hyperfine shifted β-CH_2 signals to other signals in the diamagnetic part of the spectrum of the oxidized protein.* These signals must belong to protons close to those selectively saturated, and investigation of these diamagnetic signals by 2D NMR experiments would reveal characteristic patterns of amino acid residues.

4. *Sequence-specific assignment of the four cysteines bound to the iron ions.* With the use of the primary sequence and X-ray data, in turn, this allows the assignment of the oxidation state of each iron ion within the protein frame. The structural elements derived from this analysis also constitute a first step toward the elucidation of the solution structure of these paramagnetic metalloproteins. The comparison of the various arrangements of amino acid residues around the cluster core observed in HiPIP from diverse bacterial sources may help in accounting for the different redox properties displayed.

This procedure has been applied to HiPIP from *Chromatium vinosum* and *Rhodocyclus gelatinosus*, and is in progress for HiPIP from *Ectothiorhodospira halophila II*. On the contrary, HiPIP from *Ectothiorhodospira vacuolata*, *Rhodospirillum tenue*, *Chromatium gracile*, and *Rhodocyclus globiformis*, the 1D ^1H NMR spectra of which have been performed, are still awaiting complete characterization. In the following we will

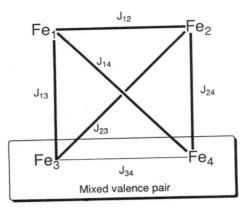

Figure 27. Magnetic coupling scheme used in the theoretical treatment of oxidized HiPIPs and reduced Fe_4S_4 ferredoxins.

thoroughly illustrate this protocol for HiPIP isolated from *C. vinosum* and *R. gelatinosus*, and we will only briefly illustrate the results obtained so far for the other HiPIP.

7.1. *Chromatium vinosum* HiPIP

The amino acid sequence (Tedro *et al.*, 1981., Przysiecki *et al.*, 1985) and high-resolution X-ray structure (Carter *et al.*, 1974) of this protein are known. The Mössbauer spectra have been discussed above (Dickson *et al.*, 1974; Middleton *et al.*, 1980), together with the assignment of the geminal methylene cysteine proton signals (Cowan and Sola, 1990a; Bertini *et al.*, 1991b).

The previously illustrated temperature dependence has been qualitatively interpreted in terms of the coupling scheme shown in Fig. 27, in which J_{nm} are the iron–iron isotropic exchange coupling constants. Considering the Fe_1–Fe_2 pair as the ferric couple, and the Fe_3–Fe_4 as the mixed-valence couple, with Fe_4 being nominally the Fe(II) ion, it has been shown that the experimentally observed pattern of down- and upfield shifts, with the relative temperature dependence, can be reproduced by a model in which J_{34} is the smallest of all J values and the removal of the degeneracy between Fe_1 and Fe_2 is achieved by differentiating the values of J_{13} and J_{14} from those of J_{23} and J_{24}, respectively (Bertini *et al.*, 1991b; Banci *et al.*, 1991c).

The smaller value of J_{34} can actually arise from an antiferromagnetic coupling partially quenched by a delocalization term B, which brings about a ferromagnetic contribution. However, the large correlation between J and B has suggested the use of simple Heisenberg Hamiltonians (Bertini *et al.*, 1991b).

The sequence-specific assignment of the cysteine signals to the various types of iron ions has been performed (Bertini *et al.*, 1992a) by first carrying out 1D NOE experiments on the oxidized protein. These spectra allow the experimenter to correlate signal f' to signals in the diamagnetic part of the spectrum, which, on the basis of T_1 values, EXSY, and TOCSY spectra, have been assigned to ring NH protons of the two tryptophan residues Trp 76 and Trp 80. X-ray data analysis shows that the Hβ1 proton of Cys 46 is very close to the ring NH proton of Trp 80, giving rise to the dipolar connectivity observed experimentally. The uniqueness of this feature suggests that f' belongs to $H\beta 1$ of Cys 46, and, in consequence, that g' belongs to $H\beta 2$ of the same cysteine. Thus Cys 46 is bound to an iron ion belonging to the ferric pair and, more precisely, to the ferric ion having a smaller coupling with the mixed valence pair, which determines its downfield shift and antiCurie behavior (Bertini et al., 1992a).

Analogous experiments involving the combination of 1D and 2D NMR techniques have allowed similar types of correlations of the remaining cysteines, thus generating the sought sequence-specific assignment. It has been thus concluded that the iron ions coordinated to cysteines 43 (signals h' and i') and 46 (f' and g') are those which show a pure ferric character in the oxidized protein, whereas cysteines 63 (a' and c') and 77 (b' and d') coordinate the iron ions forming the mixed-valence pair (Bertini et al., 1992a). An extensive assignment of the more diamagnetic signals has independently confirmed the above assignment (Nettesheim et al., 1992).

The availability of the same information for other proteins could permit the interpretation of the data in terms of the influence of the protein environment on the electronic and magnetic properties of the cluster core and possibly provide a better understanding of both the electron transfer mechanism and the redox potential differentiation among proteins from diverse microorganisms.

7.2. *Rhodocyclus gelatinosus* HiPIP

Even though the X-ray structure and Mössbauer spectra are not available for this protein, the amino acid sequence is known (Tedro et al., 1981; Przysiecki et al., 1985). Despite substantial differences between the sequences of HiPIP from *C. vinosum* and *R. gelatinosus*, the sequence positions of the cysteines bound to the cluster are invariant. The knowledge of the specific residue substitutions in the environment of the cluster has proven to be very important to predict and search for differences in the ^{1}H NMR 1D and 2D patterns between the spectra from *C. vinosum* and *R. gelatinosus* (Bertini et al., 1992c). The spectra of *R. gelatinosus* HiPIP in the reduced and the oxidized state are shown in Fig. 28A and 28B, respectively, together with the correlations obtained through EXSY (Banci et al., 1991c; Bertini et al., 1992c). These spectra are qualitatively very similar to those of the protein from *C. vinosum* (Banci et al., 1991c). In particular, the same pattern of seven downfield and two upfield-shifted signals is observed, eight assigned to β-CH$_2$ protons of bound cysteines and the last to an α-CH proton. The pairwise assignment has been carried out, and the temperature dependence of these signals has shown a pattern very similar to that observed for *C. vinosum* (Banci et al., 1991c). The conclusions drawn from the theoretical treatment suggest that the difference in isotropic shift of the corresponding pair of signals f' and g' may arise from slight differences in the coupling constants J_{13} and J_{14}, which are more similar to J_{23} and J_{24} in *R. gelatinosus* than in the case of *C. vinosum*, thus decreasing the inequivalence between the ferric ions Fe$_1$ and Fe$_2$ (Banci et al., 1991c).

^{1}H NOE and 2D NMR studies on the oxidized and reduced HiPIP have been carried out recently, and, using the same strategy as that outlined for *C. vinosum* HiPIP, the signals of the cysteines have been specifically assigned and connected with the oxidation state of each iron ion (Bertini et al., 1992c). Thus, signals a' and b' have been assigned to Cys 63, whereas signals c', d', and e' have been shown to belong to Cys 77.

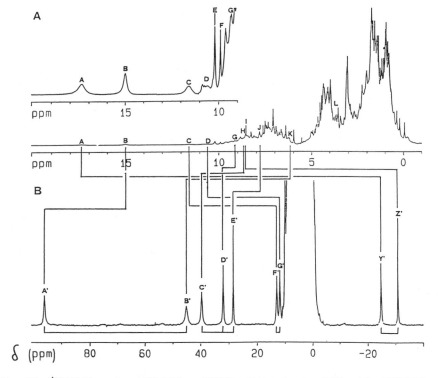

Figure 28. ¹H NMR spectra at 600 MHz and 300 K of (A) reduced and (B) oxidized HiPIP from *R. gelatinosus*. The EXSY and NOE connectivities are indicated (Bertini *et al.*, 1992c).

The signals of the methylene protons of Cys 43 are y' and z', and, finally, those of Cys 46 are f' and g'. These results show that, if the mixed valence pair is the one with larger subsite S as in *C. vinosum* HiPIP, then the iron(III) ions are bound to Cys 43 and 46, and the mixed-valence irons are bound to Cys 63 and 77 (Bertini *et al.*, 1992c).

7.3. Other HiPIP

In *E. halophila*, two HiPIP isoenzymes, I and II, can be isolated (Bartsch, 1978; Meyer, 1985; Przysiecki *et al.*, 1985). The amino acid sequences are known for both isozymes (Przysiecki *et al.*, 1985), whereas preliminary Mössbauer data are available for *E. halophila* II, indicating that, as in *C. vinosum* HiPIP, the larger subsite S is associated with the mixed-valence pair (Bertini *et al.*, 1992f). The X-ray structure of *E. halophila* I has recently been solved (Breiter *et al.*, 1991).

The ¹H NMR specra of oxidized and reduced *E. halophila* I (Krishnamoorthy *et al.*, 1986) show that the oxidized protein is qualitatively similar to that of *C. vinosum* and *R. gelatinosus*, with the only difference being the absence of two signals, corresponding to f' and g' in the other cases, which are probably hidden in the diamagnetic region. The

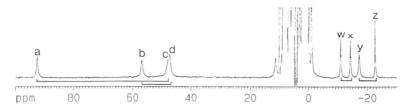

Figure 29. ^{1}H NMR spectrum at 600 MHz and 300 K of oxidized HiPIP from *E. halophila II*. The NOE connectivities are indicated (Banci *et al.*, 1991b).

temperature dependence of the signals is analogous to that of the corresponding signals in *C. vinosum* and *R. gelatinosus* (Krishnamoorthy *et al.*, 1986). No further characterization is available.

E. halophila II, on the other hand, has been studied more thoroughly, and the NMR spectra of the oxidized and reduced protein have been obtained (Krishnamoorthy *et al.*, 1986; Banci *et al.*, 1991b). In the oxidized protein spectrum (Fig. 29), eight hyperfine-shifted signals fall outside the diamagnetic region, with four signals being downfield and four upfield. The pairwise assignment of the signals to geminal couples of β-CH$_2$ cysteine protons, shown in Fig. 29, has been carried out using NOE experiments (Banci *et al.*, 1991b). The peculiarity of *E. halophila II* is the presence of two pairs of signals in the upfield region. Considering the Curie behavior of the downfield signals and the pseudoCurie temperature dependence of the upfield signals, they have been assigned to cysteines bound to the mixed-valence and to the ferric pair, respectively (Banci *et al.*, 1991b). This assignment is now confirmed by Mössbauer data (Bertini *et al.* 1992f).

The sequence-specific assignment of the signals of *E. halophila II* to the various iron ions is currently in progress (Bertini *et al.*, 1992b). Preliminary data point to a different electron distribution than that observed for *C. vinosum* and *R. gelatinosus*, thus rendering this protein more complex and interesting for an understanding of the mechanisms of redox catalysis.

The NMR spectra of *E. vacuolata* HiPIP I and II (Krishnamoorthy *et al.*, 1986), *R. tenue* (Krishnamoorthy *et al.*, 1989), *R. globiformis* (Bertini *et al.*, 1992e), and *C. gracile* (Sola *et al.*, 1989) have been obtained.

The pattern of signals in these proteins is very similar, in all cases, to that observed in *C. vinosum* and *R. gelatinosus*. The amino acid sequence is known for all proteins (Tedro *et al.*, 1981; Przysiecki *et al.*, 1985; Ambler, 1992), and, again, a low to medium sequence homology among them is probably reflected in a very similar tertiary structure arrangement around the cluster. More detailed characterization of these proteins is not available.

8. NMR OF 2[Fe$_4$S$_4$] FERREDOXINS

The large majority of Fe$_4$S$_4$ bacterial ferredoxins are characterized by the presence of two cubane clusters held in the same protein through eight cysteine residues bound to the iron ions. Each of these clusters can exist in the oxidized [Fe$_4$S$_4$]$^{2+}$ and in the reduced $[Fe_4S_4]^{1+}$ form, and they carry out electron transfer utilizing this low-potential redox couple (Howard and Rees, 1991). We will show here how considerations analogous to those seen for HiPIP can be applied to these systems and what goals are currently being pursued.

The oxidized core corresponds to the cluster present in reduced HiPIP (Antanaitis and Moss, 1975) and has a diamagnetic ground state (Thompson et al., 1974). The reduced state is paramagnetic, and it is usually present in the $S = \frac{1}{2}$ spin ground state (Middleton et al., 1978). Recently, however, deviations from this behavior have been observed, the best-characterized examples being the $[Fe_4Se_4]^{1+}$ clusters of selenium-reconstituted reduced ferredoxin from C. pasterianum, shown to have EPR and Mössbauer spectra consistent with the presence of spin states up to $S = \frac{7}{2}$ (Meyer and Moulis, 1981; Moulis and Meyer, 1982, 1984; Moulis et al., 1984a, b; Gaillard et al., 1986, 1987; Auric et al., 1987), and the $[Fe_4S_4]^{1+}$ centers of the Fe protein of A. vinelandii Mo- and V-nitrogenases, which exist as a mixture of $S = \frac{1}{2}$ and $S = \frac{3}{2}$ ground states (Lindhal et al., 1985, 1987; Hales et al., 1986). Other cases of this unusual behavior are the $[Fe_4S_4]^{1+}$ centers of amidotransferases from Bacillus subtilis (Vollmer et al., 1983) and hydrogenase II from C. pasterianum (Rusnak et al., 1987), displaying $S \geq \frac{3}{2}$. Finally, the P-clusters of nitrogenase, believed to be $[Fe_4S_4]^{1+}$ clusters, have been shown to have $S = \frac{1}{2}$ or $\frac{7}{2}$ (Zimmermann et al., 1978; Huynh et al., 1980; Johnson et al., 1981; Smith et al., 1982).

The reduced core formally contains three iron(II) ions and one iron(III) ion. Mössbauer spectroscopy under zero- and applied-field conditions indeed shows the presence of two pure ferrous ions and a Fe^{2+}–Fe^{3+} mixed-valence pair. These two pairs constitute two subsets of signals of equal intensity, one corresponding to the ferrous pair with positive internal fields ($A > 0$) and one corresponding to the mixed-valence pair with negative internal fields ($A < 0$). Thus, the ground spin state results from antiferromagnetic coupling between the pure ferrous pair (smaller S) and the mixed-valence pair (larger S) (Middleton et al., 1978).

The same protocol illustrated for HiPIP has been tentatively applied also to the [8Fe–8S] protein from C. pasterianum, utilizing advanced 1D and 2D NMR techniques developed for paramagnetic macromolecules. C. pasterianum is the most extensively characterized example of low-potential ferredoxins, its amino acid sequence (Hall et al., 1975) and Mössbauer spectra (Middleton et al., 1978) being known. Early ^1H NMR studies showed the presence of several isotropically shifted resonances (Poe et al., 1970), their assignment to β-CH_2 cysteinyl protons having been partially achieved by isotopic labelling experiments performed on an analogous ferredoxin from Clostridium acidiurici (Packer et al., 1977). Recently, systematic studies have been performed on the fully reduced ($2[Fe_4S_4]^{1+}$), partially ($1[Fe_4S_4]^{1+}$, $1[Fe_4S_4]^{2+}$) reduced, and fully oxidized ($2[Fe_4S_4]^{2+}$) forms of the protein (Bertini et al., 1990, 1991a, 1992h; Busse et al., 1991).

8.1. Fully Oxidized Ferredoxin

The spectrum shown in Fig. 30A shows thirteen isotropically shifted resonances of fully oxidized ferredoxin from C. pasterianum (a–m) in the 8–18 ppm region (Bertini et al., 1991a; Busse et al., 1991). Five additional signals belonging to protons in the immediate surroundings of the clusters have ben detected in the diamagnetic region by their very short relaxation times T_1 (Bertini et al., 1991a). To identify the eight geminal

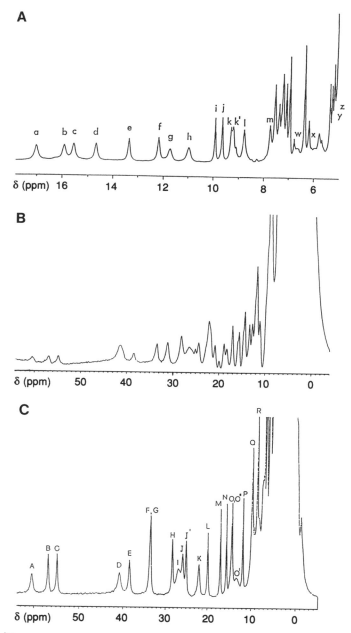

Figure 30. ¹H NMR spectra at 600 MHz of ferredoxin from *C. pasterianum*: (A) fully oxidized, 300 K (Bertini *et al.*, 1991a); (B) partially reduced, 293 K (Bertini *et al.*, 1990, 1992h); and (C) fully reduced, 293 K (Bertini *et al.*, 1992h).

pairs of β-CH$_2$ protons of cysteines bound to the iron ions of the two cores, NOE, COSY, and NOESY experiments have been successfully carried out, the signal couples being a–k', b–y, c–k, d–z, e–w, f–x, g–l, and h–m. With the same experiments, the signals i and j have ben tentatively assigned as α-CH protons of pairs c–k and a–k', respectively. As in the case of reduced HiPIP, the large delocalization present in the [Fe$_4$S$_4$]$^{2+}$ core does not allow one to distinguish the cysteines bound to Fe(II) and Fe(III).

8.2. Partially and Fully Reduced Ferredoxin

The NMR spectra of partially and fully reduced ferredoxin are shown in Fig. 30B and C (Bertini *et al.*, 1990, 1992h). In the fully reduced spectrum, the paramagnetic $S = \frac{1}{2}$ ground state causes a large chemical shift range of the isotropically shifted signals (*ca* 60 ppm). The spectrum of partially reduced ferredoxin displays a set of resonances belonging to species in which one cluster is reduced and the other oxidized, with the extra electron exchanging quickly on the NMR time scale between the two clusters, in addition to the two sets of signals of the completely reduced and oxidized proteins. EXSY experiments (Fig. 31) performed on the partially reduced protein have revealed the correlations between most of the signals of the oxidized form (already known) with those of the reduced form (Bertini *et al.*, 1990). Thus, in the spectrum of the reduced protein, the following couples of signals belong to the same geminal methylene cysteine protons: E–G, A–?, F–L, J–?, B–I, J'–?, D–C, K–H, where the question marks indicate that the partners of signals A, J, and J' have not been established. N and M would then belong to α-CH protons of pairs F–L and E–G. In addition, the β-CH$_2$ signals of the two individual clusters have been assigned, and the difference in redox potentials between them calculated on the basis of the positive or negative deviations of the isotropic shift of signals belonging to the partially reduced protein from the unweighted average isotropic shift in the fully oxidized and fully reduced ferredoxin (Bertini *et al.*, 1990, 1992h).

The temperature dependence of the shifts of the two individual sets of signals of the reduced protein is given in Fig. 32. This behavior can be interpreted in terms of the same theoretical approach utilized for HiPIP. For the cluster for which the pairwise assignment of the β-CH$_2$ protons is complete, the Heisenberg coupling scheme used is the same as in Fig. 27. As in the case of HiPIP, the experimental results can be reproduced by using a value of J_{34}, the coupling constant between the iron ions belonging to the mixed valence pair, smaller than J_{12} (Bertini *et al.*, 1990, 1992h).

The complete assignment of each iron ion in each cluster to a specific oxidation state is a goal currently being pursued.

9. COBALT(II)-SUBSTITUTED THIONEINS

Metal-deprived metallo-thioneins (MT) bind several divalent metal ions (Vasak and Kägi, 1983; Vasak, 1986). Titration of MT with up to seven moles of cobalt(II) ions per mole of protein affords the NMR spectra reported in Fig. 33 (Bertini *et al.*, 1989b, c). The development of such spectra is not linear with the amount of cobalt(II) added. It has been proposed that cobalt(II) initially binds in such a way as to avoid sharing cysteine sulfurs as ligands. Up to three cobalt(II) ions bind in this way, one in

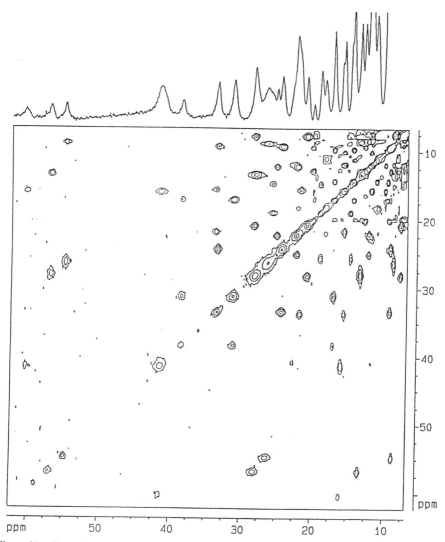

Figure 31. Downfield region of the 2D EXSY spectrum at 600 MHz and 293 K of 50%-reduced ferredoxin from *C. pasterianum* (Bertini *et al.*, 1992h).

the three-metal site and two in the four-metal site. The fourth equivalent of cobalt(II) binds in the four-metal cluster, producing a magnetically coupled trimetallic system characterized by very broad NMR lines (Fig. 33A). The fifth equivalent completes the four-metal cluster, giving rise to a large number of sharp hyperfine-shifted signals, as shown in Fig. 33B. The sixth and seventh equivalents complete the three-metal cluster.

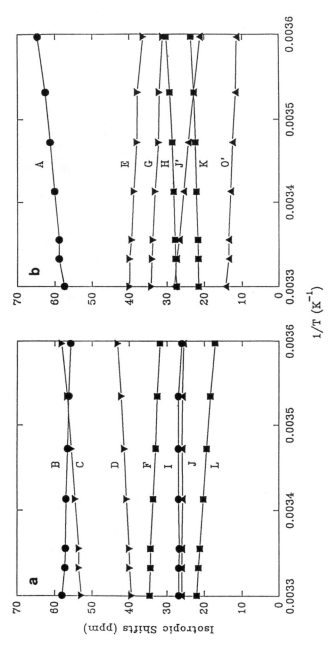

Figure 32. Isotropic shift versus $1/T$ dependencies of the hyperfine-shifted signals corresponding to the cluster of (a) lower and (b) higher potential in the case of fully reduced ferredoxin from *C. pasterianum* (Bertini *et al.*, 1992h).

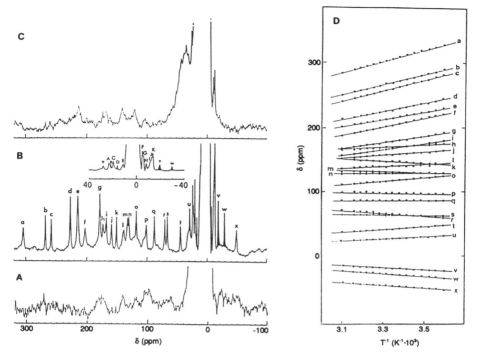

Figure 33. ^1H NMR spectra at 90 MHz and 298 K of *apo*-MT solutions reacted with (A) 3.8 or (B) 7 equivalents of cobalt(II); (C) spectrum of Co_3Cd_4MT (Bertini *et al.*, 1989b). In D the temperature dependence of the hyperfine shifted signals is shown.

The overall shape of the NMR spectrum is not much altered, as the three-metal cluster is again characterized by very broad NMR lines. This feature of the Co_3 cluster has been independently confirmed by the Cd_4Co_3MT derivative, where the four-metal cluster is filled with cadmium ions (Fig. 33C). Therefore, the spectrum of Fig. 33B essentially contains only the signals of the Co_4 cluster. We take such a spectrum as an example of how exchange coupling within a cluster can lead to unexpected temperature dependencies of the shifts even in homometallic systems.

The temperature dependence of the hyperfine-shifted signals is shown in Fig. 33D. While most of the signals follow a Curie-like behavior, there are six signals (k, m, p, q, r, s) that exhibit a more or less antiCurie behavior. From Fig. 7, it appears that this cluster is less symmetric than that of Fe_4S_4 systems: metals 1 and 2 are bound to two terminal and two bridging cysteines each, whereas metals 3 and 4 are bound to one and three bridging cysteines each. Therefore, one of the six exchange coupling constants, namely J_{12}, must be zero or at least much smaller than the others. If J_{12} in either Hamiltonian, Eq. (22) or (24), is set to zero (i.e., $\Delta J_{12} = -J$), then metals 1 and 2 are expected to be less antiferromagnetically coupled to the other metals in the cluster than metals 3 and 4, because the latter are also coupled to one another. As a result, the system has an $S = 0$ ground state, as expected for an antiferromagnetically coupled homovalent system, but paramagnetic excited states where $S_{12} > S_{34}$ lie close in energy. As in the case of Fe_4S_4 systems, one pair (the 1–2 pair in this case) forces the other pair to align antiparallel to the external magnetic field and the corresponding protons to exhibit antiCurie behavior. By referring to Fig. 7, we then expect the four terminal cysteines coordinated to metals 1 and 2 to exhibit normal Curie behavior and the two cysteines coordinated to metals 3 and 4 to exhibit antiCurie behavior. Of the bridging cysteines, the one bridging metals 3 and 4 should also exhibit antiCurie behavior, whereas the four cysteines bridging metals 1–3, 2–3, 1–4, and 2–4 should experience contrasting effects from the two metal ions and thus should have a less pronounced temperature dependence. As noted before, six signals indeed exhibit antiCurie behavior. These signals have been shown through NOE measurements to be dipolarly connected pairwise. Among the other sixteen signals, the eight showing a more pronounced Curie behavior are also connected pairwise, and the same holds for the remaining eight signals. This simple model is therefore able to explain in great detail the NMR properties of the Co_4 cluster (Bertini *et al.*, 1989b, c).

10. CONCLUDING REMARKS

This chapter has provided an overview of recent achievements in the theoretical understanding of the NMR parameters in exchange-coupled systems of biological relevance. In parallel, as reviewed in Chapters 1 and 2 of this volume, the continuous progress in NMR instrumentation and in the development of pulse sequences for multidimensional spectroscopies has almost abolished the gap between paramagnetic and diamagnetic proteins in terms of suitability for NMR studies. Pitfalls in the application of some routine 2D techniques (i.e., COSY) to paramagnetic systems have also been recognized and overcome (Bertini *et al.*, *Chem. Phys. Lett.*, in press; see also Chapter 2, this volume). As a result of both theoretical and technical achievements, we are now able to address such sophisticated issues as the identification of the oxidation states of the individual metal ions in mixed-valence polymetallic systems. We have also shown how magnetic coupling can be exploited to dramatically enhance the resolution in the NMR spectra of proteins containing slow-relaxing metals such as copper(II). The few biological systems discussed here cover only a small fraction of all the polymetallic metalloproteins and metalloenzymes described in the introductory section of this chapter, representing an overview of the field, which we hope can stimulate the interest of more NMR researchers. However, we predict that in the next few years many more of these systems will be successfully attacked by NMR spectroscopists, who have been discouraged, up to now, by the unfavorable electronic relaxation times of certain metal ions in some cases or by the large molecular weights in some others. A theoretical challenge is represented, for example, by the putative [6Fe–6S] cluster–containing proteins, in which some of the oxidation levels could involve localization or delocalization of electron density and in which NMR could contribute decisively to the elucidation of the electronic and magnetic properties of the cluster. Another challenge is represented by the understanding of the magnetic coupling and of the mechanism of action of the active site of sulfite reductase, a problem which is just being tackled by NMR (Cowan and Sola, 1990b). In the more distant future, we may hope to see NMR studies undertaken on larger and more complicated systems like nitrogenases, which are so fascinating and important for the environmental point of view. To this end, the present studies on iron–sulfur clusters will constitute a precious precedent. When the electronic relaxation times are so long as to prevent the observation of NMR signals even in the presence of magnetic coupling among paramagnetic metal centers, nuclear magnetic relaxation dispersion (NMRD) (Bertini and Luchinat, 1986; Banci *et al.*, 1991d) can still provide information on the cluster relaxation mechanisms and on the solvent accessibility to the metal site. NMRD has already been applied even to extended systems such as ferritin, which turned out to represent a very efficient relaxing agent of great potential in magnetic resonance imaging techniques (Koenig and Brown, 1990).

REFERENCES

Adman, E. T., Sieker, L. C., and Jensen, L. H., 1973, *J. Biol. Chem.* **248**:3987.
Adman, E. T., Sieker, L. C., and Jensen, L. H., 1976, *J. Biol. Chem.* **251**:3801.
Aisen, P., and Listowsky, I., 1980, *Annual Rev. Biochem.* **49**:357.
Allgood, G. S., and Perry, J. J., 1986, *J. Bacteriol.* **186**:563.
Ambler, R. P., 1992, private communication.
Anderson, P. W., and Hasegawa, H., 1955, *Phys. Rev.* **100**:675.
Anderson, R. E., Dunham, W. R., Sands, R. H., Bearden, A. J., and Crespi, H. L., 1975, *Biochim. Biophys. Acta* **408**:306.
Andréasson, L.-E., and Vänngård, T., 1970, *Biochim. Biophys. Acta* **200**:247.
Andréasson, L.-E. and Vänngård, T., 1988, *Ann. Rev. Plant Physiol. Plant Mol. Biol.* **39**:379.
Antanaitis, B. C., and Aisen, P., 1983, *Adv. Inorg. Biochem.* **5**:111.
Antanaitis, B. C., and Moss, T. H., 1975, *Biochim. Biophys. Acta* **405**:262.
Aono, S., Bryant, F. O., and Adams, M. W. W., 1989, *J. Bacteriol.* **171**:3433.
Armstrong, F. A., George, S. J., Thomson, A. J., and Yates, M. G., 1988, *FEBS Lett.* **234**:107.
Auric, P., Gaillard, J., Meyer, J., and Moulis, J.-M., 1987, *Biochem. J.* **242**:525.
Backes, G., Mino, Y., Loehr, T. M., Meyer, T. E., Cusanovich, M. A., Sweeney, W. V., Adman, E. T., and Loehr, J. S., 1991, *J. Am. Chem. Soc.* **113**:2055.
Baici, A., Luisi, P. L., Palmieri, S., and Marchesini, A., 1979, *J. Mol. Catal.* **6**:135.
Banci, L., Bertini, I., Luchinat, C., Monnanni, R., and Scozzafava, A., 1987a, *Inorg. Chem.* **26**:153.
Banci, L., Bertini, I., Luchinat, C., and Scozzafava, A., 1987b, *J. Am. Chem. Soc.* **109**:2328.
Banci, L., Bertini, I., Luchinat, C., and Hallewell, R. A., 1988a, *J. Am. Chem. Soc.* **110**:3629.
Banci, L., Bertini, I., Luchinat, C., Monnanni, R., and Scozzafava, A., 1988b, *Inorg. Chem.* **27**:107.
Banci, L., Bertini, I., Luchinat, C., Piccioli, M., Scozzafava, A., and Turano, P., 1989a *Inorg. Chem.* **28**:4650.
Banci, L., Bertini, I., Luchinat, C., and Scozzafava, A., 1989b, *J. Biol. Chem.* **264**:9742.
Banci, L., Bertini, I., Luchinat, C., Scozzafava, A., and Turano, P., 1989c, *Inorg. Chem.* **28**:2377.
Banci, L., Bencini, A., Bertini, I., Luchinat, C., and Piccioli, M., 1990a, *Inorg. Chem.* **29**:4867.
Banci, L., Bertini, I., Cabelli, D., Hallewell, R. A., Luchinat, C., and Viezzoli, M. S., 1990b, *Inorg. Chem.* **29**:2398.
Banci, L., Bertini, I., and Luchinat, C., 1990c, *Struct. Bonding* **71**:113.
Banci, L., Bertini, I., Luchinat, C., and Viezzoli, M. S., 1990d, *Inorg. Chem.* **29**:1438.
Banci, L., Bertini, I., Briganti, F., and Luchinat, C., 1991a, *New J. Chem.* **15**:467.
Banci, L., Bertini, I., Briganti, F., Luchinat, C., Scozzafava, A., and Vicens Oliver, M., 1991b, *Inorg. Chim. Acta* **180**:171.
Banci, L., Bertini, I., Briganti, F., Luchinat, C., Scozzafava, A., and Vicens Oliver, M., 1991c, *Inorg. Chem.* **30**:4517.

Banci, L., Bertini, I., and Luchinat, C., 1991d, *Nuclear and Electron Relaxation*, VCH, Weinheim, FRG.

Banci, L., Bertini, I., and Turano, P., 1991e, *Eur. Biophys. J.* **19**:141.

Banci, L., Bertini, I., Luchinat, C., Piccioli, M., and Scozzafava, A., 1992a, unpublished.

Banci, L., Bertini, I., Luchinat, C., and Viezzoli, M. S., 1992b, unpublished.

Banci, L., Carloni, P., La Penna, G., and Orioli, P. L., 1992c, unpublished.

Baranyn, V. V., Vagin, A. A., Melik-Adamyan, V. R., Grebenko, A. I., Khangulov, S. V., Popov, A. N., Andrianova, M. E., and Veinshtein, B. K., 1986, *Dokl. Akad. Nauk. SSSR* **288**:877.

Barrass, B. C., Coult, D. B., Rich, P., and Tutt, K. J., 1974, *Biochem. Pharmacol.* **23**:47.

Bartsch, R. G., 1978, *Methods Enzymol.* **53**:329.

Beinert, H., and Thomson, A. J., 1983, *Arch. Biochem. Biophys.* **222**:333.

Bencini, A., and Gatteschi, D., 1990, *Electron Paramagnetic Resonance of Exchange-Coupled Systems*, Springer-Verlag, Berlin.

Bertini, I., Lanini, G., and Luchinat, C., 1984, *Inorg. Chem.* **23**:2729.

Bertini, I., Briganti, F., Luchinat, C., Mancini, M., and Spina, G., 1985a *J. Magn. Reson.* **63**:41.

Bertini, I., Lanini, G., Luchinat, C., Mancini, M., and Spina, G., 1985b, *J. Magn. Reson.* **63**:56.

Bertini, I., Lanini, G., Luchinat, C., Messori, L., Monnanni, R., and Scozzafava, A., 1985c, *J. Am. Chem. Soc.* **107**:4391.

Bertini, I., Luchinat, C., Mancini, M., and Spina, G., 1985d, in *Magneto-Structural Correlations in Exchange-Coupled Systems* (R. D. Willett, D. Gatteschi, and O. Kahn, eds.), pp. 421–462, Reidel, Dordrecht.

Bertini, I., Luchinat, C., and Monnanni, R., 1985e, *J. Am. Chem. Soc.* **107**:2178.

Bertini, I., and Luchinat, C., 1986, in *NMR of Paramagnetic Molecules in Biological Systems*, Benjamin/Cummings, Menlo Park, CA.

Bertini, I., Luchinat, C., Owens, C., and Drago, R. S., 1987, *J. Am. Chem. Soc.* **109**:5208.

Bertini, I., Banci, L., Brown, R. D., Koenig, S. H., and Luchinat, C., 1988a, *Inorg. Chem.* **27**:951.

Bertini, I., Banci, L., and Luchinat, C., 1988b, in *Metal Clusters in Proteins* (L. Que, Jr., ed.) ACS, Washington, D.C.

Bertini, I., Luchinat, C., Viezzoli, M. S., and Wang, Y., 1988c, *Arch. Biochem. Biophys.* **269**:586.

Bertini, I., Banci, L., Bielski, B. H. J., Cabelli, D. E., Luchinat, C., Mullenbach, G. T., and Hallewell, R. A., 1989a, *J. Am. Chem. Soc.* **111**:714

Bertini, I., Luchinat, C., Messori, L., and Vasak, M., 1989b, *J. Am. Chem. Soc.* **111**:7296.

Bertini, I., Luchinat, C., Messori, L., and Vasak, M., 1989c, *J. Am. Chem. Soc.* **111**:7300.

Bertini, I., Briganti, F., Luchinat, C., and Scozafava, A., 1990, *Inorg. Chem.* **29**:1874.

Bertini, I., Briganti, F., Luchinat, C., Messori, L., Monnanni, R., Scozzafava, A., and Vallini, G., 1991a, *FEBS Lett.* **289**:253.

Bertini, I., Briganti, F., Luchinat, C., Scozzafava, A., and Sola, M., 1991b, *J. Am. Chem. Soc.* **113**:1237.

Bertini, I., Luchinat, C., Piccioli, M., Vicens Oliver, M., and Viezzoli, M. S., 1991c, *Eur. J. Biophys.* **20**:269.

Bertini, I., Capozzi, F., Ciurli, S., Luchinat, C., Messori, L., and Piccioli, M., 1992a, *J. Am. Chem. Soc.*, in press.

Bertini, I., Capozzi, F., Carloni, P., Ciurli, S., Luchinat, C., Piccioli, M. and Banci, L., 1992b, unpublished.

Bertini, I., Capozzi, F., Luchinat, C., Piccioli, M., and Vicens Oliver, M., 1992c, *Inorg. Chim. Acta*, in press.

Bertini, I., Luchinat, C., Ming, L. J., Piccioli, M., Valentine, J. S., and Viezzoli, M. S., 1992d, unpublished.

Bertini, I., Capozzi, F., Luchinat, C., and Piccioli, M., 1992e, unpublished.

Bertini, I., Texeira, M., Campos, A. P., and Luchinat, C., 1992f, unpublished.

Bertini, I., Luchinat, C., Ming, L. J., Piccioli, M., Valentine, J. S., and Viezzoli, M. S., 1992g, unpublished.

Bertini, I., Briganti, F., Luchinat, C., Messori, L., Monnanni, R., and Vallini, G., 1992h, *Eur. J. Biochem.* in press.

Beyer, W. F., and Fridovich, I., 1985, *Biochemistry* **24**:6460.

Blondin, G., and Girerd, J.-J., 1990, *Chem. Rev.* **90**:1359.

Bolin, J. T., Ronco, A. E., Mortenson, L. E., Morgan, T. V., Williamson, M., and Xuong, N.-H., 1991, in *Nitrogen Fixation: Achievements and Objectives* (R. Gresshof, A. Roth, J. Stacey, and W. E. Newton, eds.), Chapmann and Hall, New York.

Bouchilloux, S., McMahill, P., and Mason, H. S., 1963, *J. Biol. Chem.* **238**:1699.

Bovier-Lapierre, G., Bruschi, M., Bonicel, J., and Hatchikian, E. C., 1987, *Biochim. Biophys. Acta* **913**:20.

Braun, W., Wagner, G., Worgotter, E., Vasak, M., Kagi, J. H., and Wüthrich, K., 1986, *J. Mol. Biol.* **187**:125.

Breiter, D. R., Meyer, T. E., Rayment, I., and Holden, H. M., 1991, *J. Biol. Chem.* **266**:18660.

Britzinger, H., Palmer, G., and Sands, R. H., 1966, *Proc. Natl. Acad. Sci. USA* **55**:397.

Brock, J. H., 1985, *Topics Mol. Struct. Biol.* **7**:183.

Brudvig, G. W., 1987, *J. Bioenerg. Biomemb.* **19**:91.

Brudvig, G. W., 1988, in *Metal Clusters in Proteins* (L. Que, Jr ed.), pp. 221–237, ACS, Washington, D.C.

Brudvig, G. W., and Crabtree, R. H., 1986, *Proc. Natl. Acad. Sci. USA* **83**:4586.

Bruschi, M., and Guerlesquin, F., 1988, *FEMS Microbiol. Rev.* **54**:155.

Busse, S. C., La Mar, G. N., and Howard, J. B., 1991, *J. Biol. Chem.* **266**:23714.

Cammack, R., Rao, K. K., Hall, D. O., and Johnson, C. E., 1971, *Biochem. J.* **125**:849.

Cammack, R., Fernande, V. M., and Schneider, K., 1988, in *Bioinorganic Chemistry of Nickel* (J. R. Lancaster, ed.) pp. 167–190, VCH, Deerfield Beach FL.

Carter, C. W., Jr., Kraut, J., Freer, S. T., and Alden, R. A., 1974, *J. Biol. Chem.* **429**:6339.

Carter, C. W., Jr., 1977, *J. Biol. Chem.* **252**:7802.

Chan, T.-M., and Markley, J. L., 1983, *Biochemistry* **22**:6008.

Chasteen, N. D., Aisen, P., and Antanaitis, B. C., 1985, *J. Biol. Chem.* **260**:2926.

Christner, J. A., Janick, P. A., Siegel, L. M., and Munck, E., 1983a, *J. Biol. Chem.* **258**:11157.

Christner, J. A., Munck, E., Janick, P. A., and Siegel, L. M., 1983b, *J. Biol. Chem.* **258**:11147.

Christner, J. A., Munck, E., Kent, T. A., Janick, P. A., Salerno, J. C., and Siegel, L. M., 1984, *J. Am. Chem. Soc.* **106**:6786.

Ciurli, S., Yu, S.-B., Holm, R. H., Srivastava, K. K. P. and Munck, E., 1990, *J. Am. Chem. Soc.* **112**:8169.

Ciurli, S., Ross, P. K., Scott, M. J., Yu, S.-B., and Holm, R. H., 1992, *J. Am. Chem. Soc.*, in press.

Clark, P. A., and Wilcox, D. E., 1989, *Inorg. Chem.* **28**:1326.

Cline, J. A., Hoffman, B. M., Mims, W. B., LaHaie, E., Ballou, D. P., and Fee, J. A., 1985, *J. Biol. Chem.* **260**:3251.

Conover, R. C., Kowal, A. T., Fu, W., Park, J.-B., Aono, S., Adams, M. W. W., and Johnson, M. K., 1990a, *J. Biol. Chem.* **265**:8533.

Conover, R. C., Park, J.-B., Adams, M. W. W., and Johnson, M. K., 1990b *J. Am. Chem. Soc.* **112**:4562.

Cowan, J. A., and Sola, M., 1990a, *Biochemistry* **29**:5633.

Cowan, J. A., and Sola, M., 1990b, *Inorg. Chem.* **29**:2176.

Coyle, C. L., Zumft, W. G., Kroneck, P. M. H., Korner, H., and Jakob, W., 1985, *Eur. J. Biochem.* **178**:459.

Dalton, H., 1980, *Adv. Appl. Microbiol.* **26**:71.

De Paula, J. C., Beck, W. F., and Brudvig, G. W., 1986, *J. Am. Chem. Soc.* **108**:4002.

De Witt, J. G., Bentsen, J. G., Rosenzweig, A. C., Hedman, B., Green, J., Pilkington, S., Papaefthymiou, G. C., Dalton, H., Hodgson, K. O., Lippard, S. J., 1991, *J. Am. Chem. Soc.* **113**:9219.

Dickson, D. P. E., Johnson, C. E., Cammack, R., Evans, M. C. W., Hall, D. O., and Kao, K. K., 1974, *Biochem. J.* **139**:105.

Dilworth, M. J., Eady, R. R., Robson, R. L., and Miller, R. W., 1987, *Nature* **327**:167.

Ding, X.-Q., Bominaar, E. L., Bill, E., Winkler, A. X., Trautwein, S., Drüeke, S., Chauduri, P., and Wieghardt, K., 1990, *J. Chem. Phys.* **92**:178.

Doi, K., Antanaitis, B. C., and Aisen, P., 1988, *Struct. Bonding* **70**:1.

Dugad, L. B., La Mar, G. N., Banci, L., and Bertini, I., 1990, *Biochemistry* **29**:2263.

Dunham, W., Palmer, G., Sands, R. H., and Bearden, A. J., 1971, *Biochim. Biophys. Acta* **253**:373.

Eickman, N. C., Solomon, E. I., Larrabee, J. A., Spiro, T. G., and Lerch, K., 1978, *J. Am. Chem. Soc.* **100**:6529.

Ellerton, H. D., Elleton, N. F., and Robinson, H. A., 1983, *Progr. Biophys. Mol. Biol.* **41**:143.

Emptage, M. H., 1988, in *Metal Clusters in Proteins* (L. Que, Jr., ed.), pp. 343–371, ACS, Washington, D. C.

Fee, J. A., and Briggs, R. G., 1975, *Biochim. Biophys. Acta* **400**:439.

Fee, J. A., Malkin, R., Malmstrom, B. G., and Vanngard, T., 1969, *J. Biol. Chem.* **244**:4200.

Fee, J. A., Findling, K. L., Yoshida, T., Hille, R., Tarr, G. E., Hearshen, D. O., Dunham, W. R., Day, E. P., Kent, T. A., and Munck, E., 1984, *J. Biol. Chem.* **259**:124.

Freedman, T. B., Loehr, J. S., and Loehr, T. M., 1976, *J. Am. Chem. Soc.* **98**:2809.

Frey, M. H., Wagner, G., Vasak, M., Sorensen, O. W., Neuhaus, D., Worgotter, E., Kagi, J. H., Ernst, R. R., and Wüthrich, K., 1985, *J. Am. Chem. Soc.* **107**:6847.

Frieden, E., 1979, in *Copper in the Environment*, Part II (J. O. Nriagu, ed.), pp. 241–284, Wiley, New York.

Frieden, E., 1981, in *Metal Ions in Biological Systems*, Volume 13 (H. Sigel, ed.), pp. 117–142, Marcel Dekker, New York.

Fronko, R. M., Penner-Hahn, J. E., and Bender, C. J., 1988, *J. Am. Chem. Soc.* **110**:7554.

Fukuyama, K., Hase, T., Matsumoto, S., Tsukihara, T., Katsube, Y., Tanaka, N., Kakudo, M., Wada, K., and Matsubara, H., 1980, *Nature* **286**:522.

Fukuyama, K., Nagahara, Y., Tsukihara, T., and Katsube, Y., 1988, *J. Mol. Biol.* **199**:183.

Fukuyama, K., Nagahara, Y., Tsukihara, T., and Katsube, Y., 1989, *J. Mol. Biol.* **210**:383.

Furey, W. F., Robbins, A. H., Clancy, L. L., Winge, D. R., Wang, B. C., and Stout, C. D., 1986, *Science* **23**:704.

Gaillard, J., Moulis, J.-M., Auric, P., and Meyer, J., 1986, *Biochemistry* **25**:464.

Gaillard, J., Moulis, J.-M., and Meyer, J., 1987, *Inorg. Chem.* **26**:320.

Gaykema, W. P. J., Hol, W. G. J., Vereijken, J. M., Soeter, N. M., Bak, H. J., and Beintemma, J. J., 1984, *Nature* **309**:23.

Gaykema, W. P. J., Volbeda, A., and Hol, W. G. J., 1985, *J. Mol. Biol.* **187**:255.

Gibb, T. C., 1976, in *Principles of Mossbauer Spectroscopy* (A. D. Buckingham, ed.), Chapman and Hall, New York.

Gibson, J. F., Hall, D. O., Thornley, J. H. M., and Whatley, F. R., 1966, *Proc. Natl. Acad. Sci. USA* **56**:987.

Ginsberg, A. P., 1971, *Inorg. Chim. Acta Rev.* **5**:45.

Glickson, J. D., Phillips, W. D., McDonald, C. C., and Poe, M., 1971, *Biochem. Biophys. Res. Commun.* **42**:271.

Goldstein, I. M., Kaplan, H. B., Edelson, H. S., and Weissmann, G., 1979, *J. Biol. Chem.* **254**:4040.

Goodin, D. B., Yachandra, V. K., Britt, R. D., Sauer, K., and Klein, M. P., 1984, *Biochim. Biophys. Acta* **767**:209.

Gueron, M., 1975, *J. Magn. Reson.* **19**:58.

Guiles, R. D., Yachandra, V. K., McDermott, A. E., Britt, R. D., Dexhiemer, S. L., Sauer, K., and Klein, M. P., 1987, in *Progress in Photosynthesis Research* (J. Biggins, ed.) pp. 561–564, Nijhoff, Dordrecht.

Gurbiel, R. J., Batie, C. J., Sivaraja, M., True, A. E., Fee, J. A., Hoffman, B. M., and Ballou, D. P., 1989, *Biochemistry* **28**:4861.

Hagen, W. R., Pierik, A. J., and Veeger, C., 1989, *J. Chem. Soc., Faraday Trans.* **85**:4803.

Hales, B. J., Langosch, J., and Case, E. E., 1986, *J. Biol. Chem.* **261**:15301.
Hall, D. O., Rao, K. K., and Cammack, R., 1975, *Sci. Progr. Oxford* **62**:285.
Hausinger, R. P., 1987, *Microbiol. Rev.* **51**:1.
Hirasawa-Soga, M., and Tamura, G., 1981, *Agric. Biol. Chem.* **45**:1615.
Holm, R. H., Ciurli, S., and Weigel, J. A., 1990, *Progr. Inorg. Chem.* **38**:1
Holmberg, C. G., and Laurell, C. B., 1948, *Acta Chem. Scand.* **2**:550.
Howard, J. B., and Rees, D. C., 1991, *Adv. Protein Chem.* **42**:99199.
Hsieh, S. H., and Frieden, E., 1975, *Biochem. Biophys. Res. Commun.* **67**:1326.
Huynh, B. H., Henzl, M. T., Christner, J. A., Zimmermann, R., Orme-Johnson, W. H., and Munck, E., 1980, *Biochim. Biophys. Acta* **623**:124.
Johnson, M. K., Thompson, A. J., Robinson, A. E., and Smith, B. E., 1981, *Biochim. Biophys. Acta* **671**:61.
Joliot, P., and Kok, B., 1975, in *Bioenergetics of Photosynthesis* (A. Govindjee, ed.), pp. 387–412, Academic Press, New York.
Jolley, R. L., Evans, L. H., and Mason, H. S., 1972, *Biochem. Biophys. Res. Commun.* **46**:878.
Jolley, R. L., Evans, L. H., and Mason, H. S., 1973, in *Oxidases and Related Redox Systems* (T. E. King, H. S. Mason, and M. Morrison, eds.), pp. 100, University Park Press, Baltimore.
Jolley, R. L., Evans, L. H., Makino, N., and Mason, H. S., 1974, *J. Biol. Chem.* **249**:335.
Kadenbach, B., Jarausch, J., Hartmann, R., and Merle, P., 1983, *Anal. Biochem.* **129**:517.
Kanatzidis, M. G., Hagen, W. R., Dunham, W. R., Lester, R. K., and Coucouvanis, D., 1985, *J. Am. Chem. Soc.* **107**:953.
Kiick, K. L., Finnegan, M. G., Werth, M. T., Alvarez, M. L., Wilcox, D. E., Breitenbach, J., Hausinger, R. P., and Johnson, M. K., 1990, *J. Inorg. Biochem.* **43**:658.
Kissinger, C., Adman, E. T., Sieker, L. C., and Jensen, L. H., 1988, *J. Am. Chem. Soc.* **110**:8721.
Kissinger, C. R., Adman, E. T., Sieker, L. C., Jensen, L. H., and LeGall, J., 1989, *FEBS Lett.* **244**:447.
Kitajima, N., Fujisawa, K., and Moro-oka, Y., 1989, *J. Am. Chem. Soc.* **111**:8975.
Koenig, S. H. and Brown, R. D., III, 1990, *Progr. Nucl. Magn. Reson. Spectrosc.* **22**:489.
Kono, Y., and Fridovich, I., 1982, *J. Biol. Chem.* **257**:5751.
Krishnamoorthi, R., Markley, J. L., Cusanovich, M. A., and Przysiecki, C. T., 1986, *Biochemistry* **25**:60.
Krishnamoorthi, R., Cusanovich, M. A., Meyer, T. E., and Przysiecki, C. T., 1989, *Eur. J. Biochem.* **181**:81.
Kroneck, P. M. H., and Zumft, W. G., 1991, *FEMS Symp.* **56**:1.
Kroneck, P. M. H., Antholine, W. A., Riester, J. and Zumft, W. G., 1988, *FEBS Lett.* **242**:70.
Kroneck, P. M. H., Riester, J., Zumft, W. G. and Antholine, W. E., 1990, *Biol. Metals* **3**:103.
Lancaster, J. R., Vega, J. M., Kamin, H., Orme-Johnson, N. R., Orme-Johnson, W. H., Krueger, R. J., and Siegel, L. M., 1979, *J. Biol. Chem.* **254**:1268.
Lavergne, J., 1987, *Biochim. Biophys. Acta* **894**:91.
Lerch, K., 1981, in *Metal Ions in Biological Systems*, Volume 13 (H. Sigel, ed.) pp. 143–186, Marcel Dekker, New York.
Lindhal, P. A., Day, E. P., Kent, T. A., Orme-Johnson, W. H., and Munck, E., 1985, *J. Biol. Chem.* **260**:11160.
Lindhal, P. A., Gorelick, N. J., Munck, E., and Orme-Johnson, W. H., j1987, *J. Biol. Chem.* **262**:14945.
Linzen, B., Soeter, N. M., Riggs, A. F., Schneider, H.-J., Schartan, W., Moore, M. D., Yokota, E., Behrens, P. Q., Nakashima, H., Takagi, T., Namoto, T., Vereijken, J. M., Bak, H. J., Beintema, J. J., Volbeda, A., Gaykema, W. P. J., and Hol, W. G. J., 1985, *Science* **229**:519.
Makino, N., and Mason, H. S., 1973, *J. Biol. Chem.* **248**:5731.
Makino, N., McMahill, P., Mason, H. S., and Moss, T. H., 1974, *J. Biol. Chem.* **249**:6062.
Martin, C. T., Scholes, C. P., and Chan, S. I., 1988, *J. Biol. Chem.* **263**:8420.

Mauesca, J. M., Lamotte, B., and Rius, G. J., 1991, *Inorg. Biochem.* **43**:251.

McConnell, H. M., and Chestnut, D. B., 1958, *J. Chem. Phys.* **28**:107.

McRee, D. E., Richardson, D. C., Richardson, J. S., and Siegel, L. M., 1986, *J. Biol. Chem.* **61**:10277.

Messerschmidt, A., Rossi, A., Ladenstein, R., Huber, R., Bolognesi, M., Gatti, G., Marchesini, A., Petruzzelli, R., and Finazzi-Agrò, A., 1989, *J. Mol. Biol.* **206**:513.

Meyer, J., and Moulis, J.-M., 1981, *Biochem. Biophys. Res. Commun.* **103**:667.

Meyer, T. E., 1985, *Biochim. Biophys. Acta* **806**:175.

Meyer, T. E., Przysiecki, C. T., Watkins, J. A., Bhattacharyya, A., Simondsen, R. P., Cusanovich, M. A., and Tollin, G., 1983, *Proc. Natl. Acad. Sci. USA* **80**:6740.

Middleton, P., Dickson, D. P. E., Johnson, C. E., and Rush, J. D., 1978, *Eur. J. Biochem.* **88**:135.

Middleton, P., Dickson, D. P. E., Johnson, C. E., and Rush, J. D., 1980, *Eur. J. Biochem.* **104**:289.

Minami, Y., Wakabayashi, S., Wada, K., Matsubara, H., Kersher, L., and Oersterhelt, D., 1983, *J. Biochem.* **972**:745.

Ming, L. J., Banci, L., Luchinat, C., Bertini, I., and Valentine, J. S., 1988a, *Inorg. Chem.* **27**:4458.

Ming, L. J., Banci, L., Luchinat, C., Bertini, I., and Valentine, J. S., 1988b, *Inorg. Chem.* **27**:728.

Mobley, H. L. T., and Hausinger, R. P., 1989, *Microbiol. Rev.* **53**:85.

Morgenstern-Badarau, I., Cocco, D., Desideri, A., Rotilio, G., Jordanov, J., and Dupré, N., 1986, *J. Am. Chem. Soc.* **108**:300.

Moss, T. H., Bearden, A. J., Bartsch, R. G., and Cusanovich, M. A., 1968, *Biochemistry* **7**:1591.

Moulis, J.-M., and Meyer, J., 1982, *Biochemistry* **21**:4762.

Moulis, J.-M., and Meyer, J., 1984, *Biochemistry* **23**:6605.

Moulis, J.-M., Auric, P., Gaillard, J., and Myer, J., 1984a, *J. Biol. Chem.* **259**:11396.

Moulis, J.-M., Meyer, J., and Lutz, M., 1984b, *Biochem. J.* **219**:829.

Moura, J. J. G., Xavier, A. V., Bruschi, M. and LeGall, J., 1977, *Biochim. Biophys. Acta* **459**:278.

Münck, E., Papaefthymiou, V., Surerus, K. K., and Girerd, J.-J., 1988, in *Metal Clusters in Proteins* (L. Que, Jr ed.), pp. 302–325, ACS, Washington, D.C.

Murphy, M. J., and Siegel, L. M., 1973, *J. Biol. Chem.* **248**:6911.

Murphy, M. J., Siegel, L. M., Kamin, H., and Rosenthal, D., 1973, *J. Biol. Chem.* **248**:2801.

Nagayama, K., Ozaki, Y., Kyogoku, Y., Hase, T., and Matsubara, H., 1983, *J. Biochem.* **94**:893.

Nettesheim, D. G., Harder, S. R., Feinberg, B. A., and Otvos, J. D., 1992, *Biochemistry* **31**:1234.

Noodleman, L., 1988, *Inorg. Chem.* **27**:3677.

Norlund, P., Sjoberg, B.-M., and Ecklund, H., 1990, *Nature* **345**:593.

Oberley, L. W., 1982, in *Superoxide Dismutase*, CRC Press, Boca Raton, FL.

Okawara, N., Ogata, M., Yagi, H. S., and Matsubara, H., 1988, *J. Biochem.* **104**:196.

Orme-Johnson, N. R., Mims, W. B., Orme-Johnson, W. H., Bartsch, R. G., Cusanovich, M. A., and Peisach, J., 1983, *Biochim. Biophys. Acta* **748**:68.

Owens, C., Drago, R. S., Bertini, I., Luchinat, C., and Banci, L, 1986, *J. Am.Chem. Soc.* **108**:3298.

Packer, E. L., Sweeney, W. V., Rabinowitz, J. C., Sternlicht, H., and Shaw, E. N., 1977, *J. Biol. Chem.* **252**:2245.

Palmer, G., Dunham, W. R., Fee, J. A., Sands, R. H., Izuka, T., and Yonetani, T., 1971, *Biochim. Biophys. Acta* **245**:201.

Pan, R. N., Mitchenall, L. A., and Robson, R. L., 1989, *J. Am. Chem. Soc.* **171**:124.

Pan, L. P., Li, Z., Larsen, R., and Chan, S. I., 1991, *J. Biol. Chem.* **266**:1367.

Papaefthymiou, V., Girerd, J.-J., Moura, I., Moura, J. J. G., and Munck, E., 1987, *J. Am. Chem. Soc.* **109**:4703.

Papaefthymiou, V., Millar, M. M., and Munck, E., 1986, *Inorg. Chem.* **25**:3010.

Peisach, J., Orme-Johnson, N. R., Mims, W. B., and Orme-Johnson, W. H., 1977, *J. Biol. Chem.* **252**:5643.

Phillips, W. D., Poe, M., McDonald, C. C., and Bartsch, R. G., 1970a, *Proc. Natl. Acad. Sci. USA* **67**:682.

Phillips, W. D., Poe, M., Weiher, J. F., McDonald, C. C., and Lovenberg, W., 1970b, *Nature* **227**:574.

Pierik, A. J., and Hagen, W. R., 1991, *Eur. J. Biochem.* **195**:505.

Poe, M., Phillips, W. D., McDonald, C. C., and Lovenberg, W., 1970, *Proc. Natl. Acad. Sci. USA* **65**:797.

Poe, M., Phillips, W. D., Glickson, J. D., McDonald, C. C., and San Pietro, A., 1971, *Proc. Natl. Acad. Sci. USA* **68**:68.

Przysiecki, C. T., Meyer, T. E., and Cusanovich, M. A., 1985, *Biochemistry* **24**:2542.

Que, L., Jr., and True, A. E., 1990, *Progr. Inorg. Chem.* **38**:97.

Rieske, J. S., MacLennan, D. H., and Coleman, R., 1964, *Biochem. Biophys. Res. Commun.* **15**:338.

Robb, D. A., 1984, in *Copper Proteins and Copper Enzymes* (R. Lantie, ed.) pp. 207, CRC, Boca Raton, FL.

Robbins, A. H., and Stout, C. D., 1989, *Proc. Natl. Acad. Sci. USA* **86**:3639.

Robson, R. L., Eady, R. R., Richardson, T. H., Miller, R. W., Hawkins, M. H., and Postgate, J. R., 1986, *Nature* **322**:388.

Rueger, D. C., and Siegel, L. M., 1976, in *Flavins and Flavoproteins* (T. P. Singer, ed.) pp. 610, Elsevier, Amsterdam.

Rusnak, F. M., Adams, M. W. M., Mortenson, L. E., and Munck, E., 1987, *J. Biol. Chem.* **262**:38.

Salmeen, I., and Plamer, G., 1972, *Arch. Biochem. Biophys.* **150**:767.

Scaringe, R. P., Hodgson, D. J., and Hatfield, W. E., 1978, *Mol. Phys.* **35**:701.

Schimpff-Weiland, G., and Follmann, H., 1981, *Biochem. Biophys. Res. Commun.* **102**:1276.

Scott, A. I., Irwin, A. J., Siegel, L. M., and Scoolery, J. N., 1978, *J. Am. Chem. Soc.* **100**:7987.

Sheats, J. E., Czernusczewicz, R. S., Dismukes, G. C., Rheingold, A. L., Petrouleas, V., Stubbe, J., Armstrong, W. H., Beer, R. H., and Lippard, S. J. 1987, *J. Am. Chem. Soc.* **109**:1435.

Sheridan, R. P., Allen, L. C., and Carter, C. W., Jr., 1981, *J. Biol. Chem.* **256**:5052.

Siegel, L. M., 1978, in *Mechanisms of Oxidizing Enzymes* (T. P. Singer, and R. N. Ondorza, eds.), pp. 201, Elsevier-North Holland, New York.

Siegel, L. M., Rueger, D. C., Barber, M. J., Krueger, R. J., Orme-Johnson, N. R., and Orme-Johnson, W. H., 1982, *J. Biol. Chem.* **257**:6343.

Sjoberg, B.-M., and Graslund, A., 1983, *Adv. Inorg. Biochem.* **5**:87.

Skjieldal, L., Westler, W. M., Oh, B.-H., Krezel, A. M., Holden, H. M., Jacobson, B. L., Rayment, I. and Markley, J. L., 1991a, *Biochemistry* **30**:7363.

Skjieldal, L., Markley, J. L., Coghlan, V. M., and Vickery, L. E., 1991b, *Biochemistry* **30**:9078.

Smith, J. P., Emptage, M. H., and Orme-Johnson, W. H., 1982, *J. Biol. Chem.* **257**:2310.

Sola, M., Cowan, J. A., and Gray, H. B., 1989, *Biochemistry* **28**:5261.

Solomon, E. I., 1981, in *Copper Proteins* (T. G. Spiro, ed.), pp. 41–108, Wiley, New York.

Stenkamp, R. E., Sieker, L. C., and Jensen, L. H., 1984, *J. Am. Chem. Soc.* **106**:618.

Stephens, P. J., McKenna, M.-C., Ensign, S. A., Bonam, D., and Ludden, P. W., 1989, *J. Biol. Chem.* **264**:16347.

Stevens, K. W. H., 1976, *Phys. Rep.* **24c**:1.

Stevens, K. W. H., 1985, in *Magneto-Structural Correlations in Exchange-Coupled Systems* (R. D. Willett, D. Gatteschi, and O. Kahn, eds.), pp. 105, Reidel, Dordrecht.

Stiefel, E. I., Thomann, H., Jin, H., Bare, R. E., Morgan, T. V., Burgmayer, S. J. N., and Coyle, C. L., 1988, in *Metal Clusters in Proteins* (L. Que, Jr., ed.), pp. 372–389, ACS, Washington, D.C.

Stout, C. D., 1989, *J. Mol. Biol.* **205**:545.

Stout, G. H., Turley, S., Sieker, L. C., and Jensen, L. H., 1988, *Proc. Natl. Acad. Sci. USA* **85**:1020.

Sweeney, W. V., 1981, *J. Biol. Chem.* **256**:12222.

Sykes, A. G., 1991, *Met. Ions Biol. Syst.* **27**:291.

Tainer, J. A., Getzoff, E. D., Richardson, J. S., and Richardson, D. C., 1983, *Nature* **306**:284.

Takahashi, Y., Hase, T., Wada, K., and Matsubara, H., 1981, *J. Biochem.* **90**:1825.

Tan, G. O., Ensign, S. A., Ciurli, S., Scott, M. J., Hedman, B., Holm, R. H., Ludden, P. W., Korszun, R. Z., Stephens, P. J., and Hodgson, K. O., 1992, *Proc. Natl. Acad. Sci. USA*, in press.

Tedro, S. M., Meyer, T. E., Bartsch, R. G., and Kamen, M., 1981, *J. Biol. Chem.* **25**:731.

Thauer, R. K., and Schoenheit, P., 1982, in *Iron–Sulfur Proteins* (T. G. Spiro, ed.), pp. 329–341, Wiley-Interscience, New York.

Theil, E. C., 1988, in *Metal Clusters in Proteins* (L. Que, Jr., ed.), pp. 179–195, ACS, Washington, D.C.

Thompson, C. L., Johnson, C. E., Dickson, D. P. E., Cammack, R., Hall, D. O., Weser, U., and Rao, K. K., 1974, *Biochem. J.* **139**:97.

Thomson, A. J., 1985, in *Metalloproteins* (P. Harrison, ed.), pp. 79, Verlag Chemie, Weinheim, FRG.

Thomson, A. J., Brittain, T., Greenwood, C., and Springall, J. P., 1977, *Biochem. J.* **165**:327.

Thomson, A. J., Robinson, A. E., Johnson, M. K., Moura, J. J. G., Moura, I., Xavier, A. V., and LeGall, J., 1981, *Biochim. Biophys. Acta* **670**:93.

Tonge, G. M., Harrison, D. E. F., and Higgins, I. J., 1977, *Biochem. J.* **161**:333.

Tsukihara, T., Fukuyama, K., Nakamura, M., Katsube, Y., Tanaka, N., Kakudo, M., Wasa, K., Hase, T., and Matsubara, H., 1981, *J. Biochem.* **90**:1763.

Urbach, F. L., 1981, in *Metal Ions in Biological Systems* (H. Sigel, ed.) pp. 73–116, Marcel Dekker, New York.

Valentine, J. S., and Pantoliano, M. W., 1981, in *Copper Proteins* (T. G. Spiro, ed.), pp. 291–358, Wiley, New York.

Van Holde, K. E., and Miller, K. I., 1982, *Quart. Rev. Biophys.* **15**:1.

Vasak, M., 1986, in *Zinc Enzymes* (I. Bertini, C. Luchinat, W. Maret, and M. Zeppenhauer, eds.), pp. 595, Birkhauser, Boston.

Vasak, M., and Kägi, J. H., 1983, in *Metal Ions in Biological Systems* (H. Sigel, ed.), pp. 273, Marcel Dekker, New York, NY.

Vega, A. J., and Fiat, D., 1976, *Mol. Phys.* **31**:347.

Vincent, J. B., Olivier-Lilley, G. L., and Averill, B. A., 1990, *Chem. Rev.* **90**:1447.

Vollmer, S. J., Switzer, R. L., and Debrunner, P. G., 1983, *J. Biol. Chem.* **258**:14284.

Wagner, G., Frey, M. H., Neuhaus, D., Wortgotter, E., Braun, W., Vasak, M., Kägi, J. H. R., and Wüthrich, K. 1987, in *Proc. 2nd International Meeting of Metallothionein* (J. H. R. Kägi, and Y. Kojima, eds.), Birkhauser Verlag, Basel.

Wakabayashi, S., Fujimoto, N., Wada, K., Matsubara, H., Kersher, L, and Oersterhelt, D., 1983, *FEBS Lett.* **162**:21.

Walsh, C. T., and Orme-Johnson, W. H., 1987, *Biochemistry* **26**:4901.

Wilcox, D. E., Porras, A. G., Hwang, Y. T., Lerch, K., Winkler, M. E., and Solomon, E. I., 1985, *J. Am. Chem. Soc.* **107**:4015.

Wikström, M., Krab, K., and Saraste, M., 1981, *Cytochrome Oxidase, A Synthesis*, Academic Press, New York.

Wilkins, P. C., and Wilkins, R. G., 1987, *Coord. Chem. Rev.* **79**:195.

Wilkins, R. G., and Harrington, P. C., 1983, *Adv. Inorg. Biochem.* **5**:51.

Willing, A., Follman, H., and Auling, G., 1988, *Eur. J. Biochem.* **170**:603.

Winter, D. B., Bruyninckx, W. J., Foulke, F. G., Grinich, N. P., and Mason, H. S., 1980, *J. Biol. Chem.* **255**:11408.

Yang, C.-Y., Meagher, A., Huynh, B. H., Sayers, D. E., and Theil, E. C., 1987, *Biochemistry* **26**:497.

Yoch, D. C., and Carithers, R. P., 1979, *Microbiol. Rev.* **43**:384.

Zener, C., 1951, *Phys. Rev.* **82**:403.

Zimmermann, R., Münck, E., Brill, W. J., Shah, V. K., Henzl, M. T., Rawlings, J., and Orme-Johnson, W. H., 1978, *Biochim. Biophys. Acta* **537**:185.

Contents of Previous Volumes

Index